ALGORITHMS AND PROTOCOLS FOR WIRELESS SENSOR NETWORKS

WILEY SERIES ON PARALLEL AND DISTRIBUTED COMPUTING

Editor: Albert Y. Zomaya

A complete list of titles in this series appears at the end of this volume.

ALGORITHMS AND PROTOCOLS FOR WIRELESS SENSOR NETWORKS

Edited by

Azzedine Boukerche, PhD
University of Ottawa
Ottawa, Canada

WILEY
A John Wiley & Sons, Inc., Publication

Published by John Wiley & Sons, Inc., Hoboken, New Jersey
Published simultaneously in Canada

Library of Congress Cataloging-in-Publication Data

Algorithms and protocols for wireless sensor networks / edited by Azzedine Boukerche.
p. cm.
Includes bibliographical references and index.
ISBN 978-0-471-79813-2 (cloth)
1. Sensor networks. 2. Computer network protocols. 3. Computer algorithms.
I. Boukerche, Azzedine.
TK7872.D48A422 2008
681'.2–dc22

2008016869

Printed in the United States of America

This book is dedicated to my parents and my family who have always been there with me.
Love you all.

Azzedine Boukerche

CONTENTS

PREFACE

With the recent technological advances in wireless communication and networking, coupled with the availability of intelligent and low-cost actor and sensor devices with powerful sensing, computation, and communication capabilities, wireless sensor networks (WSNs) are about to enter the mainstream. Today, one could easily envision a wide range of real-world WSN-based applications from sensor-based environmental monitoring, home automation, health care, security, and safety class of applications, thereby promising to have a significant impact throughout our society. Wireless sensor networks are comprised of a large number of sensor devices that can communicate with each other via wireless channels, with limited energy and computing capabilities. However, due to the nature of wireless sensor networks, we are witnessing new research challenges related to the design of algorithms and network protocols that will enable the development of sensor-based applications. Most of the available literature in this emerging technology concentrates on physical and networking aspects of the subject. However, in most of the literature, a description of fundamental distributed algorithms that support sensor and actor devices in a wireless environment is either not included or briefly discussed. The efficient and robust realization of such large, highly dynamic and complex networking environments is a challenging algorithmic and technological task. Toward this end, this book identifies the research that needs to be conducted on a number of levels to design and assess the deployment of wireless sensor networks–in particular the design of algorithmic methods and distributed computing with sensing, processing, and communication capabilities. It is our belief that this volume provides not only the necessary background and foundation in wireless sensor networks but also an in-depth analysis of fundamental algorithms and protocols for the design and development of the next generations of heterogeneous wireless networks in general and wireless sensor networks in particular. This book is divided into 18 chapters and covers a variety of topics in the field of wireless sensor networks that could be used as a textbook for graduate and/or advanced undergraduate studies, as well as a reference for engineers and computer scientists interested in the field of wireless sensor networks.

The rest of this book is organized as follows. In Chapter 1, we address the several important algorithmic issues arising in wireless sensor networks and highlight the main differences to classical distributed algorithms. Next, an algorithmic perspective toward the design of wireless sensor networks is discussed followed by an overview of well-known algorithms for basic services (that can be used by other algorithms in WSNs), data communication, management functions, applications, and data fusion. Chapter 2 introduces heterogeneous wireless sensor networks where more than one type of sensor node is integrated into a WSN. While many of the existing civilian and military applications of heterogeneous wireless sensor networks (H-WSNs) do not differ substantially from their homogeneous counterparts, there are compelling reasons to incorporate heterogeneity into the network, such as improving the scalability of WSNs and addressing the problem of nonuniform energy drainage, among others. Chapter 2 also discusses how these reasons are interrelated and how this new dimension heterogeneity opens new challenges to the design of algorithms that run on such wireless sensor networks.

In order to develop algorithms for sensor networks and in order to give mathematical correctness and performance proofs, models for various aspects of sensor networks are needed. In the next three chapters, we focus upon the modeling, design, and analysis of algorithms and protocols for wireless sensor networks. Chapter 3 discusses how biological inspired models, such epidemic models, can be used to design reliable data dissemination algorithms in the context of wireless sensor networks. Recall that reliable data dissemination to all sensor nodes is necessary for the propagation of queries, code updates, and other sensitive WSN-related information. This is not a trivial task because the number of nodes in a sensor network can be quite large and the environment is quite dynamic (e.g., nodes can die or move to another location). Chapter 4 provides an overview and discussion of well-known sensor network models used today and shows how these models are related to each other. While the collaborative nature of the WSN brings significant advantages over traditional sensing, the spatiotemporal correlation among the sensor observations is another significant and unique characteristic of the WSN which can be exploited to drastically enhance the overall sensor network performance. Chapter 5 presents the theoretical framework to model the spatiotemporal correlation in sensor networks and describes in detail how to exploit this correlation when designing reliable communication protocols for WSN.

With the traditional TCP/IP models not suited to routing in wireless sensor networks, the network layer protocol has to be updated to be synchronized with the challenging constraints posed by WSNs. Hence, routing in these networks is a challenging task and has thus been a primary focus with the wireless networking community. The next chapters investigate the major issues to routing with the goals to devise new protocols to keep associated uncertainty under control. Chapter 6 highlights the properties of a wireless sensor network from the networking point of view, and then it presents a description of various well-known routing protocols for wireless sensor networks. The common goals of designing a routing algorithm is not only to reduce control packet overhead, maximize throughput, and minimize the end-to-end delay, but also to take into consideration the energy consumption, especially in a sensor network comprised of nodes that are considered lightweight with limited memory and battery power. In order to achieve high energy efficiency and ensure long network lifetime for routing traffic control, as well as employ bandwidth re-use for data gathering and target tracking, researchers have designed one-to-many, many-to-one, one-to-any, or one-to-all communications, routing, and clustering-based routing protocols. Chapter 7 presents different protocols developed to create clusters and select the best cluster head using Graph Theory concepts. Chapter 8 discusses the merits and challenges of algorithms and protocols that provide point-to-point services through position-based routing, where forwarding decisions are made by maximizing or minimizing some function of node locations within a coordinate system. Sensors can generally be placed in an area of interest either deterministically or randomly. However, controlled node deployment is viable and often necessary when sensors are expensive or when their operation is significantly affected by their position. Chapter 9 investigates the effect node placement strategies on the dependability of WSNs, and it presents the various sensor and base-station positioning protocols that have been developed to enhance further the performance of WSNs and extend its network lifetime.

The next generation of wireless sensor networks are envisioned to support mobile sensor devices and a variety of mobile robot sensor devices and a variety of wireless multimedia sensor services. Chapter 10 presents several techniques for exploiting the mobility of network components in large networks of resource constrained devices, such as wireless sensor networks, and improving the performance of these networks without significantly affecting data routing and end-to-end latency. A number of mobility issues in WSNs as well as the pros and cons of providing mobility to the normal nodes, relay nodes, and/or sink nodes are analyzed. Also in this chapter, solutions that use mobility to alleviate the problem of energy depletion of nodes near the sink are shown. However, this mobility as well as the random deployment of the nodes in a WSN imposes another problem to the network: how to discover the current physical position of the nodes. Chapters 11 and 12 focus on the different aspects of this problem known as the localization problem. In Chapter 11, the localization systems are divided into different components—distance estimation, position computation, and localization algorithm—and several techniques employed by these components are explained. On the other hand, Chapter 12 deals with more specific problems, such as using the signals' angle of arrival to estimate the position of the nodes.

Quality of service (QoS) provisioning in wireless sensor networks (WSNs) is an important concept to enable mission-critical and real-time applications. In Chapter 13, the necessity to support QoS in WSNs, QoS-based communication protocols, and research directions to support QoS in WSNs is discussed. Chapter 14 presents some background topics in network information theory relevant to the efficient collection, compression, and reliable communication of sensor data. Then, it discusses how a QoS perspective enables scalability in classical flat sensor networks. Finally, a number of practical QoS approaches for high-fidelity data extraction in large-scale sensor networks are explored. Chapter 15 focuses on several important aspects of energy efficiency, like minimizing the total energy dissipation, minimizing the number of transmissions, and balancing the energy load to prolong the system's lifetime. Several characteristic protocols and techniques in the recent literature that explicitly focus on energy efficiency are presented. Such techniques include clustering and probabilistic forwarding, adaptive transmission range management, and local optimization.

WSNs are supposed to be deployed in critical scenarios to be used in safety, emergency, and military applications. In these cases, security is a key technology in order to make the gathered data a reliable information. Thus, we believe that a WSN book would not be complete without a good review of the proposed techniques that aim to provide the secure operation and communication in WSNs. Thus, the next chapters of this book investigate different aspects of providing security in WSNs. Chapter 16 focuses on general aspects of the problem, showing how WSNs are vulnerable to several attacks in the different network layers. Cryptography techniques for WSNs such as cryptographic systems, authentication methods, and key distribution and management protocols are then studied and analyzed as a countermeasurement for a number of the identified attacks. Also in this chapter, secure routing protocols that are resilient to these attacks are discussed and explained. Besides securing the routing, it is also important to secure other key protocols in WSNs such as the synchronization and localization protocols. Chapter 17 provides a good overview of the proposed solutions for securing a time synchronization protocol to be used in critical applications of WSNs. This chapter shows the importance of a secure synchronization system, how current synchronization solutions are vulnerable to a number of attacks, and the proposed techniques to secure these protocols. Finally, Chapter 18 takes the security issue to the localization protocols. This chapter shows how the different components of the localization systems–distance estimation, position computation, and localization algorithm–are vulnerable to a number of attacks and then shows the proposed techniques and countermeasurements to secure these components and provide a secure localization system that are able to work in the presence of hostile nodes and compromised environments.

It is our belief that this is the first book that covers the basic and fundamental algorithms and protocols for wireless sensor networks, making their design and analysis accessible to all levels of readers.

Special thanks are due to all contributors for their support and patience, as well as to the reviewers for their hard work and timely reports, which make this book truly special. Last but not least, we wish to extend our thanks to Paul Petralia and Whitney Lesch from John Wiley & Sons for their support, guidance, and certainly their patience in finalizing this book.

AZZEDINE BOUKERCHE

University of Ottawa

ABOUT THE EDITOR

Azzedine Boukerche is a Professor and holds a Canada Research Chair position at the University of Ottawa. He is the Founding Director of Paradise Research Laboratory at the University of Ottawa. Prior to this, he held a Faculty position at the University of North Texas, and he was working as a Senior Scientist at the Simulation Sciences Division, Metron Corporation, located in San Diego. He was also employed as a faculty member at the School of Computer Science, McGill University, and he taught at Polytechnic of Montreal. He spent a year at the JPL/NASA-California Institute of Technology, where he contributed to a project centered around the specification and verification of the software used to control interplanetary spacecraft operated by JPL/NASA Laboratory. His current research interests include wireless ad hoc and sensor networks, wireless networks, mobile and pervasive computing, wireless multimedia, QoS service provisioning, large-scale distributed interactive simulation, parallel discrete event simulation, and performance evaluation and modeling of large-scale distributed and mobile systems. Dr. Boukerche has published several research papers in these areas. He was the recipient of and/or nominated for the Best Research Paper Award at IEEE/ACM PADS '97, IEEE/ACM PADS '99, IEEE ICC 2008, ACM MSWiM 2001, and MobiWac'06, and he was the co-recipient of the 3rd National Award for Telecommunication Software 1999 for his work on distributed security systems on mobile phone operations.

Dr. A. Boukerche is a holder of an Ontario Early Research Excellence Award (previously known as Premier of Ontario Research Excellence Award), an Ontario Distinguished Researcher Award, and a Glinski Research Excellence Award. He is a Co-Founder of QShine International Conference on Quality of Service for Wireless/Wired Heterogeneous Networks (QShine 2004) and has served as a General Chair for the 8th ACM/IEEE Symposium on Modeling, Analysis, and Simulation of Wireless and Mobile Systems, the 9th ACM/IEEE Symposium on Distributed Simulation and Real-Time Application, and the 6th IEEE/ACM MASCOT '98 Symposium; he has also served as the Vice General Chair for the 3rd IEEE International Conference on Distributed Computing in Sensor Systems (DCOSS '07), Program Chair for IEEE Globecom 2007 and 2008 Ad Hoc, Sensor and Mesh Networking Symposium, and a Program Co-Chair for ICPP 2008, the 2nd ACM Workshop on QoS and Security for Wireless and Mobile Networks, ACM/IFIPS Europar 2002 Conference, IEEE/SCS Annual Simulation Symposium '02, ACM WWW '02, IEEE MWCN 2002, IEEE/ACM MASCOTS '02, IEEE Wireless Local Networks 03-04, IEEE WMAN 04-05, and ACM MSWiM 98-99.

Dr. A. Boukerche is an Associate Editor for *ACM/Springer Wireless Networks*, *IEEE Transactions on Vehicular Networks*, *IEEE Wireless Communication Magazine*, *IEEE Transactions on Parallel and Distributed Systems*, Elsevier's *Ad Hoc Networks*, *Wiley International Journal of Wireless Communication and Mobile Computing*, Wiley's *Security and Communication Network Journal*, Wiley's *Pervasive and Mobile Computing Journal*, Elsevier's *Journal of Parallel and Distributed Computing*, and *SCS Transactions on Simulation*. He also serves as a Steering Committee Chair for the ACM Modeling, Analysis and Simulation for Wireless and Mobile Systems Symposium, the ACM Workshop on Performance Evaluation of Wireless Ad Hoc, Sensor, and Ubiquitous Networks, and the IEEE/ACM Distributed Simulation and Real-Time Applications Symposium (DS-RT).

CONTRIBUTORS

Mohammad Z. Ahmad, School of Electrical Engineering and Computer Science, University of Central Florida, Orlando, FL 32816-2362

Özgür B. Akan, Next generation Wireless Communications Laboratory (NWCL), Department of Electrical and Electronics Engineering, Middle East Technical University, Ankara, Turkey 06531

Kemal Akkaya, Department of Computer Science, Southern Illinois University, Carbondale, IL 62901

Liliana M. Arboleda, Department of Computing and Information Sciences, University of Guelph, Guelph, Ontario N1G 2W1, Canada

Stefano Basagni, ECE Department, Northeastern University, Boston, MA 02115

Azzedine Boukerche, School of Information Technology and Engineering, University of Ottawa, Ottawa, Ontario K1N 6N5, Canada

Alessio Carosi, Dipartimento di Informatica, Università di Roma "La Sapienza," Roma 00198, Italy

Sajal K. Das, Center for Research in Wireless Mobility and Networking (CReWMaN), Department of Computer Science and Engineering, University of Texas at Arlington, Arlington, TX 76019

Pradip De, Center for Research in Wireless Mobility and Networking (CReWMaN), Department of Computer Science and Engineering, University of Texas at Arlington, Arlington, TX 76019

Eylem Ekici, Department of Electrical and Computer Engineering, Ohio State University, Columbus, OH 43210

Marwan M. Fayed, School of Information Technology and Information, University of Ottawa, Ottawa, Ontario K1N 6N5, Canada

Bing Li, Department of Computer Science and Engineering, Arizona State University, Tempe, AZ 85287-8809

Antonio A. F. Loureiro, Department of Computer Sciences, Federal University of Minas Gerais, Belo Horizonte, Brazil, 31270-010

Angela M. Mielke, Distributed Sensor Networks Group, Los Alamos National Laboratory, Los Alamos, NM 87545

Hussein T. Mouftah, School of Information Technology and Information, University of Ottawa, Ottawa, Ontario K1N 6N5, Canada

Eduardo F. Nakamura, Research and Technological Innovation Center (FUCAPI), Brazil.

Asis Nasipuri, Department of Electrical and Computer Engineering, The University of North Carolina at Charlotte, Charlotte, NC 28223

Nidal Nasser, Department of Computing and Information Sciences, University of Guelph, Guelph, Ontario N1G 2W1, Canada

Sotiris Nikoletseas, Department of Computer Engineering and Informatics, University of Patras, Patras, Greece; and Computer Technology Institute, (CTI), Patras 26500, Greece

Horacio A. B. F. Oliveira, School of Information Technology and Engineering, University of Ottawa, Ottawa, Ontario, Canada, K1N 6N5; Federal University of Minas Gerais, Minas Gerais, Brazil, 31270-010; and Federal University of Amazonas, Amazonas, Brazil, 69077-000

Ameesh Pandya, Department of Electrical Engineering, UCLA, Los Angeles, CA 90095

Chiara Petrioli, Dipartimento di Informatica, Università di Roma "La Sapienza," Roma 00198, Italy

Gregory J. Pottie, Department of Electrical Engineering, UCLA, Los Angeles, CA 90095

Stefan Schmid, Computer Engineering and Networks Laboratory (TIK), ETH Zurich, CH-8092 Zurich, Switzerland

Violet R. Syrotiuk, Department of Computer Science and Engineering, Arizona State University, Tempe, AZ 85287-8809

Yuan Tian, Department of Electrical and Computer Engineering, Ohio State University, Columbus, OH 43210

Begümhan Turgut, Department of Computer Science, Rutgers University, Piscataway, NJ 08854-8019

Damla Turgut, School of Electrical Engineering and Computer Science, University of Central Florida, Orlando, FL 32816-2362

Serdar Vural, Department of Electrical and Computer Engineering, Ohio State University, Columbus, OH 43210

Roger Wattenhofer, Computer Engineering and Networks Laboratory (TIK), ETH Zurich, CH-8092 Zurich, Switzerland

Mohamed Younis, Department of Computer Science and Electrical Engineering, University of Maryland Baltimore County, Baltimore, MD 21250

Tanveer Zia, School of Information Technologies, The University of Sydney, Sydney, NSW 2006, Australia

Albert Y. Zomaya, School of Information Technologies, The University of Sydney, Sydney, NSW 2006, Australia

CHAPTER 1

Algorithms for Wireless Sensor Networks: Present and Future

AZZEDINE BOUKERCHE

School of Information Technology and Engineering, University of Ottawa, Ottawa, Ontario K1N 6N5, Canada

EDUARDO F. NAKAMURA

Federal University of Minas Gerais, Brazil; and FUCAPI—Analysis, Research, and Technological Innovation Center, Brazil

ANTONIO A. F. LOUREIRO

Department of Computer Sciences, Federal University of Minas Gerais, Belo Horizonte, Brazil

1.1 INTRODUCTION

Wireless sensor networks (WSNs) pose new research challenges related to the design of algorithms, network protocols, and software that will enable the development of applications based on sensor devices. Sensor networks are composed of cooperating sensor nodes that can perceive the environment to monitor physical phenomena and events of interest. WSNs are envisioned to be applied in different applications, including, among others, habitat, environmental, and industrial monitoring, which have great potential benefits for the society as a whole. The WSN design often employs some approaches as energy-aware techniques, in-network processing, multihop communication, and density control techniques to extend the network lifetime. In addition, WSNs should be resilient to failures due to different reasons such as physical destruction of nodes or energy depletion. Fault tolerance mechanisms should take advantage of nodal redundancy and distributed task processing. Several challenges still need to be overcome to have ubiquitous deployment of sensor networks. These challenges include dynamic topology, device heterogeneity, limited power capacity, lack of quality of service, application support, manufacturing quality, and ecological issues.

Algorithms and Protocols for Wireless Sensor Networks, Edited by Azzedine Boukerche

The capacity to transmit and receive data packets allows both information and control to be shared among sensor nodes but also to perform cooperative tasks, all based on different algorithms that are being specifically designed for such networks. Some of the classes of algorithms for WSNs are briefly described in the following:

- *Centralized algorithms* execute on a central node and usually benefit from a global network knowledge. This type of algorithm is not very common in WSNs because the cost of acquiring a global network knowledge is usually unfeasible in most WSNs.
- *Distributed algorithms* are related to different computational models. In a WSN, the typical computational model is represented by a set of computational devices (sensor nodes) that can communicate among themselves using a message-passing mechanism. Thus, a distributed algorithm is an algorithm that executes on different sensor nodes and uses a message-passing technique.
- *Localized algorithms* comprise a class of algorithms in which a node makes its decisions based on local and limited knowledge instead of a global network knowledge. Thus "locality" usually refers to the node's vicinity [1].

Algorithms for WSNs may also have some specific features such as self-configuration and self-organization, depending on the type of the target application. Self-configuration means the capacity of an algorithm to adjust its operational parameters according to the design requirements. For instance, whenever a given energy value is reached, a sensor node may reduce its transmission rate. Self-organization means the capacity of an algorithm to autonomously adapt to changes resulted from external interventions, such as topological changes (due to failures, mobility, or node inclusion) or reaction to a detected event, without the influence of a centralized entity.

1.2 WIRELESS SENSOR NETWORKS: AN ALGORITHMIC PERSPECTIVE

In the following, we present an overview of some algorithms for basic services (that can be used by other algorithms), data communication, management functions, applications, and data fusion.

1.2.1 Basic Services

Some of the basic services that can be employed by other algorithms in wireless sensor networks are localization, node placement, and density control.

Localization. The location problem consists in finding the geographic location of the nodes in a WSN, which can be computed by a central unit [2] or by sensor nodes in a distributed manner [3–8]. Essentially, the location discovery can be split in two stages: distance estimation and location computation [4]. Usually, the distance between two

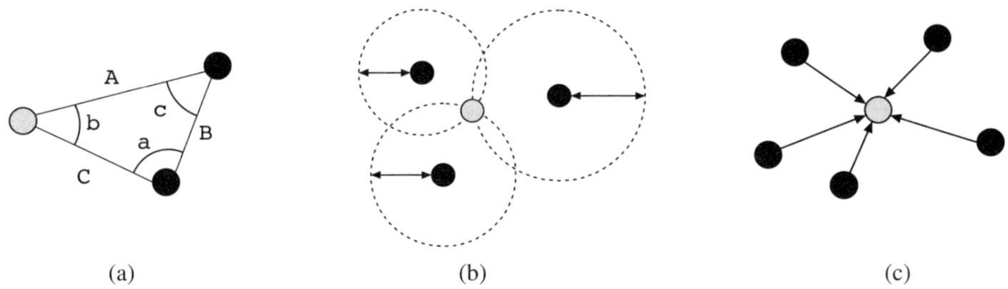

Figure 1.1. Position estimation methods: (a) triangulation, (b) trilateration, and (c) multilateration. (Adapted from reference 10.)

nodes is estimated based on different methods, such as Received Signal Strength Indicator (RSSI), Time of Arrival (ToA), and Time Difference of Arrival (TDoA) [4]. Once the distance is estimated, at least three methods can be used to compute the node location: triangulation, trilateration, and multilateration [9], as depicted in Figure 1.1. Another method to estimate the node location is called the Angle of Arrival (AoA), which uses the angle in which the received signal arrives and the distance between the sender and receiver.

Solutions for finding the nodes' location are often based on localized algorithms in the sense that every node is usually able to estimate its position. For instance, Sichitiu and Ramadurai [11] use the Bayesian inference to process information from a mobile beacon and determine the most likely geographical location (and region) of each node, instead of finding a unique point for each node location. The Directed Position Estimation (DPE) [8] is a recursive localization algorithm in which a node uses only two references to estimate its location. This approach leads to a localization system that can work in a low-density sensor network. Besides, the controlled way in which the recursion occurs leads to a system with smaller and predictable errors. Liu et al. [12] propose a robust and interactive Least-Squares method for node localization in which, at each iteration, nodes are localized by using a least-squares-based algorithm that explicitly considers noisy measurements.

Node Placement. In some applications, instead of throwing the sensor nodes on the environment (e.g., by airplane), they can be strategically placed in the sensor field according to a priori planning. In this approach, there is no need to discover the nodes' location. However, good planning depends on the knowledge of the terrain and the environmental particularities that might interfere in the operation of the sensor nodes and the quality of the gathered data.

The node placement problem has been addressed using different approaches [13–15]. However, current solutions are basically concerned with assuring spatial coverage while minimizing the energy cost. The SPRING algorithm is a node placement algorithm that also performs information fusion. In SPRING it is possible to migrate the fusion role.

Besides spatial coverage [13, 15], other aspects should be considered in a node placement algorithm, such as node diversity [14] and the fusion performance. When

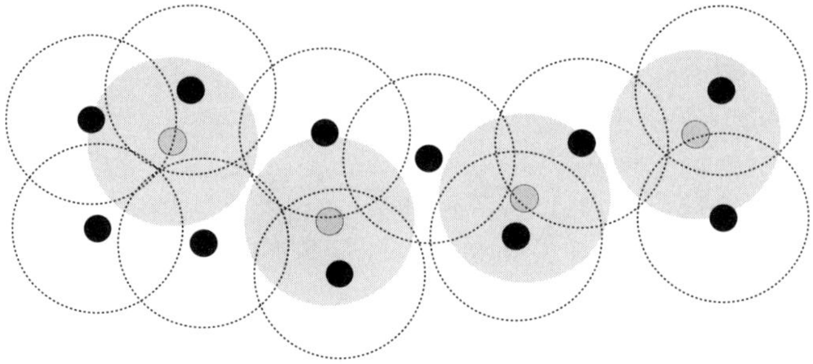

Figure 1.2. An example of node scheduling: Gray nodes are asleep and black nodes are awake.

nodes perform data fusion, an improper node placement may lead to the degradation of information fusion as illustrated by Hegazy and Vachtsevanos [16].

Density Control. The main node scheduling objective is to save energy using a density control algorithm [17–20]. Such algorithms manage the network density by determining when each node will be operable (awake) and when it will be inoperable (asleep). Figure 1.2 depicts an example of the result of a node scheduling algorithm in which gray nodes are asleep because their sensing areas are already covered by awaken nodes (in black).

Density control is an inherently localized algorithm where each node assesses its vicinity to decide whether or not it will be turned on. Some of the node scheduling algorithms, such as GAF [17], SPAN [19], and STEM [18], consider only the communication range to choose whether or not a node will be awake. Therefore, it is possible that some regions remain uncovered, and the application may not detect an event. Other solutions, such as PEAS [20], try to preserve the coverage. However, none of the current node scheduling algorithms consider the information fusion accuracy. As a result, nodes that are important to information fusion might be turned off. A key issue regarding density control algorithms is the integration with other functions such as data routing. Siqueira et al. [21] propose two ways of integrating density control and data routing: synchronizing both algorithms or redesigning an integrated algorithm.

1.2.2 Data Communication

In wireless sensor networks, the problem of data communication is mainly related to medium access control, routing, and transport protocols.

MAC Protocols. The link or medium access control (MAC) layer controls the node access to the communication medium by means of techniques such as contention [22, 23] and time division [24, 25]. Basically, the MAC layer must manage the communication channels available for the node, thereby avoiding collisions and errors in the communication.

Most solutions try to provide a reliable and energy-efficient solution. In this direction, Ci et al. [26] use prediction techniques to foresee the best frame size to reduce the packet size and save energy. To avoid transmitting packets under unreliable conditions, Polastre et al. [23] apply filter techniques to estimate ambient noise and determine whether the channel is clear for transmission. Liang and Ren [27] propose a MAC protocol based on a fuzzy logic rescheduling scheme that improves existing energy-efficient protocols. Their input variables are the ratios of nodes that (i) have an overflowed buffer, (ii) have a high failing transmission rate, and (iii) are experiencing an unsuccessful transmission.

Routing Protocols. Routing is the process of sending a data packet from a given source to a given destination, possibly using intermediate nodes to reach the final entity. This is the so-called unicast communication. In WSNs, data communication, from the point of view of the communicating entities, can be divided into three cases: from sensor nodes to a monitoring node, among neighbor nodes, and from a monitoring node to sensor nodes. Data communication from sensor nodes to a monitoring node is used to send the sensed data collected by the sensors to a monitoring application. This class includes most of the routing protocols proposed in the literature [28]. Data communication among neighbor nodes often happens when some kind of cooperation among nodes is needed. Data communication from a monitoring node to a set of sensor nodes is often used to disseminate a piece of information that is important to those nodes. Based on an efficient dissemination algorithm, a monitoring node can perform different activities, such as to change the operational mode of part or the entire WSN, broadcast a new interest to the network, activate/deactivate one or more sensor nodes, and send queries to the network.

The routing algorithms for wireless sensor networks can be broadly divided into three types: flat-based routing, hierarchical-based routing, and adaptive-based routing. Flat-based routing assumes that all sensor nodes perform the same role. On the other hand, nodes in hierarchical-based routing have different roles in the network, which can be static or dynamic. Adaptive routing changes its behavior according to different application and network conditions such as available energy resources. These routing protocols can be further classified into multipath-based, query-based, or negotiation-based routing techniques depending on the protocol operation.

A natural routing scheme for flat networks is the formation of routing trees. Krishnamachari et al. [29] provide analytical bounds on the energy costs and savings that can be obtained with data aggregation using tree topologies. Zhou and Krishnamachari [30] evaluate the tree topology with four different parent selection strategies (earliest-first, randomized, nearest-first, and weighted-randomized) based on the metrics, such as node degree, robustness, channel quality, data aggregation, and latency. Tian and Georganas [31] identify drawbacks of pure single-path and multipath routing schemes in terms of packet delivery and energy consumption. The InFRA algorithm [32] builds a routing tree by establishing a hybrid network organization in which source nodes are organized into clusters and the cluster-to-sink communication occurs in a multihop fashion. The resulting topology is a distributed heuristic to the Steiner tree problem.

For the hierarchical topology, several algorithms are provided in the literature. LEACH [33] is a cluster-based protocol that randomly rotates the cluster heads to evenly distribute the energy load among the sensors in the network. PEGASIS [34] is an improvement of LEACH in which sensors form chains, and each node communicates only with a close neighbor and takes turns to transmit messages to the sink node.

The Directed Diffusion [35] is a pioneer protocol that tries to find the best paths from sources to sink nodes that might receive data from multiple paths with different data delivery frequencies. If the best path fails, another path with lower data delivery frequency assures the data delivery. Ganesan et al. [36] propose a routing solution, which evolved from Directed Diffusion, that tries to discover and maintain alternative paths, connecting sources to sinks, to make the network more fault-tolerant.

Niculescu and Nath [37] propose the Trajectory-Based Forwarding (TBF) algorithm, a data dissemination technique in which packets are disseminated from a monitoring node to a set of nodes along a predefined curve. Machado et al. [38] extend TBF with the information provided by the energy map [39] of a sensor network to determine routes in a dynamic fashion.

In WSNs, routing protocols are closely related to information fusion because it addresses the problem of delivering the sensed information to the sink node, and it is natural to think of performing the fusion while the pieces of data become available. However, the way information is fused depends on the network organization, which directly affects how the role can be assigned. Hierarchical networks are organized into clusters where each node responds only to its respective cluster-head, which might perform special operations such as data fusion/aggregation. In flat networks, communication is performed hop-by-hop and every node may be functionally equivalent.

Transport Protocols. In general, transport protocols are concerned with the provision of a reliable communication service for the application layer. This is the main objective of the Pump Slowly, Fetch Quickly (PSFQ) protocol [40]. PSFQ is an adaptive protocol that makes local error correction using hop-by-hop acknowledgement. In this case, the adaptation means that under low failure rates, the communication is similar to a simple forward, and when failures are frequent, it presents a store-and-forward scheme. Another transport protocol that aims to provide a reliable communication is the Reliable Data Transport in Sensor Networks (RMST) [41] that also implements a hop-by-hop acknowledgment. However, RMST is designed to operate in conjunction with Directed Diffusion.

An interesting approach is introduced by the Event-to-Sink Reliable Transfer (ESRT) protocol [42, 43]. This protocol is designed for event-based sensor networks, and it changes the focus of traditional transport protocols. The authors state that for WSNs a transport protocol should be reliable regarding the event detection task. ESRT assumes that an event must be detected when the sink node receives a minimum number of event reports from sensor nodes. If this threshold is not achieved, the sink node does not recognize the event. Thus, ESRT adjusts the transmission rate of each node in such a way that the desired threshold is achieved and the event is reliably detected.

1.2.3 Management Functions

In the following, we present some high-level management functions that can be used by different monitoring applications in a WSN. We start by presenting a management architecture, followed by a discussion of data storage, network health, coverage and exposure, and security.

Architecture. A WSN management architecture can be used to reason about the different dimensions present in the sensor network. In this direction, the MANNA architecture [44] was proposed to provide a management solution to different WSN applications. It provides a separation between both sets of functionalities (i.e., application and management), making integration of organizational, administrative, and maintenance activities possible for this kind of network. The approach used in the MANNA architecture works with each functional area, as well as each management level, and proposes the new abstraction level of WSN functionalities (configuration, sensing, processing, communication, and maintenance) presented earlier. As a result, it provides a list of management services and functions that are independent of the technology adopted.

Data Storage. Data storage is closely related to the routing (data retrieval) strategy. In the Cougar database system [45], stored data are represented as relations whereas sensor data are represented as time series. A query formulated over a sensor network specifies a persistent view, which is valid during a given period [46]. Shenker et al. [47] introduce the concept of data-centric storage, which is also explored by Ratnasamy et al. [48] and Ghose et al. [49]. In this approach, relevant data is labeled (named) and stored by the sensor nodes. Data with the same name are stored by the same sensor node. Queries for data with a particular name are sent directly to the node storing that named data, avoiding the flooding of interests or queries.

Network Health. An important issue underlying WSNs is the monitoring of the network itself; that is, the sink node needs to be aware of the health of all the sensors. Jaikaeo et al. [50] define diagnosis as the process of monitoring the state of a sensor network and figuring out the problematic nodes. This is a management activity that assesses the network health—that is, how well the network elements and the resources are being applied.

Managing individual nodes in a large-scale WSN may result in a response implosion problem that happens when a high number of replies are triggered by diagnostic queries. Jaikaeo et al. [50] suggest the use of three operations, built on the top of the SINA architecture [51], to overcome the implosion problem: sampling, self-orchestrated, and diffused computation. In a sampling operation, information from each node is sent to the manager without intermediate processing. To avoid the implosion problem, each node decides whether or not it will send its information based on a probability assigned by the manager (based on the node density). In a self-orchestrated operation, each node schedules its replies. This approach introduces some delay, but reduces the collision chances. In a diffused computation, mobile scripts are used (enabled by the SINA architecture) to assign diagnosis logic to sensor nodes so they know how to perform information fusion and route the result to the manager. Although diffused computation optimizes bandwidth use, it introduces greater delay and the resultant information is less accurate. The three operations provide different levels of granularity and delay; therefore they should be used in different stages: Diffused computation and self-orchestrated operations should be continuously performed to identify problems, and sampling should be used to identify problematic elements.

Hsin and Liu [52] propose a two-phase timeout system to monitor the node liveliness. In the first phase, if a node A receives no message from a neighbor D in a given period of time (monitoring time), A assumes that D is dead, entering in the second phase. Once in the second phase, during another period of time (query time), A queries its neighbors about D; if any neighbor claims that D is alive, then A assumes it was a false alarm and discards this event. Otherwise, if A does not hear anything before the query time expires, it assumes that D is really dead, triggering an alarm. This monitoring algorithm can be seen as a simple information fusion method for liveliness detection where the operator (fuser) is a logical OR with n inputs such as input i is true if neighbor i considers that D is alive and false otherwise.

Zhao et al. [53] propose a three-level health monitoring architecture for WSN. The first level includes the *digests* that are aggregates of some network property, like minimum residual energy. The second comprises the network *scans*, a sort of feature map that represents abstracted views of resource utilization within a section of the (or entire) network [54]. Finally, the third is composed by *node dumps* that provide detailed node states over the network for diagnosis. In this architecture, digests should be continuously computed in background and piggybacked in a neighbor-to-neighbor communication. Once an anomaly is detected in the digests, a network scan may be collected to identify the problematic sections in the network. Finally, dumps of problematic sections can be requested to identify what is the problem. The information granularity increases from digests to dumps, and the finer the granularity, the greater the cost. Therefore, network scans and, especially, dumps should be carefully used.

An energy map is the information about the amount of energy available at each part of the network. Due to the importance of energy-efficiency solutions for WSNs, the energy map can be useful to prolong the network lifetime and be applied to different network activities in order to make a better use of the energy reserves. Thus, the cost of obtaining the energy map can be amortized among different network applications, and neither of them has to pay exclusively for this information itself. The energy map can be constructed using a naive approach, in which each node sends periodically only its available energy to the monitoring node. However, this approach would spend so much energy, due to communication, that probably the utility of the energy information would not compensate the amount of energy spent in this process.

Zhao et al. [55] propose a more interesting solution that obtains the energy map using an aggregation based approach. Mini et al. [39] propose another efficient solution, based on a Markov Chain mechanism, to predict the energy consumption of a sensor node in order to construct the energy map.

Coverage and Exposure. Coverage (spatial) comprises the problem of determining the area covered by the sensors in the network [13, 14, 56, 57]. Coverage allows the identification of regions that can be properly monitored and regions that cannot. This information associated with the energy map [54] can be used to schedule sensor nodes to optimize the network lifetime without compromising the quality of the gathered information.

Azzedine Boukerche [57] defines coverage in terms of the best case (regions of high observability) and the worst case (regions of low observability), and it is computed in a centralized fashion by means of geometric structures (Delaunay triangulation and Voronoi diagram) and algorithms for graph searching. Li et al. [56] extend this work considering a sensing model in which the sensor accuracy is inversely proportional to the distance to the sensed event, and they provide distributed algorithms to compute the best case of coverage and the path of greater observability. Chakrabarty et al. [14] compare coverage to the Art Gallery Problem (AGP), which consists in finding the smallest number of guards to monitor the entire art gallery. Dhillon et al. [13] consider coverage as the lowest detection probability of an event by any sensor. Exposure is closely related with coverage and it specifies how well an object, moving arbitrarily, can be observed by the WSN over a period of time [58].

Security. Security is an issue of major concern in WSNs, especially in surveillance applications, with implication to other functions. For instance, despite the fact that data fusion can reduce communication, fusing data packets makes security assurance more complex. The reason is that intermediate nodes can modify, forge, or drop data packets. In addition, source-to-sink data encryption may not be desirable because the intermediate nodes need to understand the data to perform data fusion.

Hu and Evans [59] present a protocol to provide secure aggregation for flat WSNs that is resilient to intruder devices and single device key compromises, but their protocol may become vulnerable when a parent and a child node are compromised. The Energy-efficient and Secure Pattern-based Data Aggregation protocol (ESPDA) [60] is a secure protocol for hierarchical sensor networks that does not require the encrypted data to be decrypted by cluster heads to perform data aggregation. In ESPDA, the cluster head first requests nodes to send the corresponding pattern code for the sensed data. If the same pattern code is sent to the cluster head by different nodes, then only one of them is allowed to send its data. The pattern code is generated based on a seed provided by the cluster head. No special fusion method is actually applied in the ESPDA protocol, which simply avoids the transmission of redundant data, so any information fusion must be performed by the sensor nodes, not the cluster head. Secure Information Aggregation in Sensor Networks (SIA) [61] presents a fuse–commit–prove approach in which fuser nodes need to prove that they perform fusion tasks correctly. To avoid cheating by fuser nodes, SIA adopts cryptographic techniques of commitments and provides random sampling mechanisms and interactive proofs to allow the user to verify the data given by fuser nodes, even when the fuser nodes or some sensor nodes are corrupted.

1.2.4 Applications

Two of the most basic applications for wireless sensor networks are query processing, and event and target tracking. The former is often used to answer queries posed by users outside of the network, and the latter is used to know about events happening inside the network, including specific targets. These two applications can actually be seen as *application protocols* that might be present in different monitoring applications.

Query Processing. Different solutions explore the query approach using in-network processing to filter and/or aggregate the data during the routing process. Directed Diffusion [35] introduces the concept of interests to specify which data will be delivered through a publish/subscribe scheme, but no query language is specified.

Another possibility is to model the sensor network as a database so data access is performed by declarative queries. The DataSpace Project [62] provides a means of geographically querying, monitoring, and controlling the network devices that encapsulate data. DataSpace provides network primitives to assure that only relevant devices are contacted when a query is evaluated. Sensor Information Networking Architecture (SINA) [51] is a cluster-based architecture that abstracts a WSN as a dense collection of distributed objects where users access information through declarative queries and execute tasks through programming scripts. The Cougar Project [45] handles the network as a distributed database in which each piece of data is locally stored in a sensor node and data are retrieved by performing aggregation along a query tree. Temporal coherency-aware in-Network Aggregation (TiNA) [63] uses temporal coherency tolerances to reduce the communication load and improve quality of data when not all sensor readings can be propagated within a given time constraint. The ACtive QUery forwarding In sensoR nEtworks (ACQUIRE) [64] system considers the query as an active entity that is forwarded through the network searching for a solution. In ACQUIRE, intermediate nodes, handling the active query, partially evaluate the queries by using information from nodes within d hops. Once the query is fully evaluated, a response is sent toward the querying node. TinyDB [65] provides a simple query language to specify the data of interest.

Event and Target Tracking. Event (target) tracking is one of the most popular applications of sensor systems in general. The problem consists in predicting where an event or target being detected is moving to. This is essentially a data fusion application.

Coates [66] uses filters for target tracking in cluster-based networks in which cluster heads perform computations and share information, and the other cluster members sense the environment. To track multiple targets, Sheng et al. [67] use filters that run on uncorrelated sensor cliques that are dynamically organized based on target trajectories. Vercauteren et al. [68] propose a collaborative solution for jointly tracking several targets and classifying them according to their motion pattern. Schmitt et al. [69] propose a collaborative algorithm to find the location of mobile robots in a known environment and track moving objects.

1.2.5 Data Fusion

Data fusion algorithms [70] are orthogonal to the above-mentioned problems, in the sense that these algorithms can be applied to any solution that needs to make inferences or improve estimates.

Classical data fusion techniques have been used to assist solving many problems. For instance, the Least-Squares method has been used to predict sensor data [71] and find nodes' locations [8, 12]; the moving average filter has been used to estimate link connectivity statistics [72], estimate data traffic [73] and the number of events [74], and track targets [75]; the Kalman filter has been applied to refine location and distance estimates [6, 76], track different targets [77], predict the best frame size for MAC protocols [26], and predict sensor data to reduce communication [78].

As discussed in the following, data fusion can have an important role when we design an integrated solution for a wireless sensor network.

1.3 CHALLENGE: SYNTHESIS PROCESS

One of the most important challenges in the design of wireless sensor networks is to deal with the dynamics of such networks. The physical world where the sensors are embedded is dynamic. Over time, the operating conditions and the associate tasks to be performed by the sensors can change. Some of the causes that might trigger these changes are the events occurring in the network, amount of resources available at nodes (particularly energy), and reconfiguration of nodes. Furthermore, it is important that sensors adapt themselves to the environment since manual configuration may be unfeasible or even impossible. In summary, the kind of distributed system we are dealing with calls for an entire new class of algorithms for large-scale, highly dynamic, and unattend WSN.

The complete design of a wireless sensor network, considering a particular application, should take into account many different aspects such as application goals, traffic pattern, sensor node capability and availability, expected network lifetime, access to the monitoring area, node replacement, environment characteristics, and cost. Given a particular monitoring application, the network designer should clearly identify its main goals and the corresponding QoS parameters. For instance, given a fire detection application for a rain forest, we would like to guarantee that the network will operate for the expected lifetime. However, as soon as a fire spot is detected, this information should reach the sink node as fast and reliable as possible, probably not worrying about the energy expenditure of the nodes involved in this communication.

Power-efficient communication paradigms for a given application should consider both routing and media access algorithms. The routing algorithms must be tailored for efficient network communication while maintaining connectivity when required to source or relay packets. In this case, the research challenge of the routing problem is to find a power-efficient method for scheduling the nodes such that a multihop path may be used to relay the data. But, when we consider the particular aspects of the monitoring application, we could apply, for instance, information fusion and density control algorithms to reduce the amount of data packets to be relayed and sensor nodes that need to be active, respectively.

As the sensor network starts to operate, it may be necessary to adjust the functionality of individual nodes. This refinement can take several different forms. Scalar parameters, like duty cycle or sampling rates, may be adjusted using self-configuration and self-organization algorithms. This process may occur in different ways along the operation of the network lifetime.

Ideally, a WSN designer should come up with both the hardware and software necessary to accomplish the aspects mentioned above. Unfortunately, it seems that we are far from this scenario. We are still giving the first steps in the design process of a wireless sensor network as we move toward to a more disciplined development. Most of the studies found in the literature study particular problems for a WSN. That is possibly the way we should go since we need to have more experience before we can design a complete solution in a more systematic and automated way.

Figure 1.3 depicts a possible monitoring application for a rain forest. In this case, we might be interested in detecting different events such as the presence of a rare bird, a fire spot, and different environmental variables. The operation of the sensor

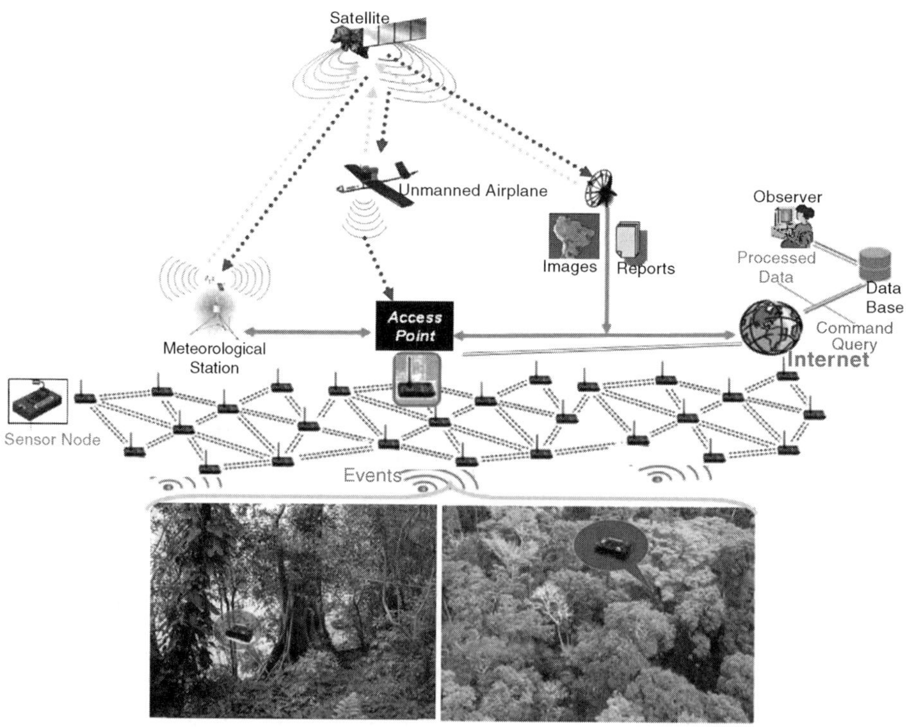

Figure 1.3. Monitoring application for a rain forest.

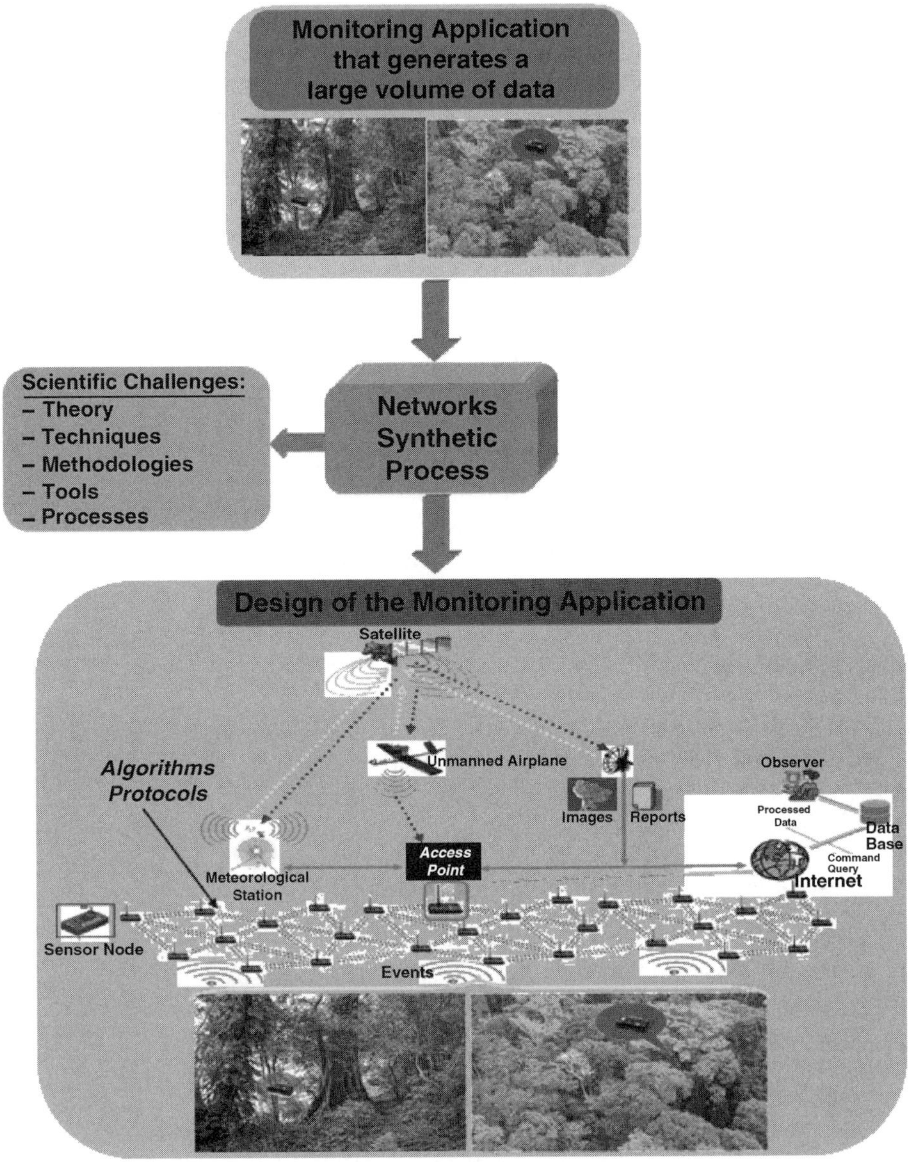

Figure 1.4. Synthesis process.

network can also be based on data received from a meteorological station, an unmanned airplane, or a satellite. Thus, given the different application requirements and data sources, what are the best algorithms and sensor nodes that should be used to accomplish the desired goals? This is a research challenge that we are starting to face once more, and more real monitoring applications are being deployed. Notice that we can even go one step further and build a specific hardware node that best fits to the proposed solution, leading to a truly hardware–software codesign.

In order to achieve this proposed solution, we need a network synthesis process, as depicted in Figure 1.4. This is similar to what happens currently in the design of an integrated circuit (IC) that starts with its high-level specification and finishes with its physical design. The synthesis process is guided by some aspects such as the testability of the IC. It is important to design a more testable IC, since a chip is tested not to check its logical correctness but to check its manufacturing process. In the case of the WSN synthesis process, there are very interesting scientific challenges that we need to overcome to have this automated development, as it happens in the synthesis of an integrated circuits.

These challenges are related to the theory, techniques, methodologies, tools, and processes. We need to propose new fundamental principles that will create a theory to synthesize both the hardware and software of a wireless sensor network. This theory will lead to techniques, methodologies, tools, and processes that will enable designers to design new sensor networks for different monitoring applications in a systematic way. In this vision, algorithms for wireless sensor networks have a fundamental role, since they will be the outcome of this synthesis process.

BIBLIOGRAPHY

1. J. Feng, F. Koushanfar, and M. Potkonjak. Localized algorithms for sensor networks. In I. Mahgoub and M. Ilyas, editors, *Handbook of Sensor Networks*, CRC Press, Boca Raton, FL, 2004.
2. L. Doherty, K. S. J. Pister, and L. El Ghaoui. Convex position estimation in wireless sensor networks. In INFOCOM [79], p. 1655–1663.
3. D. Niculescu and B. Nath. Ad hoc positioning system (APS). In *2001 IEEE Global Telecommunications Conference (GLOBECOM '01)*, Vol. 5, San Antonio, TX, Novermber 2001, IEEE, New York, pp. 2926–2931.
4. C. Savarese, J. M. Rabaey and J. Beutel. Locationing in distributed ad-hoc wireless sensor networks. In *Proceedings of the IEEE Signal Processing Society International Conference on Acoustics, Speech, and Signal Processing 2001 (ICASSP '01)*, Vol. 4, Salt Lake City, UT, May 2001, IEEE, New York, pp. 2037–2040.
5. A. Savvides, H. Park and M. B. Srivastava. The bits and flops of the n-hop multilateration primitive for node localization problems. In *Proceedings of the 1st ACM International Workshop on Wireless Sensor Networks and Applications (WSNA'02)*, Atlanta, GA, September 2002, ACM, New York, pp. 112–121.
6. A. Savvides, C.-C. Han and M. B. Strivastava The n-hop multilateration primitive for node localization. *Mobile Networks and Applications*, **8**(4):443–451, 2003, ISSN 1383-469X. doi: http://dx.doi.org/10.1023/A:1024544032357.
7. L. Hu and D. Evans. Localization for mobile sensor networks. In *Proceedings of the 10th Annual International Conference on Mobile Computing and Networking (MobiCom '04)*, Philadelphia, PA, 2004, ACM, New York, pp. 45–57, ISBN 1-58113-868-7. doi: http://doi.acm.org/10.1145/1023720.1023726.

8. H. A. B. F. de Oliveira, E. F. Nakamura, A. A. F. Loureiro, and A. Boukerche. Directed position estimation: A recursive localization approach for wireless sensor networks. In S. R. Thuel, Y. Yang, and E. K. Park, editors, *Proceedings of the 14th IEEE International Conference on Computer Communications and Networks (IC3N '05)*, San Diego, CA. October 2005, IEEE, pp. 557–562, ISBN 0-7803-9428-3.

9. J. D. Gibson, editor. *The Mobile Communication Handbook*, 2nd edition CRC Press, Boca Raton, FL, 1999.

10. A. Savvides, C.-C. Han, and M. B. Strivastava. Dynamic fine-grained localization in ad-hoc networks of sensors. In Mobicom [80], pp. 166–179, ISBN 1-58113-422-3.

11. M. L. Sichitiu and V. Ramadurai. Localization of wireless sensor networks with a mobile beacon. In *Proceedings of the 1st IEEE International Conference on Mobile Ad Hoc and Sensor Systems (MASS 2004)*, Fort Lauderdale, FL, October 2004, IEEE, pp. 174–183.

12. J. Liu, Y. Zhang, and F. Zhao. Robust distributed node localization with error management. In *Proceedings of the 7th ACM International Symposium on Mobile Ad Hoc Networking and Computing (MobiHoc'06)*, Florence, Italy, 2006, ACM, New York, pp. 250–261, ISBN 1-59593-368-9. doi: http://doi.acm.org/10.1145/1132905.1132933.

13. S. S. Dhillon, K. Chakrabarty, and S. S. Iyengar. Sensor placement for grid coverage under imprecise detections. In *Proceedings of 5th International Conference on Information Fusion (Fusion 2002)*, Vol. 2, Annapolis, MD, July 2002, IEEE, New York, pp. 1581–1587.

14. K. Chakrabarty, S. S. Iyengar, H. Qi, and E. Cho. Grid coverage for surveillance and target location in distributed sensor networks. *IEEE Transactions on Computers*, **51**(12):1448–1453, 2002.

15. E. S. Biagioni and G. Sasaki. Wireless sensor placement for reliable and efficient data collection. In *Hawaii International Conference on Systems Sciences (HICSS 2003)*, Hawaii, January 2003, IEEE, New York, pp. 127–136.

16. T. Hegazy and G. J. Vachtsevanos. Sensor placement for isotropic source localization. In F. Zhao and L. J. Guibas, editors, *IPSN '03*, Vol. 2634 of *Lecture Notes in Computer Science*, Palo Alto, CA, April 2003, Springer, New York, pp. 432–441, ISBN 3-540-02111-6.

17. Y. Xu, J. Heidemann, and D. Estrin. Geography-informed energy conservation for ad-hoc routing. In Mobicom [80], pp. 16–21, ISBN 1-58113-422-3.

18. C. Schurgers, V. Tsiatsis, S. Ganeriwal, and M. Srivastava. Optimizing sensor networks in the energy-latency-density design space. *IEEE Transactions on Mobile Computing*, **1**(1): 70–80, 2002.

19. B. Chen, K. Jamieson, H. Balakrishnan, and R. Morris. Span: An energy-efficient coordination algorithm for topology maintenance in ad hoc wireless networks. *Wireless Networks*, **8**(5):481–494, 2002.

20. F. Ye, G. Zhong, J. Cheng, S. Lu, and L. Zhang. PEAS: A robust energy conserving protocol for long-lived sensor networks. In *Proceedings of the 23rd International Conference on Distributed Computing Systems*, Providence, RI, May 2003, IEEE, New York, pp. 28–37.

21. I. G. Siqueira, C. M. S. Figueiredo, A. A. F. Loureiro, J. M. S. Nogueira, and L. B. Ruiz. An integrated approach for density control and routing in wireless sensor networks. In *Proceedings of the IEEE 20th International Parallel and Distributed Processing Symposium (IPDPS '06)*. April 2006, IEEE, New York, CD-ROM.

22. A. Woo and D. E. Culler. A transmission control scheme for media access in sensor networks. In Mobicom [80], pp. 221–235, ISBN 1-58113-422-3.

23. J. Polastre, J. Hill, and D. Culler. Versatile low power media access for wireless sensor networks. In J. A. Stankovic, A. Arora, and R. Govindan, editors, *Proceedings of the 2nd International Conference on Embedded Networked Sensor Systems (SenSys'04)*, Baltimore, MD, November 2004, ACM, New York, pp. 95–107, ISBN 1-58113-879-2.

24. W. Ye, J. Heidemann, and D. Estrin. An energy-efficient MAC protocol for wireless sensor networks. In INFOCOM, editor, *INFOCOM 2002*, New York, June 2002. IEEE, New York, pp. 1567–1576, URL http://www.isi.edu/~johnh/PAPERS/Ye02a.html.

25. V. Rajendran, K. Obraczka, and J. J. Garcia-Luna-Aceves. Energy-efficient, collision-free medium access control for wireless sensor networks. In Akyildiz et al. [81], pp. 181–192, ISBN 1-58113-707-9.

26. S. Ci, H. Sharif, and K. Nuli. A UKF-based link adaptation scheme to enhance energy efficiency in wireless sensor networks. In *Proceedings of the 15th IEEE International Symposium on Personal, Indoor and Mobile Radio Communications (PIMRC '04)*, Vol. 4, Barcelona, Spain, September 2004, IEEE, pp. 2483–2488.

27. Q. Liang and Q. Ren. An energy-efficient MAC protocol for wireless sensor networks. In *2005 IEEE Global Telecommunications Conference (GLOBECOM '05)*, Vol. 1, St. Louis, MO, November–December 2005, IEEE, New York, CD-ROM.

28. I. F. Akyildiz, W. Su, Y. Sankarasubramaniam, and E. Cyirci. Wireless sensor networks: A survey. *Computer Networks*, **38**(4):393–422, 2002.

29. B. Krishnamachari, D. Estrin, and S. Wicker. The impact of data aggregation in wireless sensor networks. In *International Workshop of Distributed Event Based Systems (DEBS)*, Vienna, Austria, July 2002, IEEE, pp. 575–578.

30. C. Zhou and B. Krishnamachari. Localized topology generation mechanisms for self-configuring sensor networks. In *2003 IEEE Global Telecommunications Conference (GLOBECOM '03)*, Vol. 22, San Francisco, CA, December 2003, IEEE, New York, pp. 1269–1273.

31. D. Tian and N. D. Georganas. Energy efficient routing with guaranteed delivery in wireless sensor networks. In *IEEE Wireless Communications and Networking Conference (WCNC 2003)*, Vol. 3, New Orleans, LA, March 2003, IEEE, New York, pp. 1923–1929.

32. E. F. Nakamura, H. A. B. F. de Oliveira, L. F. Pontello, and A. A. F. Loureiro. On demand role assignment for event-detection in sensor networks. In P. Bellavista, C.-M. Chen, A. Corradi, and M. Daneshmand, editors, *Proceedings of the 11th IEEE International Symposium on Computers and Communication (ISCC '06)*, Cagliari, Italy, June 2006, IEEE, New York, pp. 941–947, ISBN 0-7695-2588-1. doi: http://doi.ieeecomputersociety.org/10.1109/ISCC.2006.110.

33. W. Heinzelman, A. Chandrakasan, and H. Balakrishnan. Energy-efficient communication protocol for wireless microsensor networks. In *Proceedings of the 33rd Hawaii International Conference on System Sciences (HICSS '00)*, Maui, HI, January 2000, IEEE, New York, pp. 8020–8029, ISBN 0-7695-0493-0.

34. S. Lindsey, C. Raghavendra, and K. M. Sivalingam. Data gathering algorithms in sensor networks using energy metrics. *IEEE Transactions on Parallel and Distributed Systems*, **13**(9):924–935, 2002.

35. C. Intanagonwiwat, R. Govindan, and D. Estrin. Directed diffusion: A scalable and robust communication paradigm for sensor networks. In *Proceedings of the 6th Annual International Conference on Mobile Computing and Networking (MobiCom '00)*, Boston, MA, August 2000, ACM, New York, pp. 56–67.

36. D. Ganesan, R. Govindan, S. Shenker, and D. Estrin. Highly-resilient, energy-efficient multipath routing in wireless sensor networks. *ACM SIG-MOBILE Mobile Computing and Communications Review*, **5**(4):11–25, 2001.

37. D. Niculescu and B. Nath. Trajectory-based forwarding and its applications. In *Proceedings of the 9th Annual International Conference on Mobile Computing and Networking (MobiCom '03)*, 2003, pp. 260–272.

38. M. V. Machado, O. Goussevskaia, R. A. F. Mini, C. G. Rezende, A. A. F. Loureiro, G. R. Mateus, and J. M. S. Nogueira. Event-to-sink reliable transport in wireless sensor networks. *IEEE Journal on Selected Areas in Communications*, **23**(12), 2005. CD-ROM.

39. R. A. F. Mini, M. do V. Machado, A. A. F. Loureiro, and B. Nath. Prediction-based energy map for wireless sensor networks. *Ad Hoc Network Journal*, **3**:235–253, 2005.

40. C.-Y. Wan and A. Campbell. PSFQ: A reliable transport protocol for wireless sensor networks. In *Proceedings of the 1st ACM International Workshop on Wireless Sensor Networks and Applications (WSNA '02)*, Atlanta, GA, September 2002, ACM, New York, pp. 1–11.

41. F. Stann and J. Heidemann. RMST: Reliable data transport in sensor networks. In *Proceedings of the First IEEE International Workshop on Sensor Network Protocols and Applications (SNPA 2003)*, Anchorage, AK, May 2003, IEEE, pp. 102–112.

42. Y. Sankarasubramaniam, O. B. Akan, and I. F. Akyildiz. ESRT: Event-to-sink reliable transport in wireless sensor networks. In *Proceedings of the 4th ACM International Symposium on Mobile Ad Hoc Networking and Computing (MobiHoc '03)*, Annapolis, MD, June 2003, ACM, New York, pp. 177–188.

43. O. B. Akan and I. F. Akyildiz. Event-to-sink reliable transport in wireless sensor networks. *IEEE/ACM Transsactions on Networking*, **13**(5):1003–1016, 2005, ISSN 1063-6692.

44. L. B. Ruiz, J. M. S. Nogueira, and A. A. F. Loureiro. Manna: A management architecture for wireless sensor networks. *IEEE Communications Magazine*, **41**(2):116–125, 2005.

45. Y. Yao and J. Gehrke. The cougar approach to in-network query processing in sensor networks. *Sigmod Record*, **31**(3):9–18, 2002.

46. P. Bonnet, J. Gehrke, and P. Seshadri. Towards sensor database systems. In K.-L. Tan, M. J. Franklin, and J. C. S. Lui, editors, *Proceedings of the 2nd International Conference on Mobile Data Management*, Vol. 1987 of *Lecture Notes in Computer Science*, Hong Kong, China, January 2001, Springer, pp. 3–14.

47. S. Shenker, S. Ratnasamy, B. Karp, R. Govindan, and D. Estrin. Data-centric storage in sensornets. *ACM SIGCOMM Computer Communication Review*, **33**(1):137–142, 2003.

48. S. Ratnasamy, B. Karp, S. Shenker, D. Estrin, R. Govindan, L. Yin, and F. Yu. Data-centric storage in sensornets with ght, a geographic hash table. *Mobile Networks and Applications*, **8**(4):427–442, 2003.

49. A. Ghose, J. Grossklags, and J. Chuang. Resilient data-centric storage in wireless ad-hoc sensor networks. In M.-S. Chen, P. K. Chrysanthis, M. Sloman, and A. B. Zaslavsky, editors, *Proceedings of the 4th International Conference on Mobile Data Management*, Vol. 2574 of *Lecture Notes in Computer Science*, Melbourne, Australia, January 2003, Springer, New York, pp. 45–62.

50. C. Jaikaeo, C. Srisathapornphat, and C.-C. Shen. Diagnosis of sensor networks. In *Proceedings of the 2001 IEEE International Conference on Communications (ICC'01)*, Vol. 1, Helsinki, Finland, June 2001, IEEE, New York, pp. 1627–1632.

51. C.-C. Shen, C. Srisathapornphat, and C. Jaikaeo. Sensor information networking architecture and applications. *IEEE Personal Communications Magazine*, **8**(4):52–59, 2001.

52. C.-f. Hsin and M. Liu. A distributed monitoring mechanism for wireless sensor networks. In *Proceedings of the ACM Workshop on Wireless Security (WiSe'02)*, Atlanta, GA, September 2002, ACM, New York, pp. 57–66.

53. J. Zhao, R. Govindan, and D. Estrin. Computing aggregates for monitoring wireless sensor networks. In *Proceedings of the 1st IEEE International Workshop on Sensor Network Protocols and Applications (SNPA 2003)*, Anchorage, AK, May 2003, IEEE, New York, pp. 139–148.

54. J. Zhao, R. Govindan, and D. Estrin. Residual energy scans for monitoring wireless sensor networks. In *Proceedings of the IEEE Wireless Communications and Networking Conference (WCNC '02)*, Vol. 1, Orlando, FL, USA, March 2002, IEEE, New York, pp. 356–362.

55. Y. J. Zhao, R. Govindan, and D. Estrin. Residual energy scans for monitoring wireless sensor networks. In *IEEE Wilress Communications and Networking Conference (WCNC '02)*, Orlando, FL, March 2002, CD-ROM.

56. X.-Y. Li, P.-J. Wan, and O. Frieder. Coverage in wireless ad-hoc sensor networks. *IEEE Transactions on Computers*, **52**(6):753–763, 2003.

57. X. f. A. Boukerche. A coverage-preserving scheme for wireless sensor network with irregular sensing range. In INFOCOM [79], pp. 1303–1316.

58. S. Megerian, F. Koushanfar, G. Qu, G. Veltri, and M. Potkonjak. Exposure in wireless sensor networks: Theory and practical solutions. *Wireless Networks*, **8**(5):443–454, 2002.

59. L. Hu and D. Evans. Secure aggregation for wireless networks. In *Workshop on Security and Assurance in Ad Hoc Networks*, Orlando, FL, USA, January 2003, IEEE, New York, pp. 384–391.

60. H. Cam, S. Ozdemir, P. Nair, and D. Muthuavinashiappan. ESPDA: Energy-efficient and secure pattern-based data aggregation for wireless sensor networks. In *Proceedings of the IEEE Sensors*, Vol. 2, Toronto, Canada, October 2003, IEEE, New York, pp. 732–736.

61. B. Przydatek, D. Song, and A. Perrig. SIA: Secure information aggregation in sensor networks. In Akyildiz et al. [81], pp. 255–265, ISBN 1-58113-707-9.

62. T. Imielinski and S. Goel. DataSpace: Querying and monitoring deeply networked collections in physical space. *IEEE Personal Communications*, **7**(5):4–9, 2000.

63. M. A. Sharaf, J. Beaver, A. Labrinidis, and P. K. Chrysanthis. TiNA: A scheme for temporal coherency-aware in-network aggregation. In *Proceedings of the 3rd ACM International Workshop on Data Engineering for Wireless and Mobile Access*, San Diego, CA, September 2003, ACM, New York, pp. 69–76.

64. N. Sadagopan, B. Krishnamachari, and A. Helmy. Active query forwarding in sensor networks. *Ad Hoc Networks*, **3**(1):91–113, 2005.

65. S. R. Madden, M. J. Franklin, J. M. Hellerstein, and W. Hong. TinyDB: An acqusitional query processing system for sensor networks. *ACM Transactions on Database Systems*, **30**(1):122–173, 2005, ISSN 0362-5915.

66. M. Coates. Distributed particle filters for sensor networks. In K. Ramchandran, J. Sztipanovits, J. C. Hou, and T. N. Pappas, editors, *IPSN'04*, Berkeley, CA, April 2004, ACM, pp. 99–107, ISBN 1-58113-846-6.

67. X. Sheng, Y. H. Hu, and P. Ramanathan. Distributed particle filter with GMM approximation for multiple targets localization and tracking in wireless sensor network. In IPSN, editor, *IPSN'05*, Los Angeles, CA, USA, April 2005, IEEE, pp. 181–188, ISBN 0-7809-9201-9. doi: http://dx.doi.org/10.1109/IPSN.2005.1440923.

68. T. Vercauteren, D. Guo, and X. Wang. Joint multiple target tracking and classification in collaborative sensor networks. *IEEE Journal on Selected Areas in Communications*, **23**(4):714–723, April 2005, ISSN 0733-8716. doi: 10.1109/JSAC.2005.843540.

69. T. Schmitt, R. Hanek, M. Beetz, S. Buck, and B. Radig. Cooperative probabilistic state estimation for vision-based autonomous mobile robots. *IEEE Transactions on Robotics and Automation*, **18**(5):670–684, 2002.

70. A. Boukerche. In *Handbook of Algorithms for Wireless and Mobile Networks and Computing*, Chapman & Hall/CRC Press, Boca Raton, FL, 2005, pp. 841–864, ISBN 1-58488-465-7.

71. S. Santini and K. Römer. An adaptive strategy for quality-based data reduction in wireless sensor networks. In *Proceedings of the 3rd International Conference on Networked Sensing Systems (INSS 2006)*, Chicago, IL, June 2006, TRF, pp. 29–36, ISBN 0-9743611-3-5.

72. A. Woo, T. Tong, and D. Culler. Taming the underlying challenges of reliable multihop routing in sensor networks. In Akyildiz et al. [81], pp. 14–27, ISBN 1-58113-707-9.

73. E. F. Nakamura, F. G. Nakamura, C. M. S. Figueiredo, and A. A. F. Loureiro. Using information fusion to assist data dissemination in wireless sensor networks. *Telecommunication Systems*, **30**(1–3):237–254, 2005, ISSN 1018-4864.

74. C. M. S. Figueiredo, E. F. Nakamura, and A. A. F. Loureiro. An event-detection estimation model for hybrid adaptive routing in wireless sensor networks. In *Proceedings of the 2007 IEEE International Conference on Communications (ICC '07)*, Glasgow, Scotland, May 2007, IEEE, New York. IEEE CD-ROM.

75. C.-L. Yang, S. Bagchi, and W. J. Chappell. Location tracking with directional antennas in wireless sensor networks. In *2005 IEEE MTT-S International Microwave Symposium Digest*, Long Beach, CA, June 2005, IEEE, New York doi: 10.1109/MWSYM.2005.151640.

76. C. Hongyang, D. Ping, X. Yongjun, and L. Xiaowei. A robust location algorithm with biased extended Kalman filtering of TDOA data for wireless sensor networks. In *Proceedings of the International Conference on Wireless Communications, Networking and Mobile Computing (WCNM '05)*, Vol. 2, Wuhan, China, September 2005, IEEE, New York, pp. 883–886. doi: 10.1109/WCNM.2005.1544192.

77. T. Li, A. Ekpenyong, and Y.-F. Huang. Source localization and tracking using distributed asynchronous sensor. *IEEE Transactions on Signal Processing*, **54**(10):3991–4003, 2006.

78. A. Jain, E. Y. Chang, and Y.-F. Wang. Adaptive stream resource management using Kalman filters. In *Proceedings of the 2004 ACM SIGMOD International Conference on Management of Data (SIGMOD'04)*, Paris, France, 2004. ACM, New York, pp. 11–22, ISBN 1-58113-859-8. doi: http://doi.acm.org/10.1145.1007568.1007573.

79. INFOCOM, editor. *20th Annual Joint Conference of the IEEE Computer and Communications Societies (INFOCOM 2001)*, Anchorage, AK, April 2001, IEEE, New York.

80. Mobicom, editor. *Proceedings of the 7th Annual ACM/IEEE International Conference on Mobile Computing and Networking (Mobicom '01)*, Rome, Italy, July 2001, ACM, New York, ISBN 1-58113-422-3.

81. I. F. Akyildiz, D. Estrin, D. E. Culler, and M. B. Srivastava, editors. *Proceedings of the 1st International Conference on Embedded Networked Sensor Systems (SenSys '03)*, Los Angeles, CA, USA, November 2003, ACM, New York, ISBN 1-58113-707-9.

CHAPTER 2

Heterogeneous Wireless Sensor Networks

VIOLET R. SYROTIUK and BING LI

Department of Computer Science and Engineering, Arizona State University, Tempe, AZ 85287-8809

ANGELA M. MIELKE

Distributed Sensor Networks Group, Los Alamos National Laboratory, Los Alamos, NM 87545

2.1 INTRODUCTION

Wireless sensor networks (WSNs) have emerged as an important new class of computation that embeds computing in the physical world. To date, most of the work has focused on *homogeneous* WSNs, where all of the nodes in the network are of the same type. However, the continued advances in miniaturization of processors and in low-power communications combined with mass-produced sensors have enabled the development of a wide variety of nodes. When more than one type of node is integrated into a WSN, it is called *heterogeneous*. While many of the existing civilian and military applications of *heterogeneous wireless sensor networks* (H-WSNs) do not differ substantially from their homogeneous counterparts, there are compelling reasons to incorporate heterogeneity into the network. These include:

- Improving the scalability of WSNs.
- Addressing the problem of nonuniform energy drainage.
- Taking advantage of the multiple levels of fidelity available in different nodes.
- Reducing energy requirements without sacrificing performance.
- Balancing the cost and functionality of the network.
- Supporting new and higher-bandwidth applications.

As we will see, many of these reasons are interrelated. However, before discussing the new dimension that heterogeneity brings to the algorithms that run on such wireless

Algorithms and Protocols for Wireless Sensor Networks, Edited by Azzedine Boukerche

Figure 2.1. (a) Mica and (b) Stargate family of processors (not to scale).

sensor networks, we discuss the typical forms of and architectures for heterogeneous wireless sensor networks.

2.1.1 Forms of Heterogeneity

A sensor node is made up of four basic components [1]: a processing unit, a transceiver unit, a power unit, and a sensing unit. Heterogeneity may arise in each unit.

Nodes may vary significantly in their processing capability. For example, the Mica2 mote model MPR400CB [2] is based on the Atmel low-power ATmega128L microcontroller. It has only 128 Kbytes of program flash memory, 512 Kbytes of measurement (serial) flash memory, and 4 Kbytes of programmable read-only memory. In contrast, the Stargate [2] is a 400-MHz Intel PXA255 XScale processor with 64 Mbytes of synchronous dynamic random access memory and 32 Mbytes of flash memory. Figure 2.1 show the Mica and Stargate families of processors. Nodes with higher computational resources may perform more in-network processing, reducing the amount and/or frequency of sensed information that needs to travel through the network.

Often, nodes that vary in transceiver unit also vary in their power unit. For example the same Mica2 mote is a multichannel radio with four channels centered at 868 MHz. It supports a data rate of 38.4 Kbaud drawing 27 mA to transmit at maximum power, 10 mA to receive, and less than 1 μA to sleep. This mote is powered by two AA batteries. The Stargate runs a version of IEEE 801.11a/b and can run off a lithium-ion battery or an alternating-current power adaptor. Its power consumption is low, at less than 500 mA. Typically, nodes with more powerful energy resources are used to form a backbone of the network, taking the communication burden. In terms of energy consumption, the wireless exchange of data between nodes strongly dominates other node functions such as sensing and processing [3, 4].

A number of classes of sensors are available. These include light, temperature, relative humidity, barometric pressure, acceleration, seismic, acoustic, radar, magnetic, camera, and *global positioning system* (GPS) among others. In each class, the sensors vary greatly in fidelity and hence may vary significantly in accuracy and in reliability.

Higher fidelity usually comes at a higher cost, so balancing the functionality and the cost consequently impacts the network architecture.

Currently, almost all of the nodes in both the homogeneous and the heterogeneous WSNs considered are static. In the future, nodes that are mobile, such as those considered in reference 5, will inevitably be integrated into the network.

2.1.2 Architectures for Heterogeneous Wireless Sensor Networks

Two classes of architecture have emerged for heterogeneous WSNs: staged and hierarchical.

In a *staged architecture* the nodes are organized into a series of n tasks performed step-by-step. Within each stage, nodes are typically homogeneous, while successive stages are heterogeneous in their capabilities. As the nodes in stage i, $1 \leq i \leq n - 1$, complete their task, they trigger the nodes in the next stage $i + 1$ to carry out their task. Often, decreasing numbers of nodes of increasing fidelity are used in successive stages. Figure 2.2a illustrates this architecture of a heterogeneous WSN.

A *hierarchical architecture* is an organization of the nodes into a forest of trees. There are two ways in which the forest is commonly formed for n distinct node types. In the *single-hop* organization, each tree has as many levels n as node types. The root node (level zero) of each tree in the forest usually corresponds to nodes of highest fidelity. Nodes at level i, $1 \leq i \leq n - 1$, in a tree typically correspond to nodes of type i. Each leaf or intermediate node is connected directly (via an edge) to its parent. In a *multihop* organization, the key difference is that leaf nodes may traverse a multihop path to a node of higher fidelity. Therefore the number of levels in each tree does not equal n. Furthermore, each tree may have a different height.

Figure 2.2b shows a hierarchical architecture made up of two node types. The resulting forest has four trees, each of height one; each leaf node reaches its parent in a singlehop. In contrast, Figure 2.2c also shows a hierarchical architecture for two node types. Here the forest has four trees: one of height one, two of height two, and one of height three. Hence, some of the low-fidelity nodes traverse a multihop path to a node of higher fidelity.

While in each forest in Figures 2.2b and 2.2c the root of each tree forms a backbone of the network, there is an important difference between them. In Figure 2.2b the backbone is used to reach an information sink node, common in the architecture of WSNs. In Figure 2.2c the network has no sink node. The high-fidelity nodes may communicate amongst themselves and compute in a distributed manner.

Nodes physically organized into a hierarchy may be logically organized into stages to accomplish a series of tasks. However, this need not be the case. The hierarchy may exist solely to improve the scalability, functionality, or resource efficiency of the network.

2.1.3 Chapter Organization

The rest of this chapter is organized as follows. In order to motivate some of the problems that arise in heterogeneous WSNs, Section 2.2 describes two testbeds. *SensEye*

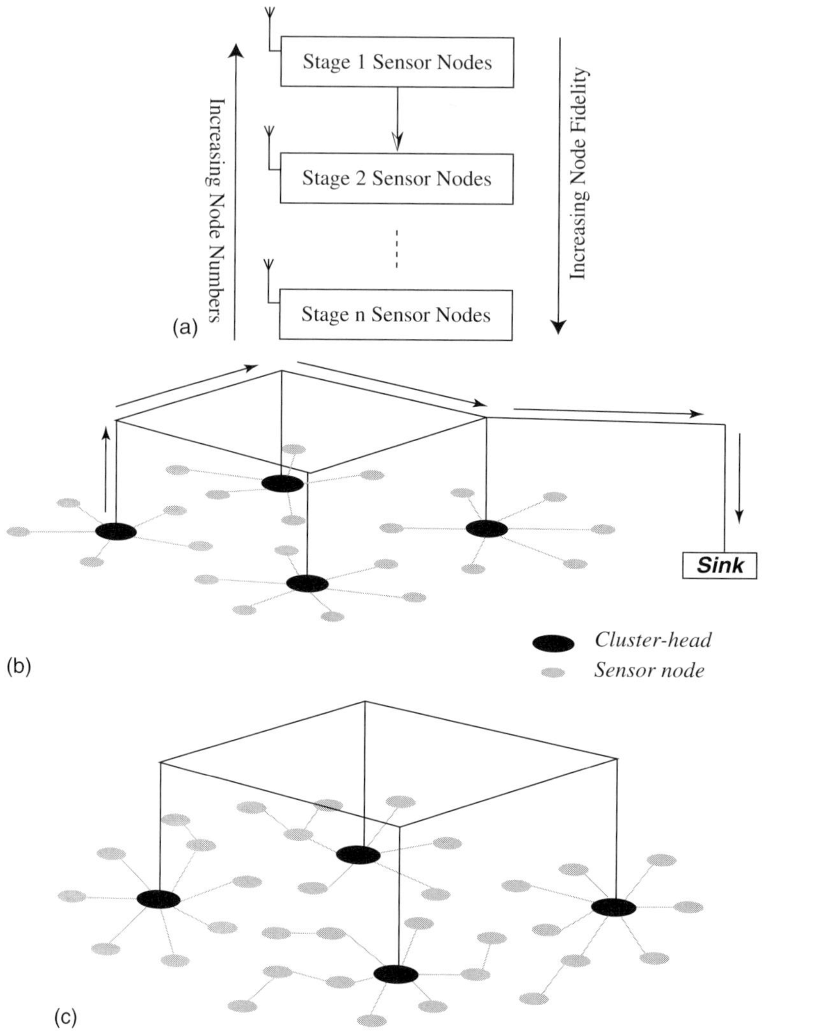

Figure 2.2. Architectures of heterogeneous WSNs: (a) Staged architecture (b) Single-hop hierarchical architecture with sink (c) Multi-hop hierarchical architecture without Sink.

is a testbed for monitoring and surveillance in which four types of cameras are used with three different hardware platforms, organized in a three-stage architecture. Radioactive source detection is the goal of the Los Alamos National Laboratory testbed. It, too, is organized in a three-stage architecture. Section 2.3 examines the problems of scalability and nonuniform energy drainage in homogeneous WSNs and how they are addressed in heterogeneous WSNs. Algorithms for topology formation and routing are presented. Coverage is a fundamental problem in homogeneous WSNs. The problems of differentiated and stochastic coverage in heterogeneous WSNs are discussed in Section 2.4. Section 2.5 examines issues in the management of heterogeneous networks. Section 2.6 presents two new applications, *live virtual reality* and *do not disturb*, that are enabled by heterogeneity. Section 2.7 provides a summary of established research projects in heterogeneous WSNs. As well, a summary of systems infrastructure (including system software, middleware, and simulators) under development for heterogeneous WSNs is provided. Finally, Section 2.8 provides many potential directions of study in this emerging area of research in heterogeneous wireless sensor networks.

2.2 HETEROGENEOUS WIRELESS SENSOR NETWORK TESTBEDS

2.2.1 A Heterogeneous Camera Sensor Network

SensEye is a heterogeneous camera sensor network motivated by two applications: (a) the monitoring of rare species in remote forests and (b) surveillance in disaster management [5]. Both applications share characteristics that involve three basic tasks:

1. *Object Detection.* The presence of a new object is detected in the monitored environment. A good algorithm minimizes the latency in detection.
2. *Object Recognition.* Once a new object is detected, it is classified by type.
3. *Object Tracking.* If the object is of interest, then tracking is warranted. This involves multiple tasks including computing the location and trajectory of the object, the ability to track the object as it moves out of the visual range of one camera sensor and into the range of another, and streaming video (or a sequence of still images) of the object to a monitoring station.

In order to achieve low latency in detection without sacrificing energy efficiency, a staged architecture is used. Low-fidelity cameras perform the simpler task of motion detection, while higher-fidelity cameras are woken up on-demand for object recognition and tracking. Figure 2.3 shows the three stages of the *SensEye* architecture. The imaging, processing, and networking capabilities improve with increasing stage.

The first stage of *SensEye* is made up of Mica2 motes [2] equipped with 900 MHz radios and low-fidelity Cyclops [6] or CMUcam [7] camera sensors. The second stage is made up of Stargate nodes [2] equipped with webcams. Each Stargate runs Linux and is equipped with a webcam that can capture higher-fidelity images than the cameras at stage one. Each stage two node has two radios: (1) an IEEE 802.11 radio that is used by the Stargates to communicate with each other and (2) a 900-MHz radio that is used to communicate with the motes in the first stage. The third stage is a sparse deployment of high-resolution *pan–tilt–zoom* (PTZ) cameras connected to embedded personal computers. These cameras are used to fill any gaps in coverage of the second-stage nodes and to provide additional redundancy for tasks such as localization. There are no base stations in the architecture; nodes are assumed to communicate in ad hoc mode in each stage and between stages.

The principles that guide the design of *SensEye* include:

- Each task is mapped to the least powerful stage with sufficient resources.
- Wake-up on demand, along with redundancy in coverage, is exploited.

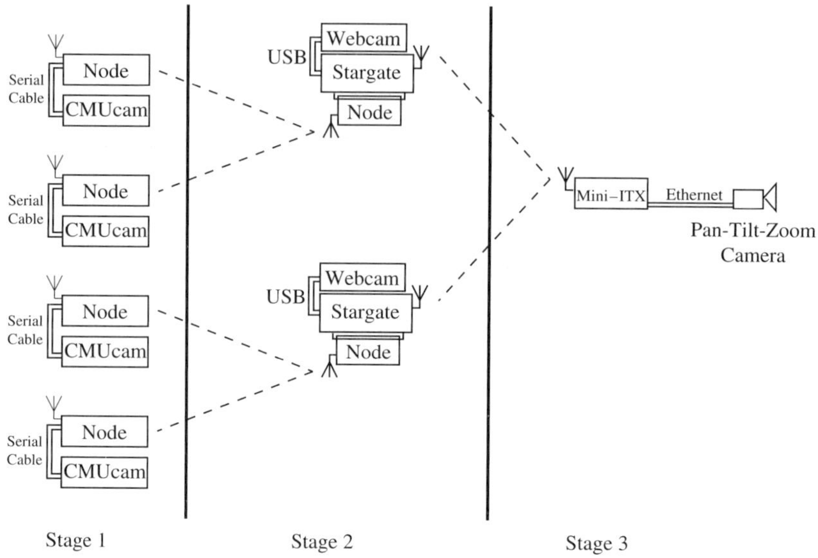

Figure 2.3. Staged architecture of *SensEye*.

In general, object detection requires few resources and it is therefore performed at stage one. The camera and the sensor node are *duty-cycled*, or woken up periodically, to detect the presence of a new object. In addition, *SensEye* uses a randomized duty-cycling algorithm where different cameras are woken up at different times to further reduce object detection latency. A frame differencing algorithm is used to detect objects. Each camera compares the newly acquired image to a background image obtained when the system is calibrated; the pixel difference is used to indicate the presence of a new object.

If a new object is detected in stage one, appropriate stage two nodes are woken up. This involves computing the coordinates of the object and determining the stage two nodes that have cameras pointing at its location; the details of object localization are provided in reference 5. Each stage one node knows the visual range of each stage two node in its vicinity and can therefore use the coordinates of the object to determine the most appropriate stage two nodes. If no appropriate stage two node is identified, a stage three camera is woken up, since it can use its pan and tilt capabilities to point to the location of the object. Localization is feasible only when at least two stage one nodes view the object. If only a single stage one node detects the object, then all stage two nodes that have overlapping coverage with it are woken up.

The separation of object detection and recognition across stages introduces latency between the execution of tasks. This latency includes the delay in receiving and processing a wake-up packet as well as the delay in waking up a stage two node. The wake-up process begins with the transmission of a wake-up packet to a stage two node similar to "wake-on-wireless" energy saving strategy of Shih et al. [8]. Upon receiving a wake-up packet, the stage two node transitions from a suspend to an awake state. By running a minimum of device drivers, this transition time is kept small.

Accurate recognition of an object requires a higher-fidelity image and significantly greater processing and memory resources than are available on a stage one node. As a result, the recognition algorithm is executed using higher-fidelity webcams and the more capable processors of stage two nodes. As a proof-of-concept, *SensEye* implements two well-known recognition algorithms from the computer vision literature in the stage two nodes [5]. Object tracking in *SensEye* involves a combination of detection, localization, and wake-up, in addition to recognition. The current system can track objects moving slowly.

An experimental evaluation of *SensEye* shows that, compared to a flat network architecture, an order of magnitude improvement in energy usage is obtained. Despite the energy reduction, similar detection performance (only 6% more missed detections) is obtained. Detection latency and energy usage at the stage one nodes is an order of magnitude less than at the stage two nodes. The mean localization errors indicate that detection can be performed by the lower-fidelity cameras of stage one while tracking is best done using higher-fidelity cameras; see Kulkarni et al. [5] for the details of the evaluation.

The *SensEye* heterogeneous camera sensor network testbed has demonstrated successfully the benefits of a staged architecture over a flat architecture with respect to energy usage. Continuing research examines tradeoffs such as system cost and coverage. Design issues that impact performance, such as (a) the number of stages in the architecture and (b) the allocation of tasks to sensors, are also under study. The problem of streaming video (or a sequence of still images) of the object to a monitoring station is not addressed in this work; providing *quality-of-service* (QoS) support for such high-bandwidth, real-time data is a challenging open problem in heterogeneous WSNs.

2.2.2 Detection of Radioactive Sources

A team of researchers at Los Alamos National Laboratory have spent the past several years focusing on the development of heterogeneous WSNs for event detection. Typical deployments of sensor networks revolve around biological or environmental monitoring applications where the emphasis is on collecting all of the data from a sensor array to be sent back to a laboratory for detailed analysis. Applications of interest to the *Distributed Sensor Networks with Collective Computation* (DSN-CC) team have instead focused on the detection, classification, and tracking of radiological materials within the sensor network. These goals are very similar to those of the *SensEye* system, requiring all processing to be performed within the network with no use of base stations. However, this work relies on multiple sensor modalities instead of a single sensor type for event detection.

The motivation for this research is to develop systems to guard against attacks from *radiological dispersal devices* (RDDs) capable of contaminating an area or population with fissile material. A potentially last line of defense for such attacks may reside in systems placed along roadways that are able to detect such material in-transit and alert the appropriate authorities before dispersal occurs [9].

One approach to such a threat employs portal monitoring equipment. Portals provide high fidelity results; however, they are large, conspicuous, and costly and require

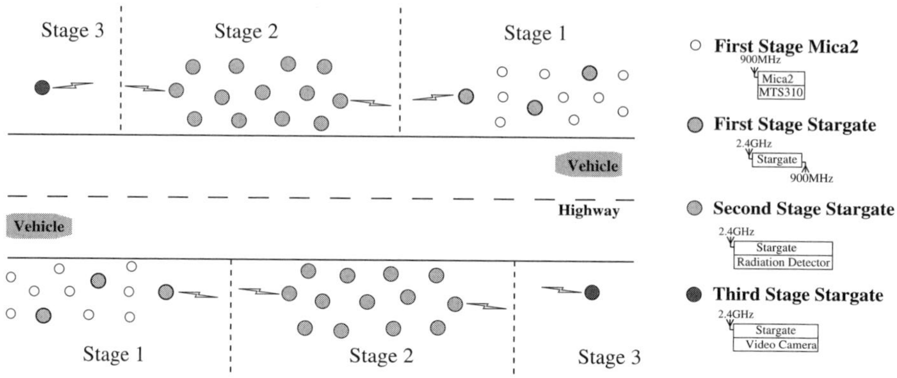

Figure 2.4. Staged architecture for radioactive source detection.

considerable time and infrastructure to set up. Instead the DSN-CC project strives to employ networks of small, low-cost, heterogeneous sensors in producing similar quality results in persistent applications where the system deployments may occur utilizing fast, low-impact methods. Such a system must be heterogeneous in sensor type, as well as in node backbone, to allow for data redundancy, in-network processing, and hypothesis validation.

In the *SensEye* system a staged architecture is employed as a means of achieving energy efficiency. While energy efficiency is an important factor for the development of the DSN-CC system, the staged architecture is utilized instead as a means of gaining confidence in a network-developed detection while decreasing system false alarms. A single radiation detector provides specific detection and false alarm rates; coupling a string of radiation detectors with seismic sensors, magnetometers, acoustic sensors, atmospheric sensors, and video cameras increases dramatically the fidelity of the decisions made within the network.

Figure 2.4 shows the three stages of the radiation detection application of the DSN-CC system. Stage one consists of both Mica2 motes equipped with the MTS310 multisensor boards and Stargate nodes. Stage two is comprised of an array of radiation detectors connected directly to the Stargate nodes, and stage three is a video camera connected to a Stargate node. In all instances the Mica2 motes communicate through an embedded 900-MHz radio while the Stargate nodes are equipped with both the 900-MHz radio and an IEEE 802.11 radio transmitting at the 2.4-GHz frequency.

Figure 2.5 shows the inexpensive Mica2 motes densely placed along the roadway in stage one to allow for hardware failures through redundancy. The goal of stage one is to detect the presence of a vehicle utilizing data from the acoustic and magnetometer sensors for cueing stage two and three sensors. Vehicle classification may also occur during this stage. The algorithms for vehicle detection and classification are embedded within the network in the stage one Stargate nodes. The stage two nodes remain in a background collection mode and stage three nodes remain inactive until a vehicle is detected within the network by the stage one nodes. This helps to minimize the false alarm rate of the radiation detection assets, as well as reducing the required energy draw of the stage three nodes.

Figure 2.5. Stage one Mica2 vehicle detection network.

When a vehicle is detected within the sensing array, the stage one Stargate nodes transmit cues to the second stage nodes. These radiation detection nodes change state from background count collection to an active, timed-count collection. Each stage two detector node takes a radiation count while the passing vehicle is directly in front of it based upon its node location and the speed of the passing vehicle. These counts are coherently added across the network and compared to the environmental background readings [10]. If the vehicle is suspected of carrying radiological material, the stage three video camera is cued to collect an image of the offending vehicle. This image, along with the corroborating event information, is then relayed to a command-and-control console similar to Figure 2.6 for monitoring personnel for interdiction. In the field, this command-and-control console is a tablet PC communicating to the network via its IEEE 802.11 radio link.

Although commercially available hardware is limited in its performance and capabilities, ongoing field experiments provide an indication that heterogeneous systems have the potential to provide low-cost, highly reliable solutions to many persistent surveillance applications. Continuing research at the Los Alamos National Laboratory is focusing on (a) the operational issues of a network such as node and network security, (b) validation studies of efficient communication protocol schemes, (c) development of additional embedded algorithms for further event classification and

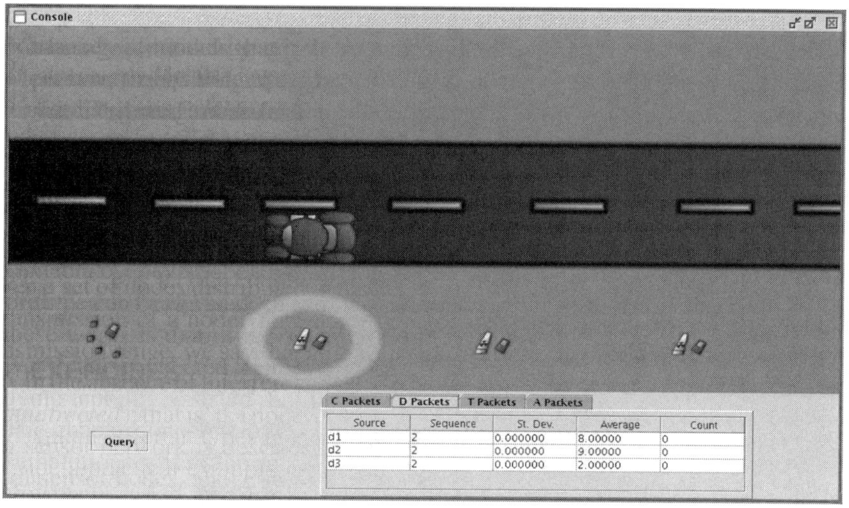

Figure 2.6. Simple command-and-control console interface.

tracking, (d) the investigation of alternate backbone hardware, and (e) the investigation of long-range exfiltration schemes.

2.3 SCALABILITY AND SYSTEM LIFETIME

It is well known that homogeneous ad hoc networks, which include homogeneous WSNs, suffer poor capacity. Gupta and Kumar [11] were the first to show that the throughput of a node is $\Theta(1/\sqrt{n \log n})$, where n is the number of nodes in the network—a very pessimistic result! In addition, as paths between nodes become longer, the probability of packets being lost becomes higher. As the number of nodes in a network grows, the successful end-to-end transmission rate drops significantly [12]. Experimentation in simulation [13] and in testbeds [14] has confirmed that the performance of homogeneous ad hoc networks does not scale with increasing n.

A primary difference between WSNs and ad hoc networks is that the traffic pattern is many-to-one, from the sensor nodes to the base station. Figure 2.7 shows a homogeneous WSN with a base station at the center. A transmission from any sensor node to the base station goes through one of the nodes within a distance of r of the base station. These *critical* nodes have the highest burden of relaying traffic. As a result, they are likely to exhaust their energy before other sensor nodes [15]. When the critical nodes die, connectivity of the network is lost. Hence the energy drainage rate of the critical nodes determine the system lifetime. Indeed, Du and Xiao [16] found that when connectivity is lost, more than half of the nodes still have more than 50% of their energy left. This energy is wasted since communication with the sink is no longer viable.

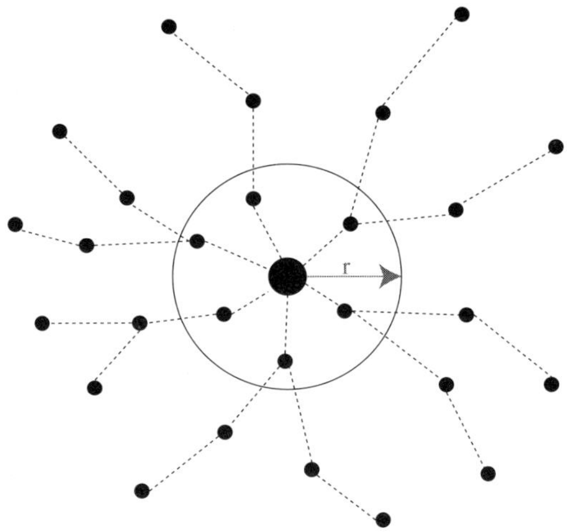

Figure 2.7. The energy of *critical* nodes (within the circle) drain nonuniformly.

Adding structure to the network can help improve scalability and also alleviate the problem of nonuniform drainage of energy. Clustering is one way to add structure in a homogeneous WSN. Heinzelman et al. propose LEACH, a clustered network in which each cluster head aggregates data and transmits it directly to the base station [17]. The cluster heads are periodically rotated for efficient load balancing and a consequent lengthening of network lifetime. In order to minimize the total energy, the number of cluster heads must scale as the square root of the total number of sensor nodes [17]. The LRS [18] and power-aware chessboard-based adaptive routing (PCAR) [19] protocols also aim to balance the energy consumption in a homogeneous sensor network. All of these protocols suffer from overhead associated with frequent cluster-head rotation.

While the problem of routing in homogeneous WSNs has been considered in flat architectures (see Directed Diffusion [20] as an example), when the architecture is hierarchical, the routing protocol makes use of the hierarchy (see TTDD [21] and LEACH [17] as examples). Hierarchy is shown to help a homogeneous WSN achieve higher total throughput and increase the network lifetime.

In heterogeneous WSNs, it is often natural to organize the nodes into a hierarchy. In this section we consider algorithms for heterogeneous WSNs to form hierarchical topologies, and address the related problem of routing, in order to tackle the problems of scalability and nonuniform energy drainage.

2.3.1 A Resource-Oriented Protocol: Topology Formation and Routing

In WSN applications in which the sensor nodes are inherently heterogeneous in energy resources, these differences should be considered in order to improve the network capacity and extend the system lifetime.

Ma et al. [22, 23] consider a wireless in-home heterogeneous sensor network. This may include devices embedded into everyday objects such as appliances, devices for climate monitoring and environmental control, and medical devices integrated into the home for monitoring medical conditions. Even mobile sensor nodes, such as those carried by people or on mobile toys, are considered.

In-home sensor nodes are heterogeneous in their power units. Some nodes are directly connected to the alternating current power supply and have, essentially, unlimited energy. Others are powered by batteries with varying capacities. To exploit the heterogeneity in power units, a *resource-oriented protocol* (ROP) is proposed. The goal is to achieve the longest system lifetime. To be precise, the *system lifetime* for a sensor network is the shortest lifetime of any participating node in the network. The *node lifetime* for a sensor is the time at which the sensor exhausts all its energy.

ROP exploits the existence of sensor nodes with unlimited resources. A topology is formed to minimize the consumption of resources of energy-constrained sensor nodes. Sensor nodes with unlimited energy serve as relays, since they can afford higher transmission power and hence longer transmission range. This saves energy in the energy-constrained sensor nodes and, as a result, extends the functional lifetime of the network.

Figure 2.8 shows an example of ROP; node U has unlimited energy resources, while the others are energy-constrained sensor nodes. Of these energy-constrained nodes, nodes E_1 and E_2 are battery-powered local cluster heads with medium energy capacity, and the rest of the nodes only have small energy capacity. When there is a message that needs to be sent, for example, from a source node 1 to a destination node 7, a traditional multihop routing protocol might route the packets from node 1 through nodes 2, 3, 4, 5, and 6 to 7. However, because of the existence of the unlimited energy node U and the medium energy capacity node E_1, ROP would route the packets from

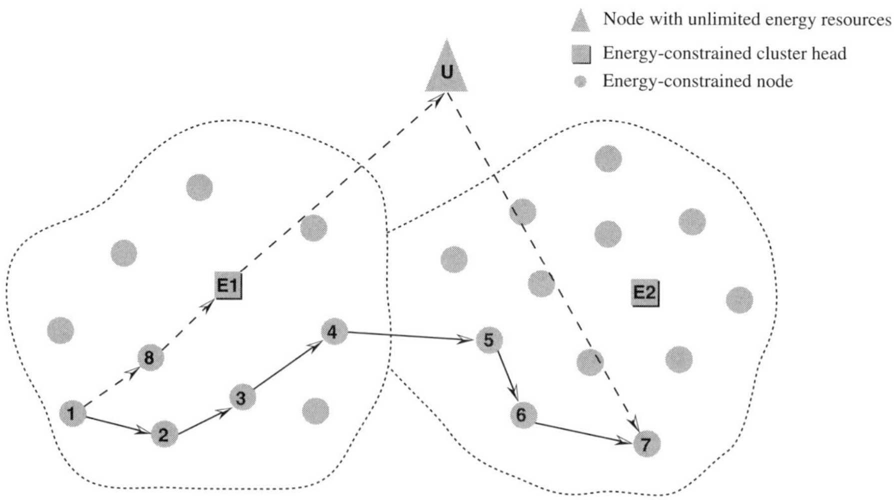

Figure 2.8. Resource oriented routing versus multihop routing.

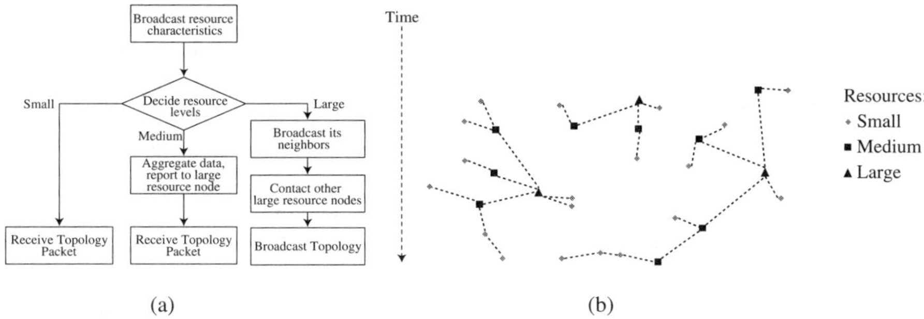

Figure 2.9. Resource-oriented protocol for heterogeneous WSNs. (a) ROP topology formation phase. (b) Sample ROP multihop hierarchical topology.

1 to 8, then to E_1 and U, and then directly to 7. Although such a path is not energy-efficient, it saves more energy in energy-constrained sensor nodes, in order to prolong the operational lifetime of these nodes.

ROP has two phases: a topology formation phase and a topology update phase. The topology formation phase, in turn, involves two steps. In the first step, each sensor node reports its characteristics and available resources to all of its neighbors. The local cluster-head aggregates these reports and sends it to the most powerful sensor nodes. In the second step, these most powerful nodes decide the topology of the network and broadcast routing information. On receipt of the topology packet, each sensor node configures its route cache based on the topology decided. Figure 2.9a shows the topology formation phase for a network with nodes at three resource levels: small, medium, and large. This phase builds a multihop hierarchical topology with the large resource nodes at level zero—that is, the root of each tree (see, for example, Figure 2.9b).

In order to reduce the energy cost of the topology formation phase, some sensors may be left isolated. The topology update phase takes care of this situation, and it establishes routes to mobile nodes. ROP is one of the few heterogeneous WSN protocols that incorporates mobile sensor nodes.

When a sensor wants to communicate with other sensors, it uses the route in its cache. If the route is outdated, it sends a *route request* (RREQ) packet. The returned route replaces the outdated route in its cache. If several routes are received, it chooses the one with the largest resources; and if two routes have the same resources, the one with fewer children is selected. The details of ROP and the reactive routing scheme are described in Ma et al. [23].

The performance of ROP is evaluated in simulation. In ROP, energy efficiency cannot always result in longer system lifetime. Rather, balancing resources among sensors and saving energy for those more resource-constrained sensor nodes contributes to lengthening system lifetime.

2.3.2 Chessboard Clustering and Routing Protocol

Du and Xiao [16] propose a chessboard clustering and routing protocol for heterogeneous WSNs to overcome the performance bottleneck and poor scalability of homogeneous WSNs and to address, at same time, the problem of nonuniform energy consumption. A good observation is made that clustering alone does not solve the problem of nonuniform energy drainage; indeed, the center node in Figure 2.7 can as well be a cluster head rather than a base station.

Two types of nodes are assumed: a small number of high-end sensor nodes and a large number of low-end sensor nodes. Each node is assumed to be aware of its location. A cluster is formed around each high-end sensor node which serves as a cluster head. Low-end sensor nodes perform the basic sensing as well as the relaying of packets within the cluster. Given its powerful energy reserve and communication ability, each high-end node performs data fusion within its cluster, and it transmits the aggregated data to the sink via a single-hop link or a multihop path. In this way, the network is divided into multiple regions, with each region assuming a smaller burden of the communication due to the smaller number of sensor nodes within the cluster. The network lifetime is therefore increased by transmitting fewer packets at low-end sensor nodes and utilizing the less power-constrained or non-power-constrained nodes as much as possible.

Figure 2.10 shows the sensor field divided into equal-sized cells with adjacent cells colored with different colors, resembling a chessboard. These nodes are assumed to be uniformly and randomly distributed in the sensor field. Since each node knows its location, it can determine if it is in a white cell or a black cell.

The basic idea is to use the underlying chessboard to define two clustered topologies, with only one clustering in use at a given time. In a white clustering, all high-end sensor nodes in white cells are active while all high-end sensor nodes in black cells are inactive. In a black clustering, all high-end sensor nodes in black cells are active while all high-end sensor nodes in white cells are inactive. Low-end sensor nodes are all active, forming multihop clusters around the currently active high-end sensor nodes. The motivation for switching colors is as follows: sensor nodes that are critical nodes in a white clustering are likely to become non-critical nodes in a black clustering and vice versa. Since critical nodes consume more energy in packet

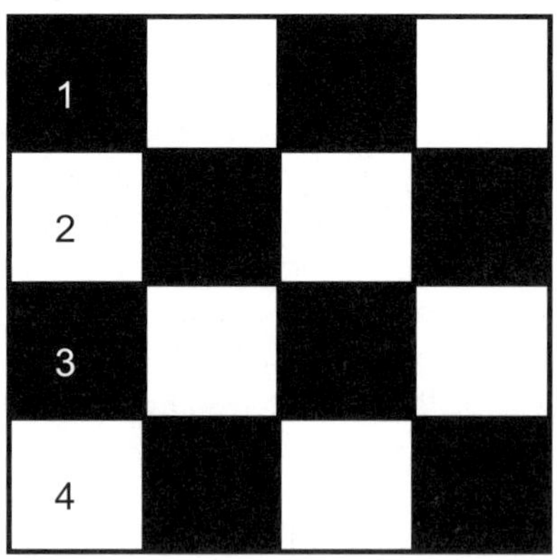

Figure 2.10. Chessboard clustering scheme.

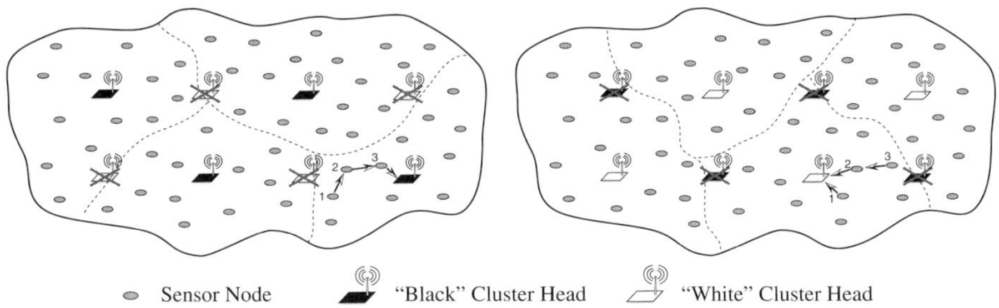

Figure 2.11. A black (left) and white (right) clustering of the heterogeneous WSN.

forwarding than do other sensor nodes, switching the color of the clustering balances the energy consumption among sensors, and prolongs the network lifetime.

Figure 2.11 shows an example of clustering based on black and on white cells, respectively. In the black clustering, sensor node 3 is a critical node forwarding packets on behalf of nodes 1 and 2. In the white clustering, nodes 1 and 2 become critical forwarding packets on behalf of other nodes in the cluster; in particular, node 2 now forwards the packets of node 3.

In order to form a black (white) clustering, each black (white) high-end sensor node broadcasts a hello packet, containing its identifier and its location. Low-end sensor nodes may receive hello packets from multiple black (white) high-end nodes. In a two-dimensional sensor field, each low-end sensor node selects the closest high-end sensor node as the cluster head; this leads to the formation of Voronoi cells where the cluster heads correspond to the nuclei of the cells [16].

The decision to switch the color of the clustering is based on the energy levels of the high-end nodes. Suppose the current clustering is black. Periodically, each black high-end sensor node exchanges packets with its neighboring white high-end nodes. The packets contain the energy remaining in the node. If the remaining energy of the black high-end node drops below a threshold, its neighboring white high-end nodes become active and initiate cluster formation. As the network runs, the black high-end nodes drain their energy and become unavailable. Gradually, the white high-end nodes become active.

Both intra- and intercluster routing protocols are proposed [13, 16]. Routing within a cluster is achieved via a greedy geographic routing protocol. Each low-end sensor node simply forwards a packet to the neighbor closest to the cluster head.

In order to support intercluster routing, after the clusters are formed, each cluster head sends its location to the sink. The sink then broadcasts the locations of all clusters heads. For a cluster head to communicate with the sink, it draws a line between itself and the sink. The line intersects some number of Voronoi cells. The packet is forwarded from the source cluster head to the sink through the cluster heads in these relay cells. The chessboard routing protocol achieves a higher delivery ratio, lower total energy consumption, smaller end-to-end delay, and better throughput than two routing protocols for homogeneous WSNs. The details of the chessboard clustering and routing protocols, as well as their performance evaluation in simulation, can be found in references 13 and 16.

2.3.3 Analyses of System Lifetime in Heterogeneous WSNs

Mhatre and Rosenberg [24] present a cost-based comparative study of (a) homogeneous WSNs and (b) heterogeneous WSNs with two types of nodes. Their model takes into account the cost of manufacturing the hardware as well as the battery energy of the sensor nodes. First, a single-hop clustered architecture is considered, with LEACH [16] selected as the representative in a single-hop homogeneous WSN. For the multihop homogeneous clustered architecture a multihop variant, called M-LEACH, is proposed and analyzed. In comparing (a) the cost of the multihop homogeneous clustered architecture with M-LEACH and (b) a multihop heterogeneous clustered architecture, the homogeneous WSN can outperform the heterogeneous one if the nonuniform energy drainage problem is not addressed.

Mhatre et al. continue their study of heterogeneous WSNs in reference 15. They consider a WSN with nodes of two types distributed over a sensor field using two-dimensional homogeneous Poisson point processes: (a) type 0 nodes with intensity (average number per unit area) λ_0 and battery energy E_0 and (b) type 1 nodes with intensity λ_1 and battery energy E_1. The type 0 nodes do the sensing while the type 1 nodes act as the cluster heads. Nodes use multihop paths to communicate with their closest cluster head. The optimum node intensities (λ_0, λ_1) and node energies (E_0, E_1) that guarantee a lifetime of at least T units, while ensuring both connectivity and coverage of the surveillance area with high probability, are determined. The overall cost of the network is minimized under these constraints. Here, the network lifetime is defined as the number of successful data gathering trips (or cycles) that are possible until connectivity and/or coverage are lost. Conditions for a sharp cutoff are taken into account. This means that it is ensured that almost all the nodes run out of energy at about the same time so that there is very little energy lost due to residual energy. The results comparing a random deployment of nodes with a deployment in which nodes are placed deterministically along grid points show that λ_1 scales approximately as $\sqrt{\lambda_0}$. The results can be extended to take into account unreliable nodes.

Duarte-Melo and Liu [25] examine the performance and the energy consumption of a heterogeneous WSN providing periodic data from a sensor field to a remote receiver. A flat homogeneous WSN is compared to one in which an overlay of fewer more powerful sensor nodes is added. The energy consumption is formulated and the estimated lifetime based on a clustering mechanism with varying parameters related to the sensor field, such as size and distance, is studied. The optimal number of clusters is quantified based on the model. Also, an allocation of energy between the two levels of the architecture is discussed.

Li and Mohapatra [26] develop an analytical model for the problem of nonuniform energy drainage. It is found that density does not affect the energy consumption rate of a node. This confirms the fact that simply deploying more nodes in a network cannot prolong its lifetime. Using the model, they investigate the effectiveness of some existing approaches toward mitigating the nonuniform energy drainage problem in a formal manner. Using a hierarchical architecture alleviates, though does not eliminate, the problem. Other approaches include investigating the impact of source bit rate and the impact of traffic compression. Simulation is used to validate the analysis.

Ai et al. [27] develop an analytical model for energy dissipation for a homogeneous WSN with a flat architecture and also for a heterogeneous WSN with a hierarchical clustered architecture. The communication cost of multihop links increases with the number of clusters, while the communication cost of forwarding messages on the backbone increases with the number of clusters. Thus, there is an optimal number of clusters that trade off the power consumption between multihop and single-hop links to minimize the energy dissipation rate.

2.4 COVERAGE IN HETEROGENEOUS WSNs

One of the fundamental issues in homogeneous WSNs is the problem of coverage, which reflects how well a sensor network is monitored by sensor nodes. Several forms of coverage have been studied.

In order to achieve *deterministic coverage*, a static network of predefined shape must be deployed. A grid-based sensor network is an example of a uniform deterministic deployment. In this case, the problem of coverage of the sensor field reduces to the problem of coverage of one cell and its neighborhood [28, 29]. A *weighted* deployment might be used in an art gallery where more valuable objects are equipped with more sensors to maximize the coverage of the security system.

In many situations, deterministic deployment of the sensor nodes is neither practical nor feasible. Instead, sensor nodes may be randomly distributed in the sensor field and *stochastic coverage* is considered [30–33]. The stochastic random distribution model may be uniform, Gaussian, or any other distribution based on the application.

In this setting, Megerian et al. [29] study the worst-case and best-case coverage problems. Informally, in the *worst-case* coverage problem, the goal is to find the closest distance to sensor nodes that an agent traveling on any path in the sensor field must encounter at least once. The idea is that the closest distance to sensor nodes is one metric by which coverage of the field may be characterized. This scheme is worst-case since the closest distances to sensor nodes is determined, even if the agent tries to avoid them. At the other extreme is the *best-case* coverage problem, where the goal is to find the farthest distance to sensor nodes that an agent traveling on any path in the sensor field must have from the nodes even if it tries to stay as close to them as possible. Provably optimal polynomial time algorithms for the best- and worst-case coverage problems are provided [29].

Yan et al. [34] examine the problem of differentiated coverage (corresponding to weighted deployments) in homogeneous WSNs. A protocol is designed in which each node is able to dynamically decide a schedule for itself in order to guarantee a degree of coverage. The schedule has an average energy consumption that is inversely proportional to the node density.

We highlight the work on coverage in heterogeneous WSNs next.

2.4.1 Differentiated Coverage in Heterogeneous WSNs

In some applications, *differentiated coverage* is necessary. This is when a different degree of coverage is applied to different parts of the network [34, 35]. For example, some areas of a battlefield are of more interest than others.

Du and Lin [35] propose an algorithm for uniform coverage that can be extended to provide differentiated coverage of a heterogeneous WSN. As in references 13 and 16, two types of nodes are assumed: a small number of high-end sensor nodes and a large number of low-end sensor nodes. Each node is assumed to be aware of its location.

A logical grid is assumed to overlay the sensor field with certain grid points requiring coverage k. In order to provide uniform coverage of the grid points, the goal is to design a node scheduling algorithm that ensures that all grid points have the required coverage while at the same time minimizes the total energy consumption and balances node energy consumption.

The high-end sensor nodes know the locations of the low-end nodes and can compute which low-end nodes cover a grid point. In Figure 2.12, the low-end sensor nodes A, B, C, and D cover grid point 1. If k sensor nodes cover a grid point, then an ideal schedule has each node awake for T/k time and asleep for $T - T/k$ time, in a round T. However, a low-end sensor node may need to cover other grid points. Therefore the high-end sensor considers the assigned slots when each low-end sensor is awake and assigns a time slot that has the maximal overlap with the existing awake slots. For example, if node D already has a slot of $[0, T/4]$ for covering grid point 1, then the high-end node can assign an awake slot of $[0, T/3]$ to D. Node D need only be active during $[0, T/3]$ in order to cover both grid points 1 and 2. If there is a conflict, then a node may require the assignment of additional slots in which it is awake. After determining the node schedule for all grid points, the high-end node broadcasts it to all low-end sensor nodes. The schedule is updated periodically to ensure the coverage algorithm is robust to sensor failure.

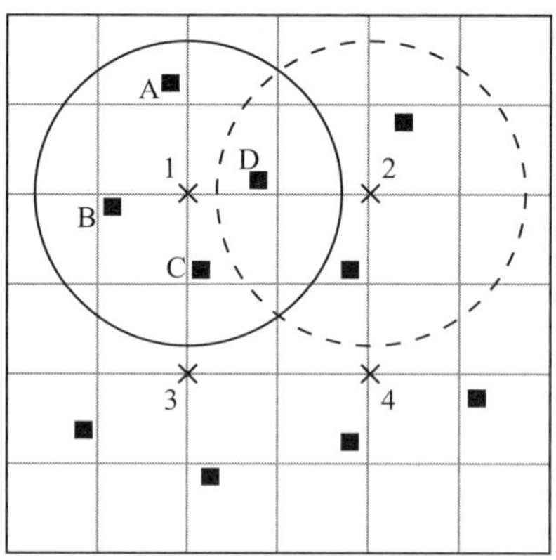

Figure 2.12. Coverage for grid points.

The uniform coverage algorithm is extended to provide differentiated coverage. To provide a coverage degree c of a certain grid point, the high-end node correspondingly adjusts the time awake for each low-end sensor in the coverage area. For a grid point covered by k sensor nodes, T/k slots in each round T are assigned to provide coverage of degree 1. For coverage of degree c, the number of slots must be $c \cdot T/k$. The complete algorithm for differentiated coverage is provided in reference 35.

While the focus of Mhatre et al. [15] is on maximizing the lifetime of the network, one of the constraints relates to coverage. Two types of node are deployed over a sensor field for the purpose of surveillance. One type of node does the sensing while the other type acts as cluster heads. An aircraft visits the area periodically and gathers data about the activity in the field from the sensor nodes. The problem is treated assuming that the base station (aircraft) receives updates from every cluster. However, if the base station is interested in receiving updates from only a few clusters (an extrasensitive region), then the analysis can be modified to accommodate this requirement. More nodes are deployed over the regions of frequent updates, and these nodes are taken into account in the overall network cost. The redundant nodes stay inactive while the battery energy of other nodes lasts; they join the cluster when the other nodes start to expire.

2.4.2 Stochastic Coverage in Heterogeneous WSNs

Lazos and Poovendran [31] study the following stochastic coverage problem in heterogeneous WSNs: Given a planar sensor field and n sensor nodes deployed according to a known distribution, compute the fraction of the sensor field that is covered by at least k sensor nodes, $k \geq 1$. This may also be viewed as a problem in k-coverage [33].

The problem is formulated as a set intersection problem arising in integral geometry. Analytical expressions for stochastic coverage as then derived. The formulation does not require the sensor nodes to have identical sensing capability, and it does not restrict the distribution according to which the sensors are deployed. In addition, the formulation is applicable to scenarios where the sensing area of each sensor node has arbitrary shape. The validity of the derived expressions are verified by simulation.

2.5 MANAGEMENT OF HETEROGENEOUS WSNs

By definition, heterogeneous WSNs have more than one type of sensor, making their management increasingly important. Management includes:

- Coordinating and scheduling tasks for sensors.
- Optimizing the use of capabilities and resources.
- Managing the sensor data aggregation and correlation.
- Assessing the situation.
- Adapting the sensor network.
- Reducing human involvement.

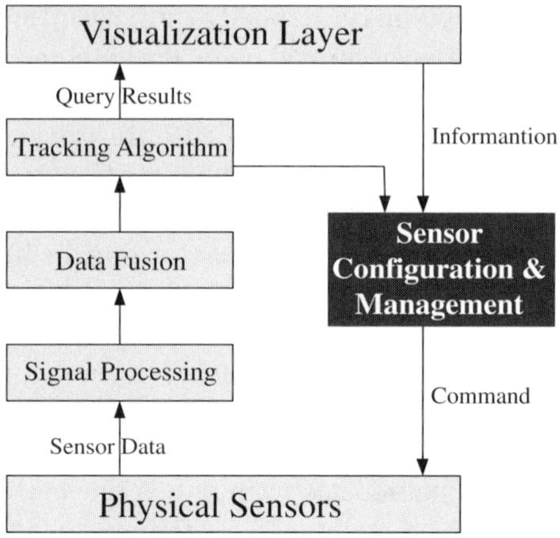

Figure 2.13. Position of management in the system model.

Vaidya et al. [36] propose a framework for sensor configuration and management to take the responsibility of making decisions in order to coordinate the assignment and scheduling of sensor nodes best suited for the application. The application considered is tracking and movement of objects in a moderately occupied confined space. Figure 2.13 shows how the management component is positioned in the unified sensing system model.

A manager is designed to operate over a heterogeneous WSN that provides sensory data from multiple types of sensor. The goal of the management system is to minimize the energy consumption and the required bandwidth while preserving the quality of tracking.

When tracking the movement of one object, the system uses a set of three sensor nodes to determine the current location of the object and to predict the next set of sensor nodes to use according to its velocity and direction. This allows the rest of the sensor nodes in the network to go to sleep for the next detection round. For multiple targets that are far away from each other, tracking is similar to a single object moving. When multiple objects move very close to each other, there is ambiguity in the data acquired from the sonar sensors. In this case, visual sensors come to the aid. Figure 2.14 shows the complete flow chart for the sensor management system. With the help of management system, a significant energy reduction is achieved compared to a randomized activation scheme.

The challenge of correlating the data gathered by several sensor nodes listening to live traffic is studied by Andersson et al. [37]. Correlating data from different types of sensor brings a number of benefits. The first is a reduction of the number of alerts that a user must address. The correlation engine should recognize when reports from multiple sensors refer to the same incident. Correlation can enhance the detection capability as a second benefit. In addition, correlation can exploit the complementary coverage from several sensors. Reports from several sensors employing diverse

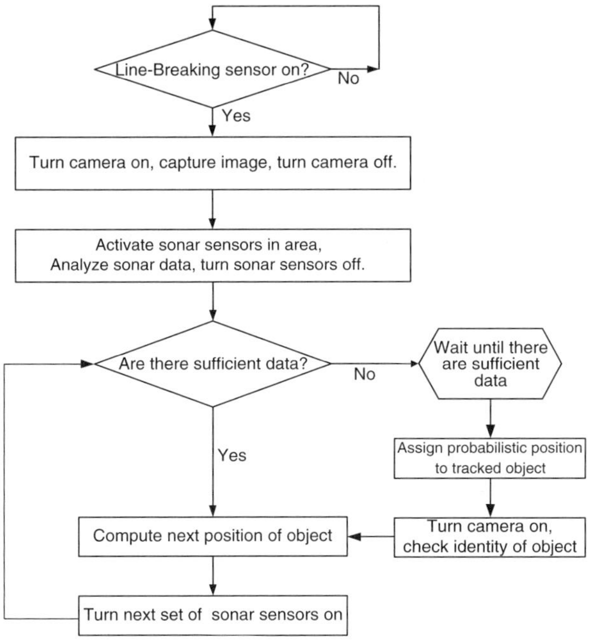

Figure 2.14. Sensor management flow chart.

analytical techniques may reinforce each other. A standard format is developed to facilitate the interoperability of diverse sensors.

Management of the WSN may also help network scalability. In reference 38, Gürgen et al. propose a hybrid approach offering scalable solutions that combine the advantages of both centralized and distributed data stream management. Their main concern is the querying and system management of large sets of sensors. In the heterogeneous WSN considered, diverse types of sensor nodes are used, each with a different data delivery rate. A three-stage architecture is proposed for distributed evaluation of queries on the network: sensor nodes, gateway, and control site. In the architecture, the load of query evaluation is distributed between stages. A mediator-wrapper [39] is applied at the gateway stage to serve as local query translator and optimizer. The wrapper proposed is an approach for heterogeneous sensor data management and provides an integrated global view of the different types of sensor. With the help of this three-stage management scheme, the query load is distributed; also, the burden at each stage is decreased. Hence, the scalability of the network is improved by manipulating the query in the heterogeneous WSN.

2.6 NEW APPLICATIONS ENABLED BY HETEROGENEOUS WSNs

Using a homogeneous WSN may not satisfy the requirements of an application [40]. The following two applications are examples requiring a heterogeneous WSN.

The ability to reliably deliver a large volume of data has opened a new range of applications for heterogeneous WSNs. This includes audio/video surveillance and the monitoring of telemetry data. Yuan et al. [40] consider a new *live virtual reality* application. The idea is to provide a user with live real-time video of a monitored sensor field together with the ability for the user to navigate virtually within the field. One example is that of a heterogeneous WSN deployed for securing a building, where a console operator can survey a building using a joystick for navigation.

Gnawali and Yarvis [41] propose a *"do not disturb"* application requiring a heterogeneous WSN. The purpose of this application is to alert people to keep the noise down in an office environment when there are people working in a nearby area.

A "do not disturb" WSN consists of nodes with motion and sound sensors. Motion detectors determine the occupancy status of a cubicle. Sound sensors in cubicles measure the loudness of the sound heard in each cubicle. Sound sensors in the hallways pick up the noise of impromptu hallway meetings. The "do not disturb" application determines the source of the noise by data fusion; this requires CPU and memory resources typically not available in sensor nodes. Figure 2.15a shows two types of sensor node deployed in an office environment, as well as actuators to alert people when it is too noisy; the network topology that corresponds to the deployment is shown in Figure 2.15b.

The use of heterogeneous nodes allows a distributed architecture using ad hoc deployment that is resilient to failure and has lower cost. In addition to the motion and sound sensors, a sensor with more powerful processing ability is required. This is where the motion and sound data are sent for processing and analysis, to determine whether an actuator signal is needed. As a result, the successful transmission rate increases, and delay decreases significantly. Moreover, as the network grows and computational demand increases, more high-end sensor nodes can be added in the network.

2.7 SUMMARY OF PROJECTS AND SYSTEMS INFRASTRUCTURE

2.7.1 Summary of Heterogeneous WSN Projects

A summary of heterogeneous WSN projects is given in Table 2.1.

2.7.2 Systems Infrastructure for Heterogeneous WSNs

A summary of systems infrastructure for heterogeneous WSNs is given in Table 2.2.

2.8 OPEN PROBLEMS

Research in heterogeneous WSNs is in its infancy and is therefore rich in open problems. Some of them include:

Inadequate Theory of Heterogeneous WSNs. Most of the models assume that a heterogeneous WSN provides data that are clock-driven (or periodic).

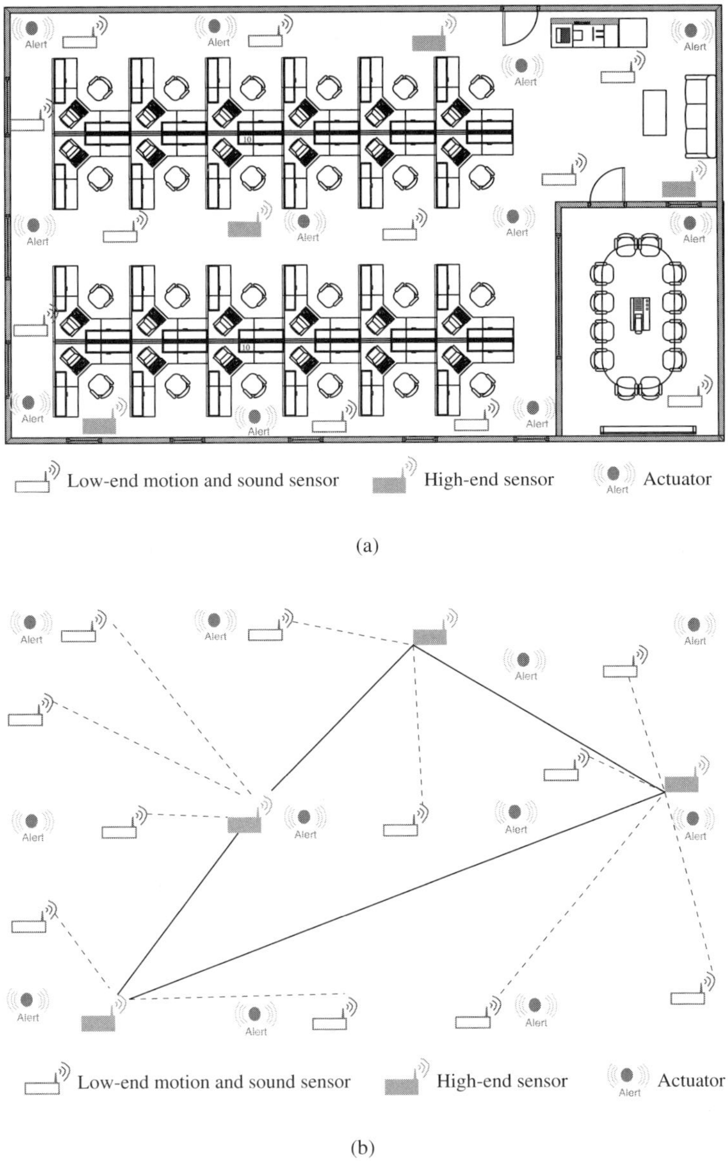

Figure 2.15. (a) A "do not disturb" application deployed in an office area and (b) its corresponding topology.

Theory for query-driven and event-driven heterogeneous WSNs needs to be explored. While Mhatre et al. [15] consider hardware cost, models that also consider energy consumed in data processing (compression, fusion, etc.) are of interest.

TABLE 2.1. Heterogeneous WSN Projects.

Project	Brief Description and Website
CENS, University of California, Los Angeles	Center for Embedded Networked Sensing A mission of CENS is to develop and demonstrate architectural principles and methodologies for deeply embedded, massively distributed, sensor-rich systems. Research areas that relate to heterogeneous WSNs include the Multiscaled Sensing and Actuation (MAS) project, the Tenet project, the EmStar family of generalized deployment software tools, and individual protocols such as the centralized (CentRoute) and distributed (Hyper) routing protocols. http://research.cens.ucla.edu/
CoSense, Palo Alto Research Center	Collaborative Sensemaking Collaborative sensemaking of distributed sensor data for target recognition and condition monitoring. http://www2.parc.com/spl/projects/cosense/
DSN-CC, Los Alamos National Laboratory	Distributed Sensor Networks with Collective Computation The goal of DSN-CC is to demonstrate in situ collective computation abilities of heterogeneous sensor networks in simulation and using inexpensive, readily available off-the-shelf platforms. One application is a staged heterogeneous wireless sensor network for the detection of radioactive sources. http://www.lanl.gov/source/orgs/isr/dsn/background.shtml
GNOMES, Rice University	Generalized Network of Miniature Environmental Sensors GNOMES is a low-cost hardware and software testbed. It is designed to explore the properties of heterogeneous wireless sensor networks, to test theory in sensor networks architecture, and to be deployed in practical application environments. http://cmclab.rice.edu/projects/sensors/
HSN, University of California, Berkeley	Heterogeneous Sensor Networks Heterogeneous sensor networks for automated target recognition and tracking in urban terrain is the focus. Issues addressed include: a new theory for distributed signal processing with random spatiotemporal sampling of complex scenes, robust design principles for sensor networks with both low- and high-bandwidth sensors, and metrics for the design and deployment of sensor networks and incorporating mobility into sensor networks. http://trust.eecs.berkeley.edu/hsn/

TABLE 2.1. (*Continued*)

Project	Brief Description and Website
Intel Research	Heterogeneous Sensor Networks To address the scalability problem in WSNs, high-end nodes (such as Intel XScale-based nodes) are overlaid on a sensor network. Goals include identifying and utilizing heterogeneous capabilities, such as links and services, for embedding local processing, imposing a database model, and enhancing routing protocols. Applications include preventive maintenance for equipment in Intel's fabs and sensor networks for theme parks. http://www.intel.com/research/exploratory/heterogeneous.htm
Microsoft Research	Networked Embedded Computing Group Microsoft is developing new service architectures, interoperation protocols, and programming models that are resource-aware and resource-efficient across heterogeneous devices that can range from extremely limited sensor nodes to more powerful servers. http://research.microsoft.com/nec/
SensEye, University of Massachusetts, Amherst	A Multitier Multimodal Camera Sensor Network Trends in technology have resulted in a spectrum of camera sensors, wireless radios, and embedded sensor platforms. SensEye is designed on the principle that multitier networks are not only scalable, but also offer a number of advantages over simpler, single-tier unimodal networks: lower cost, better coverage, higher functionality, and better reliability. http://sensors.cs.umass.edu/projects/senseye/
SensorNets, Carnegie Mellon University	Pervasive Infrastructure Sensor Networks SensorNets creates a framework for applications of networks of sensors in long-lived infrastructure systems such as buildings, bridges, and highways—a heterogeneous collection of sensors that must continue to operate even as parts of the infrastructure are changed, upgraded, or remodeled. The project has four main areas of thrust: devices, applications, systems, and data. http://www.ices.cmu.edu/sensornets/

TABLE 2.2. Systems Infrastructure for Heterogeneous WSNs

Systems	Brief Project Description and Website
Aspen, University of Pennsylvania	Abstraction-based Sensor Programming at Penn The Aspen project focuses on the challenges in developing a programming environment and runtime system for complex applications that may have heterogeneous types of sensor, confidentiality requirements, different levels of connectivity, and timing constraints. A programming model that handles heterogeneous data stream types and sensor capabilities is under development. http://www.cis.upenn.edu/~zives/aspen/
Avrora, University of California, Los Angeles	Sensor Network Simulation Avrora is an instruction-level sensor network simulator. Avrora simulates a network of AVR/Mica2 motes. The goal is to enhance Avrora with new capabilities for executing and monitoring simulations of heterogeneous sensor networks. Specifically, this includes supporting sensor code that is dynamically updated, other sensor platforms, and source-level monitoring of simulations. http://research.cens.ucla.edu/projects/ 2006/Systems/Avrora
DSS, Los Alamos National Laboratory	Distributed Sensors Simulator DSS is a simulation framework that assists in implementing and debugging wireless distributed sensor networks. The user provides data on node locations and characteristics, defines event phenomena, and plugs in the applications each node runs. DSS provides simulation of the wireless and environmental channels and was specifically designed for investigations of topological, phenomenological, networking, robustness, and scaling issues in WSNs. http://www.lanl.gov/source/orgs/isr/dsn/codes.shtml
EmStar, University of California, Los Angeles	EmStar EmStar is a family of tools, libraries, and services that provide an environment to help enable the design, development, and deployment of WSN applications. EmStar supports heterogeneous deployments consisting of both mote-class and microserver-class component systems. http://research.cens.ucla.edu/

Tradeoffs Between Local and Remote Processing. The tradeoffs regarding where processing of the sensed data should be performed are not well understood. What are the benefits of staged versus hierarchical architectures, and centralized processing at a sink node versus distributed processing by the cluster heads?

Querying, In-Network Processing, Caching. A related question to the processing tradeoff is how to support querying in a heterogeneous WSN, what to cache and where to cache, and what kind of in-networking processing can be performed.

Event Detection. A large application of heterogeneous WSNs is in event detection. How do we reliably detect events with a low false alarm rate?

Quality-of-Service Support. Heterogeneous WSNs bring the potential of of high-bandwidth sources such as audio and video. Such data streams require quality-of-service support in order to meet delay, jitter, and related constraints.

Nonuniform Energy Drainage. While hierarchical architectures have alleviated the problem of non-uniform energy drainage, the problem remains unsolved.

Mobility in Sensor Nodes. Eventually, mobile nodes will be integrated into heterogeneous WSNs. This will add another dimension of complexity to all of the problems.

ACKNOWLEDGMENTS

The work of V. R. Syrotiuk and B. Li is supported, in part, by LANL contract 13638-001-05 and NSF grant ANI-0240524. Any opinions, findings, conclusions, or recommendations expressed are those of the authors and do not necessarily reflect the views of LANL or NSF.

The work of A. M. Mielke is supported by the U.S. Department of Energy/NNSA and Los Alamos National Laboratory funds under Contract Number DE-AC52-06NA25396 and is approved for public release under LA-UR-06-5787.

BIBLIOGRAPHY

1. I. Akyildiz, W. Su, Y. Sankarasubramaniam, and E. Cayirci. A survey on sensor networks. *IEEE Communications Magazine*, **40**(8):102–114, 2002.
2. Crossbow Technology, I. hppt://www.xbow.com/.
3. I. Akyildiz, W. Su, Y. Sankarasubramaniam, and E. Cayirci. Wireless sensor networks: A survey. *Computer Networks*, **38**(4):393–422, 2002.
4. Q. Gao, D. J. Holding, Y. Peng, and K. J. Blow. Energy efficiency design challenge in sensor networks. In *Proceedings of London Communications Symposium*, September 2002.
5. P. Kulkarni, D. Ganesan, P. Shenoy, and Q. Lu. SensEye: A multi-tier camera sensor network. In *Proceedings of the 13th annual ACM International Conference on Multimedia (MM '05)*, November 2005, pp. 229–238.
6. M. Rahimi, R. Baer, O. I. Iroezi, J. C. Garcia, J. Warrior, D. Estrin, and M. Srivastava. Cyclops: In situ image sensing and interpretation in wireless sensor networks. In *Proceedings*

of the 3rd International Conference on Embedded Networked Sensor Systems (SenSys '05, (2005), pp. 192–204.

7. CMUCAM. http://www-2.cs.cmu.edu/cmucam/cmucam2/index.html.
8. E. Shih, P. Bahl, and M. J. Sinclair. Wake on wireless: An event driven energy saving strategy for battery operated devices. In *Proceedings of the 8th Annual International Conference on Mobile Computing and Networking (Mobicom '02)*, 2002.
9. A. M. Mielke, S. M. Brennan, M. C. Smith, D. C. Torney, A. B. Maccabe, and J. F. Karlin. Independent sensor networks. *IEEE Instrumentation & Measurement Magazine*, **8**(2):33–37, 2005.
10. S. M. Brennan, A. M. Mielke, D. C. Torney, and A. B. Maccabe. Radiation detection with distributed sensor networks. *IEEE Computer Magazine*, **37**(8):57–59, 2004.
11. P. Gupta, and P. R. Kumar. Capacity of wireless networks. *IEEE Transactions on Information Theory*, **IT-46**(2):388–404, 2000.
12. Intel heterogeneous sensor networks research group.
13. X. Du, and F. Lin. Improving routing in sensor networks with heterogeneous sensor nodes. In *Proceedings of the 61st IEEE Vehicular Technology Conference (VTC '05-Spring)*, May–June 2005, Vol. 4, pp. 2528–2532.
14. K. Xu, X. Hong, and M. Gerla. An ad hoc network with mobile backbones. In *Proceedings of the IEEE International Conference on Communications (ICC '02)*, Vol. 5, April–May 2002, pp. 3138–3143.
15. V. Mhatre, C. Rosenberg, D. Kofman, R. Mazumdar, and N. Shroff. A minimum cost heterogeneous sensor network with a lifetime constraint. *IEEE Transactions on Mobile Computing*, **4**(1):4–15, 2005.
16. X. Du, and Y. Xiao. Energy efficient chessboard clustering and routing in heterogeneous sensor networks. *International Journal of Wireless and Mobile Computing*, **1**(2):121–130, 2006.
17. W. R. Heinzelman, A. Chandrakasan, and H. Balakrishnan. Energy-efficient communication protocols for wireless microsensor networks. In *Proceedings of the 33rd Hawaii International Conference on System Sciences* Vol. 2, January 2000.
18. S. Lindsey, C. S. Raghavendra, and K. Sivalingam. Data gathering in sensor networks using the energy delay metric. In *Proceedings of the 15th International Parallel and Distributed Processing Symposium*, 2001, pp. 2001–2008.
19. T. Lin, Y. Chen, and S. Lee. PCAR: A power-aware chessboard-based adaptive routing protocol for wireless sensor networks. In *Proceedings of the IEEE 6th Circuits and Systems Symposium on Emerging Technologies: Frontiers of Mobile and Wireless Communication*, Vol. 1, 2004, pp. 145–148.
20. C. Intanagonwiwat, R. Govindan, and D. Estrin. Directed diffusion: A scalable and robust communication paradigm for sensor networks. In *Proceedings of the 6th Annual ACM International Conference on Mobile Computing and Networking (Mobicom '00)*, August 2000, pp. 56–67.
21. F. Ye, H. Luo, J. Cheng, S. Lu, and L. Zhang. A two tier data dissemination model for large-scale wireless sensor networks. In *Proceedings of the 8th Annual ACM International Conference on Mobile Computing and Networking (Mobicom '02)*, 2002, pp. 148–159.
22. Y. Ma, S. Dalal, M. Alwan, and J. H. Aylor. ROP: A resource oriented protocol for heterogeneous sensor networks. In *Proceeding of the 13th Virginia Tech/MPRG Symposium on Wireless Personal Communication*, June 2003, pp. 59–70.

23. Y. Ma, and J. H. Aylor. System lifetime optimization for heterogeneous sensor networks with a hub-spoke topology. *IEEE Transactions on Mobile Computing*, **3**(3):286–294, 2004.

24. V. Mhatre, and C. Rosenberg. Homogeneous vs. heterogeneous clustered sensor networks: A comparative study. In *Proceedings of IEEE International Conference on Communications (ICC '04)*, Vol. 6, 2004, pp. 3646–3651.

25. E. J. Duarte-Melo, and M. Liu. Analysis of energy consumption and lifetime of heterogeneous wireless sensor networks. In *Proceedings of the IEEE Global Telecommunications Conference (Globecom'02)*, November 2002, Vol. 1, pp. 21–25.

26. J. Li, and P. Mohapatra. An analytical model for the energy hole problem in many-to-one sensor networks. In *Proceedings of the 62nd IEEE Vehicular Technology Conference (VTC '05-Fall*, Vol. 4, September 2005, pp. 2721–2725.

27. J. Ai, D. Turgut, and L. Bölön. A cluster-based energy balancing scheme in heterogeneous wireless sensor networks. In *Proceedings of the 4th International Conference on Networking (ICN '05)*, April 2005, pp. 467–474.

28. B. Liu, and D. Towsley. A study of the coverage of large-scale sensor networks. In *Proceedings of the IEEE International Conference on Mobile Ad-hoc and Sensor Systems*, October 2004, pp. 475–483.

29. S. Megerian, F. Koushanfar, M. Potkonjak, and M. B. Srivastava. Worst and best-case coverage in sensor networks. *IEEE Transactions on Mobile Computing*, **4**(1):84–92, 2005.

30. C. Huang, and Y. Tseng. The coverage problem in a wireless sensor network. In *Proceedings of the 2nd ACM International Conference on Wireless Sensor Networks and Applications*, 2003, pp. 115–121.

31. L. Lazos, and R. Poovendran. Coverage in heterogeneous sensor networks. In *Proceedings of International Symposium on Modeling and Optimization in Mobile, Ad Hoc, and Wireless Networks*, April 2006, pp. 1–10.

32. D. Miorandi, and E. Altman. Coverage and connectivity of ad hoc networks presence of channel randomness. In *Proceedings of the 24th Annual Joint Conference of the IEEE Computer and Communications Societies (Infocom '05)*, Vol. 1, March 2005, pp. 491–502.

33. G. Xing, X. Wang, Y. Zhang, C. Lu, R. Pless, and C. Gill. Integrated coverage and connectivity configuration for energy conservation in sensor networks. *ACM Transactions on Sensor Networks*, **1**(1):36–72.

34. T. Yan, T. He, and J. A. Stankovic. Differentiated surveillance for sensor networks. In *Proceedings of the 1st International Conference on Embedded Networked Sensor Systems (SenSys '03)*, 2003, pp. 51–62.

35. X. Du, and F. Lin. Maintaining differentiated coverage in heterogeneous sensor networks. *EURASIP Journal on Wireless Communications and Networking,* **5**(4):565–572, 2005.

36. D. Vaidya, J. Peng, L. Yang, and J. W. Rozenblit. A framework for sensor management in wireless and heterogeneous sensor network. In *Proceedings of the 12th IEEE International Conference and Workshops on the Engineering of Computer-Based Systems (ECBS '05)*, April 2005, pp. 155–162.

37. D. Andersson, M. Fong, and A. Valdes. Heterogeneous sensor correlation: A case study of live traffic analysis. In *Proceedings of the 3rd Annual Information Assurance Workshop*, June 2002.

38. L. Gürgen, C. Labbe, V. Olive, and C. Roncancio. A scalable architecture for heterogeneous sensor network. In *Proceedings of the 16th International Workshop on Database and Expert Systems Applications (DEXA '05)*, 2005, pp. 1108–1112.

39. G. Wiederhold. Mediators in the architecture of future information systems. *IEEE Computer Magazine*, **25**(3):38–49, 1992.

40. L. Yuan, C. Gui, C. Chuah, and P. Mohapatra. Applications and design of heterogeneous and/or broadband sensor networks. In *Proceedings of the 1st IEEE Workshop on Broadband Advanced Sensor Networks (BaseNets '04)*, October 2004.

41. O. Gnawali, and M. Yarvis. "Do Not Disturb" an application leveraging heterogeneous sensor networks. Technical Report 03-808, University of Southern California, 2003.

CHAPTER 3

Epidemic Models, Algorithms, and Protocols in Wireless Sensor and Ad Hoc Networks

PRADIP DE and SAJAL K. DAS

Center for Research in Wireless Mobility and Networking (CReWMaN), Department of Computer Science and Engineering, University of Texas at Arlington, Arlington, TX 76019

3.1 INTRODUCTION

Sensor networks are composed of a large number of sensing devices, which are equipped with limited computing and radio communication capabilities. They have diverse application areas, ranging from tracking and intrusion detection for security purposes to environment monitoring and traffic and location systems. However, with the steady advancements in processor, memory, communication, and sensing technology, along with a drive toward a smarter environment, there is an increased interest in the development and deployment of wireless sensor networks to be used for many interesting and new applications. These applications range from real-time remote monitoring and control, military surveillance, and environmental monitoring to healthcare management, construction safety, and so on.

Within the next few years, it is very likely that the number of deployed sensors will see an exponential increase. Most of these networks will require application-specific functionalities and performance requirements [1]. However, the realization of these networks poses a lot of challenges in system and network design, algorithm and protocol design, and query language and database design. The primary issues under focus which are critical to the proper functioning of wireless sensor networks are energy consumption, connectivity, clustering techniques, data aggregation, and so on. These issues stem mostly from the stringent resource constraints of the sensor nodes. Therefore, in order to address these issues, we require efficient modeling techniques and robust algorithms and protocols before actual implementation and deployment is done.

Algorithms and Protocols for Wireless Sensor Networks, Edited by Azzedine Boukerche

Oftentimes, in the course of modeling complex entities and networks, we have taken recourse to biologically inspired paradigms. Biologically inspired modeling techniques are among the many mechanisms that have been adopted to accurately model certain phenomena in wireless sensor networks. For example, data dissemination, routing algorithms, and broadcast protocols are among the few areas that have been effectively modeled by epidemic theoretic concepts.

In this chapter, we address the modeling techniques, algorithms, and protocols proposed in wireless sensor and ad hoc networks that are primarily based on Epidemic Theoretic concepts and paradigms.

The rest of the chapter is organized as follows. In Section 3.2, we provide a general overview of Epidemic Theoretic concepts and analysis. In Section 3.3, we discuss the data dissemination models in sensor networks and their use of epidemic theory. Section 3.4 illustrates several reprogramming and code update protocols in sensor networks that adopt epidemic theoretic principles. In Section 3.5, we look into epidemic protocols in ad hoc networks. Section 3.6 looks into some security aspects and explains the propagation process modeling of malware in sensor networks. Finally, we conclude the chapter in Section 3.7.

3.2 OVERVIEW OF EPIDEMIC THEORY

In order to appreciate the epidemiological models applied in wireless sensor networks, we need to first understand the concept of epidemiology. In this section we provide a terse description of the theory and its applications. Epidemic Theory [2] is the study of the dynamics of how contagious diseases spread in a population, resulting in an epidemic. Primarily, the theory mathematically models the propagation process of an infection and measures its outcome in relation to a *population at risk*. The population at risk basically comprises of the set of people who possess a susceptibility factor with respect to the infection. This factor is dependent on several parameters such as exposure, spreading rate, previous frequency of occurrence, and so on, which define the potential of the disease causing the infection. Among the different models characterizing the infection spread, two are quite popular. They are the *Susceptible Infected Susceptible* (S-I-S) Model, *Susceptible Infected Recovered* (S-I-R) Model, and so on. In the former, a susceptible individual acquires infection and then after an infectious period (i.e., the time the infection persists) the individual becomes susceptible again. On the other hand, in the latter, the individual recovers and becomes immune to further infections.

An approach to model the propagation of an infection is to assume that the probability (per unit time) for a susceptible individual to acquire infection is equal to the average rate at which new infective partners are acquired multiplied by the probability of being infected by any one such partner. In the general deterministic S-I-R model, if $N(t)$, $X(t)$, $Y(t)$, and $Z(t)$ denote the total population, the susceptibles, the infected, and the recovered or immune individuals, respectively, at time t, we can say

$$N(t) = X(t) + Y(t) + Z(t) \quad (3.1)$$

If β denotes the infection rate and γ denotes the removal rate of infected individuals, then assuming a homogeneous mixing model (i.e., each of the susceptibles can get in contact with any of the infectives), it is simple to observe that in time Δt, there are $\beta xy\Delta t$ new infections and $\gamma y\Delta t$ removals. Therefore, the basic differential equations that describe the rate of change of susceptibles, infectives, and recovered individuals are given by

$$\frac{dX(t)}{dt} = -\beta XY,$$
$$\frac{dY(t)}{dt} = \beta XY - \gamma Y,$$
$$\frac{dZ(t)}{dt} = \gamma Y$$

The above equations can be solved either approximately or precisely based on some boundary conditions, such as, at the start of the epidemic, when $t = 0$, (X, Y, Z) can take the values $(x_0, y_0, 0)$. Note that, in particular, if y_0 is very small, x_0 is approximately equal to N. It also follows that if the *relative removal rate*, $\mu = \gamma/\beta$, is greater than x_0, only then can an epidemic start to build up as this condition will result in $[dY(t)/dt]_{t=0} > 0$, i.e. $Y(t)$ will have a positive slope. Therefore, the relative removal rate $\mu = x_0$ gives a threshold density of susceptibles.

On the other hand, the S-I-S model does not have the recovered subset $Z(t)$, and those who are infected fall back into the susceptible subset $S(t)$ after their infectivity duration.

An important aspect that is of particular interest in epidemiological studies is the phenomenon of phase transition of the spreading process that is dependent on a threshold value of the epidemic parameter; that is, if the epidemic parameter is above the threshold, the infection will spread out and become persistent; on the contrary, if the parameter is below the threshold, the infection will die out. Identification of this threshold value is critical in the study of how an epidemic spreads and how it can be controlled.

Apart from modeling technique based on the continuous differential rate equation, the study of epidemics has often been performed by treating the population as a network graph, with the nodes representing each individual and the edges their interaction. This form of analysis [3] has mainly been used in scenarios where the end result of the epidemic spread is more important than the temporal dynamics of the propagation. Several works have spawned from this formulation [3–8], where the spread of diseases have been studied by modeling the social network as a scale-free topology. Several other works also exist that model the spread of computer viruses [9, 10].

Epidemic Theory has found special attention in the design and modeling of several phenomena and protocols in sensor networks wherever there is a scope of information distribution on a large scale, preferably from a small number of sources to a large number of recipients. Among the popular phenomena in sensor and ad hoc networks where this theory has been adopted are data dissemination, broadcast protocols, and routing. We will delve into some of these areas where Epidemic Theory has been used to study and model several processes and functions of sensor networks.

3.3 DATA DISSEMINATION IN SENSOR NETWORKS: MODEL AND PROTOCOLS

The problem of reliable data dissemination in the context of wireless sensor networks is very critical. Reliable data dissemination to all nodes is absolutely necessary for the propagation of queries, code updates, and other sensitive information in a wireless sensor network. This is not a trivial task since the number of nodes in a sensor network can be quite huge and the environment is dynamic (i.e., nodes can die or move), thus making the topology change constantly.

Since data dissemination primarily deals with the transfer of messages from one node to all nodes of a network, algorithms based on epidemiological formulations are a perfect fit. Accordingly, these algorithms have been successfully used in disseminating information in sensor networks and, depending on the application, the dissemination can start at a single node, such as a base station, or at multiple sensor nodes. The decentralized and distributed nature of wireless sensor networks fits the context of epidemic algorithms aptly.

One of the prominent works of data dissemination in sensor networks is SPIN [11]. An obvious problem with normal epidemic broadcast-based dissemination is the inefficient use of bandwidth and other resources. Therefore, the basic epidemic strategy needs to be optimized for sensor networks. In reference 11, the authors proposed the concept of meta data or data descriptors to eliminate the chance of redundant transmissions in sensor networks. Their work focuses on the efficient dissemination of individual sensor observations to all the sensors in a network. Their main contribution was based on the basic deficiencies of classic flooding, namely, *Implosion*, *Overlap*, and *Resource Blindness*. Implosion is sending data redundantly to one's neighbors regardless of whether they already received it. Coverage overlap of nodes can make them gather the same data and flood it to common neighbors. Classic flooding can be blind to the availability of resources when it is flooding data across the network.

The use of metadata allows nodes to negotiate between themselves and prevent redundantly transmitting the same information. Also, in SPIN, each node has a local resource manager that keeps track of its resources and helps a node decide whether to transmit or process data. SPIN first broadcasts metadata to its neighbors. Then, if it receives a request for the data from any neighbor it sends the data to that node.

There are four protocols in the SPIN family. The first two, *SPIN-PP* and *SPIN-BC*, tackle the basic problem of data dissemination under ideal conditions. The other two, *SPIN-EC* and *SPIN-RL*, are modified versions of the first two. *SPIN-PP* is optimized for communicating in a point-to-point mode, where for each data transmission between neighbors, a three-stage handshaking (ADV-REQ-DATA) is

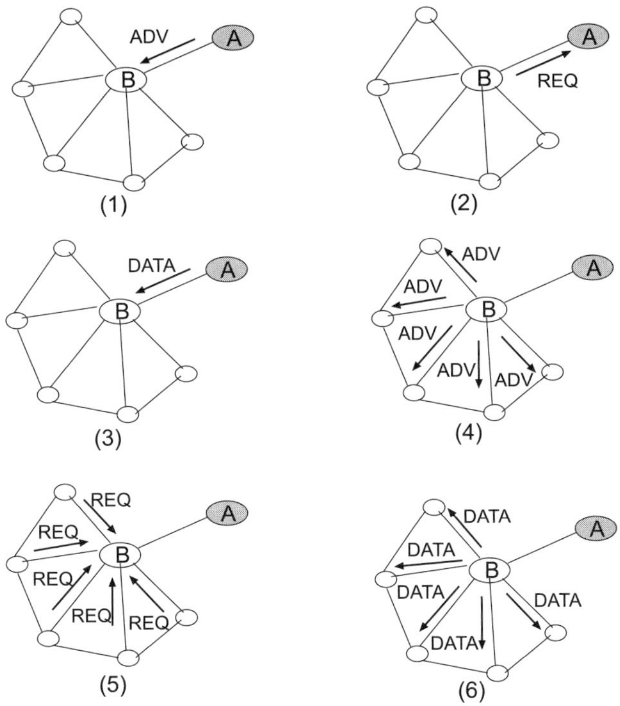

Figure 3.1. SPIN-PP protocol. Node A sends advertisement messages (*ADV*) to B. B responds with a request (*REQ*) message. Then B starts to send *ADV* to its neighbors.

performed. As illustrated in Figure 3.1, a node sends an *ADV* message whenever it has new data to advertise. Upon receiving an *ADV* message, the neighboring node verifies whether it has already received or requested the advertised data. If not, it responds by sending a *REQ* message for the missing data back to the sender. The initiator of the protocol responds to the *REQ* message with a *DATA* message containing the missing data.

Although this protocol has been designed for a lossless environment, it can be adapted for a lossy environment. Nodes can periodically send the *ADV* message to counter lost *ADV* messages. For lost *REQ* and *DATA* messages, nodes can request items that do not arrive within a fixed time period. *SPIN-EC* is a modification of *SPIN-PP* so that when a node observes that it is approaching a low-energy threshold, it reduces its participation in the protocol.

In *SPIN-BC*, which is a broadcast transmission protocol, each node transmits to the broadcast address (Figure 3.2). Every node that is in the transmission range of the sender processes the received message. This approach is justified because broadcast and unicast transmissions use the same amount of network resources in a broadcast network. The proliferation of redundant messages in the network can be curtailed by *SPIN-BC* because a node *A* suppresses its own transmission whenever it observes

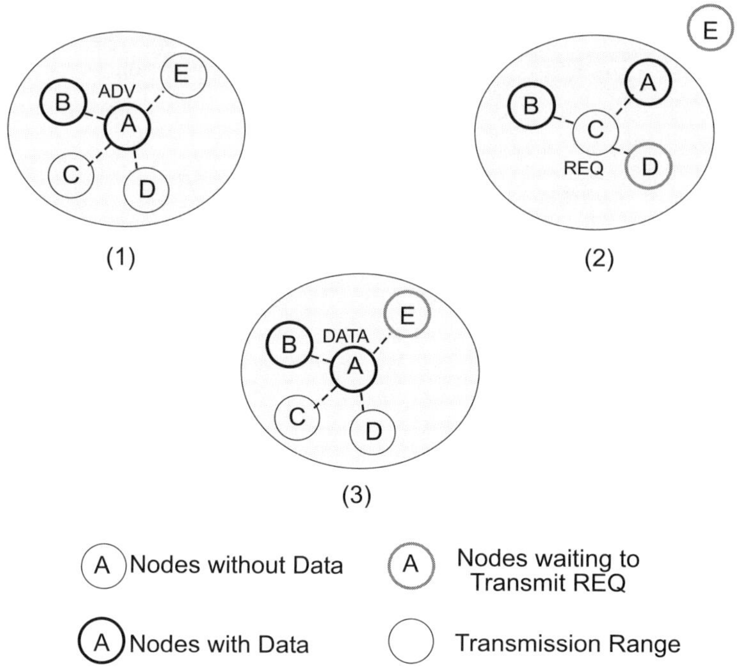

Figure 3.2. SPIN-BC protocol. (1) A sends *ADV* to all its neighbors. (2) C responds by broadcasting a request, specifying as originator of the *ADV*. It also suppresses D's *REQ*. (3) After receiving the requested data, E's *REQ* is also suppressed.

that another node *B* has transmitted the same message that *A* itself was supposed to transmit.

We observe the epidemic nature of the dissemination of data in *SPIN*, especially in *SPIN-BC*. Using the three-way handshake, a node that has the missing data passes it on to a neighbor that does not have it, thereby infecting it in the process. The working of the three-way handshaking protocol basically constitutes the contact and infecting process of the *SPIN* protocol.

3.3.1 Infuse

For the reliable dissemination of data in sensor networks, the authors of *Infuse* [12] proposed a TDMA-based data dissemination protocol for sensor networks. The primary purpose of the protocol was similar to that of *Deluge* [13]—that is, reliable dissemination of bulk data in a sensor network. We discuss Deluge later in this chapter. In Infuse, the data dissemination protocol is based on a TDMA-based medium access layer. Since TDMA ensures a deterministic slot when a sensor node should transmit its packet, it offers a degree of reliability which is used by the data dissemination strategy adopted in Infuse. The authors tackle the problem of random message losses in the presence of channel errors by considering recovery algorithms based on sliding window protocols, modified to use implicit acknowledgments.

In the ideal scenario without channel errors, the base station sends a special *Start Download* message to the sensor which contains the number of subsequent packets to follow in each TDMA slot. The sensor then reserves the necessary flash and downloads the arriving packets.

For dealing with channel errors, Infuse uses an implicit acknowledgment technique. This happens because whenever a successor sensor forwards a data packet, the predecessor node gets to hear it. This overhearing acts as an implicit acknowledgment for the predecessor node. Furthermore, Infuse forwards a received packet in the next TDMA slot, thus maintaining a pipeline effect of the transfer process which helps in reducing the total latency of the dissemination process. The use of TDMA-based data dissemination also allows Infuse to send the node to sleep except in its own transmission slot, thereby making the Infuse protocol energy-efficient.

3.3.2 Firecracker

Routing a packet from one source to a single destination is fast because forwarding nodes can retransmit without worrying about suppression or local density. At the same time, routing cannot be used to disseminate data to all the nodes in a sensor network because the nodes are not individually addressable. The Firecracker Protocol [14] uses a combination of routing and broadcast principles to rapidly disseminate data throughout a sensor network. As depicted in Figure 3.3, a data source first routes the data to be distributed to distant points in the network. Once the data reaches its destination, broadcast-based dissemination starts along the path like a string of firecrackers. Firecracker is largely designed to disseminate small pieces of data that would propagate fast, like small programs or configuration constants. While maintaining the energy efficiency of broadcasts, Firecracker can achieve dissemination rates close to routing.

From an epidemic modeling standpoint, Firecracker is fundamentally an infection propagation strategy with a predetermined set of infective nodes defined by the destination nodes of the routing protocol. Having strategically placed the infective nodes at different points of the population, the protocol starts its final broadcast to disseminate the information to the rest of the network. The dissemination strategy

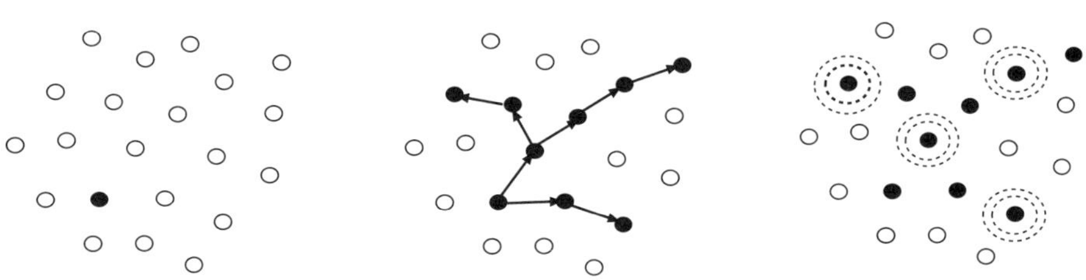

Figure 3.3. The Firecracker disseminaton mechanism.

could be Trickle [15], and the routing strategy could be any suitable one used for sensor networks.

To elucidate further, Firecracker is composed of three main parts: (a) Broadcast Protocol, (b) Routing Protocol, and (c) Seed Selection. The broadcast protocol is very important to the functioning of Firecracker. It not only must propagate data to nodes that do not have the data, but also decide when to propagate. Moreover, the protocol should minimize the cost of detection but propagate rapidly, and temporary network disconnections should not prevent reception.

The basic purpose of the routing phase is to spread data to distant points in the network so that the initial seeds are placed as deep as possible into the network. This facilitates the following broadcast process to spread the data rapidly. In this regard, a naming scheme that allows nodes to choose such points is helpful. Since wireless data, even during the routing phase, is essentially broadcast in the neighborhood, nodes along the route should be able to snoop on routed traffic to cache the data as they pass by. Moreover, reliability and nonredundancy are more important than minimum hop paths. Therefore, taking a long, winding path through different areas of the network could benefit the subsequent broadcast protocol in quickly installing the code in all the nodes.

The choice of the seed nodes is also equally important to the performance of Firecracker. The farther the seeds are from the original source and the farther they are spread apart from themselves, the faster the data would propagate to all the nodes of the network.

In general, epidemic algorithms for data dissemination follow the model of nature to spread information and define simple rules for information to flow between nodes of a network. The authors in reference 16 have done a comparative study of epidemic algorithms for data dissemination. Based on the style of communication between neighboring nodes, they have classified epidemic algorithms for data dissemination into three categories.

- *Pull-Based.* A node tries to extract new information from its neighbor.
- *Push-Based.* A node sends new information to a selected neighbor.
- *Pull–Push-Based.* A node asks its neighbors for new information as well as sends new information to its neighbors.

They have studied the performance of these three classes of epidemic algorithms on sensor networks. Their results show that both pull-based and push-based algorithms perform better than the push–pull-based epidemic algorithms in terms of delivery rate and scalability. The primary reason for this result is the restricted memory resource of sensor devices.

In reference 17, the authors performed an experimental and empirical study of the epidemic style algorithms in large-scale multihop wireless networks.

A smart tag-based data dissemination technique is explained by the authors in reference 18, where mobile individuals, equipped with smart tags disseminate data across disconnected static nodes spread across a wide area. When the mobile individuals equipped with smart tags move into a sensor field, they get updated with the latest information from the sensors. Later, when they move into another field, they disseminate the newly acquired information. The concept of using carriers, who are mobile, to carry data between connected components of the network has also been used by the authors of the epidemic routing protocol [19]. However, they did it more for the purpose of routing, whereas here the authors use smart tags to carry the sensed information to another set of output devices like display units. The authors used Bluetooth-enabled smart tags to illustrate the characteristics of their approach. As an intuitive technique, their approach is suited for applications that are delay-tolerant.

3.4 CODE UPDATE PROTOCOLS IN SENSOR NETWORKS

Several protocols have been proposed for code update and propagation in sensor networks. These protocols are mainly broadcast in nature, and tasks in sensor networks are assigned through code updates, and all the nodes in the sensor network will have the same code to execute. The propagation mechanism for the code update is basically hop by hop to all nodes in the network. Needless to say, wireless sensor nodes have limited energy, and therefore maintenance costs of the code updates must be low. Another important requirement is rapid propagation of updates, because some tasks may have to be activated as soon as possible and newly assigned tasks make the older ones obsolete. Moreover, the update process should be scalable and should work in a dynamically changing environment.

Being inherently broadcast in nature, these protocols and algorithms fundamentally transmit code updates in a manner similar to an infection spread in a susceptible population. In this subsection, we study some of these protocols and their mechanism.

3.4.1 Trickle

Trickle [15] is a broadcast algorithm for propagating and maintaining code updates in a wireless sensor network. Conceptually, Trickle borrows from epidemiological concepts and performs what the authors claim as *polite gosip*.

Sensor networks are generally deployed in remote areas and are expected to operate unattended for lengthy periods of time. Thus, there is every possibility that the requirements and environments of a sensor network evolve. As a result, users need to be able to introduce new code to retask the network. However, the large-scale and embedded nature of the network requires these code updates to propagate through the network. However, as is obvious, networking in sensor networks is very costly in terms of energy consumption, and therefore an efficient and effective reprogramming protocol is necessary.

An effective reprogramming protocol must transfer the code as fast as possible because in the transition time when the code is propagating, the network is actually in a useless state because the old and the new programs are concurrently running.

Propagation of code is costly, and nodes need to learn when they need to propagate code. Nodes, therefore, periodically communicate to learn when there is new code.

To reduce energy costs, nodes transmit metadata to determine when code is needed. However, the cost of periodically transmitting metadata consumes almost the same amount of energy as actually transmitting the code itself. Therefore, there is a crucial need for the reprogramming algorithm to be efficient in this aspect and effectively determine when nodes should propagate code. Motivated by this requirement, the authors in reference 15 have identified three main properties that a reprogramming algorithm should have. They are as follows:

- *Low Maintenance.* When a network is in a stable state, metadata exchanges should be infrequent, just enough to ensure that the network has a single program.
- *Rapid Propagation.* When the network discovers nodes that need update, it should propagate the code as fast as possible and to every node of the network.
- *Scalability.* The algorithm should obviously be scalable and be robust against any environmental changes and node failures.

Trickle tries to meet all these requirements. Its basic working principle is simple. Every so often, a mote transmits code metadata if it has not heard a few other motes transmit the same information. Trickle sends all messages to the local broadcast address. When a neighbor receives a broadcast, either it is up to date, or it detects the need for an update. Detection can be the result of either an out-of-date mote hearing someone having a new code, or an updated mote hearing someone has old code. As long as every mote communicates somehow, the need for an update is always detected. It does not matter who transmits first, but as long as some nodes communicate with each other at a nonzero rate, every node would be up to date. More formally, each node maintains a counter c, a threshold k, and a timer t in the range of $[0, \tau]$. k is a small, fixed integer (e.g., 1 or 2) and τ is a time constant. When a node hears metadata identical to its own, it increments the counter c. At the timepoint t, which is uniformly randomly chosen in the range of $[0, \tau]$, the mote broadcasts its metadata only if $c < k$. When the interval of size τ completes, c is reset to zero and t is reset to a new random value in the range $[0, \tau]$. Thus, Trickle allows each node to broadcast its metadata at most once per period τ, thus maintaining the politeness of its gossip. In each interval τ , the sum of receptions and sends of each mote is k. The random selection of t uniformly distributes the choice of who broadcasts in a given interval. This evenly spreads the transmission energy load across the network.

In Figure 3.4, the solid line represents a transmission, the broken lines represent reception, and the gray line means suppression of an advertisement. This mechanism of Trickle not only allows us to scale to high network density, but also propagates updates fast. It also distributes transmission load evenly as it spreads, and it is simultaneously robust to transient disconnections. The experimental verification by the authors shows that it imposes a maintenance overhead on the order of only a few packets per hour per node.

The epidemiological essence in the working principle of Trickle is evident. The objective is to propagate code as fast as possible to all nodes of the network. Thus,

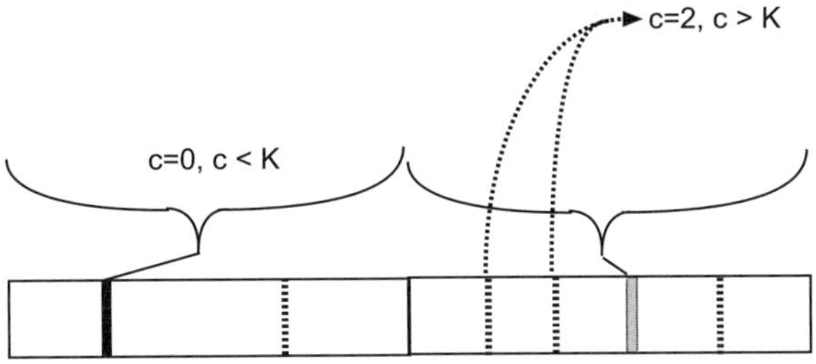

Figure 3.4. Trickle metadata advertisement.

from an epidemic theoretic standpoint, the rate at which the metadata is exchanged gives the rate at which the infection or propagation proceeds in the network. Since after a node advertises metadata every node in the neighborhood gets updated with the current code, Trickle succeeds in propagating the code update to all nodes in the network. The propagation rate is dependent on the value of τ. With a large value of τ, there is less communication overhead, but the code propagates slowly and conversely in the case of a small τ.

3.4.2 Deluge

Another type of data dissemination protocol for supporting network programming in sensor networks is Deluge (Figure 3.5) [13]. It is a reliable data dissemination protocol for propagating large data objects from a few source nodes to many other

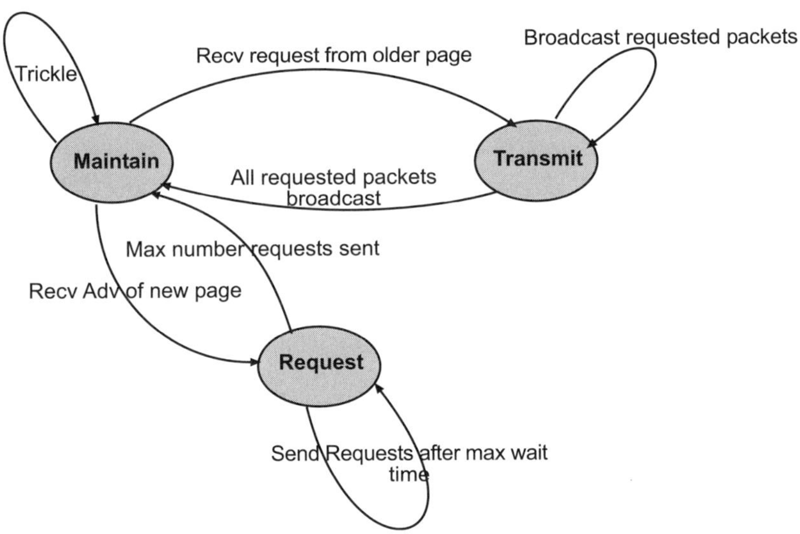

Figure 3.5. The Deluge state machine.

nodes in a wireless sensor network. Trickle's key contribution is its polite gossip that uses suppression and dynamic adjustment of the broadcast rate to limit transmissions among neighboring nodes. It only provides a mechanism for a node to decide when to propagate code. Deluge, on the other hand, though based on Trickle's principles, has the added feature of supporting the transfer of large data objects. It uses a three-phase protocol similar to SPIN-RL [11].

Deluge, being an epidemic protocol, can disseminate large data objects as quickly and reliably as possible. The basic local broadcast principle of Deluge is simple and similar to Trickle, but it also addresses several subtle issues that improve its performance. The local suppression of redundant broadcasts makes it density-aware. Its three-way handshaking mechanism ensures that there is a bidirectional link, thereby making it a reliable data dissemination protocol. Moreover, by dynamically adjusting the rate of advertisements and emphasizing on the use of spatial multiplexing to allow parallel transfers, Deluge allows quick propagation of large blocks of data.

Deluge divides the large data object into fixed-size pages for transfer. This enables efficient incremental update and also limits the amount of state that should be reserved at a time at the receiver. Each page is also divided into a fixed number of packets. Because of the epidemic nature of the page propagation, Deluge offers CRC checks at both the packet and page level to be safe from the negative effects of the epidemic nature of data transfer. The protocol resides in one of the three states, namely, *MAINTAIN*, *TX*, and *RX*. In the *MAINTAIN* state, a node uses a summary advertisement mechanism to ensure that all nodes in the communication range are up to date with the current version of the object. In the *RX* state, the node is responsible for requesting all remaining packets of a page; and while in the *TX* state, it is responsible for broadcasting all requested packets for a given page.

Another work, which was based on an epidemic style multihop reprogramming service for sensor networks, was proposed in MNP [20]. One of the basic problems in reprogramming and code update in a wireless network is the issue of message collision and the hidden terminal problem. The authors counter this problem by proposing a sender selection algorithm whereby it is guaranteed that in a neighborhood, there is at most one sensor transmitting at a time. In the basic version of the sender selection protocol, a node becomes a source node and starts advertising this fact only when it acquires the new program code entirely.

Each source node maintains a variable that indicates the number of distinct requests it has received so far, and it gets incremented each time a node receives a new download request. Two messages are used for sender selection, namely, *advertisement* and *download request*. The *advertisement* message contains information about the new program and the source node. When a node j receives the advertisement request from node i and it is in need of the new code, it sends a *download request* message to the broadcast address so that any neighboring node k becomes aware that i is a potential source.

In order to ensure that a node is aware of all the requesters who are likely to receive the code, if it is chosen to transmit the code, the node sends a download request to all senders of the advertisement messages. However, if node j loses to node k that has more requesters, then whenever j attempts to advertise again, j must reset its *request counter* value to zero and recalculate its requesters. After k finishes transmitting, it sleeps for a while, so that other sources get a better chance to send. When it wakes up and reenters the advertising state, its counter value is reset to zero, and a new round of sender selection begins.

We observe that there is considerable similarity with MNP and the SPIN family of protocols because both of them use a kind of three-way handshaking procedure for disseminating the data.

For the transfer of large-sized data, MNP tries to incorporate pipelining into the data dissemination process by breaking the large data into *segments*, each containing a constant number of packets. Thus, the protocol now operates at the segment level which helps in a node forwarding segments even if it has not received the whole data. In this aspect it is very similar to Deluge [13].

MNP is equipped to address reliability issues like loss detection and recovery. Each packet has a unique ID and each receiver is responsible for detecting its own loss. Since the size of a segment is considerably small, a bitmap of the current segment is maintained in memory, where each bit corresponds to a packet. Using this, a sensor node can receive packets in any order. This bitmap is called the *Missing Vector*. A node also maintains a *Forward Vector*, which is a bitmap of the advertised segment. Whenever a node sends a *download request*, it puts its *Missing Vector* in the request message. The advertising node marks its *Forward Vector* according to the *Missing Vector* messages it receives. A node only sends the packets indicated in the *Forward Vector*. Upon receiving all the segments of a program, the node reboots.

3.5 EPIDEMIC MODELS IN AD HOC NETWORKS

In ad hoc networks, the power supply of individual nodes, wireless bandwidths are limited, and the channel conditions can vary significantly. Moreover, since nodes can be mobile, routes may constantly change. Thus, to enable efficient communication, robust routing protocols must be developed. Several existing Mobile ad hoc routing protocols [21, 22] have been developed that allow wireless nodes to communicate with one another without any preexisting network infrastructure. In this section we look into some of the routing protocols that essentially have the flavor of epidemiology.

3.5.1 Gossip

Although flooding has been used with some optimization to route packets in an ad hoc network, many routing messages are propagated unnecessarily. The authors in reference 23 have proposed a *gossip*-based approach where each node decides to forward a message to another node based on some probability. They showed that this technique could significantly reduce the number of routing messages sent. Gossip is essentially an epidemic algorithm, where neighbors are chosen probabilistically to propagate the information in the same way as an infection spreads in a susceptible population.

In the gossip protocol a source sends the route request with probability 1. When a node first receives a route request, with probability p it broadcasts the request to its neighbors, and with probability $1-p$ it discards the request; if the node receives the same route request again, it is discarded. Thus, a node broadcasts a given route request at most once.

The problem with gossip is that if the source has very few neighbors, then the nodes will not gossip and it would die out. Basically, from the epidemiological standpoint we say that the phase transition did not happen and the propagation collapsed. In order to circumvent this problem, the authors modify gossip so that each node forwards with probability 1 for the first k hops before continuing to gossip with probability p. The modified protocol is called the GOSSIP1(p, k) protocol.

The performance study of the gossip protocol in finite networks reveals several important results. As expected, the location of the source node does not affect the fraction of the source node receiving the messages. However, it does affect the number of executions in which the gossip dies out. The number of executions in which the gossip does not die out is higher for a more central node and is lower for a corner node. The authors observe that lowering the probability significantly changes the fraction of executions in which all nodes and no nodes get the message.

The authors suggest a few optimization techniques to the basic gossip protocol. In many cases, a gossip protocol may be run in conjunction with other protocols. If the other protocols maintain fairly accurate information regarding a node's neighbors, GOSSIP1 can make use of this information effectively, by a simple optimization. In a random network, the number of neighbors of a node might not be very high. In such a case, the gossip protocol might not propagate the information and die out. To overcome such a situation, the authors proposed that the gossip probability at a node could be a function of its degree, where nodes with lower degree gossip with higher probability. The modified protocol has four parameters: p_1, k, p_2, and n. As in GOSSIP1, p_1 is the main gossip probability and k is the number of hops with which gossiping starts with probability 1. The new features are p_2 and n; the idea is that the neighbors of a node with fewer than n neighbors gossip with probability $p_2 > p_1$. Thus, if a node has fewer than n neighbors, it would instruct its neighbors to broadcast with probability p_2 rather than p_1. The modified protocol is called GOSSIP2 $(p_1; k; p_2; n)$. GOSSIP2 has significant impact in topologies that are random rather than regular.

However, GOSSIP1 and GOSSIP2 might suffer a premature death because the probability is low. In order to detect whether the gossip is dying out, a node might monitor the number of messages it is getting from its neighbors. If a node x has n neighbors and the message does not die out, then it would expect that all of its neighbors would get the message, and, if the gossip probability is p, it should get roughly pn messages from its neighbors. If it gets significantly fewer than pn messages within a reasonable time interval, then this is a clue that the message is dying out. The authors have proposed a modification to resolve this issue. If a node with n neighbors receives a message and does not broadcast it, but then does not receive the message from at least m neighbors within a reasonable timeout period, it broadcasts the message to all its neighbors. If m is chosen too large, then there may be too many messages. The experimental results show that the most significant performance improvement could be obtained with $m = 1$. Thus, in GOSSIP3 (p, k, m), if a node originally did not broadcast a received message but did get the message from at least m other nodes within some timeout period it will immediately broadcast the message after the timeout period. In what follow, we will discuss several gossip schemes.

Geographic Gossip for Efficient Aggregation. Gossip algorithms have also been used for data aggregation in sensor networks. Their forte is their simplicity in approach. However, in their basic form they may waste significant energy by essentially passing around redundant information. The authors in *Geographic Gossip* [24] propose an alternative gossiping scheme that exploits geographic information.

In a network of n sensors, a basic solution to the *averaging problem* (i.e., to compute the average of all n sensor measurements) is based on the *Gossip* algorithms where each node randomly picks a one-hop neighbor and exchanges their current values. This is performed in an iterative fashion, and ultimately all nodes converge to the global average in a distributed manner. The key issue here is the number of iterations it takes for such a gossip algorithm to converge to a sufficiently accurate estimate. Recent works [25–29] have dealt with variants of this problem. The convergence time of this algorithm is closely linked with the mixing time of the Markov Chain defined by a weighted random graph on the network. In reference 26, the authors showed how to optimize the neighbor selection probabilities for each node in order to find the fastest mixing Markov chain. However, for sensor network graphs, even an optimized gossip algorithm can result in excess energy consumption. The authors of Geographic Gossip exploit geographic information to build a completely randomized and distributed algorithm that requires substantially less communication. The idea is to include geographic routing to gossip with random nodes far away in the network. Empowered with geographic knowledge, this protocol succeeds in quickly diffusing information everywhere in the network and thus computes the average faster than the standard nearest-neighbor gossip.

Smart Gossip. The authors of *Smart Gossip* [30] propose an adaptive form of gossiping in sensor networks. They propose techniques by which a gossip-based protocol can automatically and dynamically adapt to the network topology. Smart Gossip copes well with wireless losses and unpredictable node failures that affect network connectivity. The adaptivity of the gossiping strategy also extends itself to provide reliability for disseminating messages. The authors argue that existing gossip strategies are mostly static, since there is a fixed probability for transmitting the received information. There are a few variants of the gossip protocol which are adaptive. Haas et al. [23] proposed an adaptive form of gossip which chooses its probability, based on the number of neighbors. The authors of Smart Gossip argue that simply choosing gossip probabilities based on the number of neighbors is not

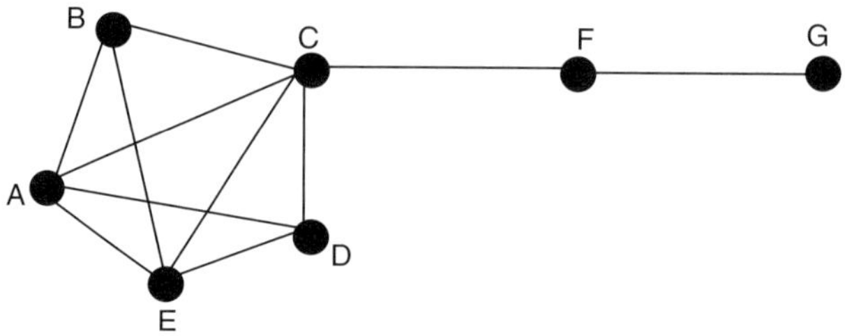

Figure 3.6. Example illustrating the argument for Smart Gossip.

correct. For example, in Figure 3.6, which is a subgraph of a random topology, node C has a high degree and therefore its probability of gossiping would be low. This could adversely affect the reception of the information at node F, which is solely dependent on C for receiving messages originating at any node to the left of C.

From the point of view of gossip percolation, the authors extract out the notion of dependence between a node X and a subset of its neighbors. These dependencies give birth to parent–child relationships between neighbors based on the direction in which the gossip can travel probabilistically. This dependency graph is just logical and also probabilistic in nature. In other words, node X does not depend on any particular parent, Y, to receive the gossip. Instead, it depends on a group of nodes, expecting at least one member of this group to probabilistically deliver the gossip to it. The gossip probabilities chosen at each node is therefore a function of the group size.

Based on this intuition, nodes promiscuously overhear broadcast messages and extract information by applying simple rules and thereby deduce whether the sender of the message is a parent, child or a sibling. A child node, on identifying its parent set, calculates the probability with which it thinks its parents are required to send, and it announces this probability by piggybacking it on every gossip it forwards. A parent node overhears such announcements and assumes its gossip probability to be the maximum of all the announced probabilities.

3.5.2 Epidemic Routing

The authors in reference 19 introduce epidemic routing for partially connected ad hoc networks where pairwise exchanges of messages among mobile hosts in a random manner ensure the eventual message delivery to the destination node. The prominent goals of epidemic routing are to (i) maximize message delivery rate, (ii) minimize message latency, and (iii) minimize the total resources consumed in message delivery.

Existing ad hoc routing protocols assume that there is a connected path from source to destination. However, with the emergence of short-range wireless communication environments (e.g., Bluetooth [31]) and the wide area over which such networks are deployed, this assumption is not always a realistic one. Unfortunately, the current ad hoc routing protocols are heavily dependent on consistent network connectivity to deliver packets between the source and the destination and generally fail in the presence of network partitions. At the same time, several applications based on a mobile sensor network exist, where there are frequent and numerous formations of network partitions.

In reference 19, the authors develop techniques for delivering application data with a high probability even when there is never a fully connected path between the source and destination. The main essence of their approach is to distribute data to connected hosts of the network, whom they call *carriers*; and depending on node mobility, the carriers can establish contact with other connected portions of the network. Through such transitive transmission of data, messages have a high probability of eventually reaching their destination. However, with basic random forwarding, the data might be transmitted to a large number of carrier hosts other than the destination that is not desirable. Since the overall goal of epidemic routing is not just to maximize message delivery rate and minimize message delivery latency, but also to minimize the aggregate system resources consumed in message delivery, the authors circumvent this problem to a reasonable extent by placing an upper bound on the message hop count and per-node buffer space (the amount of memory devoted to carrying other host's messages). Their results show that epidemic routing is able to successfully deliver messages to the destination nodes where existing ad hoc routing protocols fail because of limited node connectivity.

Although the authors of this work do not explicitly use the mathematical formulations of an epidemic model, they essentially follow the same principles of the model. The contact rate between carriers and destination or intermediate connected nodes is dependent on the mobility pattern of the carriers. However, since the notion is to route and not broadcast the message, the authors successfully constrain resources at nodes to restrict the number of messages a host is willing to carry on behalf of other hosts.

3.5.3 Epidemiology and Mobile Ad Hoc Networks

Information diffusion in mobile ad hoc networks (MANET) has been an area where epidemiological modeling concepts fit naturally. As mentioned earlier, several modeling formulations in epidemiology assume a homogeneously mixing population where each infected individual has an equal probability of having contact with any susceptible individual. Scenarios that fit this assumption can borrow the differential-equation-based formulations popular in epidemiology. Information diffusion in MANETs fit very closely in this model. Given the random mobility model, it's a fair assumption that the nodes can homogeneously mix. As a result, this phenomenon could be aptly modeled based on the differential rate equation formulations. This has been done by the authors in reference 32. Based on a simplistic S-I-S model, the authors have simplistically modeled the spread of information in a MANET. They showed that the information dissemination can be more or less accurately described by the infection rate of the model. They derived expressions that show the change of infection rate based on the node densities.

In reference 33, the authors address the issue of how to disseminate relevant information to mobile agents within a geosensor network. In their work, the authors propose an environment for simulating information dissemination strategies in mobile ad hoc geosensor networks. A geosensor network is defined as a sensor network that monitors phenomena in geographic space [34]. In the context of geosensor networks, the authors provide a decentralized location-based service that is able to disseminate relevant geospatial information to spatially dispersed mobile users that form a mobile ad hoc geosensor network. The authors explore the precise nature of efficient information dissemination strategies based on localized communication between agents in a geosensor network. Specifically, they are concerned with mobile location-aware agents who are able to sense information about their immediate geospatial environment and communicate with other agents in their neighborhood. The authors distinguish between three different strategies. The first strategy, *Flooding*, is where each geosensor node that encounters an event or receives a message about an event passes on the information to every other node within its communication range. The second approach is referred to as an *Epidemic*, in which each node only informs n other agents about the events. In the third approach which is *location-constrained*, information is only passed on in proximity to the event, and then discarded.

In reference 35, the authors propose a document oriented model for information dissemination in mobile ad hoc networks. The problem of routing messages in disconnected or partially connected mobile ad hoc networks has been dealt by previous works like references 19 and 36. The main contribution in reference 35 is the implementation of a service for document dissemination in ad hoc networks and then using this service as a building block for application level services.

Any document that is sent in the network is cached as long as possible by as many devices as possible, so that it can remain available for those devices that could not receive it at the time it was sent originally. Other than providing a caching system where documents can be maintained in mobile devices, their service also provides facilities for document advertisement, document discovery, and document transport between neighboring devices. A device can periodically advertise to its neighbors about the documents stored in its cache. It can also search for specific documents in its neighborhood and can either push documents toward or pull documents from its neighbors.

In reference 37, the authors propose a middleware for a controlled epidemic style dissemination for mobile ad hoc networks. Since traditional middleware primitives offer very little information on dissemination mechanisms and epidemic algorithms have hardly been used to control the spreading of information depending on the desired reliability and network structure, the authors present a mobile ad hoc network middleware that uses epidemic-style information dissemination techniques to tune the reliability of the communication.

The authors argue that existing epidemic algorithms have little control on the information dissemination process, and much of it is based on experimental results and not on any analytical model. In other words, the information spread cannot be accurately tuned in order to reach only a desired percentage of the hosts.

The authors, therefore, propose algorithms that rely on epidemic models and take into account the underlying network structure. They design middleware interfaces that allow programmers to set the reliability for unicasting and anycasting with a high degree of accuracy. The middleware would have primitives for epidemic dissemination and would take as control inputs the percentage of hosts to which the information is disseminated. The authors use the infectivity, which is the probability of being infected by a neighboring host, to control the reliability of the probabilistic unicast. Thus, given an expected reliability value, the middleware is able to calculate the infectivity accurately in order to obtain an infection rate proportional to the total number of hosts in the network. For constructing the analytical model, the authors adopt the simple S-I-S model of epidemiological spread to model the information dissemination in a MANET. For the analytical model, the authors assume that there is homogeneous mixing of the nodes and that the infectivity of a single host per message is constant. Using the average node degree and the probability of infection, the authors calculate the infection rate. Based on its calculation, the authors depict the epidemic spread algorithm which is executed periodically.

3.6 EPIDEMIC MODELS OF MALICIOUS CODE PROPAGATION

Computer worms have recently emerged as one of the most critical threats against information confidentiality, integrity, and service availability. Host machines in the Internet have repeatedly revealed their susceptibility to malicious intrusions like worms that have compromised millions of vulnerable hosts at an extremely fast pace [38, 39]. Given that the threat of virus and worm propagation in wireless networks is quite real, a few recent works have tried to focus on this idea and successfully utilized the concept of epidemic theory to model the spread of worms in wireless ad hoc and sensor networks.

In reference 40, the authors discuss the epidemic model of virus spreading in mobile environments. Given the increasing rise in the usage of mobile devices, it is a matter of time before viruses propagating over the air interface would be a major menace. Already, there are several viruses and worms that spread over the air. For example, the Brador virus [41] infects Pocket PCs running Windows CE; and by installing a backdoor, it allows a remote attacker access to the device. The Cabir worm [42] infects cell phones running the Symbian operating system. Identifying these examples, the authors of reference 40 stress several important factors like movement of devices and the geographic locations while formulating epidemic models for virus spread in mobile environments. In their model which they call *probabilistic queuing*, the authors investigate the behavior of malicious codes that spread via proximity-based point-to-point wireless links. They point out the drawbacks that existing epidemiological modeling of similar processes in mobile environments have, like ignoring the node velocity and the nonhomogeneous connectivity distributions that often arise in mobile environments. The Kephart–White (KW) Model [10] assumes a homogeneously connected topology, and the network is represented by a single parameter, namely, the average node degree. However, mobile environments are too dynamic in order for this model to fit. The KW model only considers mean connectivity, and it discards useful information when the underlying distribution has significant variance, which is normal in a mobile environment. Furthermore, the velocity is an important factor that influences the way a virus can spread among mobile nodes. The authors incorporate this parameter in their model, and they come up with a new epidemic threshold value when the virus spread reaches endemic state. In the KW model, which ignores node mobility, this threshold is crossed when the infecting rate is greater than the curing rate. In their model, the authors incorporate the mobility model in their derivation of a node's degree distribution, which leads to a new value for the epidemic threshold.

Several other works discuss various aspects of vulnerabilities of sensor and mobile ad hoc networks in the light of epidemic theory.

3.6.1 TWPM

In reference 43, the authors develop a *topologically aware worm propagation model* (TWPM) for wireless sensor networks. An important strategy effectively employed by many recent worms (e.g., CodeRed v2) is *localized scanning*. The local scanning worms after compromising a host machine, instead of scanning a fixed IP address space, scan neighboring hosts with a higher probability. This strategy has proven to be quite effective since the presence of a single vulnerable host implies that other hosts on the same network would also be vulnerable with a high probability.

Since general routing strategies in sensor networks have each sensor maintain a neighbor list, this procedure of *localized scanning* could be very effective for a virus spreading in a sensor network. Moreover, since the worm under consideration employs (next-hop) information from a sensor to infect other sensors, the authors refer to it as a topologically aware worm. Based on the S-I-S model of epidemic spread, the authors have constructed a differential-equation-based worm propagation model in sensor networks. Apart from simultaneously capturing both time and space propagation dynamics, TWPM incorporates physical, MAC and network layer considerations of practical sensor networks.

Dividing the sensor network into equal-sized segments and using a constant rate of infection, the authors have arrived at a closed-form expression for the number of infectives at time t that also successfully captures the spatial information in terms of the segment coordinates.

3.6.2 Compromise Propagation in Secure Sensor Networks

TWPM modeled the worm propagation process using a differential-equation-based approach. However, generally in a static network (e.g., a sensor network), the differential-equation-based approach is not feasible since it assumes a homogeneous mixing of the susceptible nodes and the infected nodes. In such a scenario, a network or graph-theoretic modeling technique is much more suitable to capture the propagation process. One such novel work has been done by the authors in reference 44, where they model the process of how a compromised node in a sensor network gradually compromises other nodes and eventually compromises the whole network.

The authors assume the nodes in the sensor network to be uniformly randomly deployed in an area and securely communicating among themselves. By secure communication, the authors assume a secret shared key-based communication paradigm. They assume that a prior random key distribution technique has distributed secret keys to each node, whereby they communicate with each other. Given such a securely communicating sensor network, the authors study how an adversary who has captured one or two sensor nodes and extracted their secret keys can possibly propagate the compromise of nodes to the whole network.

When a node is captured and its keys are known by the attacker, secure communication can be established with neighboring nodes with which the captured node shares keys. Being able to securely communicate with its neighbor, the node with the malicious code can easily attain its susceptible neighbor's trust and pass on the malicious code to the latter. Once it has passed to the susceptible neighbor, the authors assume that the malicious code has the ability to acquire the secret keys of the new node. This is when the new node also becomes infected and results in the propagation of malicious code. This process continues until the whole network gets compromised.

By constructing a random graph model of the key sharing overlay graph of the sensor network, the authors present the compromise propagation model as a Poisson process with a mean that is dependent on the *infection probability* and the *infectivity duration* at each node. The propagation process was expressed by a *transmissibility* factor of the infection, and it was basically analogous to the bond occupation probability on the graph representing the key sharing network. The size of the epidemic was equivalent to the size of the giant component formed with edge existence probability defined by the transmissibility of the compromise process.

The main focus of their work was to identify the phase transition points of the process when it attains epidemic proportions. They studied the effects of the compromise propagation under two scenarios, namely, without node recovery and with node recovery. In the event of a compromise, the network may attempt to recover the particular node. Recovery might be realized in several possible ways. For example, the keys of the nodes might be revoked and the node may be given a fresh set of secret keys. In this context, key revocation, which refers to the task of securely removing keys that are known to be compromised, has been investigated as part of the key management schemes—for example, in reference 45. Moreover, recovery can also be achieved by simply removing the compromised node from the network—for example, by announcing a blacklist—or by simply reloading the nodes programs. More sophisticated methods may include immunizing a node with an appropriate antivirus patch that might render the node immune from the same virus attack. Regardless, in their analysis, the authors studied virus spreading under the two cases, respectively, depending on whether a node can be recovered or not.

Since, contrary to the differential-rate-equation-based modeling methods, the graph theoretic model does not capture the temporal effects of an epidemic, the authors captured the temporal dynamics of the propagation process using simulation techniques.

3.7 CONCLUSION AND OPEN ISSUES

In this chapter we have delved into various epidemiological models and protocols employed in wireless ad hoc and sensor networks. Starting from data dissemination and gossip protocols to security issues in sensor networks such as propagation of compromise of sensor nodes, we observe that there have been several works inspired by this powerful concept of epidemic theory. The density and scale of a sensor network, coupled with the objective of a one-to-many data transfer from a few nodes to the rest of the network, unleash the efficiency with which this theory can effectively model and provide solutions to several problems in ad hoc and sensor networks. In this chapter, we have tried to touch most of the salient contributions that have adopted this theory and provide a concise compilation under various categories of subclassifications.

Among the open issues related to the modeling of sensor networks, we find that not many protocol models in sensor networks have been proposed with the nodes being mobile. With mobile nodes, the dynamics of the network and its properties, like connectivity, keep changing continuously, thus making it more difficult to capture in an analytical closed-form model. For example, the analysis and performance study of broadcast protocols in a mobile sensor network environment still requires considerable research. Another important aspect that needs to be dealt with while modeling a sensor network protocol is the node deployment scenario. Apart from the straightforward uniform random deployment of the sensor nodes, there could be other distributions for node deployment. A popular one among them is deploying nodes in packets such that the resident points of nodes from each packet forms a two-dimensional Gaussian distribution about the dropping point. Given such a position distribution of the sensor network, epidemic modeling of dissemination or broadcast protocols would have to deal with the change in degree distribution dependent on the location of a node in the network.

3.8 EXERCISES

1. How is epidemic modeling in sensor networks different from that used for the Internet?
2. What is the basic reproductive number in epidemiology? How is it different for scale-free networks from other networks?
3. How does a homogeneously mixing population and a heterogeneously mixing one affect the mathematical formulation of the spreading process in epidemiology?
4. How does mobility of the network nodes change the mathematical formulation of an epidemic model of the network?
5. What are the important parameters that determine the choice between a continuous and a discrete time epidemic model of a network?
6. What are the important issues while epidemic modeling of malware spread in a mobile environment? How do the location of a mobile device and the time of the day affect the model?

ACKNOWLEDGMENT

This work is supported by NSF ITR grant No. IIS-0326505 and Texas ARP grant No. 14-748779.

BIBLIOGRAPHY

1. A. Woo, S. Madden, and R. Govindan. Networking support for query processing in sensor networks. *Communications of the ACM*, **47**(6):4752, 2004.
2. R. M. Anderson and R. M. May. *Infectious Diseases of Human: Dynamics and Control*, Oxford University Press, Oxford, 1991.
3. M. E. J. Newman. Spread of epidemic disease on networks. *Physics Reviews E*, **66**: 016128, 2002.
4. R. Pastor-Satorras and A. Vespignani. Epidemic dynamics and endemic states in complex networks. *Physics Reviews E*, **63**:066117, 2001.
5. R. M. May and A. L. Lloyd. Infection dynamics on scale-free networks. *Physics Reviews E*, **64**:066112, 2001.
6. I. de S. Pool and M. Kochen. Contacts and influence. *Social Networks* **1**:148, 1978.
7. C. Moore and M. E. J. Newman. Epidemics and percolation in small-world networks. *Physics Reviews E* **61**:5678-5682, 2000.
8. P. Grassberger. On the critical behavior of the general epidemic process and dynamic percolation. *Mathematical Biosciences*. **63**:157, 1983.
9. C. Griffin and R. Brooks. A note on the spread of worms in scale-free networks. In *IEEE Transactions on Systems, Man and Cybernetics*, **36**(1):198–202, 2006.
10. J. Kephart and S. White. Directed-graph epidemiological models of computer viruses. In *Proceedings of the IEEE Computer Symposium on Research in Security and Privacy*, May 1991, pp. 343–359.
11. J. Kulik, W. Rabiner, and H. Balakrishnan, Adaptive protocols for information dissemination in wireless sensor networks. In *Proceedings of the Fifth ACM/IEEE Mobicom Conference*, Seattle, WA, August 1999. IEEE CD-ROM.
12. S. S. Kulkarni and M. Arumugam. INFUSE: A TDMA based data dissemination protocol for sensor networks. Technical Report MSU-CSE-04-46, Department of Computer Science, Michigan State University, November 2004.
13. J. W. Hui and D. Culler. The dynamic behavior of a data dissemination protocol for network programming at scale. In *2nd International Conference on Embedded Networked Sensor Systems*, 2004. IEEE CD-ROM.
14. P. Levis and D. Culler. The firecracker protocol. In *Proceedings of the 11th ACM SIGOPS European Workshop*, Leuven, Belgium, 2004. IEEE CD-ROM.
15. P. Levis, N. Patel, D. Culler, and S. Shenker. Trickle: A self-regulating algorithm for code maintenance and propagation in wireless sensor networks. In *First USENIX/ACM Symposium on Network Systems Design and Implementation (NSDI)*, 2004. IEEE CD-ROM.
16. M. Akdere, C. C. Bilgin, O. Gerdaneri, I. Korpeoglu, O. Ulusoy, U. Cetintemel. A comparison of epidemic algorithms in wireless sensor networks. *Computer Communication*, **6**:35–41, 2006.

17. D. Ganesan, B. Krishnamachari, A. Woo, D. Culler, D. Estrin, and S. Wicker. An empirical study of epidemic algorithms in large scale multihop wireless networks. Intel Research, Berkeley Technical Report IRB-TR-02-003, March 2002.
18. A. Beaufour, M. Leopold, and P. Bonnet. Smart-tag based data dissemination. In WSNA, 2002.
19. A. Vahdat and D. Becker. Epidemic routing for partially-connected ad hoc networks. Duke Technical Report CS-2000-06, July 2000.
20. S. S. Kulkarni and L. Wang. MNP: Multihop network reprogramming service for sensor networks. In *Proceedings of the 25th International Conference on Distributed Computing Systems ICDCS*, June 2005, p. 716.
21. C. E. Perkins and E. M. Royer. Ad-hoc on-demand distance vector routing. In *2nd IEEE Workshop on Mobile Computing Systems and Applications*, February 1999, pp. 90–100.
22. D. B. Johnson and D. A. Maltz. *Dynamic Source Routing in Ad Hoc Wireless Networks.* Kluwer Academic Publishers, Norwell, MA, 1996.
23. Z. J. Haas, J. Y. Halpern, and L. Li. Gossip-Based Ad Hoc Routing. In *INFOCOM 2002.* IEEE CD-ROM.
24. A. G. Dimakis, A. D. Sarwate, and M. J. WainWright. Geographic gossip: Efficient aggregation for sensor networks. In *Fifth International Symposium on Information Processing in Sensor Networks (IPSN)*, 2006. IEEE CD-ROM.
25. S. Boyd, A. Ghosh, B. Prabhakar, and D. Shah. Analysis and optimization of randomized gossip algorithms. In *Proceedings of the 43rd Conference on Decision and Control (CDC 2004)*, 2004. IEEE CD-ROM.
26. S. Boyd, A. Ghosh, B. Prabhakar, and D. Shah. Gossip algorithms: Design, analysis and applications. In *Proceedings of the 24th Conference of the IEEE Communications Society (INFOCOM 2005)*, 2005. IEEE CD-ROM.
27. J.-Y. Chen and D. X. G. Pandurangan. Robust aggregates computation in wireless sensor networks: Distributed randomized algorithms and analysis. In *Fourth International Symposium on Information Processing in Sensor Networks (IPSN)*, 2005. IEEE CD-ROM.
28. R. Karp, C. Schindelhauer, S. Shenker, and B. Vocking. Randomized rumor spreading. In *Proceedings of the IEEE Conference of Foundations of Computer Science, (FOCS)*, 2000. IEEE CD-ROM.
29. D. Kempe, A. Dobra, and J. Gehrke. Gossip-based computation of aggregate information. In *Proceedings of the IEEE Conference of Foundations of Computer Science (FOCS)*, 2003. IEEE CD-ROM.
30. P. Kyasanur, R. R. Choudhury, and I. Gupta. Smart gossip: Infusing adaptivity into gossiping protocols for sensor networks. In *ICDCS 2006*. IEEE CD-ROM.
31. J. Haartsen, M. Naghshineh, J. Inouye, O. J. Joeresson, and W. Allen. Bluetooth: Vision, goals, and architecture. *ACM Mobile Computing and Communications Review*, **2**(4):3845, 1998.
32. A. Khelil, C. Becker, J. Tian, and K. Rothermel. An epidemic model for information diffusion in MANETs. In *MSWiM, 2002*. IEEE CD-ROM.
33. S. Nittel M. Duckham, and L. Kulik. Information dissemination in mobile ad hoc geosensor networks. In *Third International Conference on Geographic Information Science*, 2004. IEEE CD-ROM.
34. S. Nittel et al. Report from the first workshop on geo sensor networks. ACM SIGMOD Record 33 2004.

35. F. Guidec and H. Roussain. Asynchronous document dissemination in dynamic ad hoc networks. In *Second International Symposium on Parallel and Distributed Processing and Applications (ISPA)*, 2004. IEEE CD-ROM.
36. A. Lindgren, A. Doria, and O. Schelen. Probabilistic routing in intermittently connected networks. In *Proceedings of the Fourth ACM International Symposium on Mobile Ad Hoc Networking and Computing (MobiHoc 2003)*, June 2003. IEEE CD-ROM.
37. M. Musolesi and C. Mascolo. Controlled epidemic-style dissemination middleware for mobile ad hoc networks. In *Mobiquitous*, 2006. IEEE CD-ROM.
38. S. Staniford, V. Paxson, and N. Weaver. How to own the Internet in your spare time. Usenix Security Symposium, 2002.
39. C. Shannon and D. Moore. The spread of the witty worm. *IEEE Security and Privacy*, **2**(4): 46–50, 2004.
40. J. W. Mickens and B. D. Noble. Modeling epidemic spreading in mobile environments. In WiSE, 2005.
41. R. Wong and I. Yap. Security information. In *Virus Encyclopedia: WINCE BRADOR.A, Technical Details*, 2004. Trend Micro Incorporated.
42. P. Ferrie, P. Szor, R. Stanev, and R. Mouritzen. Security Response: SymbOS.Cabir, 2004. Symantec Corporation.
43. S. A. Khayam and H. Radha. A Topologically-aware worm propagation model for wireless sensor networks. In *IEEE ICDCS International Workshop on Security in Distributed Computing Systems SDCS*, 2005. IEEE CD-ROM.
44. P. De, Y. Liu, and S. K. Das, Modeling node compromise spread in sensor networks using epidemic theory. In *World of Wireless, Mobile and Multimedia Networks*, WoWMoM, 2006. IEEE CD-ROM.
45. H. Chan, V. D. Gligor, A. Perrig, G. Muralidharan. On the distribution and revocation of cryptographic keys in sensor networks. In *IEEE Transactions on Dependable and Secure Computing* 2005. IEEE CD-ROM.

CHAPTER 4

Modeling Sensor Networks

STEFAN SCHMID and ROGER WATTENHOFER

Computer Engineering and Networks Laboratory (TIK), ETH Zurich, CH-8092 Zurich, Switzerland

4.1 INTRODUCTION AND MOTIVATION

In order to develop algorithms for sensor networks and in order to give mathematical correctness and performance proofs, models for various aspects of sensor networks are needed. This chapter presents and discusses currently used models for sensor networks. Generally, finding good models is a challenging task. On the one hand, a model should be as simple as possible such that the analysis of a given algorithm remains tractable. On the other hand, however, a model must not be too simplistic in the sense that it neglects important properties of the network. A great algorithm in theory may be inefficient or even incorrect in practice if the analysis is based on idealistic assumptions. For example, an algorithm that ignores interference may fail in practice since communication happens over a shared medium. Many models for sensor network have their origin in classic areas of theoretical computer science and applied mathematics. Since the topology of a sensor network can be regarded as a *graph*, the distributed algorithms community uses models from *graph theory*, representing nodes by vertices and wireless links by edges. Another crucial ingredient of sensor network models is *geometry*. Geometry comes into play as the distribution of sensor nodes in space, as well as the propagation range of wireless links, usually adheres to geometric constraints.

The chapter is organized as follows. In Section 4.2, the reader will become familiar with various models for the network's *connectivity*. Connectivity models answer the question: Which nodes are "connected" to which other nodes and can therefore directly communicate with each other. Section 4.3 then enhances these connectivity models by adding *interference* aspects: Since sensor nodes communicate over a shared, wireless medium, a transmission may disturb a nearby concurrent transmission. After having studied connectivity and interference issues, we look at

Algorithms and Protocols for Wireless Sensor Networks, Edited by Azzedine Boukerche

modeling questions related to algorithm design in Section 4.4. The reader is provided with a survey of models that influence the feasibility and efficiency of certain operations on sensor networks. We draw some general conclusions in Section 4.5, and we point out interesting areas for future research in Section 4.6.

4.2 MODELING THE SENSOR NODES' CONNECTIVITY

A first and foremost modeling question concerns the *connectivity* of sensor nodes: Given a set of nodes distributed in space, we need to specify which nodes can receive a transmission of a node. Throughout this chapter, if a node u is within a node v's transmission range, we say that u is *adjacent* to v, or, equivalently, that u is a *neighbor* of v. In the absence of interference (cf. Section 4.3), this relation is typically *symmetric* (or *undirected*); that is, if a node u can hear a node v, also v can hear u. The connectivity of a sensor network is described by a graph $G = (V, E)$, where V (*vertices*) is the set of sensor nodes, and E (*edges*) describes the adjacency relation between nodes. That is, for two nodes $u, v \in V$, $(u, v) \in E$ if v is adjacent to u. In an undirected graph, it holds that if $(u, v) \in E$, then also $(v, u) \in E$; that is, edges can be represented by sets $\{u, v\} \in E$ rather than tuples.

The classic connectivity model is the so-called *unit disk graph* (UDG) [1]. The name "unit disk graph" stems from the area of *computational geometry*; it is a special case of the so-called *intersection graph*. In this model, nodes having omnidirectional radio antennas—that is, antennas with constant gain in all directions—are assumed to be deployed in a planar, unobstructed environment. Two nodes are adjacent if and only if they are within each other's transmission range (which is normalized to 1).

Model 4.2.1 (Unit Disk Graph (UDG)). Let $V \subset \mathbb{R}^2$ be a set of nodes in the two-dimensional Euclidean plane. The Euclidean graph $G = (V, E)$ is called unit disk graph if any two nodes are adjacent if and only if their Euclidean distance is at most 1. That is, for arbitrary $u, v \in V$, it holds that $\{u, v\} \in E \Leftrightarrow |u, v| \leq 1$. Figure 4.1 depicts an example of a UDG.

The UDG model is idealistic: In reality, radios are not omnidirectional, and even small obstacles such as plants can change connectivity. Therefore, some researchers

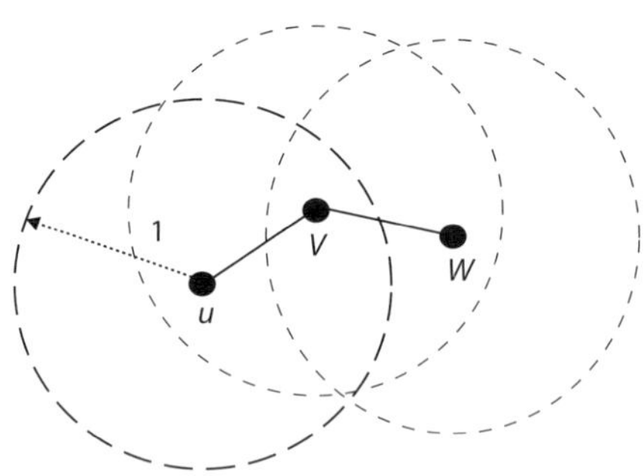

Figure 4.1. Unit disk graph: Node u is adjacent to node v (distance ≤ 1) but not to node w (distance > 1).

have proposed to study the other extreme and model the sensor network as a *general graph*; that is, each node can be adjacent to every other node.

Model 4.2.2 (General Graph (GG)). The connectivity graph is a general undirected graph G.

While a UDG is too optimistic, the GG is often too pessimistic, because the connectivity of most networks is not arbitrary but obeys certain geometric constraints. Still, in some application scenarios it might be accurate to operate either on the UDG or on the GG. Indeed, there are algorithms developed for the UDG which also perform well in more general models. Moreover, some algorithms designed for the GG are currently also the most efficient ones for UDGs (e.g., reference 2).

The research community has searched for connectivity models between the two extremes UDG and GG. For example, the *quasi unit disk graph model* (QUDG) [3, 4] is a generalization of the UDG that takes imperfections into account as they may arise from non-omnidirectional antennas or small obstacles. These QUDGs are related to so-called *civilized graphs*. The interested reader can find more information in reference 5.

Model 4.2.3 (Quasi Unit Disk Graph (QUDG)). The nodes are in arbitrary positions in $\mathbb{R}^2$. All pairs of nodes with Euclidean distance at most ρ for some given $\rho \in (0, 1]$ are adjacent. Pairs with a distance larger than 1 are never in each other's transmission range. Finally, pairs with a distance between ρ and 1 may or may not be neighboring. An example is shown in Figure 4.2.

Note that, for $\rho = 1$, a QUDG is a UDG, and therefore the following theorem holds.

Theorem 4.2.1. A UDG is a special case of a QUDG.

The QUDG model itself can be extended in several ways.

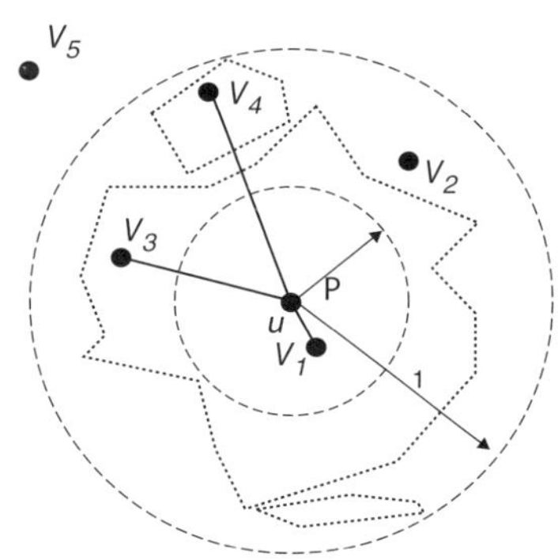

Figure 4.2. Quasi unit disk graph from the perspective of node u: Node u is always adjacent to node v_1 ($d(u, v_1) \leq \rho$) but never to v_5 ($d(u, v_5) > 1$). All other nodes may or may not be in u's transmission range. In this example, node u is adjacent to v_3 and v_4 but not to v_2.

Model 4.2.4 (QUDG Variations). The QUDG as presented in Model 4.2.3 does not specify precisely what happens if the distance is between ρ and 1. There are several options. For example, one could imagine an *adversary* choosing for each node pair whether they are in each other's transmission range or not. Alternatively, there may be a certain *success probability* of being adjacent: The corresponding probability distribution could depend on the time and/or distance [6]. For example, the QUDG could be used to study *Rayleigh fading*; that is, the radio signal intensity could vary according to a Rayleigh distributed random variable. Also, a probabilistic on/off model is reasonable, where in each round a link's state changes from good to bad and vice versa with a given probability.

Measurement studies suggest that in an unobstructed environment, and with many nodes available, $1/\rho$ is modeled as a small constant [7]. Interestingly, many algorithms can be transferred from the UDG to the QUDG at an additional cost of $1/\rho^2$ [4]. Note that while for $\rho \approx .5$ this factor is bearable, the algorithms are two orders of magnitude worse if $\rho \approx .1$. While the QUDG can be attractive to model nodes deployed in fields with few obstacles, it does not make sense for inner-city or in-building networks where obstructions cannot be ignored: Since a node may be able to communicate with another node which is dozens of meters away, but not with a third node being just around the corner, ρ would be close to 0.

However, even in such heterogeneous environments, the connectivity graph is still far from being a general graph. Although nodes that are close but on different sides of a wall may not be able to communicate, a node is typically highly connected to the nodes which are in the same room, and thus many neighbors of a node are direct neighbors themselves. In other words, even in regions with many obstacles, the total number of neighbors of a node which are not adjacent is likely to be small. This observation has motivated Model 4.2.6, see reference 8 for more details.

Model 4.2.5 (Bounded Independence Graph (BIG)). Let $\Upsilon^r(u)$ denote the set of independent nodes that are at most r hops away from node u (i.e., nodes of u's r-neighborhood) in the connectivity graph G. Thus, a set $\mathcal{S} \subseteq V$ of nodes is called independent if all nodes in the set are pairwise not adjacent; that is, for all $u, v \in \mathcal{S}$, it holds that $\{u, v\} \notin E$. Graph G has bounded independence if and only if for all nodes $u \in G$, $|\Upsilon^r(u)| = O(\text{poly}(r))$ (typically $|\Upsilon^r(u)| \in O(r^c)$ for a small constant $c \geq 2$).

The BIG model reflects reality quite well and is appropriate in many situations. Figure 4.3 shows a sample scenario with a wall; in contrast to UDG and QUDG, the BIG model captures this situation well.

Since the number of independent neighbors in a disk of radius r of a UDG is at most $O(r^2)$, we have the following fact. The proof is simple (and similar to the upcoming proof of Theorem 4.2.13) and left to the reader as an exercise.

Theorem 4.2.2. The UDG model is a special case of the BIG model. Similarly, if ρ is constant, also a QUDG is a BIG.

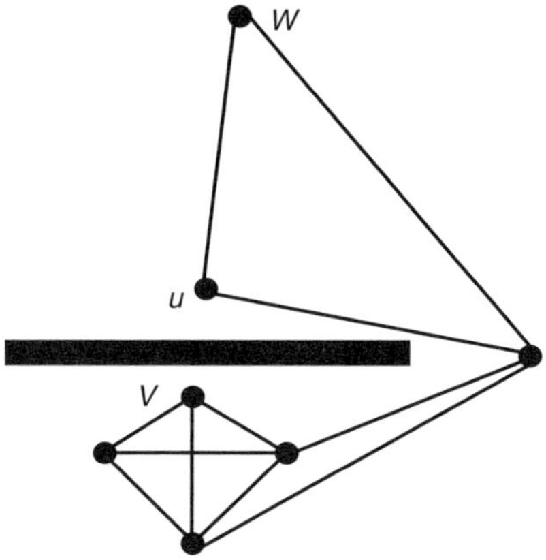

Figure 4.3. Nodes u and v are separated by a wall. Nodes on the same side of the wall are completely connected. However, due to the wall, although u can reach a distant node w, it cannot hear the close node v. Such situations can be modeled by the BIG but not by the UDG or the QUDG.

Observe that many models described so far can be generalized. For instance, the UDG and QUDG models can be studied in three dimensions rather than in the plane, or using different distance functions (*norms*). For more detailed information on the concept of norms, the reader may want to consult any introductory book on linear algebra.

Model 4.2.6 (Generalized (Q)UDG). One extension of the UDG and QUDG models is to consider nodes in $\mathbb{R}^3$. Moreover, distances between nodes could be modeled using the Manhattan norm ($\mathbb{L}_1$ norm). In the Manhattan norm, the distance between two points $u = (x_1, y_1)$ and $v = (x_2, y_2)$ in the plane is given by $d(u, v) = |x_2 - x_1| + |y_2 - y_1|$, while in the Euclidean norm ($\mathbb{L}_2$ norm), the distance is $d(u, v) = \sqrt{|x_2 - x_1|^2 + |y_2 - y_1|^2}$. Alternatively, also the *maximum norm* ($\mathbb{L}_\infty$ norm) is popular, where $d(u, v) = \max |x_2 - x_1|, |y_2 - y_1|$.

The UDG model has also been extended to more general *metric spaces*; for example, in reference 9, it was extended to *doubling metrics* [10]. Note that a metric space defines distances between all pairs of nodes while guaranteeing *non-negativity*, *identity of indiscernibles*, *symmetry*, and *triangle inequality*. A doubling metric is simply a metric space with some additional constraints which are described next.

Model 4.2.7 (Unit Ball Graph (UBG)). A doubling metric space is defined as follows: For a node u, let the ball $B_u(r)$ denote the set of all nodes at a distance at most r from u. It holds, for all nodes u and all $r \geq 0$, that the ball $B_u(r)$ can be covered by a constant number of balls of radius $r/2$; that is, $B_u(r) \subseteq \bigcup_{i=1\ldots c} B_{u_i}(r/2)$, where

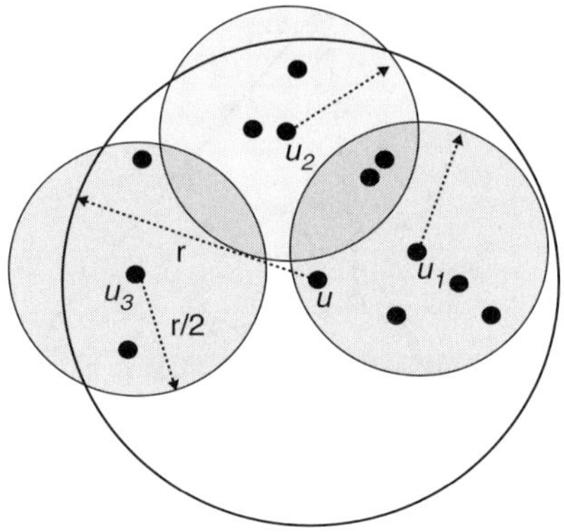

Figure 4.4. The Euclidean plane forms a doubling metric. In this example, the nodes are distributed in $\mathbb{R}^2$, and three balls of radius $r/2$ are sufficient to cover all nodes in $B_u(r)$; that is, $B_u(r) = B_{u_1}(r/2) \cup B_{u_2}(r/2) \cup B_{u_3}(r/2)$.

u_i are arbitrary nodes and c is a (usually small) constant. In the UBG model, nodes are assumed to form a doubling metric space. Two nodes u and v with $d(u, v) \leq 1$ are adjacent, whereas all other nodes are not.

The proof of the following theorem is left to the reader as an exercise.

Theorem 4.2.3. Nodes in a two-dimensional Euclidean plane (i.e., the metric space is given by the Euclidean distances) form a doubling metric. A general graph, however, does not.

Figure 4.4 shows an example for the Euclidean plane. In this setting, three balls of radius $r/2$ are enough to cover all nodes in the ball of radius r around node u. To see why a general graph may not form a doubling metric, consider a graph where all nodes have distance 1 to all other nodes. Observe that it is possible to model a UDG with a UBG by using the Euclidean distances of the UDG and connecting those node pairs which have distance at most 1. Moreover, even a QUDG can be modeled with a UBG. We have the following results.

Theorem 4.2.4. A UDG is a UBG.

Theorem 4.2.5. An undirected QUDG with constant ρ is a UBG.

Proof. The idea of the proof is as follows: First, we transform all distances in the QUDG. We then show that during this transformation, all edges are maintained; that is, the resulting graph is isomorphic to the QUDG. Moreover, it can be shown that after the transformation, the graph also fulfills the requirements of a doubling metric space.

We transform the distances between all pairs of nodes (u, v) in the QUDG as follows. Let $d_Q(u, v)$ denote the distance from node u to node v in the QUDG, and let $d_B(u, v)$ be the transformed distance in the UBG. Moreover, let $\epsilon > 0$ be an arbitrary small number.

$$d_B(u, v) := \begin{cases} d_Q(u, v)/\rho & \text{if} \quad d_Q(u, v) \leq \rho \\ 1 & \text{if} \quad \rho < d_Q(u, v) \leq 1 \text{ and } v \text{ is adjacent to } u \\ 1 + \epsilon & \text{if} \quad \rho < d_Q(u, v) \leq 1 \text{ and } v \text{ is } \textit{not} \text{ adjacent to } u \\ d_Q(u, v) & \text{if} \quad d_Q(u, v) > 1 \end{cases}$$

Observe that by this transformation, pairs of nodes that are adjacent in the QUDG are assigned distances of at most 1 and are therefore also adjacent in the UBG. Similarly, nodes that are not adjacent in the QUDG have a distance larger than 1 are therefore not neighboring in the UBG either. Also observe that the transformation increases the distance between two nodes by less than a constant factor of $\mu := (1 + \epsilon)/\rho$, but it never decreases any distances. It remains to show that after the transformation, the nodes indeed form a doubling metric space.

In order to form a metric space, the distances between the nodes are to fulfill the following properties: (1) nonnegativity, (2) identity of indiscernibles, (3) symmetry, and (4) triangle inequality. The nonnegativity and the identity of indiscernibles criteria are met trivially. The symmetry criterion, however, might not hold, because the adjacency relation can be directed in a QUDG. Therefore, in the following, we consider undirected QUDGs only. Hence, since our distance transformation maintains symmetry, Property 3 holds as well. It remains to discuss the triangle inequality.

Consider two arbitrary nodes u and v. Since in the QUDG, all distances are Euclidean, it holds that

$$\forall w : d_Q(u, v) \leq d_Q(u, w) + d_Q(w, v) \tag{4.1}$$

Let us now look at the following three cases in turn: (i) $d_Q(u, v) \leq \rho$, (ii) $\rho < d_Q(u, v) \leq 1$, and (iii) $1 < d_Q(u, v)$. In Case i, no node w with distance larger than ρ from any of the two nodes u and v can challenge the triangle inequality. For all other nodes w, however, it holds that $d_B(u, v) = d_Q(u, v)/\rho \leq (d_Q(u, w) + d_Q(w, v))/\rho = d_B(u, w) + d_B(w, v)$. Here, the equalities hold by the definition of the transformation function and the inequality is due to Eq. (4.1). Next, we tackle Case ii. Again, only nodes w with $d_Q(u, w) \leq \rho$ and $d_Q(w, v) \leq \rho$ can challenge the inequality. However, we know that $d_Q(u, v) > \rho$, and hence Eq. (4.1) yields $d_B(u, w) + d_B(w, v) = d_Q(u, w)/\rho + d_Q(w, v)/\rho > 1$. Finally, the triangle inequality also holds in Case iii, because the distance between u and v in the UBG is the same as in the QUDG, and our transformation never decreases any distances.

We conclude the proof by showing that the metric space has a constant doubling dimension. Recall that all distances are only stretched by a constant factor between 1 and μ in our transformation. Therefore, for all nodes u and arbitrary radii

r, $B_u^{QUDG}(r/\mu) \subseteq B_u^{UBG}(r)$. Thus, at most a constant factor of $O(\log \mu)$ times, more balls are needed for the UBG than for the QUDG (the Euclidean plane) in the worst case, and the claim follows.

The UBG itself has a polynomially bounded independence and is therefore a BIG.

Theorem 4.2.6. A UBG is a BIG.

Proof. Fix a node u. We have to prove that the total number of independent nodes in $B_u(r)$ grows polynomially in r. Observe that, due to the triangle inequality, in $B_u(1/2)$ there is at most one independent node. Thus, by the definition of a doubling metric, there are at most c independent nodes in $B_u(1)$, at most c^2 in $B_u(2)$, c^3 in $B_u(4)$, and so on. Generally, there are at most $c^{\log r+1}$ independent nodes in $B_u(r)$. Since $c^{\log r} \in O(r^c)$, the claim follows.

To conclude, we present two additional modeling aspects with which connectivity models are occasionally extended. The first aspect concerns the sensor nodes' antennas.

Model 4.2.8 (Antennas). Besides omnidirectional antennas, there is a wide range of more sophisticated antenna models. For example, a node can have a directional radio antenna with more gain in certain directions.

Finally, as mentioned in the discussion of the QUDG, links are not always reliable: Links may be up and down—for example, according to a probabilistic process.

Model 4.2.9 (Link Failures). Any graph-based model can be enhanced with probabilistic links.

4.3 INTERFERENCE ISSUES IN WIRELESS SENSOR NETWORKS

In wireless networks, the communication medium is shared and transmissions are exposed to interference. Concretely, a node u may not be able to correctly receive a message of an adjacent node v because there is a concurrent transmission nearby. In some sense, an interference model explains how concurrent transmissions block each other. Interference is a difficult phenomenon, with many hard-to-capture characteristics. A signal might, for example, interfere with itself due to *multipath propagation* (e.g., a direct path canceling with a longer path reflecting on an object). A discussion of these effects is beyond the scope of this overview chapter. Instead we look at models that capture reality from a worst-case perspective. The mother of all interference models is the so-called *physical* or *SINR model* [11–13], which is widely accepted by information theorists. In this model, the successful reception of a message depends on the received signal strength, the ambient noise level, and the interference caused by simultaneously transmitting nodes.

Model 4.3.1 (Signal-to-Interference Plus Noise (SINR)). Let P_r be the signal power received by a node v_r and let I_r denote the amount of interference generated by other nodes. Finally, let N be the ambient noise power level. Then, a node v_r receives a transmission if and only if $\frac{P_r}{N+I_r} \geq \beta$. Thus, β is a small constant (depending on the hardware) and denotes the minimum signal to interference ratio that is required for a message to be successfully received. The value of the received signal power P_r is a decreasing function of the distance $d(v_s, v_r)$ between transmitter v_s and receiver v_r. More specifically, the received signal power is modeled as decaying with distance $d(v_s, v_r)$ as $\frac{1}{d(v_s,v_r)^\alpha}$. The so-called *path-loss exponent* α is a constant between 2 and 6 and depends on external conditions of the medium, as well as on the exact sender–receiver distance. [1] Let P_i be the transmission power level of node v_i. A message transmitted from a node $v_s \in V$ is successfully received by a node v_r if

$$\frac{\dfrac{P_s}{d(v_s, v_r)^\alpha}}{N + \sum_{v_i \in V \setminus \{v_s\}} \dfrac{P_i}{d(v_i, v_r)^\alpha}} \geq \beta$$

In other words, in the SINR model, a node correctly receives a transmission if the received signal power—which depends on the sending power and the distance between sender and receiver—is large enough compared to the signal power of concurrent (interfering) transmissions and the ambient noise level. Sometimes a variation of this SINR model is used in literature. It has an additional requirement: For a successful reception, the received signal power must exceed a minimal threshold θ, that is, $P_r \geq \theta$. In many situations, such a threshold can also be incorporated implicitly by the ambient noise power level N. Moreover, researchers have also studied a *probabilistic* SINR model [14], where the gain of an antenna is described by a *Gaussian distribution*—independently of the distance! Apart from the interference term, and if all nodes send with the same transmission power level, the connectivity model of SINR is exactly the UDG, with path-loss exponent α and minimum ratio β such that the maximum distance for receiving a signal is 1. Hence, the SINR model can be extended similarly to the UDG model. Now, observe that the SINR model does not specify the signal power P_s used by a sender v_s to transmit data to the receiver v_r. Three models are common:

Model 4.3.2 (Power Control). CONST: All nodes use the same *constant* transmission power. DIST: The power level depends on the *distance d* between sender and receiver. Concretely, the transmission power is given by $c \cdot d^\alpha$ for some $\alpha \geq 2$ and some constant $c > 0$. GEN: A general (or *arbitrary*) power level is assumed at the

[1] In free space, α roughly equals 2. In the so-called *two-ray ground model*, it is assumed that there are two paths of the electromagnetic wave: a direct one and a ground reflected signal path; to describe this situation, $\alpha = 4$ is used. Finally, note that ever since *Marconi*'s first experiments, time has been devoted to explain radio propagation phenomena, and there is a plethora of other proposals. For example, for small urban cells, a photon propagation model has been suggested implying an *exponentially* growing path loss.

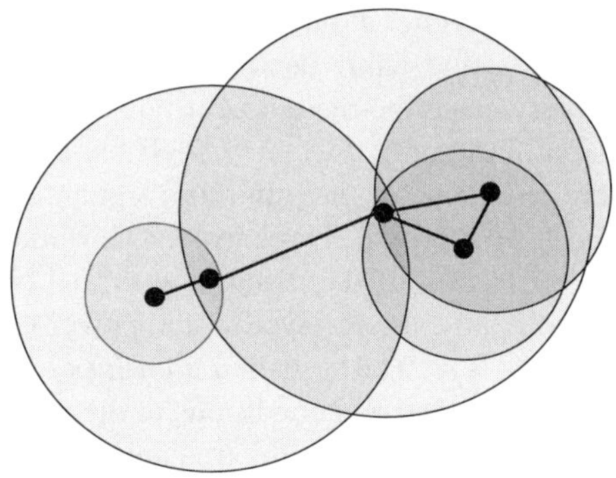

Figure 4.5. Sample network with heterogeneous transmission ranges. For instance, the node on the far left saves energy and reduces interference by using only a small power level.

sender, which may change over time. Figure 4.5 depicts a network where each node has a different power level.

Although the SINR model incorporates many important physical properties, it has not received an appropriate amount of attention from the algorithms community [12]. This can be partially explained by the fact that the SINR model is complicated. For instance, a lot of far-away transmissions sum up, and may interfere with a close-by sender-receiver pair. In practice, however, these far-away transmissions often only contribute to the ambient noise and need not be counted individually. Twiddling the knobs of the model a bit more, we might not sum up all interfering transmissions, but simply look at the worst—or, in the case of a CONST model; *closest*—disturbance: A node receives a transmission if and only if the closest simultaneously transmitting node is far enough.

Model 4.3.3 (Interfering Transmissions). SUM: All interfering transmissions are taken into account. ONE: Only the worst (or closest) interfering transmission matters. NULL: Pure connectivity models which do not consider interference aspects (cf. Section 4.2).

ONE models are quite popular because of their simplicity. The UDI—an interference-aware version of the UDG—is a prominent example (cf. Model 4.3.4). Observe that, because of the constant transmission power, the power control type of UDI is CONST. Figure 4.6 shows an example.

Model 4.3.4 (UDG with Distance Interference (UDI)). Nodes are situated arbitrarily in the plane. Two nodes can communicate directly if and only if their Euclidean distance is at most 1, and if the receiver is not disturbed by a third node with Euclidean distance less or equal a constant $R \geq 1$.

Often the constant R of the UDI model is approximated in such a way that interference can be reduced to a parameter of the UDG. For instance, some MAC

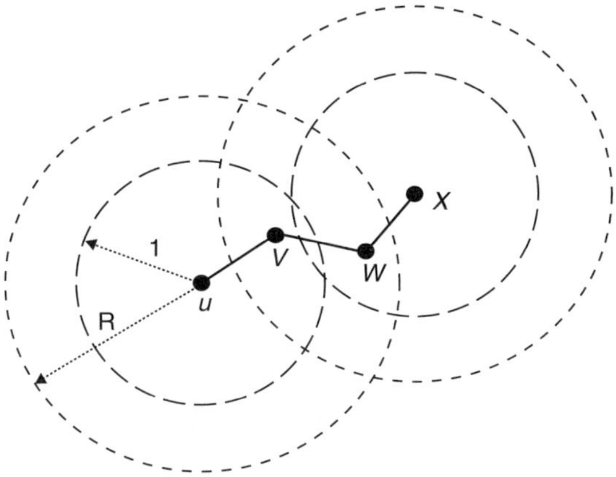

Figure 4.6. The UDI model has two radii: a transmission radius (length 1) and an interference radius (length $R \geq 1$). In this example, node v is not able to receive a transmission from node u if node x concurrently transmits data to node w—even though v is not adjacent to x.

protocols (e.g., coloring algorithms [15]) have been proposed to reduce interference by ensuring a certain hop distance between two senders. Concretely, it is assumed that only the k-neighborhood of a receiver u can interfere with u. Clearly, this is a stark simplification since in a UDG a $(k+1)$-neighbor can be close to the receiver (see Figure 4.7).

Model 4.3.5 (UDG with Hop Interference (UHI)). Nodes are located at arbitrary positions in $\mathbb{R}^2$. Two nodes are adjacent if and only if their Euclidean distance is at most 1. Two nodes can communicate directly if and only if they are adjacent, and

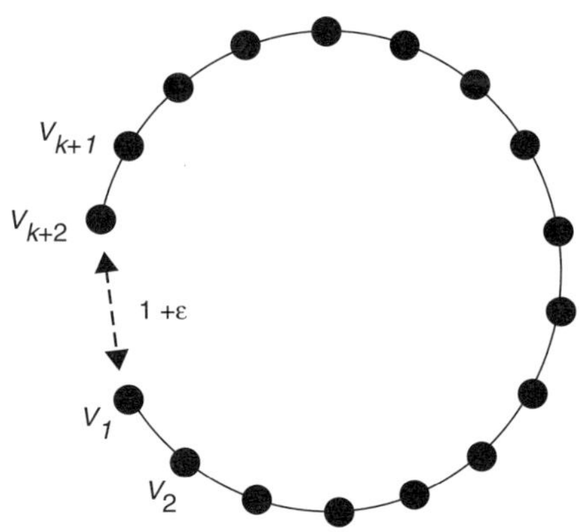

Figure 4.7. Example where UHI fails: Nodes v_1 and v_{k+2} are separated by a path of $k+1$ hops, but are close (distance $1+\epsilon$). Thus, concurrent transmissions of nodes v_2 and v_{k+2} may interfere at v_1 in spite of their large hop distance.

if there is no concurrent sender in the k-hop neighborhood of the receiver (in the UDG).

Observe that while the UHI model—for every k—sometimes overlooks interference terms which the UDI would take into account, the contrary does not hold.

Theorem 4.3.1. By choosing $R = k$, and since a hop has at most length 1, the UDI model does not overlook any interference terms that UHI would have taken into account. The contrary does not hold (cf. Figure 4.7).

Like UDI and UHI, also the *protocol model* (PM) is of type ONE (Model 4.3.3). However, the senders in the PM model adapt their transmission power according to DIST (Model 4.3.2)—that is, depending on the distance between sender and receiver. Model 4.3.7 is a variation of the model introduced in reference 11.

Model 4.3.6 (Protocol Model (PM)). Let $u_1, u_2, ..., u_k$, be the set of nodes transmitting simultaneously to receivers $v_1, v_2, ..., v_k$, respectively. The transmission of u_i is successfully received by v_i if for all $j \neq i$, it holds that $d(u_j, v_i) > \lambda \cdot d(u_j, v_j)$, where $\lambda \geq 1$ is a given constant. That is, v_i must not fall into a "guard zone" around any sender u_j which is a factor $(1 + \lambda)$ larger than u_j's transmission range.

Many interference models distinguish between senders and receivers assuming that interference arises at senders and occurs at receivers. However, often receivers acknowledge messages and are therefore also senders. If the original messages are short (e.g., control messages), then the sender/receiver distinction may not make sense. By this observation, some models (e.g., reference 16) simply consider the interference of *undirected links*. Figure 4.8 depicts an example.

Model 4.3.7 (Direction). DIR: This class of interference models distinguishes between senders and receivers (interference disks around senders). UNDIR: Interference originates from undirected links (interference "pretzels" around links).

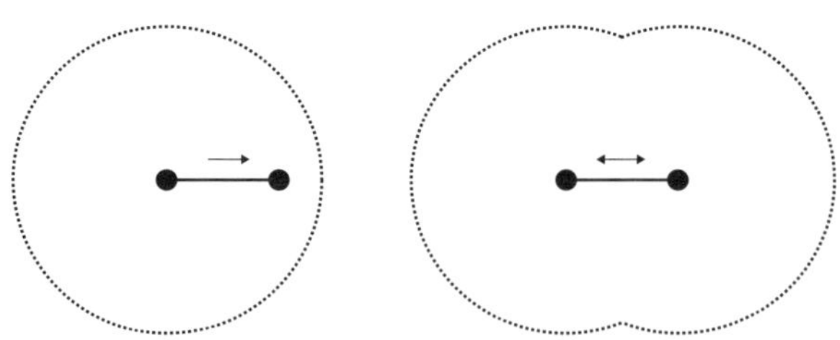

Figure 4.8. DIR vs. UNDIR: On the left, only the sender transmits data (interference disks around senders). On the right, there is no distinction between sender and receiver, and hence interference arises from the entire link ("pretzels" around links).

As in the case of connectivity models, the SINR, the UDI, and the UHI models can be extended with directional antennas and link failures, and hence Models 4.2.14 and 4.2.15 also apply here. Moreover, also the idea of quasi unit disk graphs (cf. Model 4.2.3) could be adopted. For example, the UDI can be "quasified" as follows: If two nodes are closer than a given threshold R_1, concurrent transmissions will always interfere; if the distance is larger than a second threshold R_2, there will be no interference. Finally, if the distance is between R_1 and R_2, transmissions may or may not interfere. However, these models are often too complicated to be handled algorithmically. It is sometimes simpler to study general *weighted* interference graphs instead. That is, similar to connectivity graphs, the interference model is based on *graphs*; however, the edges are now weighted. Formally, in a weighted interference graph $H = G(V, E, w)$, V represents the set of sensor nodes, E represents the set of edges, and $w : E \to \mathbb{R}^+$ is a function assigning a positive value to each edge. The weight denotes how large the interference between the corresponding nodes actually is. As in the SINR model, a transmission is received correctly if the ratio between received signal power and the amount (either the sum or the maximal interfering signal strength) of interfering traffic is smaller than a certain threshold.

Model 4.3.8 (General Weighted Graph (GWG)). A weighted interference graph H is given. A receiver v successfully receives a message from a sender u, if and only if the received signal strength (the weight of the link between u and v in H) divided by the total interference (the sum or the maximum of the weights of the links of concurrently transmitting nodes with a receiver v in H) is above the threshold given by the signal-to-interference-plus-noise ratio.

The general weighted graph model is quite pessimistic, because it allows for non-natural network topologies. Again—like in the BIG connectivity model—we need a weighted graph model that captures the geometric constraints without making too many simplifying assumptions. Again, one approach is to assume that the nodes form a *doubling metric* (cf. UBG model of Section 4.2).

Model 4.3.9 (Doubling Metric (DM)). The DM model assumes that the nodes form a doubling metric; that is, the set of nodes at a distance (which is now given by the weights of the edges) of at most r from a node u can be covered by a constant number of balls of radius $r/2$ around other nodes, for any r (cf. Model 4.2.9). Interference can be incorporated in various ways. For example, the amount of interference at a receiver u could depend on u's distance (in the doubling metric space) to the closest concurrently transmitting node (ONE model), or on the number of concurrent senders (SUM model).

As a final remark, note that so far we have only presented *binary interference models*: A message can be received either correctly or not at all. In practice, however, also the *transfer rate* at which messages can be transmitted can depend on interference: The larger the signal-to-noise ratio, the larger the available bandwidth. A WLAN 802.11, for example, exploits environments with less interference in order to transmit

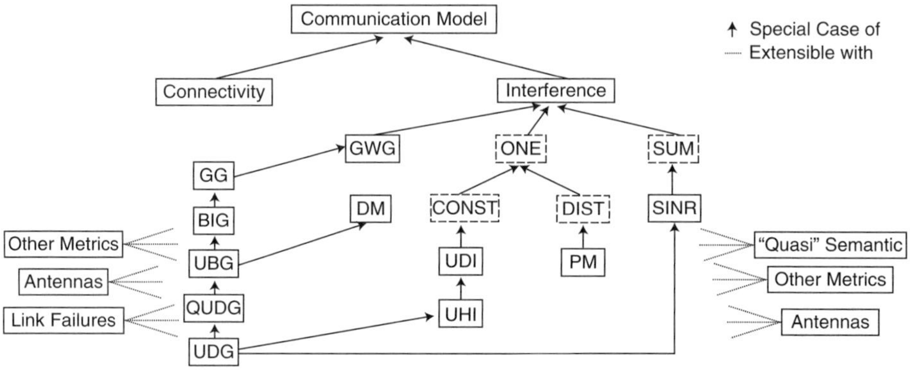

Figure 4.9. Overview of connectivity and interference models presented in Sections 4.2 and 4.3. The arrows show how the models are related.

more data per time unit. Of course, it would be possible to extend, for example, the DM model to capture this aspect as well. However, since these issues are beyond the scope of this chapter, we refer the reader to reference 17 for more details. To conclude, Figure 4.9 summarizes the connectivity and the interference models.

4.4 ALGORITHM DESIGN

The main purpose of deploying sensor networks is the collection physical data such as light intensity, sound, or temperature. In order to aggregate (e.g., compute the minimum temperature, or the average, etc.) the data that are stored at the individual nodes—and which are therefore distributed in space!— protocols or *algorithms* are needed specifying how these operations are performed. For example, due to the limited radio communication range, sensor nodes have to communicate (e.g., gather data) in a multihop manner with each other—that is, the messages have to be relayed by intermediate nodes—and hence a routing algorithm has to define which messages are to be forwarded via which other nodes.

Algorithms for sensor networks come in different flavors. In the following, we first describe the different types of algorithmic models appearing in literature today. We then discuss modeling aspects that may influence an algorithm's performance—for instance, what kind of identifiers nodes have, or how the nodes are distributed in space. Besides the classic evaluation criteria for algorithms—namely, *time complexity* and *space complexity*—algorithms for sensor networks pose additional optimization problems; for example, the number of messages that are sent should be small; or, in order to maximize the lifetime of the network, the nodes' energy consumption must be minimized. In order to facilitate a better understanding of the different algorithm types presented in the upcoming paragraphs, we will consider a sample problem: the computation of *dominating sets*.

Definition 4.4.1 (Dominating Set Problem (DS)).. The computation of a dominating set (DS) is a fundamental operation in sensor networks. For instance, such a set

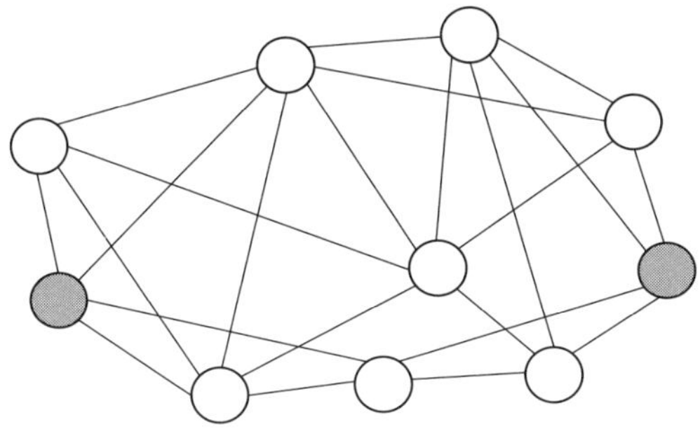

Figure 4.10. A minimum dominating set with two dominators.

can be used to build node clusters. Moreover, it may serve as a basis for constructing backbone networks that typically form a connected DS. A dominating set $D \subseteq V$ of a (undirected) network graph $G = (V, E)$ is a set of nodes such that for all nodes $u \in V$ it holds that either u is in the dominating set itself (i.e., $u \in D$), or u is adjacent to a node v in D (i.e., $\{u, v\} \in E \wedge v \in D$). It is often desirable to have small dominating sets. The minimum dominating set (MDS) is defined as the dominating set that minimizes the number of dominators $|D|$. An example of an MDS is illustrated in Figure 4.10. It can be shown that the MDS problem is NP-hard on general graphs and that a logarithmic approximation is asymptotically optimal unless $P \approx NP$ [18]. For simpler connectivity graphs such as the UDG graph, the approximation complexity of the problem may be better; for example, there is a *polynomial time approximation scheme* (PTAS) for UDG graphs!

The first category of algorithms we present here is similar to the classic (graph) algorithms appearing in the field of theoretical computer science or applied mathematics. These *global* algorithms can operate directly on the entire network or graph and can have complete information about the state of the system. For example, a system designer planning a fixed sensor network can apply a global algorithm to determine the optimal positions of the nodes in a given observation area.

Model 4.4.2 (Global Algorithms). A global algorithm can operate directly on the entire network.

Kruskal's algorithm for computing a minimum spanning tree [19] is an example of a global algorithm: The algorithm receives the entire graph as input and can sort the edges according to their weights. Kruskal's algorithm thus has a complete visibility of the entire graph and can perform arbitrary operations on it. No messages have to be sent between nodes.

Example 4.4.3. Let us tackle our dominating set problem! When faced with the task of designing an algorithm for a certain problem, it is often a good idea to start by studying greedy algorithms—that is, algorithms that in every execution step "greedily" do the currently most promising thing. Interestingly, a greedy algorithm is often optimal (see also Matroid theory [19]). So how can we greedily compute an MDS? A straightforward approach is the following (see Algorithm 1): First, we initialize the set of dominators with the empty set, that is, $D := \{\}$. We will call nodes in D black ("dominators"), nodes that are covered by nodes in D gray ("dominated nodes"), and all uncovered nodes *white*. Let $w(v)$ be the number of white nodes adjacent to v, including v itself. Then, in every step, we iterate over all nodes v (global operation!) and compute the number $w(v)$ of v's white neighbors, remembering the node x having the largest number. At the end of each step, we add node x to D. That is, we choose the node to become a dominator that covers the most new nodes, greedily reducing the number of the remaining nodes as much as possible. Obviously, the resulting dominators indeed form a DS. Moreover, it can be proven that the number of dominators is at most a logarithmic factor larger than in the optimal case. This simple approximation algorithm is therefore asymptotically optimal unless $P \approx NP$!

ALGORITHM 1. Global and Greedy MDS Algorithm

$D := \{\}$;
while $\exists$ white nodes **do**
  $x := \{x | w(x) = \max_v\{w(v)\}\}$;
  $D := D \cup x$;
od;

However, unlike global algorithms, most algorithms for sensor networks are not executed by a central designer, but rather *by the sensor nodes* themselves, for example, during the system's operation. A node a priori only knows its own state. In order to learn more about the other nodes in the network, it is bound to communicate with its neighbors *by exchanging messages*. Typically, when a node receives a message, it performs some computation, and—depending on the computation's results—sends a new message to its neighbors. By this collaboration of the nodes, global operations such as (multihop) routing between two nodes can be performed. Since the activity is distributed among the nodes, these algorithms are called *distributed algorithms* [20].

Model 4.4.4 (Distributed Algorithms). In a distributed algorithm, every sensor node runs its own algorithm. A priori, a node only knows the state of itself. In order to learn more about the rest of the network, nodes repeatedly exchange messages with adjacent nodes.

Thus, unlike in global algorithms, distributed algorithms do not see the entire graph at the beginning. If more information about the neighborhood is needed, adjacent nodes have to exchange messages. Distributed algorithms raise many interesting questions. For example: What can be computed in a distributed fashion, and what not? In contrast to global algorithms, nodes have to be coordinated somehow, and—as all nodes execute the same code—symmetries have to be broken. Besides an algorithm's correctness, execution time to perform the task (*time complexity*), and memory requirements at the nodes (*space complexity*), a new criterion becomes important, namely *message complexity*: Since distributed algorithms rely on message passing and since sending and receiving messages is an expensive operation (e.g., energy consumption, queueing delay, congestion, etc.), a distributed algorithm should minimize the total number of messages sent.

Example 4.4.5. In the following, we will see how dominating sets can be computed with distributed algorithms. Consider the following idea: Each sensor node broadcasts its own identifier plus the identifiers of all its neighbors to all other nodes in the network. Hence, each node gets a picture of the entire connectivity graph. Then, the node having the largest identifier computes the MDS using the global algorithm described in Algorithm 1, and it broadcasts the corresponding solution back to all nodes. Given this solution, each node can then decide whether it has to join the dominating set or not. Note that this algorithm is indeed distributed and yields small dominating sets: The size of the sets are again asymptotically optimal, unless $P \approx NP$. However, due to the broadcast operation, the message complexity is huge. What is more, the algorithm does not scale to a large number of sensor nodes.

Many global algorithms can be turned into distributed algorithms by just collecting the entire graph at each node and then computing the results locally. Of course, this is inefficient and inappropriate in practice. Therefore, it is often better to study a more restricted class of distributed algorithms, namely localized algorithms [21].

Model 4.4.6 (Localized Algorithms). A localized algorithm is a special case of a distributed algorithm. At the beginning, a node has only information about its own state. In order to learn more about the rest of the network, messages have to be exchanged. In a k-localized algorithm, for some constant k, each node is allowed to communicate at most k times with its neighbors. However, a node can decide to retard its right to communicate; for example, a node can wait to send messages until all its neighbors having larger identifiers have reached a certain state of their execution.

Localized algorithms are desirable in the sense that they only transmit a small number of messages. Unfortunately, however, localized algorithms can be slow: A node u might have to wait for a neighbor v to transmit all its messages, while node v in turn has to wait for its neighbor w, and so on. As a matter of fact, there can be a *linear chain of causality*, with only one node being active at any time. This yields a worst-case execution time of $\Theta(n)$, where n is the number of nodes.

Example 4.4.7. In order to compute the dominating sets of our sample problem in a localized manner, a simple algorithm can be applied (cf. Algorithm 2). Each node v waits until all its neighbors having a larger degree (or, in the case of the same degree; a larger identifier) than v have decided whether to join the dominating set or not. Then, if one of these nodes is a dominator, v decides not to join the dominating set. Otherwise, v becomes a dominator. Thus, each node has to communicate at most twice with its neighbors: once to find out their degree and once to tell them about its decision. This

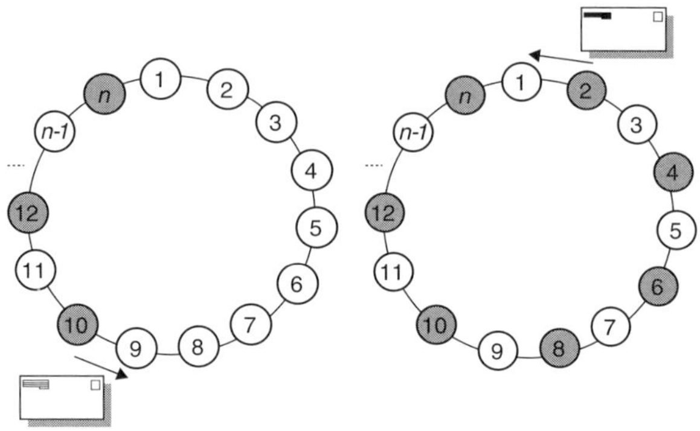

Figure 4.11. Localized algorithms can have large execution times!

localized algorithm has therefore a low message complexity. However, the execution time of this algorithm can be large. To see this, consider a cycle of nodes arranged according to their identifiers as depicted in Figure 4.11. Since all nodes have the same degree, the first node to become a dominator is node n. Only then, node $n-1$ can make its decision: It does not join the dominating set. After that, node $n-2$ decides to join the network, and so on. It's only after a linear waiting time of $O(n)$ time steps that node 1 eventually can make its decision and terminate the algorithm!

ALGORITHM 2. Localized MDS Algorithm

(* Code executed by node v *)
send degree and ID **to** all neighbors;
receive messages **from** neighbors;
while ($\exists$ undecided neighbor w with $prio(w) > prio(v)$) **do**
  wait();
od;
(* Decision *)
if ($\exists$ dominator in neighborhood) **then**
  $D := D$;
  send "I am dominated!" **to** neighbors;
else
  $D := D \cup v$;
  send "I am a dominator!" **to** neighbors;
fi;

Researchers have proposed to study yet another kind of distributed algorithm that overcomes the performance problems of localized algorithms, always terminating after a constant number of communication rounds [2].

Model 4.4.8 (Local Algorithms). Again, at the beginning, each node only knows its own state. In a k-local algorithm, for some constant k, each node can communicate at most k times with its neighbors. However, in contrast to k-localized algorithms, nodes cannot delay their decisions. In particular, all nodes process k synchronized phases,

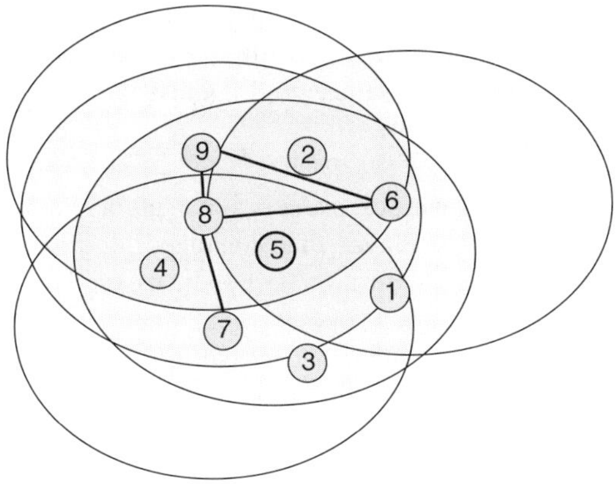

Figure 4.12. Illustration of local dominating set algorithm: Since all nodes with larger IDs are connected to each other and cover all other neighbors of node 5, node 5 does not join the dominating set—regardless of the decisions of other nodes.

and a node's operations in phase i may only depend on the information received during phases 1 to $i - 1$. The most efficient local algorithms are often randomized [22, 23], that is, the number of rounds k can vary.

Observe that in a k-local algorithm, nodes can only gather information about nodes in their k-neighborhood. In some local algorithms [22], the algorithm designer can choose an arbitrarily small constant k (at the cost of a lesser approximation ratio). This makes local algorithms particularly suited in scenarios where the nodes' environment changes frequently, because they are able to constantly adapt to the new circumstances.

Example 4.4.9. A dominating set can also be computed with a local algorithm: Each node u asks its higher-priority neighbors (with respect to degrees and identifiers) about their neighbors. If these higher-priority neighbors are connected and cover all of u's neighbors, then u does not join the dominating set and otherwise becomes a dominator. An example is illustrated in Figure 4.12. It can be shown that this algorithm even results in a connected dominating set—that is, a dominating set where any two dominators are connected by a path that only consists of other dominators. Observe that the algorithm is indeed "wait-free" or local, because a node can make its decisions only based on the identifiers of its neighbors and independently of the neighbors' decisions. Two communication rounds are sufficient. From this point of view, this local algorithm looks very appealing. Unfortunately, however, in the worst case, its approximation ratio of the optimal solution is as bad as $\Theta(\sqrt{n})$ already for simple connectivity graphs such as the UDG.[2] This indicates the existence of a challenging tradeoff: The smaller the "horizon" of a local algorithm, the more difficult it is to find good approximations of the optimal (global) solutions. In other words, there seems to be price of being

[2] For random UDGs, the performance is better: The algorithm achieves a constant approximation!

near-sighted: Similarly to online algorithms that cannot foresee the future and have to be competitive to an optimal offline algorithm, local algorithms have a restricted view of their neighborhood and are measured against the performance of a global algorithm.

Note that due to the synchronous phases, local algorithms may make greater demands on the media access sublayer than localized algorithms. In particular, in unreliable wireless networks it seems to be costly to implement a media access control scheme that allows for synchronous rounds, because messages will be lost due to interference (conflicting concurrent transmissions) or mobility (even if the nodes themselves are not mobile, the environment is typically dynamic, temporarily enabling/disabling links). A powerful concept for coping with failures is *self-stabilization* [24]. Fortunately, using a simple trick [25], every local algorithm is immediately self-stabilizing. The trick works as follows (Section 4 of reference 25): Every node keeps a log of every state transition it has taken until its current state; generally, this boils down to memorizing the local variables of each step of the main loop. If each node constantly sends its current log to all neighbor nodes, each node can check and correct every transition it has made in the past. Assuming that all inputs are correct (variable initialization and random seeds are stored in the imperishable program memory, and sensor information can be rechecked), every fault due to memory or message corruption will be detected and corrected. For details we refer to reference 25. Turning a k-local algorithm into a self-stabilizing algorithm with reference 25 blows up messages by a factor k (in the worst case); on the other hand, we immediately get an algorithm that works on a sensor network as the hardest wireless problems (messages lost due to interference and mobility) are covered by the self-stabilization model. Also, in the case of an error (such as a lost message), only the k-neighborhood of a node is affected.[3]

Having defined the most common types of algorithm, we now look at some algorithmic aspects in more detail. As mentioned, the message complexity—the total number of messages sent by an algorithm—is a main evaluation criterion of distributed algorithms. Because the number of messages typically depends on the amount of information that can be stored in a message, a model must specify the messages' sizes. A most popular model limits the message size to $O(\log n)$ bits, where n is the total number of nodes in the system. Hence, a message can store only a constant number of node identifiers (e.g., the source and destination address of a routing packet). Moreover, it is often assumed that if a node u sends a message to a neighbor v, all other neighbors of u will also receive the message (*broadcast model*). However, sometimes—for example, for lower bound proofs [26]—models are considered where the message size is *unbounded*, and where nodes can communicate with their neighbors individually (*message-passing model*). Algorithmic models also differ in their assumptions about how nodes can access the wireless medium. The concrete MAC, however, can

[3] In principle, localized algorithms can also benefit from reference 25; however, errors are not restricted to a k-neighborhood but may propagate through the entire network, resulting in a troublesome *butterfly effect*.

influence the number of retransmissions and hence an algorithm's performance. Moreover, an algorithm may be able to coordinate the medium access itself.

Model 4.4.10 (Medium Access). Some researchers assume an ideal medium access mechanism [27] where interference is impossible and where messages will always be broadcast instantaneously to all neighbors (cf. models of Section 4.2). In addition, adversarial models are used where an adversary schedules transmissions. Of course, this model only makes sense if the adversary is restricted appropriately—that is, if there are fairness guarantees. For example, the adversary might have to schedule each node at least once every $\Theta(n)$ rounds. One could also imagine an adversary that delivers a message only to a subset of a node's neighbors, because the other neighbors experience collisions. Finally, completely unstructured radio networks [28] can be considered where the algorithm designer has to implement her own medium access scheme from scratch. These models can further be classified in terms of whether collisions can be detected by a receiver or not.

As mentioned, a main objective of sensor networks is to collect physical data distributed over a given region. To achieve this, typically one or more nodes observe different sub-areas. Knowledge of the nodes' distribution, however, can be important for an algorithm designer. In a scenario where the nodes are dropped from an airplane, one might expect that the nodes are roughly randomly distributed when they reach ground.

Model 4.4.11 (Random Node Distribution). The simplest—and quite common—way to model sensor networks is to assume a UDG in combination with a *uniform node distribution* in the two-dimensional Euclidean plane. However, inspired by percolation theory, also *Poisson models* have been proposed [29]; thus, the positions of the nodes are distributed in $\mathbb{R}^2$ according to a homogeneous Poisson point process of constant density λ per unit area.

While these random models may be fine to prove the performance of an algorithm, for correctness and robustness issues, a more pessimistic model should be preferred—for example, a worst-case distribution.

Model 4.4.12 (Worst-Case Node Distribution). Nodes are distributed arbitrarily in the space given by the underlying graph (e.g., Euclidean plane, general graph, etc.).

Of course, there are again many models that lie between the two extremes. For example, random distributions with a density parameter varying over space could be considered: One can imagine that there are several nodes per square meter in areas that are "interesting" to observe, whereas in other "routing only" areas the nodes are hundreds of meters apart. Finally, speaking of node distributions, there are also models that do not allow nodes to be arbitrarily close or even assume the same position; for instance, there is such an assumption in the $\Omega(1)$ *model* [30] or in so-called *civilized graphs* [5]. Related to the distribution of nodes in space is also the issue of the distribution of node *identifiers*. Because many algorithms are based on node IDs, their performance can depend on how IDs are distributed among the nodes (and thus also in space).

Model 4.4.13 (Node Identifiers). Typically, nodes can be assumed to have unique identifiers. IDs could, for example, be generated during deployment using a random number generator. Moreover, because RFID tags already have IDs, we believe that it is reasonable to assume that sensor nodes obtain a unique ID during the production process. Finally, also note that certain tasks cannot be solved by any distributed algorithm if there are no identifiers, because there is no way to break symmetries among the nodes. Similarly to the node distribution in space, the most common models for ID distributions are random distributions and worst-case distributions. Sometimes, it also matters from which range the identifiers are chosen. Again, many variations are possible. For example, each of the n nodes can have a unique 128-bit identifier (range 0, ..., $2^{128} - 1$). Or, in a more restrictive case, the nodes may have consecutive numbers (e.g., range 1, ..., n).

Alternatively—or in addition!—node IDs can contain location information—for example, if the nodes are equipped with a *Global Positioning System* (GPS) or a *Galileo* device. Location information can boost the performance of certain operations [30]: for example, a routing algorithm can exploit geographic information to forward the message to a neighbor which lies in the direction of the message's destination (greedy routing).

Model 4.4.14 (Location Information). Sensor nodes can have access to various forms of (absolute or relative) geographic information about other nodes. For example, a node u might sense its distance to another node v, or sense in which direction (angle of arrival) u lies, or even know v's exact position.

Distributed algorithms for sensor networks are usually evaluated with respect to their time complexity, their space complexity, and their message complexity. However, in order to be successful in a real sensor network, an algorithm has to pursue additional objectives. For instance, if sensor nodes are deployed in large numbers, recharging their batteries seems out of question, in particular in adversarial territory. A node's energy supply must suffice for the whole operational phase. Therefore, the conservation of energy is of utmost importance. Basically, there are two approaches to capture the energy consumption of a node. Historically, since during the transmission of data much energy is consumed, a model has been studied which only takes the transmission energy into account [31].

Model 4.4.15 (Transmission Energy). The energy consumed by a node is calculated by the sum over all its transmissions. Thus, the energy needed to transmit one message is of the form $c \cdot d^{\alpha}$, where d is the distance between sender and receiver, α is the path-loss exponent (usually $\alpha > 2$), and c is a constant.

Although transmitting data is a costly operation, sensor nodes with short-range radios available today spend as much energy receiving or waiting for data. Therefore, techniques have been developed which allow nodes to change to a parsimonious *sleep mode* [32]. During the time periods a node is sleeping, it cannot receive any data. The idea is that if all nodes can somehow be synchronized to wake up at the same moment of time to exchange data (e.g., every minute), much energy is saved. This motivates the following model.

Model 4.4.16 (Sleeping Time). The energy consumed by a node is given by the accumulated time in which it is not in sleep mode.

If there are no external disturbances, a node is assumed to live as long as it has some energy left. The lifetime of the entire network is modeled in different ways.

Model 4.4.17 (Network Lifetime). In applications that depend on every single node, the lifetime of a network can be defined as the time until the first node runs out of battery power [33]. Alternatively, a network might be able to tolerate certain node failures; for example, the network might live as long as all live nodes are still connected to each other.

4.5 FINAL REMARKS

This chapter has given an overview and discussion of many sensor network models in use today. It has been shown how the models are related to each other. Therefore, we have assumed an algorithmic point of view and have concentrated on models of higher levels of abstraction. Of course, it does not make sense to argue about which model is "better" and which is "worse." For example, a large warehouse has different physical characteristics and signal propagation paths than an office building; or GPS might not work indoors and hence algorithms based on coordinates are not be feasible; and so on. A good model also depends on the question studied. A media access study might need a detailed model capturing several low-level aspects; for example, it has to be taken into account that a message might not be received correctly due to a nearby concurrent transmission. Hence, it is crucial that the model appropriately incorporates interference aspects. For a transport layer study, however, a much simpler model that assumes random transmission errors might be sufficient. This chapter helps to compare the different options.

Clearly, it is always desirable to have algorithms for sensor networks which can be proved correct in the most general possible model that covers all possible characteristics of a real environment. Only then can we be sure that the algorithm will actually work in practice. However, for efficiency considerations, a more idealistic model that does not yield overly conservative results might be fine. Moreover, we believe that when developing algorithms for sensor networks, it is often useful to study idealistic models first, because these models are simpler and may provide helpful insights into the given problem. After having found algorithms for these models, it is still possible to tackle the more general cases.

4.6 FUTURE RESEARCH DIRECTIONS

Models for sensor networks have developed quickly and are now much more sophisticated than they were some years ago. However, we believe that the quest for new models is still of prime importance. In particular, with the wider deployment of sensor networks the experiences with complex issues such as connectivity and interference will increase; consequently, models will evolve as well. Interestingly, many problems are still unexplored for the models presented in this chapter. For many models, there exist no algorithms with provable performance for fundamental operations such as the computation of dominating sets. For example: How well can the minimum dominating set be approximated in bounded independence connectivity graphs (BIG)? Such questions are exciting because they are interdisciplinary and require knowledge of various mathematical fields. We want to encourage the more advanced readers to address these problems!

4.7 EXERCISES

The following exercises are based on this chapter only, but sometimes require some mathematical background.

1. Name some scenarios where the QUDG model is not appropriate. Which alternative model might capture the situation better? Under what circumstances may it still be useful to study the QUDG?

2. Prove that both the UDG and the QUDG have a bounded independence (i.e., that they are a BIG). *Hint:* The proof is similar to the proof of Theorem 4.2.13.

3. **(a)** Prove that the two-dimensional Euclidean plane has a constant doubling dimension.

(b) *Formally* prove that not every general graph has a doubling dimension.

4. A basic operation in sensor networks is the distributed computation of a (minimum) *connected* dominating set (CDS). A CDS is a dominating set with the additional requirement that any two dominators are connected to each other by a path that only consists of other dominators. For instance, such a CDS can be useful to establish a *routing backbone network*.

(a) Can you come up with a *local* algorithm which computes a CDS in a UDG if all nodes are equipped with a GPS device (i.e., if they know their position)?

(b) Does your algorithm yield the optimal solution, or just an approximation? Can you prove its quality?

(c) How efficient is your algorithm, in terms of number of messages transmitted and in terms of communication rounds needed?

5. The connected dominating set problem is well-studied.

(a) Surf the web to find the currently best-known algorithm for the UDG.

(b) Can you find an algorithm that works on *general* connectivity graphs and that does not require nodes to have position information?

(c) Compare the complexity and efficiency of the best-known UDG algorithm to the best-known GG algorithm: What is the "price" of the more pessimistic model?

6. Is the UBG model equivalent to the BIG model?

BIBLIOGRAPHY

1. B. N. Clark, C. J. Colbourn, and D. S. Johnson. Unit disk graphs. *Discrete Mathematics*, **86**:165–177, 1990.
2. F. Kuhn and R. Wattenhofer. Constant-time distributed dominating set approximation. In *Proceedings of the 22nd Annual Symposium on Principles of Distributed Computing (PODC)*, 2003, pp. 25–32.
3. L. Barrière, P. Fraigniaud, and L. Narayanan. Robust position-based routing in wireless ad hoc networks with unstable transmission ranges. In *Proceedings of the 5th International Workshop on Discrete Algorithms and Methods for Mobile Computing and Communications (DIALM)*, 2001, pp. 19–27.
4. F. Kuhn, R. Wattenhofer, and A. Zollinger. Ad-hoc networks beyond unit disk graphs. In *Proceedings of the 1st ACM Joint Workshop on Foundations of Mobile Computing (DIALM-POMC)*, 2003.
5. S. O. Krumke, M. V. Marathe, and S. S. Ravi. Models and approximation algorithms for channel assignment in radio networks. *Wireless Networks*, **7**(6):575–584, 2001.
6. I. Stojmenović, A. Nayak, and J. Kuruvila. Design guidelines for routing protcols in ad hoc and sensor networks with a realistic physical layer. *IEEE Communications Magazine*, **6**:36–42, 2005.
7. D. Ganesan, D. Estrin, A. Woo, D. Culler, B. Krishnamachari, and S. Wicker. Complex behavior at scale: An experimental study of low-power wireless sensor networks. Technical Report, UCLA Computer Science Technical Report UCLA/CSD-TR 02-0013, 2002.
8. F. Kuhn, T. Nieberg, T. Moscibroda, and R. Wattenhofer. Local approximation schemes for ad-hoc and sensor networks. In *Proceedings of the Joint Workshop on Foundations of Mobile Computing (DIALM-POMC)*, 2005, pp. 97–103.
9. F. Kuhn, T. Moscibroda, and R. Wattenhofer. On the locality of bounded growth. In *Proceedings of the 24th ACM Symposium on Principles of Distributed Computing (PODC)*, 2005, pp. 60–68.
10. A. Gupta, R. Krauthgamer, and J. R. Lee. Bounded geometries, fractals, and low-distortion embeddings. In *Proceedings of the 44th Annual IEEE Symposium on Foundations of Computer Science (FOCS)*, 2003, p. 534.
11. P. Gupta and P. R. Kumar. The capacity of wireless networks. *Transactions of Information Theory*, **46**(2):388–404, 2000.
12. T. Moscibroda and R. Wattenhofer. The complexity of connectivity in wireless networks. In *Proceedings of the IEEE Infocom*, 2006. CD-ROM.
13. T. Rappaport. *Wireless communications: Principles and practices*," Prentice Hall, Englewood Cliffs, NJ, 1996.

14. D. Bliò and G. Proietti. On the complexity of minimizing interference in adhoc and sensor networks. Technical Report, Dipartimento di Informatica TRCS 009/1006, 2006.

15. T. Moscibroda and R. Wattenhofer. Coloring unstructured radio networks. In *Proceedings of the 17th Annual ACM Symposium on Parallelism in Algorithms and Architectures (SPAA)*, 2005, pp. 39–48.

16. F. Meyer auf de Heide, C. Schindelhauer, K. Volbert, and M. Grünewald. Energy, congestion and dilation in radio networks. In *Proceedings of the 14th Annual ACM Symposium on Parallel Algorithms and Architectures (SPAA)*, 2002, pp. 230–237.

17. A. El Fawal, J.-Y. Le Boudec, Ruben Merz, B. Radunovic, J. Widmer, and G. M. Maggio. Trade-off analysis of PHY-aware MAC in low-rate low-power UWB networks. *IEEE Communications Magazine*, **43**(12):147, 2005.

18. R. Raz and S. Safra. A sub-constant error-probability low-degree test, and sub-constant error-probability PCP characterization of NP. In *Proceedings of the 29th Annual ACM Symposium on Theory of Computing (STOC)*, 1997, pp. 475–484.

19. T. H. Cormen, C. E. Leiserson, R. L. Rivest, and C. Stein, *Introduction to Algorithms*, MIT Press, Cambridge, MA, 2001.

20. D. Peleg, *Distributed Computing: A Locality-sensitive Approach*, Society for Industrial and Applied Mathematics, Philadelphia, PA, 2000.

21. Y. Wang, X.-Yang Li, P.-J. Wan, and O. Frieder. Sparse power efficient topology for wireless networks. *Journal of Parallel and Distributed Computing*, 2002.

22. F. Kuhn, T. Moscibroda, and R. Wattenhofer. The price of being near-sighted. In *Proceedings of the ACM-SIAM Symposium on Discrete Algorithms (SODA)*, 2006.

23. M. Luby. "A simple parallel algorithm for the maximal independent set problem. *SIAM Journal of Computing*, **15**(4):1036–1053, 1986.

24. E. W. Dijkstra. Self-stabilizing systems in spite of distributed control. *Communications of the ACM*, **17**(11):643–644, 1974.

25. B. Awerbuch and G. Varghese. Distributed program checking: A paradigm for building self-stabilizing distributed protocols. In *Proceedings of the 32nd Annual IEEE Symposium on Foundations of Computer Science (FOCS)*, 1991. CD-ROM.

26. F. Kuhn, T. Moscibroda, and R. Wattenhofer. What cannot be computed locally!. In *Proceedings of the 23rd ACM Symposium on the Principles of Distributed Computing (PODC)*, 2004. CD-ROM.

27. P.-J. Wan, K. M. Alzoubi, and O. Frieder. Distributed construction of connected dominating set in wireless ad hoc networks. *Mobile Networks and Applications*, **9**(2):141–149, 2004.

28. T. Moscibroda, P. von Rickenbach, and R. Wattenhofer. Analyzing the energy-latency trade-off during the deployment of sensor networks. In *Proceedings of the IEEE Infocom*, 2006. CD-ROM.

29. O. Dousse, P. Thiran, and M. Hasler, Connectivity in ad-hoc and hybrid networks. In *Proceedings of the IEEE Infocom*, 2002. CD-ROM.

30. F. Kuhn, R. Wattenhofer, and A. Zollinger. Worst-case optimal and average-case efficient geometric ad-hoc routing. In *Proceedings of the 4th ACM International Symposium on Mobile Ad Hoc Networking & Computing (MobiHoc)*, 2003, pp. 267–278.

31. A. E. F. Clementi, G. Huiban, and P. Penna. On the approximation ratio of the MST-based heuristic for the energy-efficient broadcast problem in static ad-hoc radio networks. In

Proceedings of the 17th International Symposium on Parallel and Distributed Processing (IPDPS), Washington, DC, 2003. CD-ROM.

32. V. Raghunathan, C. Schurgers, S. Park, and M. Srivastava. Energy aware wireless microsensor networks. *IEEE Signal Processing Magazine*, **19**(2):40–50, 2002.
33. J.-H. Chang and L. Tassiulas. Energy conserving routing in wireless ad-hoc networks. In *Proceedings of the IEEE Infocom*, 2000, pp. 22–31.

CHAPTER 5

Spatiotemporal Correlation Theory for Wireless Sensor Networks

ÖZGÜR B. AKAN

Department of Electrical and Electronics Engineering, Middle East Technical University, Ankara, Turkey 06531

5.1 INTRODUCTION

Wireless sensor networks (WSNs) are generally composed of densely deployed sensor nodes that collaboratively observe and communicate their readings of a certain physical phenomenon [1]. In general, the main objective of the WSN is to reliably detect/estimate event features from the collective information provided by sensor nodes. Therefore, the energy and hence processing constraints of small wireless sensor nodes are overcome by this collective sensing notion that is realized via their networked deployment.

To this end, accurate and efficient operation of any WSN deployment requires that we maintain sufficient network and sensing coverage in the deployment field. To assure network and sensing coverage, WSN applications require sensor nodes to be densely deployed in the field. Dense deployment of sensor nodes makes the sensor observations highly correlated in space domain. Similarly, in periodic sensing applications, all consecutive sensor readings are temporally correlated.

While the collaborative nature of the WSN brings significant advantages over traditional sensing including greater accuracy, larger coverage area, and extraction of localized features, the spatiotemporal correlation among the sensor observations is another significant and unique characteristic of the WSN which can be exploited to drastically enhance the overall network performance. In general, and based on the application, the physical phenomenon to be observed can be modeled as *point source* (e.g., target detection/tracking) or *field source* (e.g., monitoring of magnetic field and seismic activities) [2]. Events generating a signal that originates from a single point in the field can be modeled as a point source. The cases where the physical phenomenon

Algorithms and Protocols for Wireless Sensor Networks, Edited by Azzedine Boukerche

is dispersed over the field can be modeled as field source. The characteristics of the correlation in the WSN can be summarized as follows [3]:

- *Spatial Correlation.* Typical WSN applications require spatially dense sensor deployment in order to achieve satisfactory coverage [4]. As a result, multiple sensors record information about a single event in the sensor field. Due to high density in the network topology, spatially proximal sensor readings are correlated with the degree of correlation increasing with decreasing internode separation.
- *Temporal Correlation.* Some of the WSN applications such as event tracking may require sensor nodes to periodically perform observation and transmission of the sensed event features. The nature of the energy-radiating physical phenomenon constitutes the temporal correlation between each consecutive observation of a sensor node [5]. The degree of correlation between consecutive sensor measurements may vary according to the temporal variation characteristics of the phenomenon.

The existence of the above-mentioned spatial and temporal correlations bring significant potential advantages for the development of efficient communication protocols well-suited for the WSN paradigm. For example, intuitively, due to the spatial correlation, data from spatially separated sensors is more useful to the sink than highly correlated data from nodes in proximity. Therefore, it may not be necessary for every sensor node to transmit its data to the sink. Instead, a smaller number of sensor readings may suffice to communicate the information on the sensed phenomenon to the sink within a certain reliability/fidelity level. Similarly, for a certain target tracking application, the measurement reporting frequency, at which the sensor nodes transmit their observations, can be adjusted such that a temporally correlated phenomenon signal is captured at the sink within a certain distortion level and with minimum energy expenditure.

There has been some research effort to study the correlation in WSNs [2, 3, 6–11] most of which mainly investigate the information and coding theoretical aspects of the correlation. Clearly, it is of great importance to capture the spatiotemporal correlation characteristics to be able to design energy-efficient communication protocols that can exploit the potential advantages of correlation in WSNs. In this chapter, a theoretical analysis of spatiotemporal correlation in WSNs is presented. The objective of this analysis is to capture the spatiotemporal correlation characteristics of sensor networks to reveal its potential advantages and possible approaches to exploit correlation in designing efficient communication techniques for WSNs. More specifically, in Section 5.2, the theoretical framework is developed to model the spatiotemporal correlation in sensor networks. Based on the developed correlation model, the spatial and temporal correlations are separately captured in Section 5.3.1 and 5.3.2, respectively. In order to understand the joint effects of spatial and temporal correlation, spatiotemporal correlation characteristics of point and field sources are then investigated in Section 5.3.3. In Section 5.4, based on this analysis, possible approaches are also discussed to exploit spatiotemporal correlation for efficient communication in WSNs. Finally, the concluding remarks, open research problems, and potential directions are discussed in Section 5.5.

5.2 COMMUNICATION ARCHITECTURE AND CORRELATION MODEL FOR WIRELESS SENSOR NETWORKS

In a sensor field, each sensor observes the noisy version of a physical phenomenon. The sink is interested in observing the physical phenomenon using the observations from sensor nodes with the highest accuracy. The physical phenomenon in interest can be modeled as a spatiotemporal process $s(t, x, y)$ as a function of time t and spatial coordinates (x, y).

Depending on the specific sensor application, the physical phenomenon may be a spatiotemporal process generated by a point source in case of applications such as object tracking. In this case, the sink is interested in reconstructing the source signal at a specific location (x_0, y_0) based on sensor observations. In other applications, the spatiotemporal process may be a combination of multiple point sources where the sink is interested in reconstructing the signal in multiple locations or over an event area. Although the reconstruction is application specific, the properties of the observations can be modeled based on the spatiotemporal process $s(t, x, y)$.

The model for the information gathered by N sensors in an event area is illustrated in Figure 5.1. The sink is interested in estimating the event source, S, according to the observations of the sensor nodes, n_i, in the event area. Each sensor node n_i observes

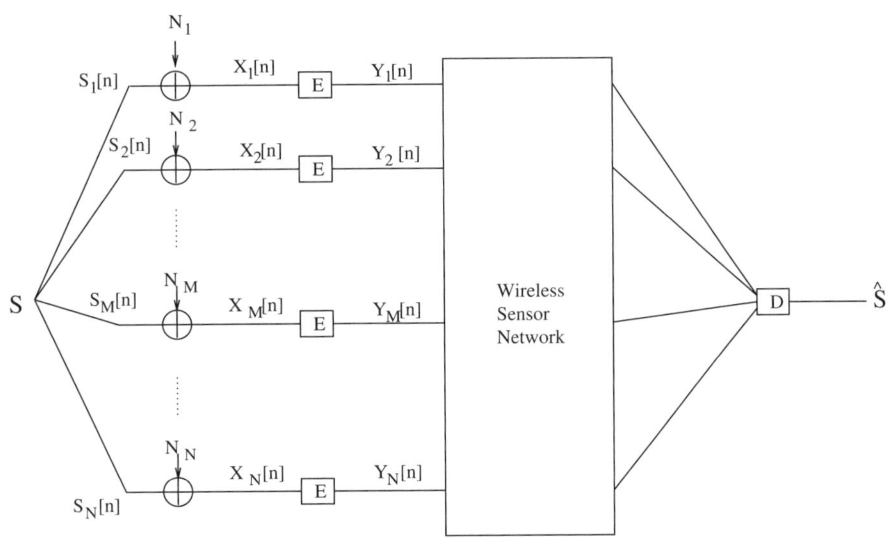

Figure 5.1. Correlation model and architecture.

$X_i[n]$, the noisy version of the event information, $S_i[n]$, which is spatially correlated to the event source, S. In order to communicate this observation to the sink, each node has to encode its observation. The encoded information, $Y_i[n]$, is then sent to the sink through the sensor network. The sink, at the other end, decodes this information to get the estimate, $\hat{S}$, of the event source S. The encoders and the decoders are labeled as E and D, respectively, in Figure 5.1. Using this model, we will exploit various aspects of correlation among sensor readings in terms of both time and space.

Each observed sample, $X_i[n]$, of sensor n_i at time n is represented as

$$X_i[n] = S_i[n] + N_i[n] \tag{5.1}$$

where the subscript i denotes the spatial location of node n_i [i.e., (x_i, y_i)], $S_i[n]$ is the realization of the space–time process $s(t, x, y)$ at time $t = t_n$[1] and $(x, y) = (x_i, y_i)$, and $N_i[n]$ is the observation noise. $\{N_i[n]\}_n$ is a sequence of i.i.d Gaussian random variables of zero mean and variance σ_N^2. We further assume that the noise each sensor node encounters is independent of each other; that is, $N_i[n]$ and $N_j[n]$ are independent for $i \neq j$ and $\forall n$.

As shown in Figure 5.1, each observation $X_i[n]$ is then encoded into $Y_i[n]$ by the source-coding at the sensor node as

$$Y_i[n] = f_i(X_i[n]) \tag{5.2}$$

and then sent through the network to the sink. The sink decodes the received data to reconstruct an estimation $\hat{S}$ of the source S

$$\hat{S} = g(Y_1[n_1], ..., Y_1[n_\tau]; ...; Y_N[n_1], ..., Y_N[n_\tau]) \tag{5.3}$$

based on the data received from N nodes in the event area over a time period $\tau = t_{n_\tau} - t_{n_1}$. The sink is interested in reconstructing the source S according to a distortion constraint

$$D = E\left[d(S, \hat{S})\right] \tag{5.4}$$

In the next sections, the general distortion function in (5.4) will be used to independently obtain the distortion functions for spatial and temporal correlation in the WSN, which will be further extended to capture the joint spatiotemporal correlation analysis of point and field sources as well.

[1] Note that we use a discrete-time model since each node is assumed to sample the physical phenomenon synchronously after the initial wake-up.

5.3 SPATIOTEMPORAL CORRELATION IN WIRELESS SENSOR NETWORKS

5.3.1 Spatial Correlation in Wireless Sensor Networks

In this section, based on the communication architecture and the theoretical correlation model presented in Section 5.2, the spatial correlation between observations of each sensor node is modeled. The information gathered by N sensors in an event area can be modeled as shown in Figure 5.1. The sink is assumed to be interested in a point source S. Since we only consider the spatial correlation between nodes, in this analysis we assume that the samples are temporally independent. Hence, by dropping the time index n, (5.1) can be restated as

$$X_i = S_i + N_i, \qquad i = 1, ..., N \tag{5.5}$$

The sink is interested in reconstructing the source S according to observations of nodes n_i which observe the spatially correlated version of S at (x_i, y_i), that is, S_i. The physical phenomenon is modeled as joint Gaussian random variables (JGRVs) at each observation point as

$$\begin{aligned} E\{S_i\} &= 0, \qquad i = 1, ..., N \\ \text{var}\{S_i\} &= \sigma_S^2, \qquad i = 1, ..., N \\ \text{cov}\{S_i, S_j\} &= \sigma_S^2 \,\text{corr}\{S_i, S_j\} \\ \text{corr}\{S_i, S_j\} &= \rho_{i,j} = K_\vartheta(d_{i,j}) = \frac{E[S_i S_j]}{\sigma_S^2} \end{aligned}$$

where $d_{i,j} = \| \mathbf{s}_i - \mathbf{s}_j \|$ denotes the distance between nodes n_i and n_j located at coordinates $\mathbf{s}_i$ and $\mathbf{s}_j$, respectively and $K_\vartheta(\cdot)$ is the correlation model. The covariance function is assumed to be nonnegative and to decrease monotonically with the distance $d = \| \mathbf{s}_i - \mathbf{s}_j \|$, with limiting values of 1 at $d = 0$ and of 0 at $d = \infty$. Generally, covariance models can be classified into four groups [12]:

- *Spherical*:

$$K_\vartheta^S(d) = \begin{cases} 1 - \frac{3}{2}\frac{d}{\theta_1} + \frac{1}{2}\left(\frac{d}{\theta_2}\right)^3 & \text{if} \quad 0 \le d \le \theta_1 \\ 0 & \text{if} \quad d > \theta_1 \end{cases}; \quad \theta_1 > 0$$

  In this model, two observations taken more than θ_1 distance apart are uncorrelated.

- *Power Exponential*:

$$K_\vartheta^{PE}(d) = e^{(-d/\theta_1)^{\theta_2}}; \quad \theta_1 > 0, \theta_2 \in (0, 2]$$

For $\theta_2 = 1$ the model becomes exponential, while for $\theta_2 = 2$ it becomes squared exponential.

- *Rational Quadratic*:

$$K_{\vartheta}^{RQ}(d) = \left(1 + \left(\frac{d}{\theta_1}\right)^2\right)^{-\theta_2}; \quad \theta_1 > 0, \theta_2 > 0$$

- *Matérn*:

$$K_{\vartheta}^{M}(d) = \frac{1}{2^{\theta_2 - 1}\Gamma(\theta_2)}\left(\frac{d}{\theta_1}\right)^{\theta_2}\mathcal{K}_{\theta_2}\left(\frac{d}{\theta_1}\right); \quad \theta_1 > 0, \theta_2 > 0$$

where $\mathcal{K}_{\theta_2}(\cdot)$ is the modified Bessel function of second kind and order θ_2.

The correlation model can be chosen according to the properties of the physical phenomenon the sink is interested in. Since we are interested in S, which is also a JGRV, we use a special notation with

$$\text{var}\{S\} = \sigma_S^2$$

$$\text{corr}\{S, S_i\} = \rho_{s,i} = K_{\vartheta}(d_{s,i}) = \frac{E[SS_i]}{\sigma_S^2}$$

where $d_{s,i}$ denotes the distance between the source S and the node n_i. The observation noise N_i of each node n_i is modeled as i.i.d. Gaussian random variable with zero mean and variance σ_N^2, that is, $N_i \sim \mathcal{N}(0, \sigma_N^2)$.

As each sensor node n_i observes an event information X_i, this information is encoded and then sent to the sink through the WSN. In traditional point-to-point communication, the optimum performance is obtained by compressing the information according to the source statistics and then adding redundant information to accommodate the errors introduced in the wireless channel. This technique is known as the separation principle. In WSNs, where multiple nodes try to send information about the same event, however, it is known that joint source-channel coding outperforms separate coding [10, 13]. In addition, for Gaussian sources, if the source is Gaussian and the cost on the channel is the encoding power, then uncoded transmission is optimal for point-to-point transmission [14]. Furthermore, for sensor networks with finite number of nodes, uncoded transmission outperforms any approach based on the separation paradigm leading to the optimal solution for infinite number of nodes [10]. Hence, we adopt uncoded transmission for the sensor observations in this model. Each node n_i sends to the sink, a scaled version, Y_i, of the observed sample X_i according to encoding power constraint P_E.

$$Y_i = \sqrt{\frac{P_E}{\sigma_S^2 + \sigma_N^2}}X_i, \qquad i = 1, \ldots, N \tag{5.6}$$

where σ_S^2 and σ_N^2 are the variances of the event information S_i and the observation noise N_i, respectively.

The sink needs to calculate the estimation of each event information, S_i, in order to estimate the event source S. Since uncoded transmission is used, it is well known that minimum mean square error (MMSE) estimation is the optimum decoding technique [15]. Hence, the estimation, Z_i, of the event information S_i is simply the MMSE estimation of Y_i, which is given by

$$Z_i = \frac{E[S_i Y_i]}{E[Y_i^2]} Y_i \tag{5.7}$$

Note that the estimated values of Z_i's are spatially correlated since the actual event information S_i's are spatially correlated. This spatial correlation results in redundancy in each event information sent to the sink. Although the sink is interested in estimating the event source, S, with a distortion constraint, intuitively, this constraint can still be met by using a smaller number of sensor nodes rather than all the nodes in the event area. In order to investigate the distortion achieved with a smaller number of nodes sending information, we assume that only M out of N packets are received by the sink, where N is the total number of sensor nodes in the event area. Since the sink decodes each Y_i using the MMSE estimator, the event source can simply be computed by taking the average of all the event information received at the sink. Then, $\hat{S}$, the estimate of S, is given as

$$\hat{S}(M) = \frac{1}{M} \sum_{i=1}^{M} Z_i \tag{5.8}$$

The distortion achieved by using M packets to estimate the event S is given as

$$D(M) = E[(S - \hat{S}(M))^2] \tag{5.9}$$

where we use the mean-squared error as the distortion metric. Using (5.5) and (5.6) in (5.7), the estimate Z_i of each event information S_i can be written as

$$Z_i = \frac{E[S_i Y_i]}{E[Y_i^2]} \sqrt{\frac{P}{\sigma_S^2 + \sigma_N^2}} (S_i + N_i) \tag{5.10}$$

denoting $\alpha = \sqrt{\frac{P}{\sigma_S^2 + \sigma_N^2}}$,

$$E[S_i Y_i] = \alpha \sigma_S^2$$

$$E[Y_i^2] = \alpha^2 \left(\sigma_S^2 + \sigma_N^2 \right)$$

Then, (5.10) is restated as

$$Z_i = \frac{\sigma_S^2}{\sigma_S^2 + \sigma_N^2}(S_i + N_i) \tag{5.11}$$

Using (5.11) and (5.8) in (5.9), the distortion function $D(M)$ is found to be [3]

$$D(M) = \sigma_S^2 - \frac{\sigma_S^4}{M(\sigma_S^2 + \sigma_N^2)}\left(2\sum_{i=1}^{M} \rho_{(s,i)} - 1\right) + \frac{\sigma_S^6}{M^2(\sigma_S^2 + \sigma_N^2)^2}\sum_{i=1}^{M}\sum_{j \neq i}^{M} \rho_{(i,j)} \tag{5.12}$$

$D(M)$ shows the distortion achieved at the sink as a function of number of nodes M that send information to the sink and correlation coefficients $\rho_{(i,j)}$ and $\rho_{(s,i)}$ between nodes n_i and n_j and between the event source S and node n_i, respectively. Based on the distortion function, we discuss possible approaches that can be used in the design of efficient medium access control in sensor networks in Section 5.4.1.

5.3.2 Temporal Correlation in Wireless Sensor Networks

As mentioned in Section 5.1, the energy-radiating physical phenomenon constitutes the temporal correlation between each consecutive observation of a sensor node [5]. For the periodic sensing applications such as event tracking, all consecutively taken sensor observations are temporally correlated to a certain degree. In this section, we establish the theoretical analysis for this temporal correlation, which will be further elaborated in the context of correlation-based reliable event transport approach discussed in Section 5.4.2.

Here, we consider the temporal correlation between the sensor observations and hence we omit the spatial variation in this analysis. We are interested in estimating the signal $s(t)$ in a decision interval of τ. In our theoretical analysis, we model an event-to-sink distortion metric, where all the information coming from the sensor nodes in the event area is considered as if it is generated by a single source node during the decision interval τ.

Assume that the sensed information from the sensors are sent to the sink using a reporting frequency of f. In this case, we seek to control the reporting frequency f such that a desired distortion level is not exceeded in the estimation of the event features at the sink. The event signal $s(t)$ is assumed to be a Gaussian random process with $\mathcal{N}(0, \sigma_s^2)$. The sink is interested in finding the expectation of the signal $s(t)$ over the decision interval τ, that is, $S(\tau)$. Assuming that the observed signal $s(t)$ is wide-sense stationary (WSS), the expectation of the signal over the decision interval τ can be calculated by the time average of the observed signal [16], that is,

$$S(\tau) = \frac{1}{\tau} \int_{t_0}^{t_0+\tau} s(t)\, dt \tag{5.13}$$

where t_0 is the time the sensor node wakes up for the sampling of the signal. With a change of variables, $S(\tau)$ can be shown as

$$S(\tau) = \frac{1}{\tau} \int_{0}^{\tau} s(t_0 + \Gamma)\, d\Gamma \tag{5.14}$$

We define the value of the signal at each sampling interval as

$$S[n] = s\left(t_0 + \frac{n}{f}\right) \tag{5.15}$$

where f is the sampling frequency and $S[n]$ are JGRV with $\mathcal{N}(0, \sigma_s^2)$.[2] For the derivation of the distortion function, the following definitions are needed:

$$\begin{aligned} E\{S[n]\} &= 0 \\ E\{(S[n])^2\} &= \sigma_S^2 \\ E\{S[n]S[m]\} &= \sigma_S^2 \hat{\rho}_S(n, m) \\ E\{s(t)s(t+\delta)\} &= \sigma_S^2 \rho_S(\delta) \end{aligned}$$

where $\hat{\rho}_S(n, m) = \rho_S(|m - n|/f)$ is the covariance function that depends on the time difference between signal samples. Among the covariance models introduced in Section 5.3.1, we use the power exponential model in the derivation since the physical event information (such as an electromagnetic wave), is modeled to have an exponential autocorrelation function [17]. Hence, the covariance function becomes

$$\rho_S(\delta) = e^{-|\delta|/\theta_1} \tag{5.16}$$

Each sensor node observes the noisy version of the signal given as

$$X[n] = S[n] + N[n] \tag{5.17}$$

and the transmitted signal is expressed by

$$Y[n] = \sqrt{\frac{P_E}{\sigma_S^2 + \sigma_N^2}}\, X[n] \tag{5.18}$$

[2] Note that the samples of a Gaussian random process are jointly Gaussian [16].

based on the discussion in Section 5.3.1. Using the MMSE estimator at the sink, each sample is estimated as

$$Z[n] = \frac{E\Big[S[n]Y[n]\Big]}{E\Big[Y^2[n]\Big]} Y[n] \tag{5.19}$$

Hence, each estimated sample from the sensor nodes can be represented as

$$Z[n] = \frac{\sigma_S^2}{\sigma_S^2 + \sigma_N^2}\Big(S[n] + N[n]\Big) \tag{5.20}$$

After collecting all the samples of the signal in the decision interval τ, the sink estimates the expectation of the signal over the last decision interval by

$$\hat{S}(\tau) = \frac{1}{\tau f}\sum_{k=1}^{\tau f} Z[k] \tag{5.21}$$

where τf is the total number of sensor samples taken within a decision interval with duration of τ. As a result, the distortion achieved by using τf samples to estimate the event is given as

$$D = E\left[\Big(S(\tau) - \hat{S}(\tau)\Big)^2\right] \tag{5.22}$$

Using the definitions above and substituting (5.14), (5.20), and (5.21) into (5.22), the distortion function can easily shown to be [3]

$$\begin{aligned} D(f) = \sigma_S^2 &+ \frac{\sigma_S^4}{\tau f(\sigma_S^2 + \sigma_N^2)} + \frac{\sigma_S^6}{\tau^2 f^2(\sigma_S^2 + \sigma_N^2)^2}\sum_{k=1}^{\tau f}\sum_{l \neq k} e^{-(\frac{|k-l|}{f})/\theta_1} \\ &- \frac{2\sigma_S^4\theta_1}{\tau^2 f(\sigma_S^2 + \sigma_N^2)}\sum_{k=1}^{\tau f}\left(2 - e^{-\frac{k}{f\theta_1}} - e^{-(\tau - \frac{k}{f})/\theta_1}\right) \end{aligned} \tag{5.23}$$

It is observed from (5.23) that the distortion in the estimation decreases with increasing f. Note that a distortion level D for the estimation of event features from the sensor observations corresponds to a certain reliability level of the event-to-sink communication in the WSN.

5.3.3 Joint Spatiotemporal Correlation in Wireless Sensor Networks

In the previous sections, the spatial and temporal correlations were separately captured. In order to understand the joint effects of spatial and temporal correlation, spatiotemporal correlation characteristics of point and field sources are investigated next.

Spatiotemporal Characteristics of Point Sources. In many WSN applications such as target detection and fire detection, the goal is to estimate the properties of an event generated by a single point source, through collective observations of sensor nodes. In this section, we first introduce our model for the point source and formulate its spatiotemporal characteristics. Next, we derive the distortion function for the estimation of the point source.

Here, we are interested in observing the joint behavior of spatial and temporal correlation. Therefore, in order to capture the spatiotemporal correlation characteristics, we follow a different approach than in Section 5.3. Here, the point source is assumed to generate a continuous signal that is modeled by a random process $f_S(s, t)$, where s denotes the outcome and t denotes time. For ease of illustration, we use $f_S(t)$ in the remainder of the chapter. We model the point source, $f_S(t)$, as a Gaussian random process such that $f_S(t)$ is first-order stationary; that is, $\mu_S(t) = \mu_S$ and has a variance σ_S^2. Without loss of generality, we assume $\mu_S = 0$.

For ease of illustration, we assume that the coordinate axis is centered at the point source. As a result, the received signal, $f(x, y, t)$, at time t at a location (x, y) can be modeled as

$$f(x, y, t) = f_S\left(t - \frac{\sqrt{x^2 + y^2}}{v}\right) e^{-\frac{\sqrt{x^2+y^2}}{\theta_s}} \tag{5.24}$$

which is the delayed and attenuated version of the signal $f_S(t)$. In this model, we assume that the event signal travels with the speed, v, and is attenuated based on an exponential law, where θ_s is the attenuation constant. Note that the function $f(x, y, t)$ is also a Gaussian random process and that the samples taken by the sensors are jointly Gaussian random variables (JGRVs). Since $\mu_S = 0$, the mean of the received signal is given by $\mu_E = 0$.[3] The variance of the received signal is also given as follows:

$$\sigma_E^2(x, y) = E\left[f^2(x, y, t)\right] = \left(\sigma_S\, e^{-\sqrt{x^2+y^2}/\theta_s}\right)^2 \tag{5.25}$$

An interesting result from (5.25) is that the variance of the signal observed at location (x, y) depends on the distance between the observation location and the point source. As in Figure 5.1, the received signal at time t_k by a sensor n_i at location (x_i, y_i) is given by

$$S_i[k] = f(x_i, y_i, t_k) \tag{5.26}$$

[3] The subscripts S and E that are used here represent the *source* and *event*, respectively.

Assuming wide-sense stationarity, the *spatiotemporal correlation function* for two samples of a point source taken at locations (x_i, y_i) and (x_j, y_j) and at times t_k and t_l, respectively, is given by

$$\rho_P(i, j, k, l) = \frac{E\left[S_i[k]\, S_j[l]\right]}{\sigma_E(x_i, y_i)\,\sigma_E(x_j, y_j)} = \rho_S(\Delta_t) \tag{5.27}$$

where $\Delta_t = |t_k - t_l - (d_i - d_j)/v|$, $d_i = \sqrt{x_i^2 + y_i^2}$ is the distance of the sensor n_i to the point source, and $\rho_S(\Delta_t) = E[f_S(t)\, f_S(t + \Delta_t)]/\sigma_S^2$ is the correlation function of the point source which is given by $\rho_S(\Delta_t) = e^{-\Delta_t/\theta_t}$, where θ_t is a constant governing the degree of correlation. Note that the spatiotemporal correlation between two samples, $\rho_P(i, j, k, l)$, depends mainly on the difference between sample times t_k and t_l since generally $v \gg (d_i - d_j)$.

In WSNs, we are interested in estimating the signal generated by the point source using the samples collected by the sensor nodes. The expectation of the generated signal, $f_S(t)$, over an interval τ is given by

$$S(\tau) = \frac{1}{\tau}\int_0^{\tau} f_S(t)\, dt \tag{5.28}$$

Each sensor node, n_i, receives the attenuated and delayed version of the generated signal $f_S(t)$, that is, $S_i[k]$. Due to the impurities in the sensor circuitries, the sampled signal is the noisy version of this received signal which is given by

$$X_i[k] = S_i[k] + N_i[k] \tag{5.29}$$

where the subscript i denotes the location of the node n_i, that is, (x_i, y_i), k denotes the sample index which corresponds to time $t = t_k$, $X_i[k]$ is the noisy version of the actual sample $S_i[k]$, and $N_i[k]$ is the observation noise, that is, $N_i[k] \sim \mathcal{N}(0, \sigma_N^2)$. $S_i[k]$ is given by (5.24) and (5.26).

The observed information, $X_i[k]$, is then encoded and sent to the sink through the WSN. It has been shown that joint source-channel coding outperforms separate coding. Moreover, as discussed in Section 5.3, for WSNs with a finite number of nodes, uncoded transmission outperforms any approach based on the separation paradigm leading to the optimal solution for infinite number of nodes [3]. In the light of these results, we assume that uncoded transmission is deployed in each node. Hence, the transmitted observation, $Y_i[k]$, is given by

$$Y_i[k] = \sqrt{\frac{P_E}{\sigma_S^2 + \sigma_N^2}}\, X_i[k], \qquad i = 1, \ldots, N \tag{5.30}$$

where σ_S^2 and σ_N^2 are the variances of the event information $S_i[k]$ and the observation noise $N_i[k]$, respectively.

The transmitted information is decoded at the sink. Since uncoded transmission is used, it is well known that minimum mean square error (MMSE) estimation is the optimum decoding technique [3]. Hence the estimation, $Z_i[k]$, of the event information $S_i[k]$ is simply the MMSE estimation of $Y_i[k]$, which is given by

$$Z_i[k] = \frac{\sigma_E^2(x_i, y_i)}{\sigma_E^2(x_i, y_i) + \sigma_N^2}\left(S_i[k] + N_i[k]\right) \tag{5.31}$$

The sink is interested in estimating the expected value of the event during a decision interval τ that is given by (5.28). Assuming each sensor node sends information at a rate of f samples/sec, this estimation can simply be found by

$$\hat{S}(\tau, f, M) = \frac{1}{\tau f M}\sum_{i=1}^{M}\sum_{k=1}^{\tau f} Z_i[k] \tag{5.32}$$

where M is the number of sensor nodes that send samples of the observed point source. M nodes are chosen among the nodes in the network to represent the point source and, hence, are referred to as *representative nodes*. Consequently, the distortion achieved by this estimation is given by [2]

$$D_p(\tau, f, M) = E\left[(S(\tau) - \hat{S}(\tau, f, M))^2\right] \tag{5.33}$$

where the subscript p denotes the point source. Using (5.24), (5.25), (5.28), (5.31), and (5.32), (5.33) can be expressed as

$$\begin{aligned} D_p(\tau, f, M) = \sigma_S^2 &- \frac{2}{\tau^2 f M}\sum_{i=1}^{M}\sum_{k=1}^{\tau f}\frac{\sigma_S^4\, e^{-3d_i/\theta_s}}{\sigma_S^2\, e^{-2d_i/\theta_s} + \sigma_N^2} \\ &\theta_t\left[2 - e^{-\left(t_k + d_i/c\right)} - e^{-\left(\tau - t_k - d_i/c\right)/\theta_t}\right] \\ &+ \frac{\sigma_N^2}{\tau f M^2}\sum_{i=1}^{M}\frac{\sigma_S^4\, e^{-2d_i/\theta_s}}{\left(\sigma_S^2\, e^{-d_i/\theta_s} + \sigma_N^2\right)^2} \\ &+ \frac{1}{\tau^2 f^2 M^2}\sum_{i=1}^{M}\sum_{j=1}^{M}\sum_{k=1}^{\tau f}\sum_{l=1}^{\tau f}\alpha\,\rho(i, j, k, l) \end{aligned} \tag{5.34}$$

where

$$\alpha = \frac{\sigma_S^8\, e^{-2(d_i + d_j)/\theta_s}}{\left(\sigma_S^2\, e^{-d_i/\theta_s} + \sigma_N^2\right)\left(\sigma_S^2\, e^{-d_j/\theta_s} + \sigma_N^2\right)}$$

and $d_i = \sqrt{(x_i + y_i)}$, and $\rho(i, j, k, l)$ is the spatiotemporal correlation function given in (5.27).

Spatiotemporal Characteristics of Field Sources. In some WSN applications such as temperature monitoring and seismic monitoring, the physical phenomenon is dispersed over the sensor field and, hence, can be modeled as a field source. Thus, here we explore the spatiotemporal characteristics of observing such a phenomenon in WSNs.

As in Section 5.3.3, the event signal $f(x, y, t)$ is assumed to be a Gaussian random process with $\mathcal{N}(0, \sigma_s^2)$. The sink is interested in estimating the signal $f(x_0, y_0, t)$ over the decision interval τ at location (x_0, y_0). Assuming that the observed signal $f(x, y, t)$ is wide-sense stationary (WSS), the expectation of the signal over the decision interval τ [i.e., $S(\tau)$] can be calculated by the time average of the observed signal as

$$S(\tau) = \frac{1}{\tau} \int_0^\tau f(x_0, y_0, t)\, dt \tag{5.35}$$

where (x_0, y_0) is the event location. The signal $S_i[k]$ received at time t_k by a sensor node at location (x_i, y_i) is defined as in (5.26), and the $S_i[k]$'s are JGRV with $\mathcal{N}(0, \sigma_s^2)$. The covariance of two samples, $S_i[k]$ and $S_j[l]$, is given by

$$\text{cov}\{S_i[k], S_j[l]\} = \sigma_S^2\, \rho_s(i, j)\, \rho_t(\delta) \tag{5.36}$$

where

$$\rho_s(i, j) = e^{-d_{i,j}/\theta_s} \quad \text{and} \quad \rho_t(\delta) = e^{-|\delta|/\theta_t} \tag{5.37}$$

are spatial and temporal correlation functions, respectively, $\delta = (k - l)/f$, f is the sampling rate, $d_{i,j} = \sqrt{(x_i - x_j)^2 + (y_i - y_j)^2}$ is the distance between two nodes n_i and n_j, and θ_s and θ_t are spatial and temporal correlation coefficients, respectively.

Following the discussion and derivations in Section 5.3.3, the noisy version of the signal, $X_i[k]$, and the transmitted signal, $Y_i[k]$, are given by (5.29) and (5.30), respectively. The estimation $Z_i[k]$ can be found as

$$Z_i[k] = \frac{\sigma_S^2}{\sigma_S^2 + \sigma_N^2} \Big(S_i[k] + N_i[k] \Big) \tag{5.38}$$

After collecting the samples of the signal in the decision interval τ from M nodes, the sink estimates the expectation of the signal over the last decision interval as given in (5.32). As a result, the distortion achieved by this estimation is given as in (5.33). Using the definitions above and substituting (5.35), (5.38), and (5.32) into (5.33), the distortion function can be derived as [2]

$$D_f(\tau, f, M) = \sigma_S^2 - \frac{2\sigma_S^2}{\tau^2 fM(\sigma_S^2 + \sigma_N^2)} \sum_{i=1}^{M} \rho_s(i, s)$$

$$\sum_{k=1}^{\tau f} \theta_t \left[2 - e^{-k/(f\theta_t)} - e^{-\left(t - \frac{k}{f}\right)/\theta_t}\right]$$

$$+ \frac{\sigma_S^4 \sigma_N^2}{\tau fM(\sigma_S^2 + \sigma_N^2)^2} + \frac{\sigma_S^6}{(\tau fM(\sigma_S^2 + \sigma_N^2))^2}$$

$$\sum_{i=1}^{M} \sum_{j=1}^{M} \sum_{k=1}^{\tau f} \sum_{l=1}^{\tau f} \rho_s(i, j)\rho_t(|k - l|/f) \tag{5.39}$$

In order to provide further insight into the spatiotemporal correlation characteristics and distortion analysis derived in this section, next we discuss possible approaches that can be used in the design of efficient communication techniques exploiting the spatiotemporal correlation observed in the wireless sensor networks.

5.4 COROLLARIES AND EXPLOITING CORRELATION IN WIRELESS SENSOR NETWORKS

Spatiotemporal correlation, in addition to the collaborative nature of the WSN, bring significant potential advantages for the development of efficient communication protocols well-suited for the WSN paradigm. In this section we discuss possible approaches exploiting spatiotemporal correlation to achieve energy-efficient medium access and reliable event transport in WSN, respectively.

5.4.1 Spatial Correlation and Medium Access Control

The shared wireless channel between sensor nodes and energy considerations of the WSN make the medium access control (MAC) a crucial part of the wireless sensor networking. Furthermore, the scarce energy sources of sensor nodes necessitate energy-aware MAC protocols. Hence, MAC protocols for WSN should be tailored to the physical properties of the sensed phenomenon and the specific network properties so that the access to the channel is coordinated with minimum collisions without affecting the connectivity throughout the network.

In WSNs, many individual nodes deployed in large areas sense events and send corresponding information about these events to the sink. As discussed in Section 5.3.1, due to the physical properties of the event, this information may be highly correlated according to the spatial distribution of the sensor nodes. Intuitively, data from spatially separated sensors are more useful to the sink than highly correlated data from closely located sensors. Hence, it may not be necessary for every sensor node to transmit its data to the sink. Instead, a smaller number of sensor measurements might be adequate to communicate the event features to the sink within a certain reliability constraint. As a result, the MAC protocol can *reduce the energy consumption of the network by exploiting spatial correlation in the WSNs without compromising on the access latency as well as the distortion achieved.*

In order to gain more insight to these intuitions, a case study is performed using the distortion function (5.12). In a 500-by-500 grid, 50 sensor nodes are randomly deployed. The *Power Exponential* model is used with $\theta_2 = 1$ and $\theta_1 = \{10, 50, 100, 500, 1000, 5000, 1000\}$ as the covariance model for the covariance function, $K_\vartheta(\cdot)$ in (5.36). The parameter θ_1 controls the relation between the distance of the nodes and the correlation coefficient. For each value of θ_1, the distortion function (5.12) is calculated varying the number of sensor nodes sending information. Starting from 50 nodes, we decrease the number of nodes that send event information to the sink. We refer to these nodes as the *representative nodes*.

Representative nodes are selected randomly among the 50 nodes for each trial, and the distortion function is calculated according to the locations of these nodes. The average distortion calculated from these simulations and the distribution of the distortion for each number of representative nodes is shown in Figure 5.2.

As shown in Figure 5.2, the achieved distortion stays relatively constant when the number of representative nodes is decreased from 50 to 15. This behavior is due to the highly redundant data sent by the sensor nodes that are close to each other. In addition, with increasing θ_1, the observed event distortion decreases because close nodes become less correlated with increasing θ_1. Based on the results shown in Figure 5.2 and the distortion function (5.12), the following discussions about the observed distortion at the sink can be made:

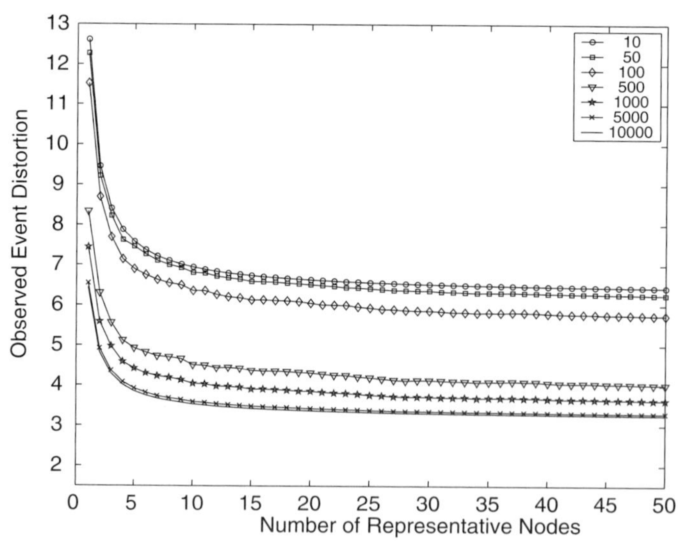

Figure 5.2. Observed event distortion for different θ_1 values according to changing number of representative nodes.

Remark 1. The minimum distortion is achieved when all the nodes in the event area send information to the sink. However, the achieved distortion at the sink can be preserved even though the number of the representative nodes decreases. As a result, significant energy saving is possible by allowing a smaller number of nodes to send information to the sink about an event.
Remark 2. Based on (5.12), there are two factors affecting the distortion other than the number of representative nodes.

1. *The correlation coefficient,* $\rho_{(s,i)}$, between a node n_i sending information and the event source S affects the distortion function negatively. The distortion increases as the distance between the event source S and the node n_i increases. Intuitively, if a representative node is chosen apart from the source, it observes relatively inaccurate data resulting in higher distortion at the sink.
2. *The correlation coefficient,* $\rho_{(i,j)}$, between each representative node n_i and n_j affects the distortion positively. As the distance between nodes increases, distortion decreases. Since nodes that are further apart observe less correlated data, the distortion is decreased if these nodes are chosen as the representative nodes.

Consequently, due to the spatial correlation between sensor observations, significant energy saving can be achieved by choosing representative nodes among the nodes in the event area without degrading the achieved distortion at the sink. It is clear that smaller number of nodes transmitting information reduces contention in the wireless medium, resulting in decreased energy consumption. Energy consumed from both transmission of packets and collision penalties can be reduced drastically if the spatial correlation is exploited. As a result, it is important to find the minimum number of representative nodes that achieve the distortion constraint given by the sensor application. This minimum number can be given as

$$M^* = \arg_M (\min \{D(M) < D_{\max}\})$$

where $D_{\max}$ is the maximum distortion allowed by the sensor application.

It is important to note that the minimum number of representative nodes, M^*, depends on the locations of the representative nodes. It follows from our previous discussions that, for a fixed number of representative nodes, the minimum distortion can be achieved by choosing these nodes such that (i) they are located as close to the event source as possible and (ii) they are located as farther apart from each other as possible.

As an example, as illustrated in Figure 5.3, choosing representative nodes such that they are spread over the event area results in a decrease in distortion, due to less redundant data sent by these nodes. Note that such a formation also improves the medium access performance during the transmission of the information. Since the representative nodes are not located close to each other, the probability of collision in the wireless medium decreases. As a result, exploiting spatial correlation not only improves the distortion but also utilizes the wireless channel due to the spatial reuse property of the wireless medium.

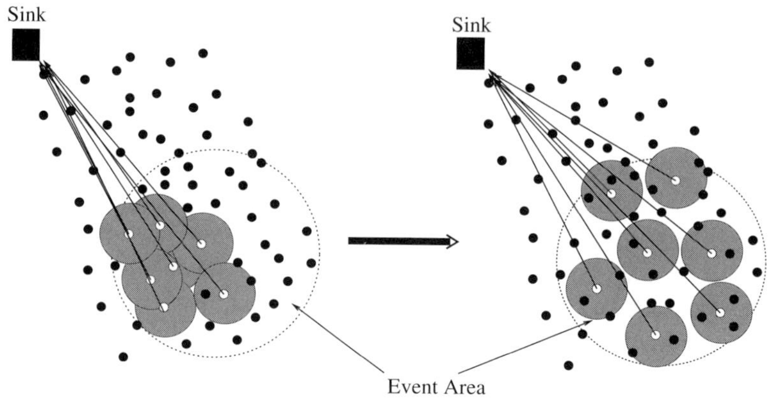

Figure 5.3. Spatial re-usage in sensor networks [18].

In a recent work [18], a correlation-based MAC protocol has been proposed which exploits the spatial correlation between closely located sensor nodes that regulate medium access and prevent redundant transmissions from closely located sensors. Based on the spatial correlation among sensor nodes, the MAC protocol collaboratively regulates medium access so that redundant transmissions from correlation neighbors are suppressed. In addition, necessary mechanisms for the efficient transmission of the information from the sensor nodes to the sink have also been incorporated into the proposed correlation-based MAC solution. The experimental results in reference 18 reveal that significant performance gains and energy savings are obtained by exploiting spatial correlation in sensor networks.

5.4.2 Spatiotemporal Correlation and Reliable Event Communication

In order to realize the potential gains of the WSN, it is imperative that desired event features are reliably communicated to the sink. Unlike traditional communication networks, the sensor network paradigm necessitates that the event features are estimated within a certain distortion bound (i.e., required reliability level) at the sink as discussed in Section 5.2. Reliable event detection at the sink is based on collective information provided by source nodes and not on any individual report. Hence, conventional end-to-end reliability definitions and solutions are inapplicable in the WSN regime and would only lead to overutilization of scarce sensor resources. On the other hand, the absence of reliable transport altogether can seriously impair event detection, which is the main objective of WSN deployment. Hence, the WSN paradigm necessitates a collective *event-to-sink* reliability notion rather than the traditional end-to-end notion [19]. Such *event-to-sink reliable transport* notion based on collective identification of spatially and temporally correlated data flows from the event to the sink is illustrated in Figure 5.4 and depends on the following definitions:

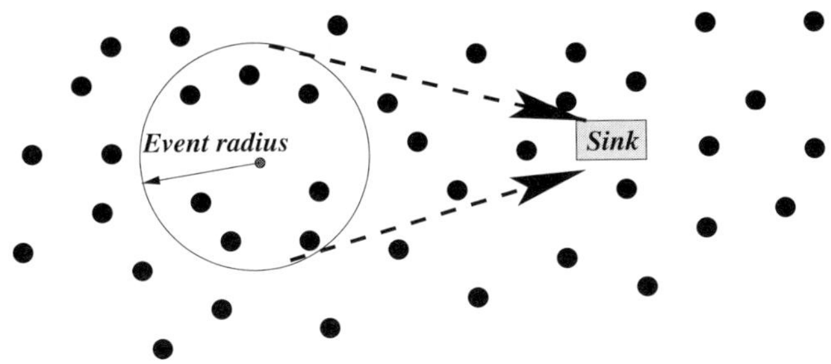

Figure 5.4. Typical sensor network topology with event and sink. The sink is only interested in collective information of sensor nodes within the event radius and not in their individual data.

Definition 1. The *observed event distortion* D_i is the distortion achieved [i.e., as in (5.23)] when the sink performs estimation of the signal S being tracked in decision interval i.

Definition 2. The *desired event distortion* D^* is the maximum distortion allowed to assure reliable event detection in the estimation performed by the sink. This upper bound for the distortion level is determined by the application and based on the physical characteristics of the signal S being tracked.

Based on the packets generated by the sensor nodes in the event area, the sink estimates the event features to determine the necessary action and observes D_i at each decision interval i. Note that a distortion level D for the estimation of event features from the sensor observations corresponds to the reliability level of the event-to-sink communication in the WSN. If observed event distortion is less than the distortion bound, i.e., $D_i < D^*$, then the event is deemed to be reliably detected. Else, appropriate action needs to be taken to assure the desired reliability level in the event-to-sink communication.

The main rationale behind such *event-to-sink reliability* notion is that the data generated by the sensors are temporally correlated, which tolerates individual packets to be lost to the extent where the desired event distortion D^* is not exceeded. Let f be the reporting frequency of a sensor node defined as the number of samples taken and hence packets sent out per unit time by that node for a sensed phenomenon. This reporting frequency can be attributed to increase in sampling rate as in Section 5.3.2. Hence, the reporting frequency f controls the amount of traffic injected to the sensor field while regulating the number of temporally correlated samples taken from the phenomenon. This, in turn, affects the observed event distortion—that is, event detection reliability. Thus, the reliable event transport problem in WSN is to *determine the reporting rate (f) of source nodes so that the maximum event estimation distortion bound D^* is not exceeded; that is, required event detection reliability is achieved at the sink, with minimum resource utilization.*

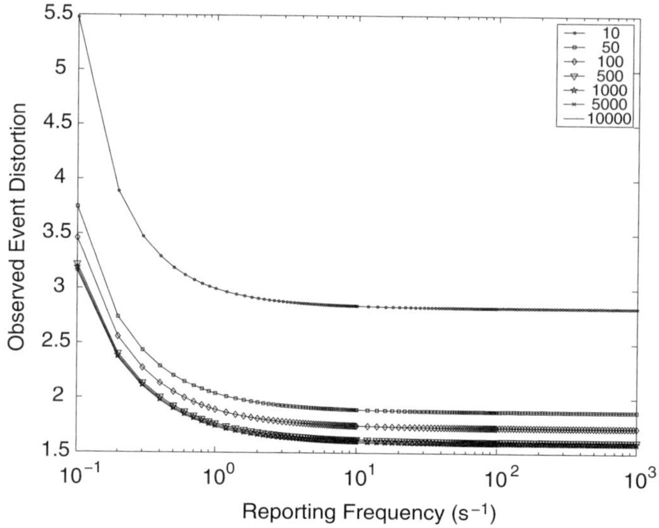

Figure 5.5. Observed event distortion for varying normalized reporting frequency.

The determination of an appropriate reporting frequency f in order to assure the desired event distortion with minimum energy expenditure and without causing congestion is a challenging issue. As derived in Section 5.3.2, the distortion D_i observed in the estimation of the signal S being tracked depends on the reporting frequency f used by the sensor nodes sending their readings to the sink in the decision interval i. A case study with the same network configuration and parameters in Section 5.4.1 is also performed to observe the variation of the observed event distortion at the sink for varying reporting frequency f—that is, distortion function $D(f)$ in (5.23). It is observed from (5.23) and Figure 5.5 that the observed event distortion at the sink decreases with increasing f. This is because the number of samples received in a decision interval i increases with increasing f conveying more information to the sink from the event area. Note that after a certain reporting frequency f, the observed event distortion cannot be further reduced. Therefore, a significant energy saving can be achieved by selecting small enough f which achieves desired event distortion D^* and does not lead to an overutilization of the scarce sensor resources.

On the other hand, any f chosen arbitrarily small to achieve a certain distortion bound D^* using (5.23) may not necessarily achieve the desired distortion level and hence assure the event transport reliability. This is mainly because all of the sensor samples generated with this chosen reporting frequency may not be received because of packet losses in the sensor network due to link errors and network disconnectivity. Similarly, very high values of f do not bring any additional gain in terms of observed event distortion as shown in Figure 5.5; on the contrary, they may endanger the event transport reliability by leading to congestion in the sensor network. Let $f_{\max}$ be the maximum reporting frequency which the network capacity can accommodate. Thus, $f > f_{\max}$ leads to congestion and hence packet losses, resulting in an increase in the observed event distortion.

Therefore, the main objective would be to operate the network at optimal operating region—that is, achieve required event reliability level (desired distortion level in the estimation) with minimum energy expenditure. To achieve this objective and address the reliable event transport problem, an event-to-sink reliable transport (ESRT) protocol has been developed in reference 19 based on the *event-to-sink reliability* notion for WSN. The objective of this scheme is to achieve reliable event transport with minimum energy expenditure and congestion control by exploiting the correlation and the collaborative nature of the WSN. To help accomplish this, the protocol uses a congestion control mechanism that serves the dual purpose of reliable detection and energy conservation. For example, in the states where the observed reliability is greater than that required (i.e., very low observed event distortion) and there is no congestion, the protocol conservatively reduces the reporting frequency f to conserve energy, while not compromising on the event estimation distortion. On the other hand, the protocol pursues more aggressive update policies in the network states with congestion and low event reliability—that is, high observed event distortion [19]. As a result, the spatiotemporal correlation conveyed in the physical characteristics of the phenomenon and deployment of the sensor network can be exploited in addressing the energy-efficient reliable event communication problem in wireless sensor networks.

5.5 CONCLUSION AND OPEN RESEARCH ISSUES

In addition to its collaborative nature, the existence of spatiotemporal correlation among the sensor observations is a significant and unique characteristic of the WSN. In this chapter, a theoretical analysis of spatiotemporal correlation characteristics in WSN is presented. It has been shown via mathematical analysis, their results, case studies, and discussions that correlation in WSNs can be exploited to significantly improve the energy-efficiency in WSNs. Therefore, this theoretical framework provides tools for finding the feasible operating region in terms of spatial and temporal resolution for a specific distortion constraint considering spatiotemporal correlation, signal properties, and network variables in WSNs.

Although there exists a considerable amount of existing research and studies on the correlation characteristics of sensor networks, there are many open research issues and new directions on this topic. One important research direction would be to obtain real sensor data and capture the spatiotemporal correlation behavior observed in different practical sensor network applications in order to enhance the correlation models developed so far. This will improve the accuracy of the correlation and distortion analysis derived in the current literature.

On the other hand, the effects of the network parameters such as topology, node distribution, radio, and sensing ranges of sensor nodes, heterogeneity of events occurring in the network need to be carefully investigated in order to reveal other important characteristics of spatiotemporal correlation in WSNs.

The correlation in WSNs can be considered in developing new energy-efficient networking protocols specifically tailored for the WSN paradigm. These protocols utilizing the correlation to conserve energy resources may drastically enhance the overall network performance. Furthermore, spatiotemporal correlation could be a baseline for cross-layer design of energy-efficient communication techniques for wireless sensor networks.

Moreover, the spatiotemporal correlation characteristics can also be exploited for efficient distributed source coding and information processing techniques. To this end, new spatiotemporal correlation modeling analysis can be developed for wireless multimedia sensor networks [20] that involve the communication of event data in the form of multimedia such as still image, video, and audio. Based on the spatiotemporal correlation models that can capture the unique communication paradigm of multimedia over WSNs, energy-efficient multimedia processing and communication algorithms can be devised for wireless multimedia sensor networks as well.

5.6 EXERCISES

1. Explain different types of correlation observed in wireless sensor networks. Provide a practical example for each case and discuss the main reasons for the observed correlation in the examples you provide.
2. What is the relation between the correlation and the distortion observed in the estimation of event features in a given sensor network? Propose another simple (yet practical) definition for distortion [as in (5.9)] which could be used as a reliability indicator in wireless sensor networks.
3. Derive the distortion function in (5.12) using the Spherical Covariance Model instead of the Power Exponential Model. Obtain the distortion as a function of number of nodes M that send information to the sink and correlation coefficients and comment on the results comparing them with Figure 5.2.
4. In the analysis of temporal correlation in WSNs, it is assumed that the observed signal $s(t)$ is wide-sense stationary (WSS). What is the effect of this assumption? How could this assumption be relaxed without losing the validity of the results of the analysis?
5. Note that the distortion analysis in this chapter does not explicitly incorporate the effects of channel and network capacity on the successful reception performance of the transmitted samples. Propose a modification to the model presented in Section 5.2 in order to include the effects of channel error rate and network congestion on the derived distortion functions.
6. Based on the spatiotemporal correlation characteristics and the distortion functions derived in this chapter, propose a new routing protocol for WSNs which can find the minimum energy and minimum distortion paths from the event field to the sink. Clearly state the assumptions you make and outline the algorithm you propose by explaining the rational behind your idea.

BIBLIOGRAPHY

1. I. F. Akyildiz, W. Su, Y. Sankarasubramaniam, and E. Cayirci. Wireless sensor networks: A survey. *Computer Networks (Elsevier) Journal*, **38**(4):393–422, 2002.

2. M. C. Vuran and O. B. Akan. Spatiotemporal characteristics of point and field sources in wireless sensor networks. In *Proceedings of IEEE ICC 2006,* Istanbul, Turkey, June 2006. IEEE CD-ROM.
3. M. C. Vuran, O. B. Akan, and I. F. Akyildiz. Spatiotemporal correlation: Theory and applications for wireless sensor networks. *Computer Networks Journal (Elsevier)*, **45**(3):245–261, 2004.
4. S. Meguerdichian, F. Koushanfar, M. Potkonjak, and M. B. Srivastava. Coverage problems in wireless ad-hoc sensor networks. In *Proceedings of IEEE INFOCOM 2001*, Anchorage, Alaska, April 2001.
5. J. Kusuma, L. Doherty, and K. Ramchandran. Distributed compression for sensor networks. In *Proceedings of IEEE Image Processing 2001,* Vol. 1, October 2001, pp. 82–85.
6. S. Bandyopadhyay and E. J. Coyle. Spatiotemporal sampling rates and energy efficiency in wireless sensor networks. In *Proceedings of the IEEE INFOCOM 2004*, Vol. 3, 2004, pp. 1728–1739.
7. R. Cristescu, B. Beferull-Lozano, and M. Vetterli. On network correlated data gathering. In *Proceedings of IEEE INFOCOM 2004*, Vol. 4, March 2004, pp. 2571–2582.
8. R. Cristescu and B. Beferull-Lozano. Lossy network correlated data gathering with high-resolution coding. In *Proceedings of IPSN 2005*, April 2005, pp. 218–224.
9. R. Cristescu and M. Vetterli. On the optimal density for real-time data gathering of spatiotemporal processes in sensor networks. In *Proceedings of IPSN 2005*, April 2005.
10. M. Gastpar and M. Vetterli. Source-channel communication in Sensor Networks. In *Proceedings of the 2nd International Workshop on Information Processing in Sensor Networks (IPSN'03)*, 2003.
11. M. Gastpar and M. Vetterli. Power, spatiotemporal bandwidth, and distortion in large sensor networks. *IEEE Journal on Selected Areas in Communications*, **23**(4):745–754, 2005.
12. J. O. Berger, V. de Oliviera, and B. Sanso. Objective bayesian analysis of spatially correlated data. *Journal of the American Statistical Association* **96**, 1361–1374, 2001.
13. S. S. Pradhan, R. Puri, and K. Ramchandran. n-Channel symmetric multiple descriptions, Part I: (n,k) source-channel erasure codes. *accepted for publication in IEEE Transactions on Information Theory,* 2003.
14. T. J. Goblick. Theoretical Limitations on the Transmission of Data from Analog Sources. *IEEE Transformation on Information Theory*, **IT-11**(4):558–567, 1965.
15. V. Poor. *An Introduction to Signal Detection and Estimation*, 2nd edition, Springer-Verlag, Berlin, 1994.
16. S. Haykin. Communication systems. 3rd edition, John Wiley & Sons, New York, 1994.
17. G. L. Stuber, *Principles of Mobile Communication*, Kluwer Academic Publishers, Norwell, MA, 2001.
18. M. C. Vuran and I. F. Akyildiz. Spatial correlation-based collaborative medium access control in wireless sensor networks. *IEEE/ACM Transactions on Networking,* **14**(2): 316–329, 2006.
19. O. B. Akan and I. F. Akyildiz. Event-to-sink reliable transport in wireless sensor networks. *IEEE/ACM Transactions on Networking,* **13**(5):1003–1016, 2005.
20. E. Gurses and O. B. Akan. Multimedia Communication in Wireless Sensor Networks. *Annals of Telecommunications,* **60**(7–8):799–827, 2005.

CHAPTER 6

A Taxonomy of Routing Protocols in Sensor Networks

AZZEDINE BOUKERCHE

School of Information Technology and Engineering, University of Ottawa, Ottawa, Ontario K1N 6N5, Canada

MOHAMMAD Z. AHMAD and DAMLA TURGUT

School of Electrical Engineering and Computer Science, University of Central Florida, Orlando, FL 32816-2362

BEGUMHAN TURGUT

Department of Computer Science, Rutgers University, Piscataway, NJ

6.1 INTRODUCTION

Sensors, in the sense of devices that perform the measurement of certain quantities and transform them into a computer readable digital format, have been around for at least a few decades. These sensors were either connected to a data collection device or connected directly to computers, using traditional wired communication such as a serial interface. The development of highly integrated computer devices, wireless radios, and miniaturization allowed the development of wireless sensor nodes, which are miniature computers with integrated sensing and wireless communication capabilities [1, 2]. The continuing miniaturization effort allowed the development of nodes with a physical size of several millimeters. Although many of these devices can act as a general-purpose computer, with the ability to perform computation as well as sensing, there are obvious limitations. First, the limited memory and computational power does not allow us to run a full-featured operating system. Most of the time, the networking stack needs to be simplified as well. The wireless transmission range is limited by the small size of the antennas. The power resources of the sensor nodes are limited by the physical size of the batteries; moreover, some of the proposed deployment models do not allow the sensor nodes to recharge their batteries.

Algorithms and Protocols for Wireless Sensor Networks, Edited by Azzedine Boukerche

Let us now consider the deployment models of the wireless sensor nodes. Classical sensors are typically used in pre-engineered deployment, being placed at a carefully chosen locations. For instance, the space shuttle uses several dozen temperature sensors, carefully positioned in the structure of the spacecraft and reporting through a wired connection to a central computer. Naturally, wireless sensor nodes can be deployed similarly, simply by replacing the wired connection with a wireless one. However, we can also choose a radically different deployment method: Blanket the desired area with a large number of nodes. Instead of careful positioning, we need to worry only about making sure that every area of interest is covered by one node (or preferably, several). The deployed nodes might not have the transmission range to reach the central computer, but they can transmit the collected information on a hop-by-hop basis to collection points called *sinks*. We call the resulting structure a "wireless sensor network (WSN)." In our example application, a WSN can provide several advantages: Due to the large number of sensors, it can collect more data than sensors with pre-engineered deployment. As several sensors cover the same area, they provide fault tolerance. In addition, having computational as well as sensing capabilities, the wireless sensor network can provide preliminary processing of the collected data concomitantly with the sensing and forwarding.

Let us now investigate the properties of a wireless sensor network from the networking point of view. First of all, there is no infrastructure available. Because the sinks are accessible only to a limited subset of nodes, the sensor nodes need to participate in the forwarding of the packets. Due to the random deployment, the routing architecture cannot be pre-established; the network needs to be set up through self-configuration. In these respects, wireless sensor networks are similar to ad hoc wireless networks.

There are, however, several important differences. The sensor nodes have significantly lower communication and computation capabilities than do the full-featured computers participating in ad hoc networks. The problem of energy resources is especially difficult. Due to their deployment model, the energy source of the sensor node is considered nonrenewable (although some sensor nodes might be able to scavenge resources from their environment). Routing protocols deployed in sensor networks need to consider the problem of efficient use of power resources.

An additional difference between ad hoc and sensor networks refers to the uniqueness of the nodes. Ad hoc nodes have a hard-wired unique MAC address, which forms the basis of node identification on the higher levels of the networking stack. The cheap, disposable sensor nodes usually come without any pre-wired identifiers; they acquire a unique identity only after deployment, by virtue of their position in the environment.

In addition to these, several other factors such as the large number of nodes in sensor networks, the high failure rates of the sensor nodes, and the frequent use of broadcasting in sensor networks as opposed to the typically unicast communication in ad hoc networks [3] require new types of MAC [4, 5] and routing protocols, specifically targeted toward the requirements of WSNs.

In this chapter, we succintly present the major applications of the WSNs, describe some of the design issues associated with routing algorithms for WSN, and finally present a survey of the state of the art in WSN routing protocols.

6.2 APPLICATIONS

Sensor networks can be deployed in a wide variety of applications. One of the main classification criteria is whether the sensor nodes are mobile or immobile. The data collection might be either continuous or periodic; the latter can lead to bursty traffic patterns. Naturally, every application requires a specific set of sensor types. Some of the most popular sensor types are: light, sound, magnetic field, accelerator, temperature, humidity, chemical composition such as soil makeup, mechanical stress levels on an object, and many others [6]. Some of the primary application domains for sensor networks are the following:

Environmental. Environmental sensors can be used to detect and track natural disasters such as forest fires or floods. They can also be used to track the movement of birds and other animals.

Military. The sensor networks will be an integral part of the future C4ISRT systems (command, control, communications, computing, intelligence, surveillance, reconnaissance, and targeting.) They can, for instance, be used to track the movement of the enemy in the battlefield. The main advantage of WSNs is that they can be deployed and operated remotely, without putting human lives at risk. Naturally, military deployments bring their own challenges of security and confidentiality.

Health. Sensor networks can be used in hospitals and clinics for patient monitoring and tracking of various systems and humans. Sensors can be also used to track and monitor the drug doses prescribed to patients and prevent situations where the drugs are administered to the wrong patient. Sensors can be deployed for telemonitoring of patients, a promising new direction for at-home monitoring and care for the elderly.

Home. The various home appliances can be sensor enabled and interconnected with each other and a central control system of the home. These sensor-enabled sensor homes might not only offer additional conveniences, but will also be safer and more energy-efficient.

6.3 DESIGN ISSUES

The challenges posed by the deployment of sensor networks is a superset of those found in wireless ad hoc networks. Sensor nodes communicate over wireless, lossy lines with no infrastructure. An additional challenge is related to the limited, usually nonrenewable energy supply of the sensor nodes. In order to maximize the lifetime of the network, the protocols need to be designed from the beginning with the objective of efficient management of the energy resources [3]. Let us now discuss the individual design issues in greater detail.

Fault Tolerance. Sensor nodes are vulnerable and frequently deployed in dangerous environment. Nodes can fail due to hardware problems or physical damage or by exhausting their energy supply. We expect the node failures to be much higher than the one normally considered in wired or infrastructure-based wireless networks. The protocols deployed in a sensor network should be able to detect these failures as soon as possible and be robust enough to handle a relatively large number of failures while maintaining the overall functionality of the network. This is especially relevant to the routing protocol design, which has to ensure that alternate paths are available for rerouting of the packets. Different deployment environments pose different fault tolerance requirements.

Scalability. Sensor networks vary in scale from several nodes to potentially several hundred thousand. In addition, the deployment density is also variable. For collecting high-resolution data, the node density might reach the level where a node has several thousand neighbors in their transmission range. The protocols deployed in sensor networks need to be scalable to these levels and be able to maintain adequate performance.

Production Costs. Because many deployment models consider the sensor nodes to be disposable devices, sensor networks can compete with traditional information gathering approaches only if the individual sensor nodes can be produced very cheaply. The target price envisioned for a sensor node should ideally be less than $1.

Hardware Constraints. At minimum, every sensor node needs to have a sensing unit, a processing unit, a transmission unit, and a power supply. Optionally, the nodes may have several built-in sensors or additional devices such as a localization system to enable location-aware routing. However, every additional functionality comes with additional cost and increases the power consumption and physical size of the node. Thus, additional functionality needs to be always balanced against cost and low-power requirements.

Transmission Media. The communication between the nodes is normally implemented using radio communication over the popular ISM bands. However, some sensor networks use optical or infrared communication, with the latter having the advantage of being robust and virtually interference free.

Power Consumption. As we have already seen, many of the challenges of sensor networks revolve around the limited power resources. The size of the nodes limits the size of the battery. The software and hardware design needs to carefully consider the issues of efficient energy use. For instance, data compression might reduce the amount of energy used for radio transmission, but uses additional energy for computation and/or filtering. The energy policy also depends on the application; in some applications, it might be acceptable to turn off a subset of nodes in order to conserve energy while other applications require all nodes operating simultaneously.

6.4 SENSOR NETWORKS ROUTING PROTOCOLS

Routing has been carried out in sensor networks with the emphasis on conserving energy. Power efficiency is the most important design metric because it is an

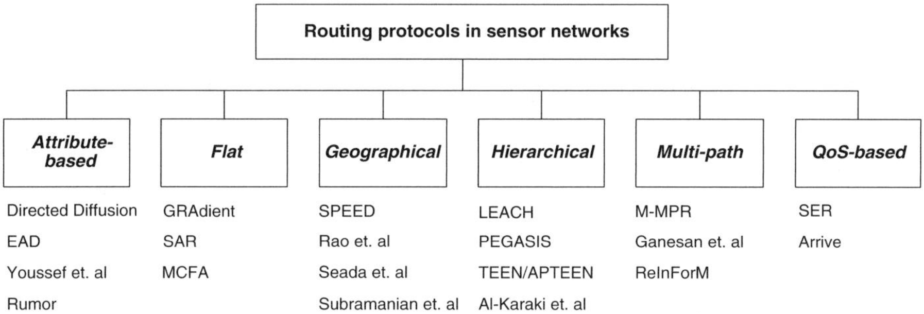

Figure 6.1. Categories of sensor routing protocols.

important design goal of any sensor network. Routing algorithms are also data-centric and employ attribute-based addressing strategies along with location awareness. These can be used with various clustering and hierarchical approaches to make an efficient routing algorithm for sensor networks. A robust and scalable stategy is required in designing a routing protocol that is also energy-efficient with minimal control overhead. Different routing protocols in sensor networks can be divided into six categories based on their underlying architectural framework. These different categories are as follows (see Figure 6.1).

- Attribute-based (Section 6.4.1)
- Flat (Section 6.4.2)
- Geographical (Section 6.4.3)
- Hierarchical (Section 6.4.4)
- Multipath (Section 6.4.5)
- QoS-based (Section 6.4.6)

Some prior studies discussing the various sensor network routing protocols have been published [3, 7–9]. We list the classifications of sensor network routing protocols mostly based on the type of deployment of these networks. This classification helps the advent of newer ideas with special focus on the application being serviced by the sensor network and hence develops efficient algorithms to facilitate better routing techniques in these networks.

6.4.1 Attribute-Based Protocols

Attribute-based routing protocols in sensor networks concentrate on routing data packets based on the content of the packets and are not device-specific. Hence, these are also known as data-centric routing approches. Since each node within the network is engaged in the routing mechanism, these nodes can make any decision and apply any routing rules to packets—that is, either forward or drop the packets. In this class of routing algorithms, contents of the transmitted data are evaluated at each hop in the network.

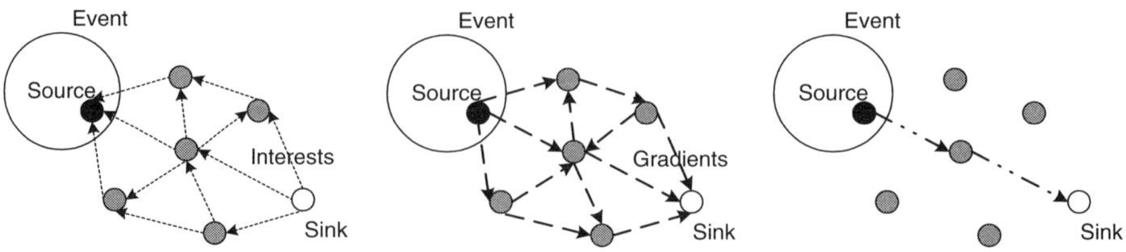

Figure 6.2. Directed diffusion [10].

***Directed Diffusion* [10].** Directed diffusion (Figure 6.2) is a data-centric approach, which means that all communication is for *named data*. It can be categorized under both attributed-based routing and flat routing protocols. Data generated by the application-aware sensor nodes is named using attribute–value pairs. A node requests data by sending interests for named data. A sensing task is disseminated via sequence of local interactions throughout the network as an interest for named data. Nodes diffusing the interest set up their own caches and gradients within the network to which the data delivery is carried out. During the data transmission, reinforcement and negative reinforcement techniques are used to converge to efficient distribution. Intermediate nodes fuse interests and aggregate, correlate, or cache data. Algorithm 1 presents a pseudocode of the directed diffusion algorithm.

ALGORITHM 1. Pseudocode Describing the Directed Diffusion Algorithm

```
Setup phase:
1.    The base station broadcasts its set of interests
2.    Do
3.     ForEach network node N receiving an interest from node M
4.      N forwards the received interest to its neighbors
           (other than M)
5.      N sets up a gradient with M
6.     EndFor
7.    Until all gradients are set up
8.    Check for loops in the paths and remove them

Operating phase:
1.   ForEach node
2.     Colect sensor data.
3.     Receive messages containing sensor data readings.
4.     Aggregate, correlate or fuse data (if necessary)
5.     If data maches an interest
6.      Forward the data according to the gradient associated
          with the interest
7.     EndIf
8.   EndFor
```

Energy-Aware Data-centric Routing (EAD) **[11].** This protocol proposes to build a virtual backbone which contains all the active sensors in the network to facilitate energy aware routing. A routing heuristic allows us to build a broadcast tree rooted at a gateway to enable the data-centric approach. The tree spans all sensors within the network and has a large number of leaves that save power by turning off their radios. The active sensors continue working as relays for traffic generated in the network.

In data-centric routing, backbone senders are in charge of data processing and information dissemination throughout the network. At each individual sensor, the local raw data is initially combined/aggregated with data from other sensors located farther away from the sink. This aggregated data are then sent to a sensor closer to the sink or to the sink itself. EAD consists of two main components: the *neighboring broadcast scheduling* and the *distributed competition* among neighbors. These components ensure that the final tree has many leaves, and sensors with relatively higher residual power also have a greater chance of being part of the virtual backbone. EAD basically makes certain the formation of a specially rooted broadcast tree designed for data-centric routing. This protocol is suitable for applications requiring frequent queries and events.

Constrained Shortest-Path Energy-Aware Routing **[12].** The distance from a source to a destination can be used as a metric for energy consumption and estimation of the propagation delay between them. It is shown that by changing the transmission power level and thereby changing the network topology graph, the algorithm can be optimized to ensure higher throughput and energy efficiency while maintaining lower end-to-end delay.

The nodes are grouped into clusters with each cluster having a clusterhead or gateway node. Routing decisions are determined and maintained at the clusterhead. Such a centralized approach is more efficient than a distributed approach since it entails less control packet overhead maintainance. The network operates in two main cycles: *data* and *routing*. The data cycle consists of the nodes sending data to the gateway nodes, whereas during the routing cycle the routing state of each node is determined by the clusterhead and the routing information is sent to all the nodes accordingly.

The constrained shortest-path algorithm uses the distance between any two nodes to determine transmission power required to send packets from one to another. The transmission energy varies inversely with d^n, where d is the distance between the transmitter and receiver and n is a value based on the system and application in use. The connectivity between nodes in a cluster can be maintained by this energy parameter, making the network topology dynamic. If no constraints are enforced for the transmission energy, then each node uses its maximum energy to transmit directly to the destination. This is certainly not maintainable in the long run; therefore, a constraint has to be put in place. Rerouting is carried out by the clusterhead if (i) the sensors within the clusters are reorganized, (ii) the battery level of the active nodes falls under a certain threshold, or (iii) there are some changes required in the energy model of the nodes.

***Rumor* [13].** Rumor allows the routing of queries to nodes that have observed an event of interest. As a result, retrieval of data is based on events and not on an addressing scheme. An event is an activity related to the phenomena being sensed (e.g., increased movement in an area being monitored). Events are assumed to be localized phenomena that occur in fixed regions of space. A query is issued by the sink node for one of two reasons: as an order to collect more data or as a request for information. Once a query arrives at its destination, data are issued to the originator of the query. Depending on the amount of data (whether it is more or less) being issued to the originator of the query, shorter paths from the source to the sink are discovered.

Any node may generate a query for a particular event. If it knows the route to the event, it transmits the query. Otherwise, the query is sent in a random direction, and this continues until the query reaches a node that has a route to the event.

Each node has its neighbor list and an events table with forwarding information to all the events it knows. After a node witnesses an event, an agent may be created, which is a long-lived packet and travels around the network. Each agent contains an events table, including the routing information for all events it knows. Since an event happens in a zone, composed of several or many nodes, it is possible that multiple agents are created from the zone and moving in the network. When an agent travels in the network, its routing table is updated if there exist a shorter path to an event within the routing table of the node it is visiting. In a similar way, the routing table of the currently visited node is updated if its route to an event is more costly than the agent's route. See Figure 6.3.

Rumor routing uses agents that have a limited life determined by a TTL field; these agents create paths in the direction of any events they may come across. If an agent crosses a path to an event that it has not yet come across in the network, it creates a path that leads to both events.

If flooding was to happen on a regular basis, network resources would be consumed quickly, thus Rumor routing was created to be an alternative to flooding queries and events. When a query is generated, it is sent randomly through the network until it finds the event path instead of flooding it. When the query finds the event path, it is

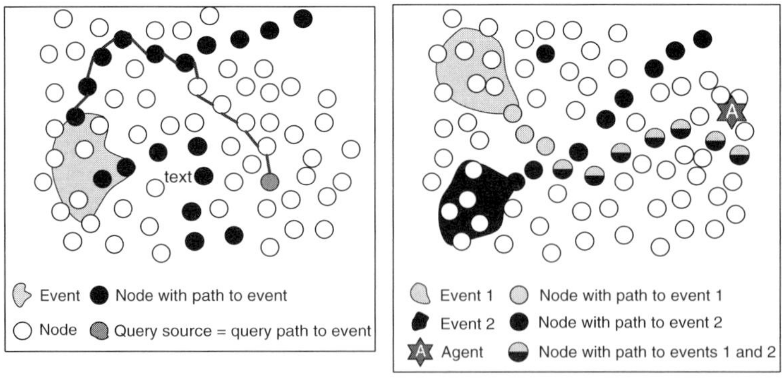

Figure 6.3. Rumor [13].

routed directly to the event. Only if the path cannot be found, it will be flooded as a last resort.

6.4.2 Flat Protocols

In flat sensor networks, there is a large number of nodes that collaborate together to sense the environment. These nodes are similar to each other in all respects; and due to their sheer number, they cannot be assigned specific global IDs. Routing protocols servicing such flat networks fall under this category of sensor networks.

***Gradient Broadcast (GRAB)* [14].** This protocol is designed for efficient data forwarding in large-scale dense sensor networks where particular objects or events called *stimuli* are monitored by all the sensor nodes in the neighborhood. An earlier version of this work has been reported in reference 15. These sensors pick up the same stimuli simultaneously but collectively elect a leader that creates a sensing report on behalf of the entire group. The election is carried out by the signal strength of the field created by the stimulus. All the nodes picking up the stimulus broadcast their respective signal strengths with a random delay to avoid collisions. Whenever a node hears this signal from another node, the node checks the strength of the signal and compares it with its own. If it is lower than the received signal, it stops; otherwise it rebroadcasts its own signal strength. This helps in rolling the messages to the center of the stimuli (CoS), where ultimately the leader is chosen. This is the node with the largest signal strength, and then it generates the report. This leader is named as the *data source*. On the other hand, each node always maintains a *cost* value, a metric representing the cost of sending data from the node to the destination sink. It is the job of the sink to build and maintain this cost field associated with each node, but the nodes also know their respective costs. Nodes located closer to the sink obviously have lower cost values than ones that are located farther away. The main difference of GRAB from other protocols is that instead of the sender deciding the path for the data packet, it is the intermediate nodes who decide whether or not they should be forwarding the packet to the sink. This decision is made by the receiver nodes simply by comparing their respective cost values with that of the sender; and if it is lower, then they accept forwarding the packet. Due to this inherent mechanism, data will always be forwarded on the shorter path—that is, the path of descending cost to the sink.

However, some small issues come up if there are multiple paths to the destination. To solve some of these problems, the source also assigns a *credit* value along with the report it generates and sends out for transmission. With this credit value, a report may be sent out on a mesh of interleaved paths where none of the paths have a total cost greater than the cost of the source in addition to its credit. This simply makes use of the multiple paths if available and hence introduces a certain degree of robustness into the protocol. This degree of overhead and robustness is the "width" of the mesh, which is ultimately dependent on the credit of the report. This credit value also helps in controlling the robustness degree, which may be useful for different scenarios and events occurring within the sensor network. No state information is required to be maintained at the individual nodes, thereby eliminating overhead due to network complexity and path maintenance.

***Sequential Assignment Routing (SAR)* [16–18].** This is the very first sensor routing algorithm to include QoS concepts. It is mainly a table-driven multipath approach toward routing and concentrates on energy efficiency and fault tolerance. Path selection is made by considering the energy resources of the nodes on the path along with the priority levels of each packet. The packet priority is calculated using an innovative technique. There is an energy cost and delay value associated with each link, which provides a *resistance* to packet flow. It is against this metric that a packet will have credits so that it achieves a certain priority in using these paths. A weighted QoS metric, which is the product of the additive QoS metric mentioned earlier and a certain weight coefficient, *measures the QoS provided to each packet relative to the priority level of the packet* [16]. The main objective of the SAR algorithm is to minimize this average weighted QoS metric.

The SAR algorithm creates trees that have roots at the one-hop neighbors of the sink by taking the priority level of each packet, the QoS metric, and the energy resources for the nodes on each path. With these trees, multiple paths are created and maintained from source to destination. When a path fails, the recovery procedure is carried out by forcing updates of the routing table between upstream and downstream nodes. Hence, there is always an available path for sending the packet, and any path failure always leads to a local path restoration being carried, which mainly means updating the respective routing tables for the nodes.

The major disadvantage of this protocol is that each node has to maintain entire routing tables of not only one path between a set of sources and destinations, but many available paths. This leads to greater overhead and memory requirements and also leads to more energy waste, especially in case of dense networks.

***Minimum Cost Forwarding Algorithm (MCFA)* [15].** This paper presents a cost-field approach to minimum cost forwarding in sensor networks. The authors propose a "cost field," which is defined as "the minimum cost from that node to the sink on the optimal path." Since the direction of the routing is always known, the sensor node is not required to have unique ID or maintain a routing table, but rather each node maintains the minimum cost estimate from itself to the sink. Each sensor node broadcasts the message to be forwarded to its neighbors. When the neighbor node receives the message, it verifies if it is on the minimum cost path toward the sink. If so, it broadcasts the message to its neighbors. This forwarding scheme continues until the message reaches to the sink.

6.4.3 Geographical Routing

Geographical routing uses location devices to estimate the location of a node prior to forwarding the packets to the destination region. In wireless sensor networks, geographical forwarding (GF) is a widely used approach due to its low overhead and localized interactions. In GF, location information is exchanged with the neighbors, and the neighbor closest to the destination is selected as a result of forwarding decisions.

***Speed* [19].** This is the stateless protocol for soft real-time communication in sensor networks. Each node maintains information about all of its neighbors and uses a geographic forwarding technique to find the paths. The SPEED protocol tries to ensure a certain speed for each packet sent such that the application has an idea of the average end-to-end delay between the nodes. If the routing protocol makes the estimated speed of the packet available, the application can calculate this delay beforehand, simply dividing the distance between the source and the sink by the estimated speed. The protocol also provides congestion avoidance mechanism when needed.

The routing component, called the stateless geographic nondeterministic forwarding (SNFG), in SPEED uses four modules listed below to function effectively.

1. *Beacon Exchange*: Similar to any other geographic routing algorithm, location information is updated periodically by nodes using specific beacon packets broadcast. Two other types of on-demand beacons in SPEED include *delay estimation* and *backpressure* beacons. These are mainly used to quickly identify traffic changes in the network.
2. *Delay Estimation*: This is carried out at each node by calculating the time elapsed since the acknowledgment receipt of the last data packet sent. By observing these delay values, the SNFG selects the best node conforming to the speed requirement of the application.
3. *Neighborhood Feedback Loop*: If no node with an appropriate delay value is obtained, this module is consulted for the *relay ratio* of the node. This ratio is simply the miss ratio of all the neighbors of the node—that is, the nodes that were unable to provide the desired speed for the application. A random number between zero and one is generated; and if the relay ratio is less than the random number generated, the packet is dropped.
4. *Backpressure Rerouting*: This module helps to reduce congestion by sending messages to the source nodes such that alternate routes can be discovered.

Due to the stateless architecture of SPEED, it requires minimal memory and minimal MAC layer support. It also provides traffic load balancing and also enables QoS routing and congestion management in the sensor network applications.

***Geographic Routing with No Location Information* [20].** This is a geographic routing algorithm that does not use any location information. The previous geographic protocols presented always assume the prescence of location identification techniques; however, this requirement cannot always hold. This scheme assigns *virtual* *coordinates* in such a scenario in which any other geographic routing protocols can be deployed without any location information.

The protocol has three stages. The initial assumptions are gradually removed as the protocol proceed further. It is essential to point out that the virtual coordinates assigned to the nodes do not have to accurately match to the geography of the places, but they should reflect the underlying connectivity. The local connectivity information is used to set these coordinates. The three stages of the protocol are described below in detail.

1. *Perimeter Nodes Know Location*: Initially, it is assumed that all perimeter nodes know their locations and the nonperimeter nodes are assigned virtual coordinates. The analogy used, which is borrowed from graph theoretic approaches, states that "each neighbor relation is represented by a force that pulls the neighbors together." Thus, the force in the x and y direction is proportional to the difference in x and y coordinates of the nodes, respectively. Iteratively, each node updates its virtual coordinates by calculating the average of x and y coordinates of the neighbor set of each node. As the number of iterations increase, the nodes tend to move toward the perimeter nodes closest to them, and ultimately the algorithm converges to a steady state where the nodes are spread throughout the region.
2. *Perimeter Nodes Are Known*: The perimeter nodes are aware that they are on the perimeter; but unlike the first step, they do not know their location. Each perimeter node broadcasts a HELLO packet to the entire network to discover the hop distance to every other perimeter node. These distance pairs are stored in a *perimeter vector* in which each perimeter node in turn broadcasts its perimeter vector to the entire network. As a result, every perimeter node becomes aware of its hop distance from other perimeter nodes. A triangulation algorithm is executed to enable the nodes to compute the coordinates of all the perimeter nodes in the network. The triangulation algorithm could fail under some conditions. As a part of the solution to this problem, two nodes are designated as *bootstrap beacons*. These nodes flood the network with HELLO messages. The perimeter nodes include these beacons in their triangulation algorithm to compute the coordinates of all the perimeter nodes.
3. *No Location Information*: The nodes use the following criteria as stated in reference 20 to decide if they are perimeter nodes: *If a node is the farthest away, among all its two-hop neighbors from the first bootstrap node, then the node decides that it is on the perimeter.*

***Energy-Efficient Forwarding Strategies for Geographic Routing in Lossy Wireless Sensor Networks* [21].** The main aspect of geographic forwarding includes the greedy forwarding of message packets by a node, usually to the neighbor located closest to the destination. This strategy works only under the following assumptions: (i) sufficiently dense network; (ii) accurate localization; and (iii) high link reliability irrespective of the distance within the physical radio ranges. It has been observed that even though the first two assumptions may hold for most types of systems, the final assumption regarding high reliable links will not be fully satisfied in a realistic scenario. This is because wireless links are inherently unreliable in nature and are subject to various environmental conditions causing the weak links to drop a high number of packets. These packets are then retransmitted, which leads to high energy consumption. In order to alleviate the problem, the neighbors are classified based on link reliability. Not only could some neighbor links be weaker than the others, their loss characteristics could also be different. Hence, a *blacklisting/neighbor selection* scheme is devised to avoid the weak and unreliable links. The *distance-hop energy tradeoff* performance metric is proposed for geographic forwarding. If the forwarding scheme aims at minimizing hops as in greedy forwarding, a significant energy consumption can occur due to transmitting and/or retransmitting the packets on long and possibly unreliable weak links. On the other hand, if smaller distances are covered across stronger links, more hops would be visited again, causing increased energy usage. The optimal choice is generally in the transitional region between these two strategies. However, let us note that not too many links should be blacklisted since this would cause greater route disconnections and lower delivery rates.

***Geographic Routing with Limited Information in Sensor Networks* [22].** This protocol shows that even for instances where there is erroneous or limited location information due to faulty GPS, the order of routing delays is within a constant factor of straight-line greedy routing strategies. The limited information means that nodes only know the relative quadrant or zone where the destination node is situated. Nodes can be forwarded in the wrong direction if the GPS at nodes are faulty or biased even when the nodes have the correct destination coordinates. The four basic scenario models are discussed below.

1. *Location Errors (Imprecise GPS)*: An angular error is introduced within greedy straight-line routing so that the packet is forwarded anywhere in the zone within the angle.
2. *Limited Destination Information*: It is assumed that the node has only a coarse estimate of the destination location such as the quadrant or half-plane information.
3. *Small Fraction of Nodes with Routing Information*: In this scenario, only a small fraction of nodes know about the destination quadrant. If a node that has no information about the destination has a packet to send, it simply selects a neighbor randomly to forward the packet.
4. *Throughput-Capacity Networks*: The packet is forwarded in the right direction but not in a straight line along the shortest path. It uses specific *progressive* routing strategies to selectively forward the packets toward the destination. This leads to spatial "hot spots" within the network because many paths may intersect due to suboptimal selection of routes.

6.4.4 Hierarchical Protocols

As the name suggests, this class of routing techniques create a virtual hierarchy among the nodes of the sensor network. Such a hierarchy may be based on zones which comprise of divisions of the entire network area, node functionality or node location. Irregardless of the system creating the network hierarchy, the routing protocol must be designed to make maximum use of this hierarchical pattern.

***Low-Energy Adaptive Clustering Hierarchy (LEACH)* [23].** LEACH is based on a simple clustering mechanism by which energy can be conserved since cluster heads are selected for data transmission instead of other nodes in the network (Algorithm 2). By the received signal strengths, local cluster heads are selected to serve as the routers to the data sinks. A sensor node sends its data to the local cluster head in turn transmitted to the nearest cluster head on the way to the sink. Since the cluster heads are only responsible for bulk of the data transfer, the overhead is minimized; however, if the cluster heads are chosen beforehand and remain fixed throughout the network lifetime, they will easily die out, thereby ending the lifetime of the member nodes of the particular cluster as well. To solve this problem, LEACH performs a periodic randomized rotation of the cluster head to enable all the nodes within the cluster to take on a collective responsibility in order not to drain the battery of a single node. The optimal number of cluster heads is considered 5 percent of the total number of nodes.

LEACH also performs local data fusion and aggregation to compress the data received from each cluster. Sensor nodes are selected as cluster heads by the node choosing a random number between zero and one. The node is selected as a cluster head for the current round if the number is less than the following threshold values:

$$T(n) = \frac{p}{1 - p * (r \bmod \frac{1}{p})} \quad \text{if} \quad n \in G$$

where p is the desired percentage of cluster heads, r is the current round, and G is the set of nodes that have not been cluster heads in the last $\frac{1}{p}$ rounds.

Once all the nodes are organized into clusters, the cluster head will create a schedule for the nodes in its cluster which enables the radio components of each cluster node to be turned off for most of the time. Each node transmits its data to the cluster head according to its schedule. On completion, the cluster head aggregates and sends all the data to the sink.

LEACH achieves a significant reduction in energy dissipation when compared with direct communication and other minimum energy routing protocols. Properties of LEACH includes (i) dynamic clustering to increase network lifetime; (ii) single-hop routing from node to cluster head, hence saving energy; (iii) distributiveness; (iv) additional overhead due to cluster head changes and calculations leading to energy inefficiency for dynamic clustering in large networks.

ALGORITHM 2. Pseudo-code Describing the Operation of the LEACH Protocol

```
Setup Phase:
In this phase clusters are created---cluster heads (CHs) are
chosen

ForEach (node N)
  N selects a random number r between 0 and 1
  If (r < Threshold value)
    N becomes a CH
    N broadcasts a message advertising its CH status
  Else
    N becomes a regular node
    N listens to the advertising messages of the CHs
    N chooses the CH with the strongest signal as its
      cluster head
    N informs the selected CH and becomes a member of
      its cluster
  EndIf
ForEach (clusterhead CH)
  CH creates a TDMA schedule for each node to transmit
    data
  CH communicates the TDMA schedule to each node in the
    cluster
EndFor

Steady State Phase:
ForEach (regular node N)
  N collects sensed data
  N transmits the sensed data to the CH in the
     corresponding TDMA time slot
EndFor
ForEach (cluster head CH)
  CH receives data from the nodes of the cluster
  CH aggregates the data
  CH transmits the data to the base station
EndFor
```

***Power-Efficient Gathering in Sensor Information Systems (PEGASIS)* [24].** PEGASIS forms chains of the sensor nodes instead of forming multiple clusters as performed in LEACH protocol. Each node in the chain can transmit and receive data from its neighbors. In the entire chain, one node is selected to transmit all the data received to the sink or base station. The chain construction follows a greedy approach. The problem of building a chain to minimize the total length is similar to the traveling salesman problem.

The elected local leader in the chain waits for data from its closest neighbors. These neighbors first receive data from their own respective closest neighbors and aggregate the data before transmitting to the leader. The leader then sends the data received from its closest neighbors to the sink.

Even though PEGASIS is similar to LEACH, it differs in the following ways. It uses multihop routing while only one node is selected to transmit to the base station. The reduction in overhead due to dynamic clustering as in LEACH leads to a performance gain of almost 100 to 300 percent in PEGASIS. This overhead is reduced when the following occur: (i) Transmission distances of non-leader nodes are minimized. (ii) One transmission is made to the sink per round by aggregating all the data. This reduces local energy consumption but introduces a large delay for nodes farther away from the leader node of that chain. It also results in a single point of failure by the bottleneck created at the chain leader. (iii) The number of transmission among the nodes are reduced leading to overall energy efficiency.

Threshold-Sensitive Energy-Efficient Sensor Network Protocol (TEEN) [25]. The TEEN protocol is designed to respond to sudden changes in the sensed attributes and uses a hierarchical model along with a data-centric mechanism. Clusters are formed in a hierarchical fashion at different levels with elected clusterheads serving as communication links between each other and the data sink.

Initially, the clusters are formed after which each clusterhead broadcasts two threshold values to all the nodes. These are *hard* and *soft* thresholds for the sensed attributes. The hard threshold is the minimum possible value of an attribute based on which the sensor will be transmitting data to the sink. When the sensed value of the attribute is greater than this threshold, the data are sent to the clusterhead. This enables the nodes to transmit only relevant data. Once a value above the hard threshold is sensed, the node checks if the difference in the current and earlier values is greater than the soft threshold; if so, the new data are transmitted. Hard and soft threshold values can be adjusted per requirements allowing to control the packet transmissions.

This protocol is succeeded by the **adaptive threshold-sensitive energy-efficient sensor network protocol** (APTEEN) [26], which aims at capturing periodic data collections and respond to time-critical events. While the architecture remains the same as TEEN, APTEEN supports three types of data query: (1) *historical*, to analyze and monitor past data values and take decisions based on these recorded values; (2) *one-time*, to take a snap view of the current network situation and visualize it at a particular time instant; (3) *persistent*, to monitor the network over a continuous time interval especially during an event taking place.

Data Aggregation—Exact and Approximate Algorithms [27]. The aim is to develop better data aggregation and in-network data processing schemes for energy savings in sensor networks. These lead to lesser packet transmissions and reduce redundancy, thereby helping in increasing the network lifetime.

The protocol employs a hierarchical model that uses data aggregation and in-network processing at two different levels of the network hierarchy. A set of nodes, called the *local aggregators* (LA), are elected. These form a backbone routing architecture on which the first instance of data aggregation and routing is performed. From the set of LAs, another set of aggregators, called *master aggregators* (MA), are selected to form the second level of nodes to carry out the second level of data aggregation. Since choosing the optimal set of MAs is an NP hard problem, a separate integer linear program (ILP) is developed to identify the optimal MA set. The goal of maximizing network lifetime must be considered during the selection process of MAs.

The time required to solve the ILP increases exponentially with the number of LAs. Three near-optimal approximation algorithms, which serve dual purposes of selecting MAs and employing routing to the external traffic sink, are proposed. The differences from conventional data aggregation approaches are as follows: (i) Data aggregation is performed at optimal network points rather than at arbitrary points in other approaches. The selected points maximize the network lifetime. (ii) Routing and data aggregation are carried out simultaneously with the proposed algorithms. (iii) There are no additional overheads due to dynamic clustering and topologies. The architecture is simple and uses a fixed hierarchy of nodes with no additional complications.

6.4.5 Multipath Routing

Multipath routing techniques compute multiple paths from source to destination to effectively route around failed nodes or invalid links. In single-path routing protocols, if a link fails, additional control packets have to be generated and broadcast to discover newer paths. Multipath routing avoids such pitfalls, and there is always a secondary route ready in the event of a route failure.

***Meshed Multipath Routing (M-MPR)* [28].** Two ways of affecting disjoint multipath routing (MPR) include (i) *Disjoint (or split) MPR (D-MPR) with selective forwarding (SF)* in which case each packet is sent along different disjoint routes and the decision of path selection is made by the source on packet-by-packet basis and (ii) *D-MPR with packet replication (PR) (or limited flooding)* where multiple copies of a data packet are transmitted simultaneously along multiple disjoint routes from a source to a destination.

A meshed multipath is set up in three steps: (i) acquiring neighborhood information, (ii) route discovery, and (iii) route reply as described below.

- *Acquiring Neighborhood Information*: Each active node broadcasts its ID, residual battery power, and location information to local neighbors. For each active neighbor i, a node maintains the following information in its database: ID_i, $location_i$, $residual\_power_i$. Since the sensor nodes are considered stationary, period update on neighborhood status is not needed unless the node is

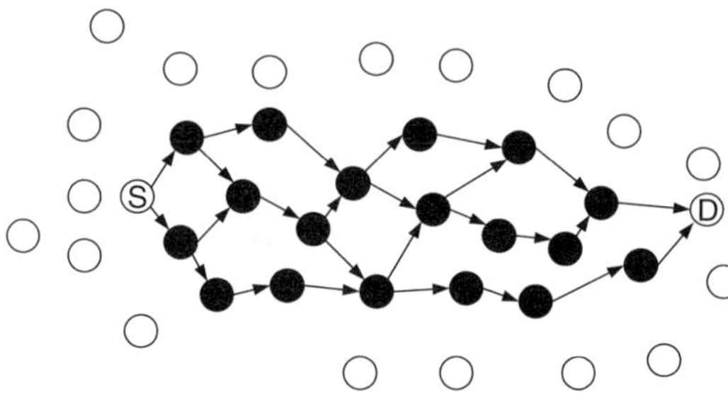

Figure 6.4. M-MMPR: Source-to-destination meshed multipath [28].

entering to sleep mode or has just woken up. In this case, the nodes status is locally broadcast based on which of the neighborhood tables of nearby nodes are updated.

- *Route Discovery*: Each node attempts to form a meshed multipath based on the neighborhood database and location information of the controller node. So far, the intermediate node is allowed to accept multiple discovery packets. During source-to-destination route discovery process, at most two copies of a discovery packet are accepted by an intermediate node, and the first arrived packet is forwarded to maximum two downstream neighbors nodes to ensure the reduction of the receiver complexity and power consumption of a node as can be seen from Figure 6.4. A maximum two forwarding node is chosen since this allows an alternate route with minimum possible extra control overhead.

  The route packet has the following fields: *source_ID*, *source_location*, *intermediate_node_ID*, *next_node_ID*1, *next_node_ID*2, *destination_ID*, *destination_location*, *and TTL*, where the IDs of forwarding nodes (*next_node_IDi*, $i = 1, 2$), *intermediate_node_ID*, and *TTL* values are updated at each intermediate stage.

  Each intermediate node maintains the following information in its routing database: *previous_node_IDi*, *previous_node_IDn*, *next_node_ID*1, *next_node_ID*2. Since several nodes are targeting the same destination, an intermediate node can have more than two "previous_node" entries in its routing table although there will be no more than two "next_nodes" as can be seen from Figure 6.5. The list of "previous_node" is bounded since the number of local

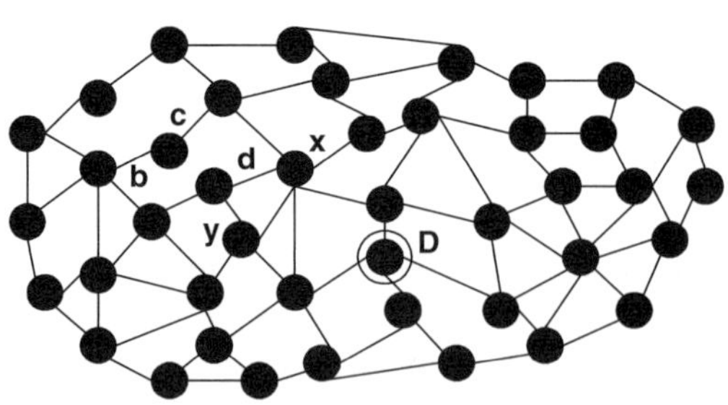

Figure 6.5. M-MPR. Meshed topology formed by many-sources-to-a-destination routes [28].

neighbors are finite and no entry is created in the routing table for discovery packets coming from an upstream neighbor which is already listed in the list. If an intermediate node that has already forwarded a discovery packet receives another discovery packet, it updates the "previous_node" list in its routing table and drops the packet.

Entry in the routing table at each node is maintained as a soft state that is deleted after a timeout unless a reply is received from a controller node. Since most sensor applications are data-centric, delay differences (jitter) between packet arrivals is not a big concern. No other resource reservation apart from storing and maintaining upstream and downstream nodes information is made during this phase; therefore, the route discovery phase can be considered as a *topology construction* process.

- *Route Reply*: Route reply message identifies the nodes comprised the meshed path. The controller node, upon receiving the discovery packets from a single source, selects the first two and sends a route reply following the original links by the route discovery packets in reverse direction with the following fields: *source_ID*, *source_location*, *intermediate_node_ID*, *previous_node_ID*1, *previous_node_ID*2. Each intermediate node changes the states of its corresponding entries from soft to permanent for the duration of its active participation, updates the fields of the reply packet other than the source information, and forwards the reply packet to its upstream node. When forwarding the route reply message, the node does not require the knowledge of source information.

  In case of the discovery packets arriving to controller node from several sensor nodes, multicast reply is used. If an intermediate node is out of service or goes to sleep mode, the upstream nodes select necessary neighbors to sustain connectivity. Intermittent "link breakage" will not trigger reconfiguration of meshed multipath, instead it is handled using selective forwarding. In the constructed meshed topology, the number of downstream links is no more than two, whereas the number of upstream nodes can be more. As can be observed from Figure 6.5, node n has three upstream nodes (a, b, and c) and two downstream nodes (x and y).

After the meshed multipath is constructed, the information packet is forwarded from the source to the destination using either the packet replication (PR) or selective forwarding (SF) variants. The methods are explained below in a nutshell.

- In PR, a source packet is copied from along all possible paths to its destination. To reduce power consumption due to transmission of multiple copies of the same packet, there is a provision of discarding the packets if a node receives more than one correct copy of the packet from one of the upstream nodes to be forwarded to the downstream node.

- In the SF variant of M-MPR, if more than one downstream node is available at either the source or an intermediate node, the information packet is forwarded to only one of the nodes based on local conditions.

In addition to a fault tolerance objective, the selective forwarding approach along meshed multipath offers more efficiency than PR in terms of resource utilization and congestion avoidance. The main difference between PR and SF is that in PR, the packet is intended for multiple neighbors in which each of them will receive and forward the packet whereas in SF, only one receiver will receive and forward the packet. Due to broadcast nature, M-MPR requires less transmission energy than D-MPR; MPR provides more flexibility in selective forwarding decisions than D-MPR, resulting in more successful packet delivery rate.

ALGORITHM 3. Pseudocode Description of the Meshed Multipath Routing Algorithm

```
Setup phase:

ForEach (node N)
   N broadcasts node information
   N listens to broadcast packets of neighboring nodes
   N adds the received information to the neighborhood
     database
EndFor
ForEach (node N)
     N forwards a route discovery packets to a maximum of
       2 hops
     N listens to route discovery packets and sends route
       reply packets
     N listens to route reply packets and sets up the
       meshed multipaths.
EndFor

Operating phase:

ForEach (node N)
   Select forwarding technique T.
   If T is packet replication
     N forwards data to all possible paths to the
       destination
   ElseIf T is selective forwarding
     N chooses a path to destination based on some criteria
     N forwards the data on the selected path
   EndIf
EndFor
```

***Highly Resilient, Energy-Efficient Multipath Routing in Wireless Sensor Networks* [29].** Energy-efficient multipath routing defines a set of localized algorithms to construct multiple paths from the source to the destination (Algorithm 3). These multiple paths may come in handy if one particular route fails during the network operation. Multipaths are constructed between nodes using two different approaches. The first is the original node *disjoint* paths where none of the paths created intersect with each other, meaning that they are completely disjoint. This ensures that for k paths constructed, no k node failures can eliminate all the paths. The second technique defines the *braided* paths in which there are generally no completely disjoint paths, but instead numerous partially disjoint alternative paths.

Conventional approaches to energy-efficient and robust routing have resulted in periodic flooding of the network, which in turn lead to energy inefficiency. By constructing alternate paths initially, there may be no need for periodic flooding of the data packets. In the event of the primary path failure along with all the other alternate paths simulatenously, there is no other option left but to flood the network to ensure that the data reach the destination. However, such a condition should not occur very frequently. The two mechanisms for identifying disjoint paths and braided paths are as follows:

- *Disjoint Multipaths*: These can be constructed by identifying a few alternate paths that are node disjoint with the primary path. Here, initially the primary path is selected between the source and the sink. This is followed by selecting the first alternate path as the best path, which is node disjoint with the primary path. Other secondary paths are selected based on the first disjoint path, and this process continues until the number of alternate paths reach the maximum specified limit. These disjoint paths are resilient, but generally energy-inefficient. See Figure 6.6 [29].
- *Braided Multipaths*: In this technique, the alternate paths created are not completely node disjoint from the primary path, but instead have a few nodes in common. Selecting a braided path is similar to selecting disjoint paths; however, each node on the primary path receiving a *reinforcement* propagates it to its most preferred neighbors, who in turn carry out the same operation. This ultimately leads to multiple alternate paths, some of which can be parts of the primary path and are not completely node disjoint. See Figure 6.7 [29].

In this system, two different modes are considered: *isolated* and *patterned* node failures. In the first case, each node has an individual probability of failure whereas in the second case, all nodes within a particular radius fail simultaneously.

***ReInForM* [30].** The reliable information forwarding using multiple paths (ReInForM) protocol brings to the forefront issues relating to information aware data delivery in sensor networks. It may happen that essential data are sent along a lossy and unreliable route. Let us look at an example in which a sensor network senses temperature at some points in the forest. On a given day, a sensor may sense a temperature of 60°F at a particular point that is normal. On the other hand, at that very

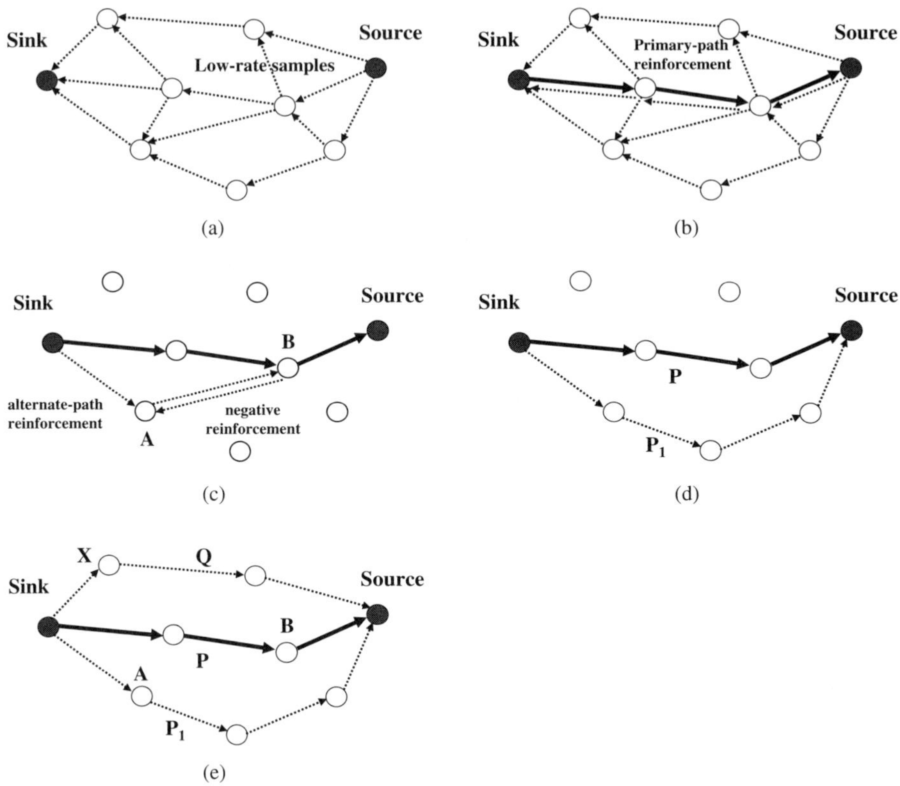

Figure 6.6. Construction of localized disjoint paths. Diagram (a) presents the low-rate date sample. Primary path P and the alternate path negative reinforcement are shown in diagrams (b) and (c), respectively. The alternate path P_1 and the idealized algorithm are given in diagrams (d) and (e) [29].

moment, another sensor placed some distance away may be sensing a temperature of 1000°F. This packet containing the abnormal temperature is the more important packet and should be sent to the sink by more reliable links to ensure delivery with the lowest latency. Basically, the proposed mechanisms support delivery of data at

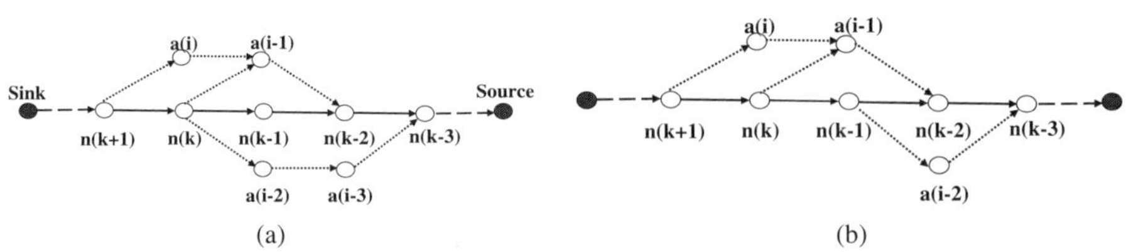

Figure 6.7. Braided multipaths. Diagrams (a) and (b) present idealized and localized braided multipaths [29].

any desired reliability. However, with the increase in reliability, one expects an increase in the overhead required to transfer the packet. Multiple paths from source to destination are discovered while reliability is ensured by redundant copies of a packet. The desired reliability, the local channel error conditions, and the neighborhood information at each node determines the degree of redundancy needed. Another key feature is that data caching is not required at intermediate nodes, thereby saving memory.

In multihop sensor networks where the source is located far away from the sink with high number of channel error occurences, a simple packet forwarding strategy may become unreliable. To improve the reliability, two solutions have been proposed: (i) *No acknowledgments*—sending data from source to sink along the shortest path with no acknowledgments. Multiple copies of the packet are sent along the same path to increase the probability of packet delivery. (ii) *With acknowledgments*—sink sends back an acknowledgment to the source after receiving the data packet. The simulation results show that the acknowledgement-based scheme has similar overhead to the non-acknowledgment-based scheme. Hence, the non-acknowledgment scheme is used in the protocol description.

ReInForM is based on the Dynamic Packet State (DPS) [31] approach used for data networks. The source uses local knowledge of the network such as channel error, hop count to sink, and so on, and sends multiple copies of the data packets through multiple paths. With every hop, network conditions are recorded in the packet header and forwarded to the next node. Every intermediate node takes forwarding decisions based on this stored information. This strategy enables the forwarding node to route the packets according to local network conditions, but without maintaining the state information on its own. It also helps in making localized decisions on packet forwarding since network characteristics at an earlier point on the path may be much different from those at the current point. This mechanism adapts to channel errors and topological deviations while maintaining a reasonable overhead.

6.4.6 QoS-Based Protocols

Routing is generally carried out based on metrics, prominent among which are control packet overhead and energy efficiency. However, additional quality of service (QoS) parameters may also be considered to facilitate more efficient routing in sensor networks.

***Stream-Enabled Routing (SER)* [32].** SER protocol allows sources to choose the routes based on the instruction (or task) provided by the sinks. An *instruction* is a predefined identifier value. Therefore, only an identifier is sent rather than the attribute list, resulting in memory conservation. It takes into account the available energy of the sensor nodes, QoS requirements of the instruction, memory limitation of nodes, and the localized effect of dense nodes. Sinks can give new instructions to the sources without establishing another path. SER uses four types of messages: (i) scout message (`S-message`), (ii) information message (`I-message`), (iii) neighbor–neighbor message (`N-message`), and (iv) update message (`U-message`).

An `S-message` is used during the source discovery to determine sources that will process the instruction (or task) specified in the `S-message`. Sources decide the type and level of the routes needed by the instruction. There are four types of routes, each with two levels (i.e., level 1 and level 2). The μ value is the radius of the level 2 routes.

Stream is identified by both the type and the level of route. Each level 2 stream includes level 1 stream. In level 2, the size of the radius μ of the stream can be determined based on QoS specified in the instruction. Combination of types and levels creates different kinds of QoS for a stream. After streams are chosen, the source sends an `N-message` to establish the streams back to the sink. The repairs of streams are accomplished via `N-` and `S-messages`. Once the streams are accomplished, data travels from sources to the sink through either level 1 or level 2 stream with an `I-message`. The sink can update the instruction (or task) at the sources through either level 1 or level 2 stream using a `U-message`. Both sources and sink can terminate the streams using `U-message`.

SER has seven phases:

1. **Source Discovery.** A sink broadcasts an `S-message` to find routes from sink to source. `S-message` contains the following fields: TID, NAP, LID, NH, AE. Table 6.1 represents the parameters used throughout the algorithm description.

   The average energy of a node is given by the following formula:

$$AE = \frac{NH_{i-1} \times AE_{i-1} + E_i}{NH_{i-1} + 1}$$

   The TID field is further composed of LI, MT, INS, and $TLOC$ fields. When a sensor node receives an `S-message`, it determines if the instruction, (INS), is intended for the node. If the INS in the `S-message` is not intended for the node, the node stores the fields of `S-message` in a *connection-tree (C-tree)*. A C-tree is a logical tree that represents possible connections through the node. It maintains the node's neighbors that can participate in a routing back to sink.

   A sensor node in an established route knows the LID values of both uplink and downlink nodes Initially, $DLID$ and $ULID$ values are not set, and DSP and NS values are set to OFF. Updated values of AE, NH, and LID fields of an `S-message` are broadcast to the neighbors.

   If a sensor node receives the same `S-message` from its neighbors, it dismisses it. The sources store an `S-message` in a *task-tree (T-tree)*. T-tree has x $DLID$ values since source can select up to x LIDs to route `I-message` back to the sink based on QoS requirement. The max value of x is the number of neighbor nodes. Each $DLID$ value corresponds to a DSP indicator. In the T-tree, the leaf nodes has no $ULID$ and NS indicator since the sources are the

TABLE 6.1. SER Protocol Parameters

TID	Task ID
NAP	Network access point (represents a unique sink)
LID	Local ID
NH	Number of hops from the sink
AE	Average energy of a route
LI	Length indicator
MT	Message type, (MT = 0 (S-message); MT = 1 (I-message); MT = 2 (U-message); MT = 3 (N-message)
INS	Instruction
TLOC	Targeted location
DSP	Downlink sensor problem
NS	Node selected
DLID	LID of downlink sensor node
ULID	LID of uplink sensor node
SLID	Selected ID
MES	Message
FI	Flow indicator (message going uphill or downhill)
CNH	Current number of hops
Payload	Description
NINS	New INS

destination of an `S-message`. A source can receive an x *S-message* since it has x neighbors. The route associated with the first received `S-message` is considered the shortest route. Sources select a neighbor node to send an `I-message` back to sink based on the QoS requirement of INS.

2. **Route Selection.** Once the sources receive the `S-message`, they determine the QoS requirement of task in the `S-message`. There are four types of stream for communication between sources and sinks, and each stream can be either level 1 or level 2:
   - Type 1: Time critical but not data critical
   - Type 2: Data critical but not time critical
   - Type 3: Not time and data critical
   - Type 4: Not time and data critical

   After the sources select the neighbor nodes, the sources broadcast an `N-message` to their neighbors indicating the level and size of the stream. `N-message` contains TID, NAP, LID, $SLID$, and MES fields. If the stream is level 1, $\mu = 0$ (width of the stream). At level 1, messages are routed back to the sink via hop-by-hop communication. Messages are sent to only one node. The Level 2 stream contains the level 1 stream, which serves as a backbone in setting up the level 2 stream. The value of μ is the number of hops away from the nodes in the level 1 stream. Messages can flow downhill to the sink or uphill to the sources by flooding through only the nodes that are part of the stream. An `I-message` flows downhill from sources to sink by using the *NH* value stored in each node in the C-tree. The nodes near the sources have higher *NH* values. *U-message* flows uphill from sink to sources by using the negative of the *NH* value. The nodes near the sources have higher negative *NH* values.

3. **Route Establishment.** An `N-message` is used by a sensor node to inform neighbors about its local information. The source sends an `N-message` to establish a stream back to the sink. Sensor nodes that are not part of a stream delete all data associated with the `N-message` from the C-tree. If intermediate nodes between the sources and sinks have not received an `N-message` in response to an `S-message` in a set time interval, the sensor node deletes the C-tree branch that is associated with the `S-message`. After the `N-message` arrives to the sink, the minimum delay or maximum average energy stream is established. Sources can start sending `I-messages` to the sink. An `I-message` contains *TID*, *FI*, *CNH*, and *Payload* fields.
4. **I-Message Transmission.** The neighbor nodes can determine if they need to route the `I-message` by the *TID* since each neighbor nodes maintain a C-tree. When a source broadcasts an `I-message`, it sends a *CNH* field with the value from a T-tree. Intermediate nodes between sources and sink use a C-tree. *FI* and *CNH* fields are only used when the stream is level 2. Each node only rebroadcasts once to avoid a node from broadcasting the same message over again. After an `I-message` is received, the sensor nodes turn off the receiver for some amount of time if the sleep mode operation is on such that the node can avoid listening neighbors broadcasting the same `I-message`. The C-tree indicates which instructions the sensor nodes need to route.
5. **Route Reconnection.** If a sensor node is low on energy or there is too much noise around when transmitting at level 1, it can broadcast an `N-message` by setting up a reconnect message indicator. Once the neighbors receive an `N-message`, they check their C-tree to decide if there are possible alternate routes. An `N-message` will be broadcasted until the alternate route is found.

   Sudden death of route: If the stream suddenly terminates, the sink cannot get the `I-messages`. The sink sends out a new `S-message` with higher QoS requirement version of the same instruction (higher QoS *INS* value). New streams can be found to avoid broken paths. Multiple streams of level 2 can be set up between source and sink to improve robustness of `I-message` routing.
6. **Instruction Update.** A `U-message` allows a sink to update its instruction to the sources. The `U-message` from the sink to the sources flows uphill while it flows downhill from sources to the sink when streams are level 2. A `U-message` contains *TID*, *FI*, *CNH*, and *NINS* fields.
7. **Task Termination.** A task at the sources are terminated in two ways:

   (a) Sources have finished the task associated with the instruction given by sink. A `U-message` with the task completed instruction indicator is broadcast by sources.

(b) A sink decides to terminate the instruction. A `U-message` with the task termination instruction indicator is set by the sink.

The streams are torn down by removing C-tree branches at the intermediate nodes and removing a T-tree at the sources.

Algorithm for Robust Routing in Volatile Environments (ARRIVE) [33]. This is a probabilistic algorithm and makes packet forwarding decisions based on localized information, and it has a tree-like topology rooted at the sink of the network. The forward approach is employed to achieve end-to-end reliability. The packet loss is avoided by sending multiple packets of the single event. Basically, the three sources of packet loss expected are (i) isolated link, (ii) patterned node failures, and (iii) malicious or misbehaving nodes. Figure 6.8 presents the overview of the ARRIVE algorithm.

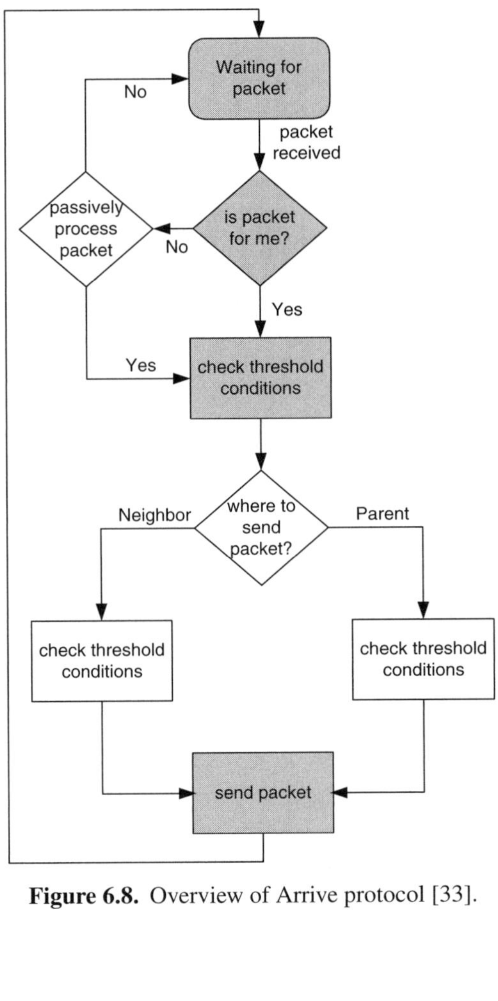

Figure 6.8. Overview of Arrive protocol [33].

- **Event**: Identified by [SourceID, EventID]
- **Level**: Each node has unique level indicating distance from source to sink (in terms of hops)
- **Parents**: Nodes one level closer to the sink
- **Neighbors**: Nodes on the same level and be able hear each other
- **Push**: Push packet to one of the neighbors
- **Forward**: Forward packet to one of the parents
- **Forwarding Probability**: Included in the packet header and used to probabilistically select whether to push or forward
- **Reputation History**: Each node keeps this information for each of its parents and neighbors
- **Convergence**: Prevents multiple packets of the same event being sent to same source of failure

ARRIVE achieves diversity in paths in two ways: (i) Upon receiving a packet, the next hop is selected probabilistically based on link reliability and node reputation and (ii) when more than two or more packets of the same event are processed, these packets are ensured to follow different outgoing links. Each keeps the following information: level, neighbors list, parents list, reputation history of neighbors and parents, and convergence history of specific events.

Several assumptions are made about the network and the algorithm. The network, which is almost considered a static network, is assumed to be dense enough for sufficient multiplicity of paths to exist between sources and sink for the algorithm to perform well. Sensors used are considered to have a low per-node cost. Routes used by the packets are unlikely to be optimal due to the probabilistic nature of the algorithm. Messages flow only from nodes to one and only sink.

The breadth first search rooted at the sink is used to initialize level, parents, and neighbors state information at each node. When a node hears a packet, it checks to see if the packet is addressed for it. If so, threshold processing takes place. Nodes are filtered by their reputation and convergence history of the neighbors and parents. A decision needs to be made to either choose to forward the packet to a parent or push it to one of its neighbors with the probability value found in the packet header. This is randomly determined by the forwarding probability function $\Pr(f)$. Each node is weighed by their reputation. The destination is randomly selected from the rest of the nodes (since bad reputation nodes are eliminated). If the the packet is forwarded to one of the parents, $\Pr(f)$ is not changed; however, its value is increased.

6.5 CONCLUSIONS

In this chapter, we have discussed various routing protocols in sensor networks. The common goals of designing a routing algorithm is not only to reduce control packet overhead, maximize throughput, and minimize the end-to-end delay, but also to take

TABLE 6.2. Comparison of Sensor Network Routing Protocols

Protocol	Classification	Mobility	Power usage	Negotiation based	Data Aggregation	Scalability
Directed diffusion	Attribute-based	Restricted	Limited	Yes	Yes	Restricted
EAD	Attribute-based	Virtual backbone	Limited	No	Yes	Restricted
RUMOR	Attribute-based	Very restricted	—	No	Yes	High
Youssef et al.	Attribute-based	Cluster based	Low	No	Yes	Limited
LEACH	Hierarchical	Fixed base station	Highest	No	Yes	High
PEGASIS	Hierarchical	Fixed base station	Highest	No	No	High
TEEN	Hierarchical	Fixed base station	Highest	No	Yes	High
APTEEN	Hierarchical	Fixed base station	Highest	No	Yes	High
Al-Karaki et al.	Hierarchical	No	—	Yes	Yes	High
M-MPR	Multipath	No	—	No	No	Restricted
Ganesan et al.	Multipath	No	High	No	No	High
ReInForM	Multipath	Limited	—	No	Yes	Restricted
SPEED	Geographical	No	—	No	No	Restricted
Seada et al.	Geographical	No	High	—	Yes	Restricted
Subramanian et al.	Geographical	No	—	—	No	Restricted
Rao et al.	Geographical	Yes	—	No	—	High
SER	QoS	No	—	Yes	No	Restricted
ARRIVE	QoS	No	—	Yes	Yes	Restricted
SAR	Flat	No	—	Yes	Yes	Restricted
GRAdient	Flat	No	High	No	No	High
MCFA	Flat	No	Low	No	No	High

into consideration the energy consumption, especially in a sensor network comprised of nodes that are considered lightweight with limited memory and battery power. We have divided the sensor routing protocols into six categories: (i) attribute-based, (ii) flat, (iii) geographical, (iv) hierarchical, (v) multipath, and (vi) QoS-based. We have then compared the protocols belonging to the same category based on various characteristics.

6.6 EXERCISES

1. List the characteristics of the directed diffusion protocol which makes it considered both a flat and attribute-based sensor routing protocol (see Table 6.2).
2. For which type of sensor applications is the EAD protocol most suited? Detail some of the factors by which EAD helps in providing better services to these applications.
3. On what basis are the local cluster heads selected in the LEACH protocol? Describe the cluster-head selection procedure in detail. Why are cluster heads not selected at the initial network deployment time itself?
4. How do PEGASIS and LEACH protocols differ?
5. In the ReInForM protocol, determine the reasons why the No ACK scheme has a similar control packet overhead to the ACK scheme, since at first glance it would logically seem that the scheme without the ACK packets would result in much less control overhead.
6. The SPEED protocol is characterized by minimal memory requirements. What are the probable reasons for such a characteristic?
7. The ARRIVE protocol uses a tree-based routing approach. Discuss various advantages derived by using such a routing scheme.

BIBLIOGRAPHY

1. D. E. Culler, D. Estrin, and M. B. Srivastava. Guest editors' introduction: Overview of sensor networks. *IEEE Computer*, **37**(8):41–49, 2004.
2. G. J. Pottie and W. J. Kaiser. Wireless integrated network sensors. *Communications of ACM*, **43**(5):51–58, 2000.
3. I. F. Akyildiz, W. Su, Y. Sankarasubramaniam, and E. Cayirci. Wireless sensor networks: A survey. *Computer Networks*, **38**(4):393–422, 2002.
4. W. Ye, J. S. Heidemann, and D. Estrin. Medium access control with coordinated adaptive sleeping for wireless sensor networks. *IEEE/ACM Transactions on Networking*, **12**(3):493–506, 2004.
5. V. Rajendran, K. Obraczka, and J. Garcia-Luna-Aceves. Energy-efficient collision-free medium access control for wireless sensor networks. In *Proceedings of SenSys*, 2003, pp. 181–192.

6. D. Estrin, R. Govindan, J. S. Heidemann, and S. Kumar. Next century challenges: Scalable coordination in sensor networks. In *Proceedings of MOBICOM*, 1999, pp. 263–270.
7. S. Tilak, N. B. Abu-Ghazaleh, and W. R. Heinzelman. A taxonomy of wireless micro-sensor network models. *ACM Mobile Computing and Communications Review*, **6**(2):28–36, 2002.
8. J. N. Al-Karaki and A. E. Kamal. Routing techniques in sensor networks: A survey. *IEEE Communications*, **11**(6):6–28, 2004.
9. K. Akkaya and M. Younis. A survey on routing protocols for wireless sensor networks. *Ad Hoc Networks Journal*, **3**:325–349, 2005.
10. C. Intanagonwiwat, R. Govindan, and D. Estrin. Directed diffusion: A scalable and robust communication paradigm for sensor networks. In *Proceedings of MOBICOM*, 2000, pp. 56–67.
11. A. Boukerche, X. Cheng, and J. Linus. Energy-aware data-centric routing in microsensor networks. In *MSWIM '03: Proceedings of the 6th ACM International Workshop on Modeling Analysis and Simulation of Wireless and Mobile Systems*, 2003, pp. 42–49.
12. M. A. Youssef, M. F. Younis, and K. Arisha. A constrained shortest-path energy-aware routing algorithm for wireless sensor networks. In *Proceedings of WCNC*, 2002, pp. 794–799.
13. D. Braginsky and D. Estrin. Rumor routing algorithm for sensor networks. In *Proceedings of WSNA*, 2002, pp. 22–31.
14. F. Ye, G. Zhong, S. Lu, and L. Zhang. Gradient broadcast: A robust data delivery protocol for large scale sensor networks. *Wireless Networks*, **11**(3):285–298, 2005.
15. F. Ye, A. Chen, S. Liu, and L. Zhang. A scalable solution to minimum cost forwarding in large sensor networks. In *Proceedings of the Tenth International Conference on Computer Communications and Networks (ICCCN)*, 2001, pp. 304–309.
16. K. Sohrabi, J. Gao, V. Ailawadhi, and G. J. Pottie. Protocols for self-organization of a wireless sensor networks. *IEEE Personal Communications Magazine*, **7**(5):16–27, 2005.
17. K. Sohrabi. On low power wireless sensor networks. Ph.D. thesis, Electrical and Computer Engineering Department, UCLA, June 2000.
18. J. L. Gao. Energy efficient routing for wireless sensor networks. Ph.D. thesis, Electrical and Computer Engineering Department, UCLA, June 2000.
19. T. He, J. A. Stankovic, C. Lu, and T. F. Abdelzaher. SPEED: A stateless protocol for real-time communication in sensor networks. In *Proceedings of ICDCS*, 2003, pp. 46–58.
20. A. Rao, C. Papadimitriou, S. Shenker, and I. Stoica. Geographic routing without location information. In *Proceedings of MOBICOM*, 2003, pp. 96–108.
21. K. Seada, M. Zuniga, A. Helmy, and B. Krishnamachari. Energy-efficient forwarding strategies for geographic routing in lossy wireless sensor networks. In *Proceedings of SenSys*, 2004, pp. 108–121.
22. S. Subramanian and S. Shakkottai. Geographic routing with limited information in sensor networks. In *Proceedings of IPSN*, 2005, pp. 269–276.
23. W. R. Heinzelman, A. Chandrakasan, and H. Balakrishnan. Energy-efficient communication protocol for wireless microsensor networks. In *Proceedings of HICSS*, 2000.
24. S. Lindsey and C. Raghavendra. PEGASIS: Power-efficient gathering in sensor information systems. In *Proceedings of IEEE Aerospace Conference*, 2002, pp. 1125–1130.
25. A. Manjeshwar and D. P. Agrawal. TEEN: A routing protocol for enhanced efficiency in wireless sensor networks. In *Proceedings of IPDPS*, 2001, pp. 2009–2015.

26. A. Manjeshwar and D. P. Agrawal. APTEEN: A hybrid protocol for efficient routing and comprehensive information retrieval in wireless sensor networks. In *Proceedings of IPDPS*, 2002, pp. 195–202.
27. J. N. Al-Karaki, R. Ul-Mustafa, and A. E. Kamal. Data aggregation in wireless sensor networks - exact and approximate algorithms. In *Proceedings of IEEE Workshop on High Performance Switching and Routing (HPSR)*, 2004, pp. 241–245.
28. S. De, C. Qiao, and H. Wu. Meshed multipath routing with selective forwarding: An efficient strategy in wireless sensor networks. *Computer Networks*, **43**(4):481–497, 2003.
29. D. Ganesan, R. Govindan, S. Shenker, and D. Estrin. Highly-resilient, energy-efficient multipath routing in wireless sensor networks. *ACM Mobile Computing Communication Review*, **5**(4):10–24, 2001.
30. B. Deb, S. Bhatnagar, and B. Nath. ReInForM: Reliable information forwarding using multiple paths in sensor networks. In *Proceedings of LCN*, 2003, pp. 406–415.
31. I. Stoica, S. Shenker, and H. Zhang. Stateless fair queueing: Achieving approximately fair bandwidth allocations in high speed networks. In *Proceedings of SIGCOMM*, 1998, pp. 118–130.
32. W. Su and I. F. Akyildiz. A stream enabled routing (SER) protocol for sensor networks. In *Proceedings of Med-hoc-Net*, 2002.
33. C. Karlof, Y. Li, and J. Polastre. ARRIVE: Algorithm for robust routing in volatile environments. Technical Report UCB/CSD-03-1233, Computer Science Division, UC Berkley, May 2003.

CHAPTER 7

Clustering in Wireless Sensor Networks: A Graph Theory Perspective

NIDAL NASSER and LILIANA M. ARBOLEDA

Department of Computing and Information Sciences, University of Guelph, Guelph, Ontario N1G 2W1, Canada

7.1 INTRODUCTION

Wireless Sensor Networks (WSNs) consist of tiny sensing devices that are spread over a large geographic area and can be used to collect and process environmental data such as temperature, humidity, light conditions, seismic activities, images of the environment, and so on. These data can be used to detect certain events and to trigger activities. For example, sensors distributed over large woodland could automatically raise an alarm if a fire has broken out somewhere, or sensors distributed over a large farmland could trigger irrigation if the ground of a field is not moist enough.

A WSN is usually comprised of a large number of sensors that are physically small, communicate wirelessly among each other, and are deployed without prior knowledge of the network topology. Due to the limitation of their physical size, the sensors tend to have storage space, energy supply, and communication bandwidth so limited that every possible means of reducing the usage of these resources is aggressively sought. The use of the WSN potential will provide efficient and costs-effective solutions for several problems. However, it is necessary to implement mechanisms or procedures to deal with the sensor constraints. The hierarchical organization of the sensors, grouping them and assigning those specific tasks into the group before transferring the information to higher levels, is one the mechanisms proposed to deal with the sensors limitations and is commonly referred to as *clustering*.

The creation of clusters in a WSN field is generally done by taking into account the proximity between the sensors, measured through the radio-frequency signal they

Algorithms and Protocols for Wireless Sensor Networks, Edited by Azzedine Boukerche

emit. Each cluster has a cluster head, which is the node that directly communicate with the sink (base station) for the user data collection. By assuming roles within a cluster hierarchy, the nodes in a WSN can control the activities they perform and therefore reduce their energy consumption. However, the election of when to act as a simple data provider (saving energy) and when to act as a gateway (cluster head) between the nodes and the base station is a difficult problem. To make this decision, it is necessary to take into account several aspects including power level signal, transmission schedules, nodes localization, and networking function. Clustering helps in solving some of the sensors' constraints by reducing the cost of transmitting data to base stations, reducing the power consumption in the devices, facilitating the gathering of sensed data, maximizing the routing process execution, and allowing scalability.

One of the techniques used to create the clusters and represent the topology of the WSN involves the use of Graph Theory concepts to represent the topology of the sensors in the field. Graph theory can be used to create the sensor clusters and can help in identifying the cluster head. Sensor network can be represented by a graph $G = (V, E)$, where the vertices (V) represent the sensors and the set of links (E) represents the connections between vertices if the they are within the transmission range of each other, as shown in Figure 7.1.

In this chapter we will study different existing approaches to (a) create clusters and select the cluster head using Graph Theory concepts and (b) identify if they can be used in either static WSN or mobile WSN. In mobile WSN, the dynamic changes in the topology of the clusters creates additional challenges. In this situation it is necessary to implement mechanisms to control the changes in the cluster graph.

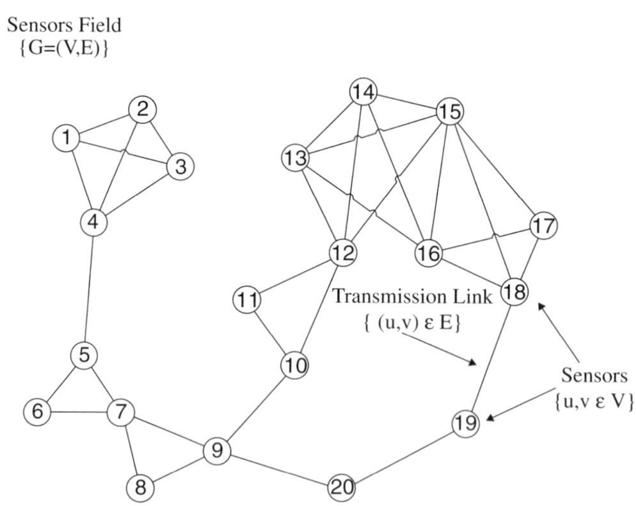

Figure 7.1. Graph representation of a WSN.

7.2 FUNDAMENTAL CORRESPONDING BETWEEN WIRELESS SENSOR NETWORKS AND GRAPH THEORY

A wireless sensor network is a set of sensors deployed in a sensor field to monitor specific characteristics of the environment, to measure those characteristics, and to collect the data related to those phenomena. The sensors are small devices with limited resources: limited battery power, low memory, little computing capability, very low data rates, low bandwidth processing, variable link quality, and so on. However, despite their constraints, when the sensors are deployed in large numbers, they can provide us with a very real picture of the field being sensed. WSN can provide an area coverage that was not possible with other wired and wireless networks. They can be deployed in different environments and can be permanently attended or can be left unattended once they have been deployed in the field.

The use of the WSN potential will provide efficient and cost-effective solutions for many problems. However, it is necessary to implement mechanisms or procedures to deal with the sensor constraints. The use of clustering techniques has been proposed to help solve some of those constraints, by allowing the organization of the sensors in a hierarchical manner, grouping them into clusters and assigning a specific task to the sensor in the clusters, before moving the information to higher levels. The concept of clustering is very useful in different contexts of WSN. Clustering is a fundamental mechanism to design scalable sensor network protocols. In general terms, clustering is the classification of similar objects into different groups or subsets. The formed subsets in some sense belong together, because they share one or more similar characteristics or behaviors. Examples of such common characteristics could be: proximity according to some defined distance measure, similar behaviors, common data patterns, and so on. In the most general problem the number of clusters or groups is unknown, as are the properties that make them similar.

Clustering techniques have been proposed in wireless networks in order to achieve high energy efficiency and assure long network lifetime, for bandwidth reuse, for data gathering [1] and target tracking [2], one-to-many, many-to-one, one-to-any, or one-to-all communications, routing [3–6], and so on. Clustering is particularly useful for applications that require scalability to hundreds or thousands of nodes. Scalability in this context implies the need for load balancing, efficient resource utilization, and data aggregation [7]. Also, many routing protocols can use clustering to create a hierarchical structure and minimize the path cost when communicating with the base station. In many sensors network applications where data collection and processing can be done in situ, this hierarchical approach is a promising method for efficiently organizing the network. Also, many signal processing algorithms used for extraction of final information from the data gathered by the sensors are well-suited for local processing of data within the clusters.

Graph Theory concepts can be used to describe, analyze and represent a WSN in a very clear way. Several of these concepts refer to structures and algorithms that had been previously used to address other aspects like topology management, localization techniques and routing, not only in WSN but also in other types of wireless and wired networks [8–11]. For example, the OSPF (Open Shorted Path First) routing algorithm used for routing in wired LANs implements the SPF (Shorted Path First) algorithm or Dijkstra's algorithm, which solves the single-source shortest path problem for a directed graph with nonnegative edge weights.

Graph data structures and algorithms can easily represent the network with stationary wireless sensor. The construction of basic graph structures like trees, cliques or dominating sets, and the use of graph algorithms like Breadth First Search (BFS) or Depth First Search (DFS) will help in the construction of clusters to improve the communication in the WSN. However, in mobile WSN, the dynamic change of the network topology due to the sensor movement creates additional challenges when forming the clusters. In this situation it is necessary to implement additional mechanisms to control the changes in the cluster graph.

In this section, we give the basic concepts and definitions that provide a detailed overview of the corresponding relation between clustering in WSN and graph theory, focusing on concepts and definitions that are important for the understanding of the material to follow.

7.2.1 Wireless Sensor Networks and Graph Theory Concepts

A cluster in WNS consists of three main different elements: sensor nodes (SNs), base station (BS), and cluster heads (CH); see Figure 7.2. The SNs are the set of sensors present in the network, arranged to sense the environment and collect the data. The main task of an SN in a sensor field is to detect events, perform quick local data processing, and then transmit the data. But the greatest constraint it has is the power consumption, which usually is caused when the sensor is observing its surroundings,

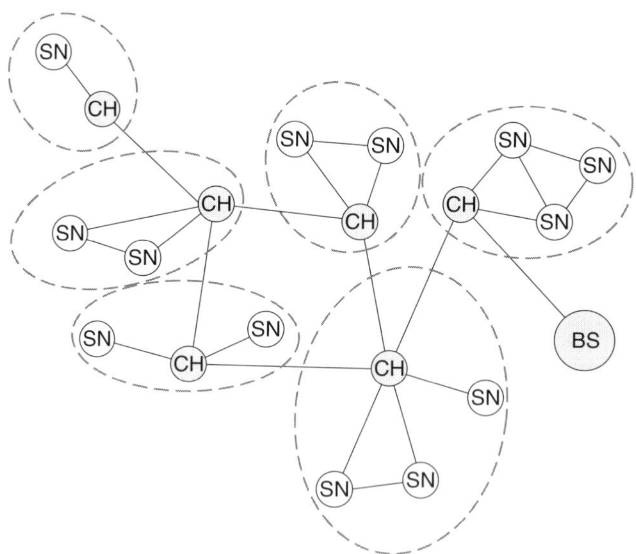

Figure 7.2. Elements in a clustered WSN.

and communicating (sending and receiving) data. The BS is the data processing point for the data received from the sensor nodes, and where the data are accessed by the end-user. It is generally considered fixed and at a far distance from the sensor nodes. The CH acts as a gateway between the SNs and the BS. The function of the cluster head is to perform common functions for all the nodes in the cluster, like aggregating the data before sending it to the BS. In some way, the CH is the sink for the cluster nodes, and the BS is the sink for the cluster heads. This structure formed between the sensor nodes, the sink, and the base station can be replicated as many times as it is needed, creating the different layers of the hierarchical WSN.

The SNs and the communication links between them can be represented by an undirected graph $G = (V, E)$, where each vertex $v \in V$ (the set of vertices in the graph) represents a sensor node with a unique ID. An edge $(u, v) \in E$ (the set of edges in the graph) represents a communication link if the corresponding nodes u and v are within the transmission range of each other.

The graph is formed by defining the neighborhood of each node. The neighborhood of each one of the nodes in the network and the k-neighborhood of the nodes can be defined as follows:

Definition 1. Node's Neighborhood. The neighborhood $N(v)$ is the set of nodes (neighbors) that reside within the circular transmission range of node v, which means the vertices adjacent to v. If v is included into the neighborhood, it is called closed neighborhood of v and it is represented $N[v]$.

Definition 2. k-Neighborhood. The k-neighborhood of v, $N_k(v)$, is the set of nodes with distance at most k from v.

$$N_k(v) = \{u | u \in V \wedge d(u, v) \leq k\}$$

Following the creation of the graph, by defining the adjacent nodes in the network and the communication links between nodes, it is possible to determine which nodes are reachable from a specific node and it allows us to calculate the correspondent hop distance (see Definition 3) between any source and target nodes. It also allows us to determine the k_{th} power graph of a node (see Definition 4), to limit the number of nodes that will be considered among the node's transmission radius when creating the clusters.

Definition 3. Hop Distance. The shortest path between two nodes u and v is the path with the minimum number of hops between them. The distance $d(u, v)$ is the number of hops in the shortest path between u and v.

Definition 4. k_{th} Power Graph. The k_{th} power graph of G, $G_k = (V, E_k)$, is the graph between the nodes in V and an edge between every pair of nodes $u, v \in V$, such that $d(u, v) \leq k$ in G.

Knowing the existent links between the nodes, the next process involved in the communication is determining the routes available to send information through the WSN. One of the most common mechanisms used for message delivery inside the network is based on a spanning trees structure, defined below, that allows the use of algorithms like Depth First Search (DFS) and Breadth First Search (BFS) to send messages along the created graph in linear time. We use T to represent the spanning tree of G_c, rooted at y. The i_{th} level of the tree is the set of nodes with hop distance equal to i from y. The depth of the tree, $depth(T)$, is the index of the farthest level in the tree.

Definition 5. Spanning Tree. A spanning tree is a connected and undirected graph, with no cycles. Each tree has with n vertices. Each pair of vertices has exactly one path connecting them, creating $n - 1$ edges in the tree.

Definition 6. Depth First Search (DFS). Depth First Search (DFS) is an uninformed tree search that works by expanding the first child node that appears on the graph and goes deeper each time until a goal node is found, or until it hits a node that has no children. After this the search backtracks, returning to the most recent node it hadn't finished exploring.

Definition 7. Breadth First Search (BFS). Breadth First Search (BFS) is a graph search algorithm that begins at the root node and explores all the neighboring nodes. After this, it explores the unexplored neighbor nodes of each of the previously found nodes, repeating this procedure until it finds the goal node.

Cluster Representation. There are many different graph concepts used for the creation of clusters in a WSN. Among them is the following definition for cluster:

Definition 8. Cluster. A cluster is any subset of nodes $C \subseteq V$. $y \in V$ is the cluster head and $G_c = (C, E_c)$ is the cluster graph.

$$E_c = \{(u, v) | u, v \in C \wedge (u, v) \in E\}$$

If G_c is connected, then the cluster is connected. $d_c(u, v)$ is the shortest path inside the cluster, and the cluster radius is the maximal distance between y and any other node $v \in C$.

$$max_{v \in C} d_c(y, v)$$

After determining the nodes' neighbors and the distance between them, it is possible to use additional graph concepts like the node's weight to use it as one of the parameters in the definition and functionality of the clusters.

Definition 9. Graph Weight. The nodes in the network graph can have a positive weight w_v. The total weight of a cluster is given by

$$W_{sum}(C) = \sum_{v \in C} w_v$$

Dominating Sets and Covers. Another group of graph definitions which are very useful when trying to model the clusters is the concept of sets. Set definition in the context of WSN are listed below. By being able of defining these sets in the network, the nodes can calculate their real coverage and establish the stability of the communication paths between them. Generally, the clusters are defined using the vertex cover of the graph (see Definition 10) and the nodes that belong to that vertex cover are selected as the cluster heads. The remaining nodes should calculate the stability of their communication link to the neighboring CHs and join the cluster corresponding to the link with the best connection stability.

Definition 10. Vertex Cover. A vertex cover is a subset $S \subset V$, such that for all edges in E, $S \cap e \neq 0$, which means that every edge has at least one edge in S.

Definition 11. Independent Set. The independent set IS_G is the subset $S \subset V$, such that there is no edge between any pair of nodes in S. The maximum independent set in G is equal to $|V|$, the size of the minimum vertex cover of G.

Definition 12. Dominating Set. The dominating set DS_G is the subset $S \subset V$ such that every node is in S or if it is in $V - S$ and has at least one neighbor (adjacent vertex) in S. A vertex of S is said to dominate itself and all adjacent vertices. An edge is dominated if either of its endpoints is in S. The nondominated edges are called free edges.

Definition 13. Minimum Dominating Set. The Minimum Dominating Set (MDS) problem is the problem of finding a dominating set of minimum size. This is an NP-complete problem.

Definition 14. Independent Dominating Set. The independent dominating set IDS_G is the set $S \subset V$ that is both dominating and independent.

Definition 15. Connected Dominating Set. The connected dominating set CDS_G is a dominating set whose induced subgraph S' is connected.

Definition 16. Weakly Induced Subgraph. For any subset $S \subset V$, the subgraph weakly induced by S, S_w, is the graph $(N[S], E \cap (N[S] \times S)$). This means that S_w contains the vertices of S, their neighbors, and all edges with at least one endpoint in S. A vertex subset S is a **weakly connected dominating set** if S is dominating and S_w is connected.

7.3 GRAPH-BASED APPROACHES FOR CLUSTERING IN WSN

In this section, we present a classification of proposed clustering techniques based on the application of Graph Theory concepts. Although many of the references are related with general wireless and ad hoc networks and not specifically WSN, we include them either because such approaches can be applied to WSN, especially static WSN, or because we need to clarify why they are not suitable for WSN. We summarize the main characteristics for each one of the algorithms and protocols, with emphasis on describing the ones suitable for WSN.

Once it has been decided that the creation of clusters inside the WSN is an appropriate solution to support other network and application functions, it is necessary to clearly define the characteristics desired in the clusters to obtain a "well-formed" structure. Some of those characteristics are as follows:

1. Every node should be in exactly one cluster. The objective of this is to maximize the average cluster sizes while maintaining full coverage.
2. Guarantee the total coverage of the network.
3. Minimize the number of CH to provide an efficient network coverage while minimizing the cluster overlap. A minimum cluster overlap reduces the amount of channel contention between clusters and improves the efficiency of algorithms that execute at the level of the CH.
4. Create a highly uniform, balanced clustering.

As an example of the importance of highly uniform clustering with low overlap, consider the clustered broadcast protocol described by Ni et al. [12]. In this protocol, the broadcast message is relayed from CH to CH, which then broadcast the message to their associated nodes in each cluster, called *followers*. In a clustering with few CH and large cluster sizes, the clusters have minimal overlap and provide the best coverage of the network with the fewest clusters. In this configuration, the number of repeated broadcast transmissions over any area will be small, thus reducing the amount of transmission collisions and channel contention. This allows for faster, more efficient and more reliable communications. On the other hand, a poor clustering containing a lot of cluster overlap and a large number of CHs lose most of the benefits of clustering because transmissions will be repeated in areas of overlap with significant channel contention

In the following subsections we make a classification of the graph theory based clustering protocols for static and mobile WSN. This classification contains two main groups as shown in Figure 7.3: static-based WSN algorithms and mobile-based WSN algorithms.

7.3.1 Centralized Algorithms and Self-Elective Protocols

Many centralized algorithms often use graph theoretic properties for clustering. These algorithms deal with the topology of the entire network as a whole, creating structures

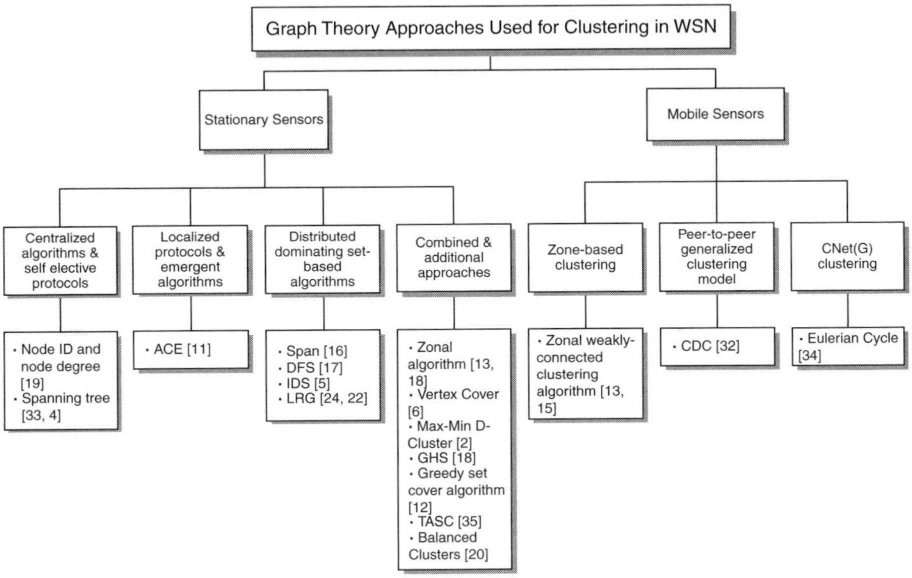

Figure 7.3. Graph theory approaches used for clustering in WSN.

that often present an increased vulnerability to node failure in certain key parts of the network, usually near the root of the tree (if they use a spanning tree) or near the BS. Additionally, these protocols may require significant communications or computation overhead for very large networks.

It is necessary to clarify that not all the clustering algorithms and protocols that have been proposed using graph theory concepts are suitable for WSN. For example, in reference 13, the authors propose a technique where each cluster forms a clique. However, they do not select a CH that makes the protocol unfeasible for WSN. Additionally, their *"createClusters()"* function has a relatively large overhead of $O(d^3)$, where d is the density of the network (number of nodes per area unit).

A group of clustering protocols called self-elective protocols use the Node ID and Node Degree to select the CH. In reference 14, Gerla and Tsai proposed two weight-based clustering algorithms, where each vertex v selects the node with optimal weight within $N(v)$ as CH. In the first algorithm, the optimal vertex is the one with lowest node ID as shown in Figure 7.4. The neighborhood of the node selected as CH is the cluster. A node that can hear two or more CHs is a "gateway" or border node.

In the second clustering algorithm, the highest-degree node in a neighborhood is selected as the optimal node to be the CH and the neighbors are covered by it (see Figure 7.5). Although the algorithm is expected to perform well on many randomly defined graphs, it may not produce any CH for graphs that do not have any node with the highest number of neighbors (like interval graphs). Thus, the algorithm must be completed by adding nontrivial tie resolution rules.

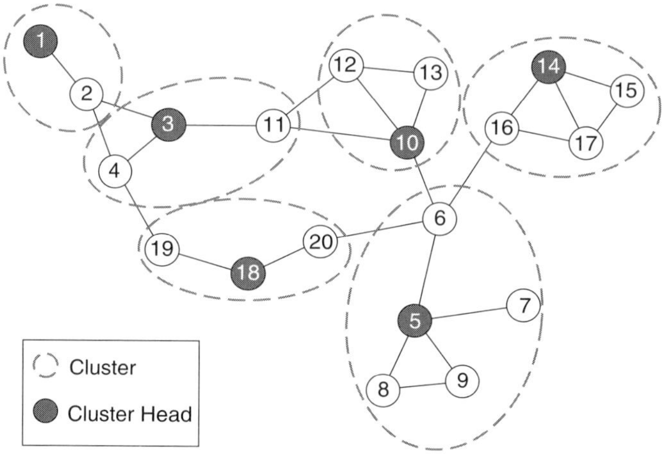

Figure 7.4. Node ID clustering.

In references 15 and 16 the authors propose algorithms that control the size of each cluster and the number of hierarchical levels. However, these algorithms are not suitable for WSN, since they add a significant computation overhead to the nodes in the networks and incorrectly assume that the topology can be controlled and deployed in a regular organized basis. Additionally, they are unable to prevent two nodes that are just over one cluster radius apart from simultaneously electing themselves as CH, causing a large overlap in the network's clusters. This problem occurs sufficiently frequently to make the resultant cluster packing inefficient.

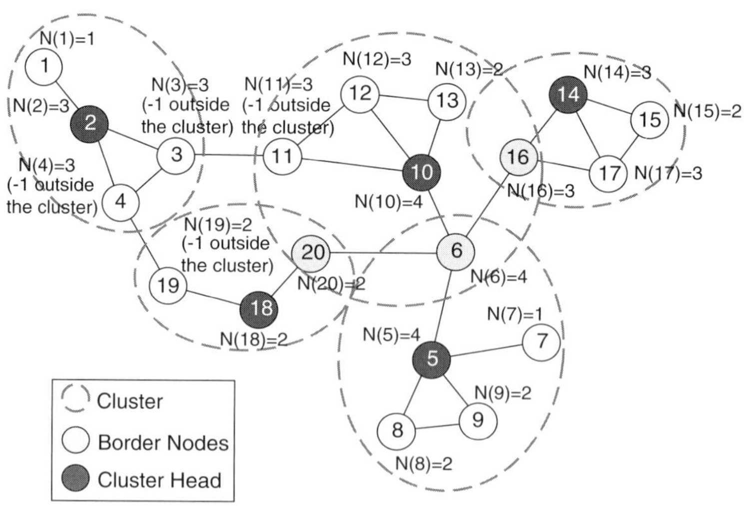

Figure 7.5. Node degree clustering.

Zhang and Arora [17] present a centralized scheme to produce an approximate hexagonal close packing. However, they assume that each node knows their precise location, which may be difficult to address in WSN. The clustering algorithms that use a centralized approach are generally based on the use of Spanning Trees to associate the vertices and to have several parent nodes in charge of the interactions with their child nodes and with their own parents in the superior level. A generalized clustering technique creating a tree-based construction for network partitioning is presented in reference 18. Thaler and Ravishankar propose to construct a top-down hierarchy based on an initial root node. However, the level of broadcasting messages used to create the tree is too large.

Another tree-based clustering protocol is proposed by Banerjee and Khuller in reference 19. They use a Breadth First Search (BFS) algorithm to partition the network. However, a drawback for using their algorithm in WSN is that only one node initiates the clustering process, which can provoke the loss of an entire subtree if one of the higher-level nodes in the tree suffers a failure. Additionally, this protocol still requires $O(n)$ time in linear networks, where n is the entire size of the network, which is outperformed by other protocols as we will show ahead.

7.3.2 Localized Protocols and Emergent Algorithms

In contrast with centralized algorithms, *localized algorithms* are characterized by reducing the amount of central coordination necessary and only require each node to interact with its local neighbors. Emergent algorithms are a class of localized algorithms. The emerging localized algorithms have the additional characteristic that the individual agents (i.e., sensor nodes) only encode simple local behaviors and do not explicitly coordinate on the global scale. A localized protocol for a sensor network is a protocol in which each sensor node only communicates with a small set of other sensor nodes within close proximity in order to achieve a desired global objective. Locality reduces the chances for protocol failure due to transmission errors and node failure.

In reference 20, the authors present an emergent algorithm called ACE (Algorithm for Cluster Establishment). The goal of ACE is to select the smallest set of cluster heads such that all nodes in the network belong to a cluster. The problem is similar to the minimum dominating set problem in graph theory. In ACE, the sensors are considered stationary and with uniformly random coordinates in the sensor field. The proposed algorithm is completed in constant time regardless of the size of the network, and it consists of two logical parts: The first part controls how clusters can spawn (by having a node elect itself to be leader), and the second part controls how clusters "migrate" dynamically to reduce overlap. The sensors continually compute the clusters, without the necessity of a specific event. This makes the protocol suitable for proactive networks. ACE results in a highly uniform cluster formation that can achieve a packing efficiency close to hexagonal close-packing, which minimizes the overlap between uniform circular clusters while ensuring full coverage. ACE is scale-independent (it is completed in constant time regardless of the size of the network) and operates without requiring geographic knowledge of node positions or any kind of distance or direction estimation between nodes.

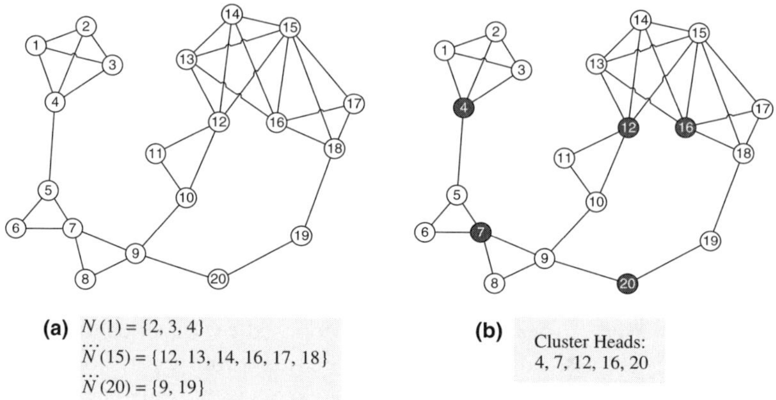

Figure 7.6. ACE cluster formation. (a) Initial state. (b) CHs dominating set.

The nodes initiate actions at random intervals to avoid collisions. Each time that an action can be initiated for a node is called a node's iteration. The communication between nodes is local and do not require time synchronization. ACE distinguishes between three possible node states: unclustered (still during the cluster formation process), follower (a node that belongs to a cluster with a designated CH), or cluster head. The iteration interval, or time interval between iterations among nodes, is uniformly random distributed. All nodes in the sensor field are initially in the unclustered state as shown in Figure 7.6a. When a node is unclustered at the beginning of its iteration, it assesses its surroundings and counts the number of loyal followers it would receive if it declared itself a CH of a new cluster. A loyal follower is a node that can belong only to the cluster that would be formed by the current node sensing its signal.

If the number of loyal followers for the node is greater than or equal to its spawning threshold function, the node will span a new cluster. The "Spanning Threshold Function" ($f_{\min}$) is given by the formula: $f_{\min} = (e^{-k_1(\frac{t}{cI})} - k_2)d$, where l is the number of loyal followers in the network, c is the desired number of algorithm iterations, I is the expected length of the iteration interval, t is the time passed since the protocol began, d is the estimated average degree of a node in the network, and k_1 and k_2 are constants to determine the shape of the exponential graph. $f_{\min}$ is an exponentially decreasing function, where the nodes at the beginning of the algorithm have less probability of becoming a cluster head, since the number of loyal followers is usually going to be less than $f_{\min}$, but the probability grows as the cluster formation process progresses. Based on the $f_{\min}$ function, the node will decide if it is becoming a cluster head or not. The set of CHs is a minimum dominating set in G (see Figure 7.6b).

If at the beginning of the iteration interval a CH already exists, this will send a POLL message to all its followers to determine which of them is the best candidate to become the new leader of the cluster. The best candidate to become CH is the node that has the greatest number of possible followers while minimizing the amount of overlap with existing clusters. Once the best candidate is determined by the current cluster head, it will promote the best candidate as the new cluster head and abdicate its position as the old cluster head. In some cases, the best candidate can be the same node that was already acting as CH. When a node becomes a CH, it conserves the previous ID if it was promoted by a previous CH or generates a random cluster ID if it declared itself as CH. After generating the cluster ID, the CH broadcasts a RECRUIT message to its neighbors, who will become followers of the new cluster. A node can be a follower of more than one cluster while the protocol is running (it picks a single cluster for membership only at the end of the protocol).

When the cluster head promotes a new node to CH, the position of the cluster will appear to migrate in the direction of the new CH because some of the former followers of the old CH are no longer part of the cluster, while some new nodes near the new CH become new followers of the cluster. Nodes that are not within one-hop radius of any CH can pick a clustered neighbor to act as their bridge to the CH, becoming two-hop followers. In the worst-case scenario, the loss of a CH in ACE would leave at most one cluster of the nodes unclustered.

Listing 1 shows and outline of the algorithm used by the nodes during ACE's first iteration, according to the pseudocode presented in reference 10. The number of iterations that the algorithm should execute is a nonfixed parameter that should be determined by the system using the algorithm, in order to determine a good tradeoff between communication overhead and cluster size. This algorithm only covers the aspects related to the clusters formation and does not include aspects related to data transmission after that. According to the simulations presented in the paper, the authors determined that 3 iterations were good for its execution, but they do not guarantee that this number is going to be the correct one for all executions of the protocol.

LISTING 1. Pseudocode for the ACE Algorithm

```
public class ACENode {

int ID, clusterID, bestLeader;
String state;
//time since the node started the protocol
double time;
int expectedIterationLenght;
int numLoyalFollowers, bestFollowerCount;

  public ACENode( ){
     //All nodes initiate in an UnclusteredState
     state = ''Unclustered'';
  }
  public int ScaleOneIteration( ){
      if (time > 3*expectedIterationLenght) {
```

(Continued)

(LISTING 1. *Continued*)

```
        if (state==''ClusterHead''){
            return 1; //DONE  }
        if (state==''Clustered''){
            //wait for cluster-head to terminate
            waitClusterHead();
            //pick one as cluster-head
            pickNewClusterHead();
            return 1; //DONE  }
        if (state==''Unclustered''){
            //pick a random clustered node to act as
              proxy
            //after it terminates
            pickRandomProxy();
            //wait for random proxy to terminate
            waitRandomProxy();
            return 1; //DONE  }
    }else{
        if (state==''Unclustered''){
            numLoyalFollowers = countLoyalFollowers();
            if (numLoyalFollowers >= fmin(time)){
                clusterID = generateNewRandomID();
                locallyBroadcast(''Recruit'',ID,
                  clusterID);  }
        }else{
            if (state==''ClusterHead''){
                bestLeader = ID;
                bestFollowerCount = numLoyalFollowers;

            //all nodes where the node is a potential
               new cluster-head
               ArrayList potCH = definePotential-
                 ClusterHeads();
               Iterator iter = potCH.iterator();
               ACENode potential=null;
               while (iter.hasNext()){
                 potential = (ACENode) iter.next();
            followerCount = PollForNumLoyalFollowers
              (potential,clusterID);
                  if (followerCount > bestFollower-
                    Count){
                      bestLeader=potential.ID;
                      bestFollowerCount=
                        followerCount;  }
            }
            if (bestLeader!= ID) {
               send(bestLeader,''Promote'',clusterID);
               //waits for bestLeader to broadcast
```

```
                        it's ''Recruit''
                        //message
                        while(!recruitMessageReceived());
                        locallyBroadcast(''Abdicate'', ID,
                          clusterID);  }
                  }
              }
          }
      }
}
```

Because of being an strictly localized algorithm, ACE uses a fixed amount of time $O(d)$ for its execution, regardless of the total amount of nodes in the network. The estimated average degree d of a node in the network is calculated prior to the network deployment.

7.3.3 Distributed Dominating Set-Based Algorithms

One of the problems that arise in a number of distributed network applications is how to locate a small number of centers in the network such that every node is close to at least one center. Clearly, this is the same objective targeted by *clustering*, and one of the main objectives when doing this is to define the cluster concentration points while creating the minimum number of clusters in the network and guaranteeing the total coverage of the network.

Graph Theory approaches this problem by calculating a dominating set DS_G in the graph. Computing a DS_G is also a way of increasing the network's lifetime. The idea is to build a network backbone (the DS) consisting of nodes with higher energy. In reference 21, the nodes in the dominating set (called *coordinators*) change from time to time in order to maximize their lifetime. An approach similar to the ones presented in reference 14 is proposed in reference 22 using both the node ID and the node degree of the vertices to select the CH. After that, they use clustering algorithms designed based on a Depth First Search (DFS) traversal of nodes in the network. These algorithms are initialized by any node and are fully distributed. They create and maintain k-clusters, in which any node is at distance at most k hops from the CH. In the clustering process, each node is either a CH, a covered nodes or an undecided node. In the basic k-cluster algorithm, each undecided visited node in DFS declares itself a CH and covers all its k-hop neighbors. In the highest connectivity k-cluster algorithms, each undecided visited node checks all its undecided k-hop neighbors and chooses one with the largest connectivity to be its CH. Chen and Stojmenovic consider the mobility of nodes and describe algorithms for modifying cluster structure in the presence of topological changes. The efficiency of the clustering algorithms is tested by measuring the average number of created clusters, the number of border nodes, and the cluster size in random unit graphs. Basagni [23], also uses the IDS_G approximation but generalizes the idea of using "weight," such that any meaningful parameter can be used in order to best exploit the network properties.

The fast distributed approximation algorithms proposed in reference 24 use a synchronous model of computation using a variant of DS_G called the k-dominating set where every node is within a distance k of some server or directory copy—in this case, of some CH. The proposed algorithm is called Local Randomized Greedy Algorithm (LRG) and is similar to the greedy algorithm for Minimum Set Covering Computation presented in reference 25, but instead of only considering the span of the nodes, LRG considers how the nodes covered by an individual node are also covered by other nodes.

The algorithm creates the clusters in three steps. First each node v calculates its span $d(v)$; this is how many of the current adjacent uncovered nodes it will cover if it becomes a CH. This span $d(v)$ plus the node ID is sent to all nodes within a distance equal to 2 around u. Each neighbor computes the maximum and forwards it one more step. With this information, every node can determine whether it is a candidate or not. In the third step, called the *selection step*, a node u calculates its support $s(u)$, which is the number of candidates that cover u. The node u sends its $s(u)$ to the candidates in its $N(u)$. Each node v adds itself to the dominating set lexicographically higher than that of any node within its 2-hop neighbors. One drawback of this approach for WSN is that once a node and it neighbors are covered, the node does not participate further in the cluster formation process, which will cause a higher rate of energy consumption in the node if it is a CH, leading it to its death because of battery depletion.

7.3.4 Combined and Additional Approaches

In this section we present some other clustering mechanisms that combine some of the previously explained approaches and therefore cannot be classified in only one of the categories. We include the Zonal Algorithm, the Max-Min D-Cluster formation, and the Topology Adaptive Spatial Clustering algorithm (TASC).

The Zonal Algorithm proposed in reference 26 aims to find weakly connected dominating sets in the network. The algorithm consist of three phases: (1) An input graph representing the network is partitioned into regions of approximately size x. (2) The distributed algorithm for weakly connected sets is run in each region. (3) Some additional border vertices are added.

In phase 1, graph partitioning using minimum spanning forests, a sender v of degree $d(v)$ can broadcast to its neighbors in constant time with $O(d(v))$ messages. A vertex can also send a message to any adjacent vertex in constant time. For finding a Minimum Spanning Tree (MST), all edge weights are distinct, breaking ties using the vertex IDs of the endpoints. The MST is unique for a given graph with distinct edge weights. The algorithm maintains a spanning forest. Initially, the spanning forest is a collection of trees of single vertices. At each step the algorithm merges two trees by including an edge in the spanning forest. During the process of the algorithm, an edge can be in any of the three states: tree edge, rejected edge, or candidate edge. All edges are candidate edges at the beginning of the algorithm. When an edge is included in the spanning forest, it becomes a tree edge. If the addition of a particular edge would create a cycle in the spanning forest, the edge is called a rejected edge and will not be considered further. In each iteration, the algorithm looks for the

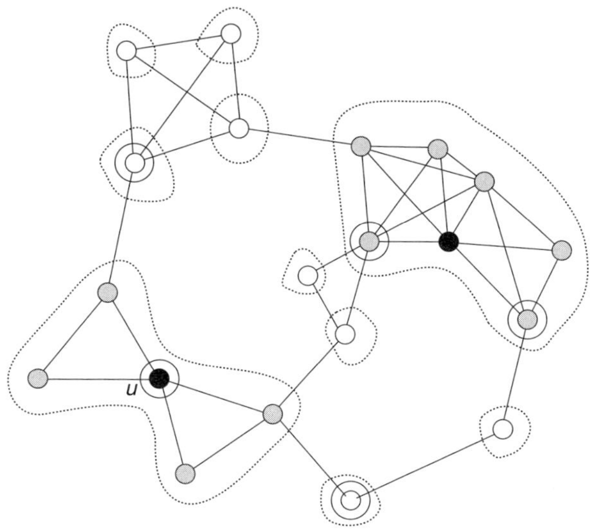

Figure 7.7. GHS (Gallager, Humblet, and Spira) algorithm for finding an MST (Minimum Spanning Tree).

candidate edge with minimum weight, and the algorithm changes it to a tree edge merging two trees into one. During the algorithm, the tree edges and all the vertices form a spanning forest. The algorithm terminates when the forest becomes a single spanning tree. This partitioning process is based on the GHS algorithm proposed in reference 27, however, any distributed algorithm for constructing spanning trees can be used. Figure 7.7 shows an example of the zonal division obtained after applying the GHS algorithm.

In phase 2, for computing the weakly connected DS_G of the regions, they use a color-based algorithm. This is a distributed implementation of the centralized greedy algorithm for finding small weakly connected DS_G in graphs. The root of the spanning tree is used as an arbitrator. The root starts an iteration by broadcasting a request through the tree to search for the vertex with the maximum improvement value. Once the root has determined the global maximum, it sends a message to that vertex and colors it. Phase 3 is used to fix the cluster borders by constructing a bipartite graph. Two regions R_i and R_j joined by a dominated edge can comprise a single region with a dominating set $S_i \cup S_j$ and do not need to have their shared border fixed. If neighboring regions R_i and R_j are not joined by a shared dominated edge, the region with the lower subscripts adds a new vertex from the $\frac{R_i}{R_j}$ border into the dominating set. Figure 7.8 shows an example where R_3 and R_4 share a dominated edge and would be fixed into one only region R_3.

In reference 28, the author presents a polynomial-time approximation algorithm to minimize the number of clusters in the network. First, it uses the concept of vertex cover to create the clusters and then creates a tree inside each cluster to improve the transmission processes in the network. This approach uses the community access network concept to connect static wireless networks with a wired backbone network.

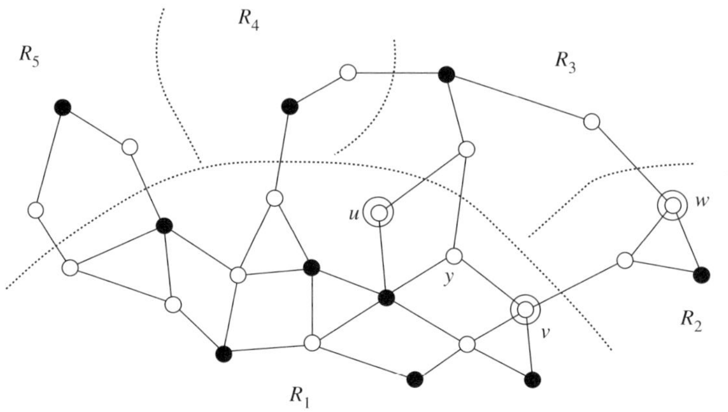

Figure 7.8. Border fixing in the zonal clustering algorithm.

It divides the network into clusters and selects a single access point at each cluster. This access point has a direct link to the wired network and acts as a gateway between the wireless nodes and the wired network.

For its execution, the proposed algorithm assumes that the upstream and downstream bandwidth requirements of every node in the network are the same. It also assumes that there is a maximal weight $W \in \mathbb{R}^+$ for the graph. The algorithm performs an initial operation to create the clusters and in the process takes into account the following constraints: (a) The resulting clusters should be disjoint, which means that every node is included in a single tree; (b) they should contain all network nodes; (c) each cluster must induce a connected graph with fixed bound R, which is the radius for multihop routing with bounded propagation delay and therefore is the maximum value for the depth of every tree; and (d) a spanning tree T is created inside each cluster to simplify the routing process and maximize the network utilization. The delivery trees inside the clusters must cover all the nodes in the network. Each node knows its parent and its descendant nodes in T. The total weight of all the nodes in a tree is at most W. A graph decomposition that satisfies the previous requirements is called a feasible partition, and every tree in the partition is called a feasible delivery tree.

In the first stage of the clustering algorithm, it looks for the minimal number of cluster heads such that they cover their R-neighbor nodes. The author proposes three different ways to select the CHs: (1) to use the greedy set cover algorithm proposed in reference 29, (2) to use a shifting strategy approach, which is based on the *"divide-and-conquer"* method and performs a series of shifting operations to subdivide the networks in a progressive manner, finding a near-optimal solution but with the expense of a high running time, or (3) to use a dominating independent set approach, where each iteration of the algorithm selects the node with the maximal number of uncovered nodes in the R-neighborhood of each node v. The second stage of the algorithm performs a cluster refinement by associating each node with its closest $y_i \in Y$ (the CHs set), and it breaks ties by selecting the CH with the lowest ID. The third stage creates the tree inside the cluster and executes a post-order tour of the tree to calculate its total weight. Finally, the algorithm reduces the relay load of the cluster nodes by selecting as an access point a node that is closer to the tree center. The root of every tree T is the access point to the wired network. This root node coordinates all the transmissions inside the clusters by using slotted channels and polling mechanisms. The delivery mechanism attempts to maximize the clusters' throughput.

According to the simulations presented in reference 28, this clustering technique yields results close to optimal. Since the weight partitioning algorithm allows the average cluster size to be as small as $W/3$, it produces clusters that are small in comparison to other clustering techniques. However, there is a drawback in this proposed technique: The access point constitutes a single point of failure in each cluster. If the access node is a cute node (if its removal leaves the graph disconnected), the system attempts to attach the disconnected nodes to an adjacent cluster. Otherwise, the algorithm should send the proper recovery messages to modify T for reconnecting the detaching nodes, which can be very costly for a WSN. In the Max–Min D-Cluster formation proposed in reference 30, the CHs are selected such that they form a d-hop DS_G, where $d > 1$ distance hops form the vertices in the DS_G away from any other vertex in C. Since d is an input value to the heuristic, it enables control over the density of cluster heads in the network.

The Topology Adaptive Spatial Clustering (TASC) algorithm presented in reference 31 is a distributed algorithm that partitions the network into a set of locally isotropic, nonoverlapping clusters without prior knowledge of the number of clusters, cluster size, and node coordinates.

The spatial grouping of nodes with respect to regions of close proximity and similar deployment density does the following: (a) It promotes efficient data aggregation, (b) it allows the use of different data compression rates in each cluster, (c) it improves the overall compression rate in the whole network, and (d) as a consequence, it also helps reducing the propagation of redundant data inside the network. Spatial clustering would also assist transmission power control, since intracluster communication requires less transmission power in dense clusters. TASC is useful for proactive networks, where the clusters provide the information needed for the cluster formation process. The nodes in the sensor field are assumed to be stationary. Each node can measure distances to its 1-hop neighbors and has knowledge of its 2-hop neighborhood, as shown in Figure 7.9.

TASC operates on a combination of node weights and a dynamic density reachability criterion. The main objective of TASC is to cluster nonuniform sensor networks. In this type of network, the node density variations are globally large but there exist subgroups of nodes such that density variations are locally small. The idea of TASC is to partition those networks in a way such that relative node density variation in individual clusters is smaller than relative node density variation in the whole network, obtaining a distribution similar to the one shown in Figure 7.10.

For the cluster formation, two different parameters must be previously specified: required minimum cluster size and density reachability parameter. The latter

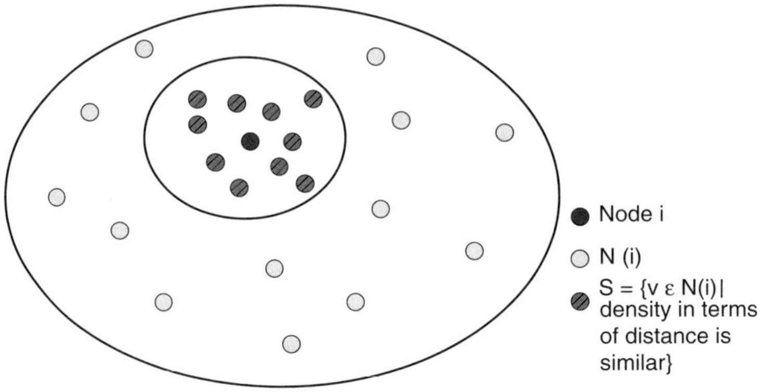

Figure 7.9. TASC neighbor's information.

parameter allows each node to further limit the number of nodes that it can potentially nominate as CH by considering only density reachable nodes as nomination candidates. This effect pulls cluster leaders toward the most dense groups in the cluster, but nomination among density reachable candidates is still based on weights.

Listing 2 shows an outline of the algorithm used by the nodes during TASC's execution, according to the pseudocode presented in reference 31. Initially, each node in the WSN computes its own weight based on shortest Euclidean paths in its

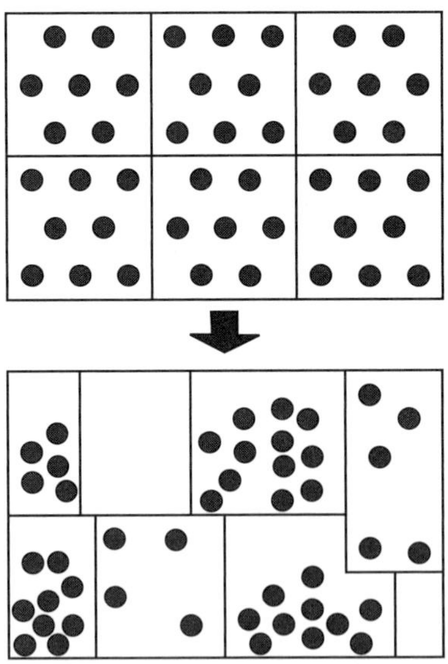

Figure 7.10. TASC cluster's nodes distribution.

2-hop environment. In a nonuniform deployment WSN, the node that tends to be the midmost related to all shortest communication paths (in terms of hops) gets the largest weight. Instead of incrementing the weight by one each time a node is used in a path, we increment the weight as a function of the distance a node contributes to the path.

LISTING 2. Pseudocode for the TASC Algorithm

```
public class TASC {

int ID;
double weight;
TASC nominee, leader;
int clusterSize;
ArrayList clusterMembers;

 /*
 * 2HNeighborhood = Euclidean paths in the 2-hop
   neighborhood of the node
 * internode = Inter-node distance measurements
 * minClustSize = Minimum cluster size
 * Dr = Density reachability parameter
 */

  public TASC(2HNeighborhood,internode,minClustSize,Dr){

        //The node computes its own weight based on
        //2HNeighborhood
        weight = computeWeight(2HNeighborhood);
        broadcastToNeighborhood(weight);

        //all weights received
If (receiveWeights()){
            //finds the heaviest density reachable node
            nominee = findHDRNode(Dr);
            broadcastToNeighborhood(nominee);
        }

        //all nominations have been received
        If (receiveNominations()){
            //Selects the closest nominee
            leader = findClosestNominee();
            broadcastToNeighborhood(leader.ID, this.ID);
        }

        //this node is leader
```

(Continued)

(LISTING 2. *Continued*)

```
        If (leader.ID == this.ID){
            //Wait until election timeout
            while (!electionTimeOut);

            broadcastToNeighborhood(clusterMembers,
            clusterSize);
        }

        //cluster size is received
        If ((clusterSize = receiveClusterSize())!=0){
            If (clusterSize < minClustSize){
                //Selects the closest neighbor for which
                //clustersize >= minimum cluster size
                leader=selectClosestCluster
                  (2HNeighborhood);
                //joins previous node's cluster;
                this.joinCluster(leader);
            }
            this.broadcastToNeighborhood(leader.ID,
            clustersize);
        }
    }
}
```

The weight value calculated by the node is broadcasted to its 2-hop neighborhood. Since this is done by all nodes, each node also receives the weights of its 2-hop neighbors. By comparing the weights, each node nominates the node having biggest weight in the density-reachable subset of its 2-hop neighbors and broadcasts its nominee to its 2-hop neighborhood. After receiving all nominees in its 2-hop neighborhood, each node elects the closest nominee to its leader. Each node that ends up in a cluster where the total number of nodes is smaller than a prespecified minimum cluster size joins the closest cluster, where the number of nodes exceeds the required minimum cluster size.

After creating the clusters, TASC uses an "all-pairs-shortest path routing." Instead of trying to forward traffic to the neighboring node that is closest to the destination, TASC routing is based on distance measurements to extract information about the network topology. More specifically, node weight is a measurement of two key quantities: (1) the frequency a node is found on the shortest path between pairs of nodes, and (2) the distance contribution of the edges of that node with respect to the total length of the path.

As mentioned at the beginning of the section, creating balanced clusters is one of the main objectives for WSN. In reference 32, the authors propose an algorithm for clustering the sensor nodes such that each cluster (and its corresponding CH) is balanced and the total distance between sensor nodes and CH is minimized. Balancing the clusters is needed for evenly distributing the load on all master nodes. If there is no balance constraint, the authors propose the utilization of Voronoi Diagrams [7], by constructing it based on the number of CH deployed previously in the sensor field. However, each CH can only manage a certain number of communication channels. In this case, trying to solve the k-clustering problem optimally helps to maintain the network's operation. This k-clustering problem attempts to group the sensor nodes such that each cluster is balanced and has exactly one CH. To solve the problem, the authors transform it into a min-cost flow instance by adding a source node s and a sink node t to G, both with infinite capacity. Each edge has a weight equal to the message transmission energy dissipation between the two end vertices. There are n directed edges from s to all vertices corresponding to sensors nodes. Similarly, there are k directed edges from vertices corresponding to CH to t. All edges incident to s or t have weight 0. Finally, nodes corresponding to sensors have capacity 1, while nodes corresponding to CH have capacity $\frac{n}{k}$. Each flow solution is corresponding to a k-clustering solution. Constructing G and the corresponding k-clustering solution can be done in $O(n.k)$ time. Hence, the k-clustering problem can be optimally solved in $O((n+k)^3)$ time. However, the major drawback of this proposal is that it assumes that the nodes are *not randomly deployed* in the sensor field. This is not a valid assumption for the majority of WSN application.

7.4 CLUSTER'S CONSTRUCTION AND MAINTENANCE IN MOBILE ENVIRONMENTS

Maintaining clusters as the topology changes is an issue that arises in mobile wireless sensor networks. A local change in the weight may result in a different vertex becoming a cluster head. As an example, if we are using IDS_G, when two CH become adjacent, one of them has to abdicate. This and other event *occurrences* can generate a total reordering of the nodes in the WSN and create the necessity to have maintenance operations, which in turn may cause other changes to propagate to the network.

Despite the evident impact of mobility issues in WSN, very few clustering proposals take it into account. Most proposals consider the nodes as stationary since designing mechanisms to solve the issue is a very challenging task. During the development of this survey, we only found three research papers involving mobility in the clustering techniques proposed for WSN using Graph Theory. References 33–35 are the subject of our next analysis in this section. The purpose is to analyze how those proposals approach the creation and maintenance of clusters in a mobile environment.

7.4.1 Zone-Based Clustering

Chen and Liestman [33], present a modification to their "Zonal weakly connected clustering algorithm," presented in reference 26, including cluster maintenance in the presence of network topology changes. In this new proposal, the cluster maintenance is divided into two layers: intrazonal (inside the zones) and interzonal (between zones, at the borders). One of the assumptions made to manage mobility is that all the changes occur sequentially and that the network can be restructured before the next topology change occurs. As a result of the nodes motion, the original weakly connected graph can become disconnected. This creates induced disconnected subgraphs that are controlled using some timeout mechanism that allows the formation of a new zone inside the network. If when the timeout interval is completed no other zones have annexed the nodes to its cluster, a new cluster is created.

The cluster formation is performed in three phases. Phase 1 divides the network into zones, creating a set of trees. Phase 2 executes a greedy algorithm to find a small weakly connected dominating set within each zone. Phase 3 fixes the border vertex of each zone. A *border vertex of a zone* is a vertex that is adjacent to vertices of other zones. In this last phase, if in the border between two zones a dominating edge exists between the border vertices, there is no need of new vertices, but if there is no dominating edge, it is necessary to add new vertices to "fix" the border. All the cluster formation process is supposed to be executed at the beginning of the network's life. After this process, the following operations are *maintenance* operations. There are four different events to be considered during this maintenance stage: (1) edge-down, (2) vertex-down, (3) edge-up, and (4) vertex-up.

In the **edge-down** event, if the lost edge is a free, nondominated edge, the graph remains the same and no action is taken, but if the edge (u, v) is a *dominating edge*, its loss must be reported by the vertices incident upon it. u and v broadcast an *edge-loss* message to determine if the graph has been broken into two pieces. If so, the root node in the cluster performs the *breach suturing* procedure to add a dominated vertex to the zone and fix the breach. The dominated vertex added is the one with the greatest *improvement value*, the node that is going to allow the current zone to connect with the greatest number of other zones. In the example shown in Figure 7.11 a, the initial clustering scheme has the edges (2,13) and (4,5). If (2,13) is down, nothing happens because it is not a dominating edge. If (4,5) is down, the root has to perform the breach suturing to create the new edge (3,5) and must calculate the new dominating set in G (Figure 7.11b).

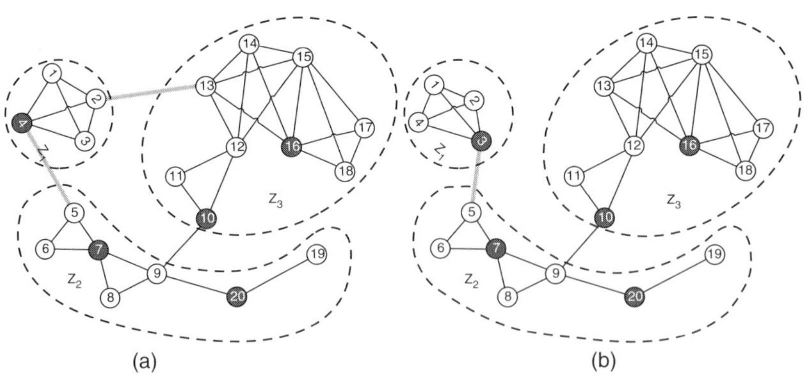

Figure 7.11. Breach suturing procedure in an edge-down event.

The **vertex-down** event is treated very similar to the edge-down event, since a detecting vertex cannot easily distinguish between the loss of an edge and the loss of a neighboring vertex. If the vertex that detects the loss still has another dominated edge incident to the lost vertex's zone, it does nothing; otherwise, it will send a message to its root for it to fix the border. However, a special operation must be performed in case the vertex down is the root vertex for the zone. In this case, another vertex must take on the role of the root. To achieve this, every root has a neighbor as a backup. If the loss is discovered by the backup, it assumes the role of the root. However, it may be the case that what is lost is the edge between the backup and the root, not the nodes. In this case, when the duplicated roots detect the redundant condition, the backup must give up on its new role and the original root joins the regions of itself and its backup neighbor.

The **edge-up** event is caused by one node moving closer toward another. In this case, if the new edge is a free edge, no changes need to be made, but if the new edge is a dominated edge, some vertices in the DS_G can become redundant and can be eliminated from the weakly dominating subgraph.

When a new vertex is added in an **vertex-up** event, it is dominated if it has a neighbor in the dominating set; otherwise, it changes its status to be dominating.

As can be observed in the previous operations, it is necessary to add or remove vertices from the zones every time the network topology changes. To accomplish this, a bipartite graph is formed between the dominating vertices of the zones.

Although Chen and Liestman [33] show in their results that the weakly connected dominating set remains roughly the same size over time, it is not clear from the paper how they add or remove vertices to the zones. If these operations involve the additional deployment of the nodes in the field, the application of this algorithm in many sensor network applications will not be feasible, since they consider a unique node deployment at the beginning of the WSN operation.

7.4.2 Peer-to-Peer Generalized Clustering Model

Ramaswamy et al. [34], present **CDC: Connectivity-based decentralized node clustering scheme**, a clustering scheme where every node only requires local knowledge about its neighbors, providing the ability to cluster the network automatically or to discover clusters around a given set of nodes. CDC is able to handle nodes' dynamics without resorting the whole network at each entry and exit.

CDC is based on the idea of flow in the network, where every edge is considered a road between two points. Every time a message is passed through an edge, it decrements its TTL (Time to Live) counter until it becomes negligible and is discarded, but it also accumulates a communication weight. Nodes with greater communication weight are those that lie inside a cluster. To calculate the accumulated communication weight for each node, a group of messages is sent from some specific nodes called the *originators*.

To select the originators, two properties are used: (1) The set of originators should be spread out in all regions of the graph, and (2) a node v_l is considered to be a good originator if it acquires more weight due to messages initiated by it than the weight acquired by messages initiated by any other originator: $\text{TotalWeight}(V_l, V_l) \geq \text{TotalWeight}(V_l, V_i) \forall V_i \in V$. To evaluate the first property, each node checks if it has already received any clustering messages from other nodes in its vicinity. If so, it means that there are other neighbor nodes that have already chosen to be originators and, hence, the node opts not to become an originator. Otherwise, the node moves to evaluate the second property. This property, however, can only be evaluated if the CDC algorithm has already been executed, but to execute the CDC algorithm we need to select the originators. To solve this situation, the authors use the **THP (two-hop return probability)**. THP, calculated as

$$\text{TwoHopProb}(V_l) = \sum_{V_i \in N(V_l)} \frac{1}{\text{Degree}(V_l) \times \text{Degree}(V_i)}$$

determines the probability of returning to a node V_l in the graph in two hops if random messages were sent starting at V_l. If the THP is higher than a predefined threshold, then the node chooses to be an originator.

The selected originators use the same TTL value for their messages. Messages are sent to the network, and each node that receives the messages calculates (after an stabilization time) which originator has a higher weight and joins it.

To manage with the dynamics of nodes, two mechanisms are considered: **node entry** and **node exit**. In the **node entry** mechanism, the entering node calculates its *attraction* to the vertices on an specific cluster. The *Neighbor-Attraction* function toward another node is defined as

$$\text{NbrAttraction}_{V_{N+1}}(V_j) = \frac{1}{\text{Degree}(V_j)}$$

where V_{N+1} is the entering node and $V_j \in N(V_{N+1})$. The node joins the cluster where the sum of the attractions to the nodes in that cluster is the highest. In the **node exit** mechanism, the neighbors of the exiting node have two possibilities. If they are not in the same cluster as the exiting node, they only update their connectivity information. If they are in the same cluster as the exiting node, they have to perform the *Node attraction* computation; to define it after the node leaves, they are going to stay in the same cluster or they are going to "move" to a new one. If the leaving node is an originator, the neighbors have to be informed and they elect the neighbor with the highest number of in-cluster edges as the new originator. Special cases may occur when a node is entering the network while one of its neighbors is leaving it, or vice versa. In these cases, the computation of the attraction has to be done more than once and increases the computation overhead of the clustering algorithm.

Although the authors state that this model is suitable for any P2P network, including sensor networks, we consider that this proposal is not suitable for WSN. Without considering the "special cases," the amount of information that the nodes have to handle is large and the computation overhead would cause a rapid depletion of the sensor's energy.

7.4.3 Cluster-Based Graph Network

In the schema for the *construction and maintenance of a cluster-based architecture for a sensor network* proposed in reference 35, the network is treated as a bidirectional graph G and $CNet(G)$ is a cluster-based network of G. The $CNet(G)$ is a spanning tree of G. The nodes in the WSN are grouped into disjoint clusters, where the backbone is a tree consisting of the cluster heads and gateway nodes. A cluster of G is a star subgraph of G, where the cluster head has an edge to each other cluster member and no edges exist between two cluster members in the cluster.

To minimize the number of clusters, two cluster heads cannot be neighbors with each other. Thus, they form a maximal independent set in G. To guarantee the communication between cluster heads, any two cluster heads are joined through one special cluster member called a *gateway node*, which is an intersection of neighbors in G of two cluster heads. The cluster members that do not act as gateways are called *pure cluster members*. A transmission between a cluster head and its members is called *local transmission*, and a transmission between cluster heads is called *backbone transmission*.

Figure 7.12 shows an example of a clustering scheme, where nodes 3, 7, 10, and 16 are CH, nodes 6 and 13 are gateway nodes, and the remaining nodes are pure cluster members. CH and gateway nodes form the network backbone. Bold edges and the vertices they connect form the $CNet(G)$ in G.

The broadcasting technique used here is called ***Eulerian broadcasting procedure***. In this technique, a message called *token* starts from the source node, visits every node in depth-first order, and turns to the source node, forming an Eulerian cycle. When a node v gets the token, it sends the token with the message and its ID to one of its neighbors which has not been visited by the token yet. If v has no unvisited

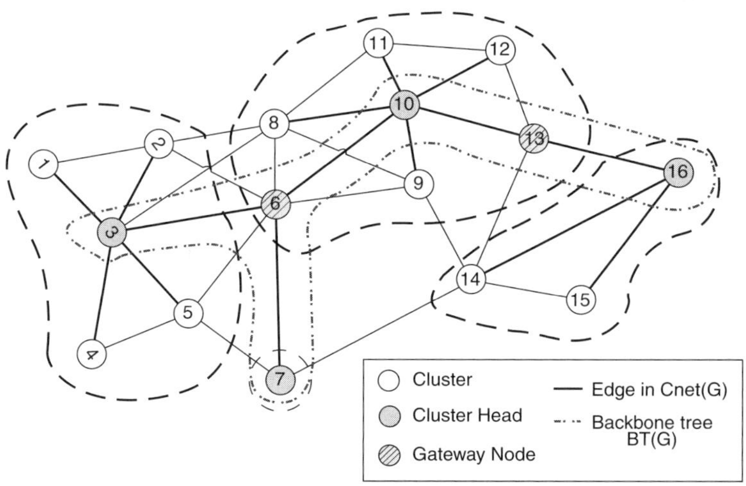

Figure 7.12. *CNet*(*G*) clustering scheme.

neighbors, it returns the token to the node from which it got the token for the first time. Two operations are defined to treat the movement of nodes: **node-move-in** and **node-move-out**. These operations are supposed to be atomic, and they can not be performed simultaneously.

In this clustering architecture, a new node u announces itself by sending a message to join the existing $CNet(G)$ and the network reorganizes itself. When u performs the **node-move-in** operation, it checks the status of its neighbors. If there exist cluster heads in $N(u)$, u selects one to be its head and itself becomes a pure cluster member. Otherwise, if there are gateway nodes in its neighbors, u selects one of them as its gateway and becomes a cluster head of a new cluster. If there are no cluster heads and no gateway nodes in $N(u)$, u becomes a cluster head and sets one neighboring pure cluster member to be the gateway of itself. After this process, the nodes in $N(u)$ update their information. According to the authors, a $CNet(G)$ can be formed statically and dynamically in $O(n)$ and $O(\sum_{i=1}^{n} T_i)$, respectively, where $T(i)$ is the number of rounds that a node-move-in operation requires for an i-node graph.

When a node u wishes to leave the network, it must perform the **node-move-out** operation. If u is a pure cluster member, it sends an *"I'm leaving"* message and simply leaves from the network. Otherwise, the node-move-out algorithm works as follows: If u is a cluster head or a gateway node, $CNet(G)$ is divided into two subtrees. One is the tree T with u as the root (not including the root in $CNet(G)$) and one is the tree H with the root of $CNet(G)$ as the root. u is removed from T and the other nodes of T are added to H by using the node-move-in operation on each node, so that the resulted tree is $CNet(G)$. If u is the root of the whole network and $N(u) \neq \varnothing$, a cluster head that is connected with u by a gateway is elected and is set as the new root of $CNet(G)$; if $N(u) = \varnothing$ and u is the only cluster head in $CNet(G)$, one of the nodes in $N(u)$ becomes the new root and calls the Eulerian procedure to create a new $CNet(G)$.

Uchida et al. also propose a second version of the node-move-in and node-move-out operations, considering networks where the first operation is performed more frequently than the second one. In this case, the list of neighbors for every node is not maintained. For the **node-move-in** operation, one leader is first elected from the neighbors of the new node u. If $S = \{$the leader (not a cluster head)$\} \cup \{$neighbor cluster heads of $u\}$, every node $v \in S$ sends its ID at some round t. If u receives an ID at round t, it means $|S| = 1$ and S contains only the leader and there are no cluster-head neighbors of u. If u receives no ID at round t, it means $|S| \geq 2$ and there exists at least one cluster head from which u can elect its cluster head. Otherwise, it can check for gateways in order to select one of them as its gateway when it becomes a CH. For the **node-move-out** operation, u recognizes its neighbors (except its parent and its children) by calling the Eulerian procedure described above, and after that it performs the same operations described in the former **node-move-out** operation.

Although this clustering architecture is the one we consider the most appropriate between the tree approaches considered in this section, there are still some issues remaining—for example, if one sensor is aware when it is leaving the network in case of energy depletion or malfunction, instead of just moving out of the cluster range.

7.5 SUMMARY AND OPEN RESEARCH PROBLEMS

In this chapter we study the fundamental concepts of clustering process in WSN from the Graph Theory perspective. To this end, we survey current existing works in the literature which use graph theory concepts to perform clusters in WSN. Most of the algorithms and protocols presented in this chapter meet the design criteria mentioned in Section 7.3 for clustering in WSN; however, the inclusion of mobility as a new criterion for the clustering creation and maintenance adds new challenges for these clustering techniques. Research for the maintenance of the cluster organization under this new scenario is seldom seen. To cope with the typical dynamics of mobile WSN, some of the existing clustering protocols, like the ones presented in Section 7.4, include new operations that allow new nodes to be incorporated into appropriate existing clusters and also allow the properly management of the departure of nodes from the clusters. However, those mechanisms still have to be further modified to handle situations where the nodes cannot give any warning to the networks about their abandon of the network, like in the case of energy depletion.

Likewise, there are still several aspects to improve the performance of clustering mechanisms in WSN. One of those aspects is related to exploiting the redundancy and spatial correlation of the nodes in the sensor field. This can allow the implementation of fault tolerance techniques during the data transmission phase of the clustering algorithms, taking into account that sensors will measure similar data as long as they remain located in a relatively close position among them. In the same way, this aspect can help reduce the energy consumption in the nodes, thus making their lifetime longer.

The inclusion of security techniques also affects the functioning of the clustering methods, when the information exchanged between the nodes is critical for the network operation. Integrating dynamic clustering with secure information quality-driven clustering and routing protocols will allow a more efficient communication inside the network and between the cluster heads and the base station.

In the mobility scenarios, there is still some work to do to (a) determine the best mechanisms to trigger the update routines in the WSN, (b) look for a better stability in the network, and (c) maintain a consistent state of the initially created clusters. Issues such as how to adapt the clusters very quickly to the continuous topology changes or how to make the network react when those changes take effect during an update period are still to be considered. Also, the inclusion of mobility patterns in the simulation will allow us to create test environments more similar to the real world and to investigate how the complex set of interactions occur between the nodes deployed in the field.

7.6 EXERCISES

1. Why do you think creating the minimum number of clusters (and also with smaller sizes) in the network is one of the main objectives of clustering techniques?

2. In the Local Randomized Greedy Algorithm (LRG) proposed in reference 24, the authors define the clusters using the calculation of the set covering in the graph, taking into account the span of the nodes and how the nodes covered by an individual node are also covered by other nodes. How does this last aspect improve the performance of the clustering process?
3. How do spatial correlation and data correlation affect clustering mechanisms in WSN? Is it feasible to exploit redundancy by taking these aspects into account when developing clustering mechanisms?
4. In topology adaptive spatial clustering (TASC), the authors use the 2-hop neighborhood of the nodes. Why use the 2-hop instead of reducing it to only the 1-hop neighborhood or extending it to the n-hop environment of the nodes?
5. How would you modify the node-move-out operation in the cluster-based graph technique proposed in Section 7.4.3, so that the nodes that remain in the network could react even if the moving-out node does not announce that it is leaving the network?
6. Clustering techniques should remain as a network layer problem or should be integrated with the lower and upper layers (MAC, Application)? For example, should they allow the management of administrative policy constraints during the clusters creation?
7. What is the relation between mobility and topology changes and energy constraints in WSN?
8. The Algorithm for Cluster Establishment (ACE) presented in Section 3.2 uses a localized emergent algorithm to create the clusters in the network. What are the advantages or disadvantages of this kind of algorithm, compared to those that rely on global interaction or information?
9. How does the in-network processing required for the creation of clusters affect the security mechanisms in WSN?
10. How often should the topology change mechanisms be triggered in Mobile WSN to ensure the consistency of the clusters and maintain the network functions?
11. An extension to the mobility management in WSN considers the problem of "group mobility," where the nodes do not move using an individual pattern but following a group pattern. How do you think this affects the proposals presented in Section 7.4?

BIBLIOGRAPHY

1. M. Ye, C. Li, G. Chen, and J. Wu. EECS: An energy efficient clustering scheme in wireless sensor networks. In *National Laboratory of Novel Software Technology, Nanjing University—China, Department of Computer Science and Engineering, Florida Atlantic University, USA*, 2005.

2. W.-P. Chen, J. C. Hou, and L. Sha. Dynamic clustering for acoustic target tracking in wireless sensor networks. In *Proceedings of the 11th IEEE International Conference on Network Protocols (ICNP 2003)*, November 2003, pp. 284–294.
3. J. N. Al-Karaki and A. E. Kamal. Routing techniques in wireless sensor networks: A survey. Department of Electrical and Computer Engineering, Iowa State University, December 2004.
4. W. R. Heinzelman, A. Chandrakasan, and H. Balakrishnan. Energy-efficient communication protocol for wireless microsensor networks. In *Proceedings of the 33rd Hawaii International Conference on System Sciences*, Vol. 8, January 2000, pp. 3005–3014.
5. C. Liu, K. Wu, and J. Pei. A dynamic clustering and scheduling approach to energy saving in data collection from wireless sensor networks. In *Proceedings of the Second Annual IEEE Communications Society Conference on Sensor and Ad Hoc Communications and Networks (SECON 2005)*, Santa Clara, California, September 2005.
6. A. Manjeshwar and D. P. Agrawal. TEEN: A routing protocol for enhanced efficiency in wireless sensor networks. In *Proceedings of the 15th International and Distributed Processing Symposium*, April 2001, pp. 2009–2015.
7. F. Aurenhammer. Voronoi diagrams—A survey of a fundamental geometric data structure. *ACM Computing Surveys*, **23**(3):345–405, 1991.
8. A. Boukerche. Performance evaluation of on-demand routing protocols for ad-hoc wireless networks. *ACM/Kluwer Mobile Networks and Applications*, **9**:333–342, 2004.
9. A. Boukerche and S. K. Das. Congestion control performance of R-DSDV protocol for multihop wireless ad hoc networks. *ACM/Kluwer Wireless Networks*, **9**:261–270, 2003.
10. A. Boukerche and S. Rogers. GPS query optimization in mobile and wireless ad hoc networks. In *Proceeedings of the 6th IEEE Symposium on Computers and Communications*, July 2001, pp. 198–203.
11. A. Boukerche and S. Vaidya. A performance evaluation of a dynamic source routing discovery optimization using GPS system. *Kluwer Telecomunication Systems*, Vol. **22**:337–354, 2003.
12. S.-Y. Ni, Y.-C. Tseng, Y.-S. Chen, and J.-P. Sheu. The broadcast storm problem in a mobile ad hoc network. In *Proceedings of the Annual ACM/IEEE International Conference on Mobile Computing and Networking (MOBICOM 1999)*, August 1999, pp. 151–162.
13. P. Krishna, N. H. Vaidya, M. Chatterjee, and D. K. Pradhan. A cluster-based approach for routing in dynamic networks. *ACM SIGCOMM Computer Communication Review*, **27**(2):49–65, 1997.
14. M. Gerla and J. T.-C. Tsai. Multicluster, mobile, multimedia radio network. *ACM/Kluwer Journal of Wireless Networks*, **1**(3):255–265, 1995.
15. R. Ramanathan and M. Steenstrup. Hierarchically-organized, multihop mobile wireless networks for quality-of-service support. *ACM/Baltzer Mobile Networks and Applications*, **3**(1):101–119, 1998.
16. R. Krishnan, R. Ramanathan, and M. Steenstrup. Optimization algorithms for large self-Structuring networks. In *Proceedings of the Conference on Computer Communications, Eighteenth Annual Joint Conference of the IEEE Computer and Communications Societies (INFOCOM 1999)*, Vol. 1, March 1999, pp. 71–78.
17. H. Zhang and A. Arora. GS^3: Scalable self-configuration and self-healing in wireless networks. In *21st ACM Symposium on Principles of Distributed Computing (PODC 2002)*, July 2002, pp. 57–68.

18. D. G. Thaler and C. V. Ravishankar. Distributed top-down hierarchy construction. In *Proceedings of the Conference on Computer Communications, Seventeenth Annual Joint Conference of the IEEE Computer and Communications Societies (INFOCOM 1998)*, March 1998, pp. 693–701.
19. S. Banerjee and S. Khuller. A clustering scheme for hierarchical control in wireless networks. In *Proceedings of the Conference on Computer Communications, 20th Annual Joint Conference of the IEEE Computer and Communications Societies (INFOCOM 2001)*, Vol. 2, April 2001, pp. 1028–1037.
20. H. Chan and A. Perrig. ACE: An emergent algorithm for highly uniform cluster formation. In *Proceedings of the First European Workshop on Sensor Networks (EWSN 2004)*, January 2004, pp. 154–171.
21. B. Chen, K. Jamieson, H. Balakrishnan, and R. Morris. Span: An energy-efficient coordination algorithm for topology maintenance in ad hoc wireless networks. *ACM Wireless Networks Journal*, **8**(5):481–494, 2002.
22. G. Chen and I. Stojmenovic. Clustering and routing in mobile wireless networks. Technical Report TR-99-05, SITE, June 1999.
23. S. Basagni. Distributed clustering for ad hoc networks. In *Proceedings of the IEEE International Symposium on Parallel Architectures, Algorithms, and Networks (I-SPAN)*, June 1999, pp. 310–315.
24. L. Jia, R. Rajaraman, and T. Suel. An efficient distributed algorithm for constructing small dominating sets. In *Proceedings of the 20th Annual ACM Symposium on Principles of Distributed Computing (PODC 2001)*, August 2001, pp. 33–42.
25. Z. J. Haas and B. Liang. Virtual backbone generation and maintenance in ad hoc network mobility management. In *Proceedings of the IEEE Conference on Computer Communications (INFOCOM 2000)*, Vol. 3, March 2000, pp. 1293–1302.
26. Y. P. Chen and A. L. Liestman. A zonal algorithm for clustering ad hoc networks. *International Journal of Foundations of Computer Science*, **14**(2):305–322, 2003.
27. R. G. Gallager, P. A. Humblet, and P. M. Spira. A distributed algorithm for minimum-weight spanning tree. *ACM Transactions on Programming Languages and Systems*, **5**(1): pp. 66–77, 1983.
28. Y. Bejerano. Efficient integration of multihop wireless and wired networks with QoS constraints. In *Proceeedings of the eighth Annual International Conference on Mobile Computing and Networking (MOBICOM 2002)*, September 2002, pp. 215-226.
29. V. Chvatal. A greedy heuristic for the set-covering problem. *Mathematics of Operations Research*, **4**(3):233–235, 1979.
30. A. D. Amis, R. Prakash, T. H. P. Vuong, and D. T. Huynh. Max–min D-cluster formation in wireless ad hoc networks. In *Proceedings of the IEEE Conference on Computer Communications (INFOCOM 2000)*, Vol. 1, March 2000, pp. 32–41.
31. R. Virranoski and A. Savvides. TASC: Topology adaptive spatial clustering for sensor networks. In *Proceedings of the Conference on Computer Communications, 24th Annual Joint Conference of the IEEE Computer and Communications Societies (INFOCOM 2005)*, March 2005.
32. S. Ghiasi, A. Srivastava, X. Yang, and M. Sarrafzadeh. Optimal energy aware clustering in sensor networks. *Sensors 2002*, **2**(7):258–269, 2002.
33. Y. P. Chen and A. L. Liestman Maintaining. weakly-connected dominating sets for clustering ad hoc networks. *Ad Hoc Networks*, **3**(5):629–642, 2005.

34. L. Ramaswamy, B. Gedik, and L. Liu. A distributed approach to node clustering in decentralized peer-to-peer networks. *IEEE Transactions on Parallel and Distributed Systems*, **16**(9):814–829, 2005.

35. J. Uchida, I. A. K. M. Muzahidul, Y. Katayama, W. Chen, and K. Wada. Construction and maintenance of a cluster-based architecture for sensor networks. In *Proceedings of the 39th Hawaii International Conference on System Sciences*, January 2006.

CHAPTER 8

Position-Based Routing for Sensor Networks: Approaches and Obstacles

MARWAN M. FAYED and HUSSEIN T. MOUFTAH

School of Information Technology and Information, University of Ottawa, Ottawa, Ontario K1N 6N5, Canada

8.1 INTRODUCTION

A growing number of sensor network applications require point-to-point services. Intuitively, pure sensing applications, where environment is monitored or events are tracked, require some geographical or geometric context for successful operation. In such a context, data interpretation and management is often tied to node positions. In addition to traditional sensing applications are a growing number of proposed applications that require no knowledge of geography yet do require advanced point-to-point services [1–4]. Traditional Internet routing techniques are unsuitable in either setting [5]. In this chapter we focus on the merits and challenges of algorithms and protocols that provide point-to-point services through position-based routing, where forwarding decisions are made by maximizing or minimizing some function of node locations within a coordinate system. The focus of discussion is on those protocols suited to static sensor networks where node positions within the network are known and unknown.

The remainder of this chapter is organized as follows: In Section 8.2 we give some context to our presentation as well as some background information relevant to its content. Our discussion of position-based routing protocols suitable to sensor networks begins in Section 8.3 with those protocols where knowledge of sensor positions is assumed to exist prior to protocol execution; while the presentation is cumulative, each section may be read independently of the others. In Section 8.4 we present protocols where no a priori knowledge of position exists; amongst such protocols the routing element is coupled to the localization mechanism. We summarize our discussion in

Algorithms and Protocols for Wireless Sensor Networks, Edited by Azzedine Boukerche

Section 8.5; then we follow with open problems in Section 8.6 and, finally, exercises in Section 8.7.

8.2 BACKGROUND AND OVERVIEW

In sensor networks, IP-like routing techniques face scalability issues since node identifiers, if available, share no topological or geographical similarities and, hence, no addressing information. In addition, IP-like routing requires global cooperation and dissemination of information which places undue demand on the energy constraints that are inherent in such devices.

Current point-to-point routing solutions under investigation in the research community fall into one of two classes, on-demand routing [6–8], and position-based (often referred to as geometric or geographic) routing [9–12]. Among the former class, routes are found as needed, often without any prior communications. These methods consume network resources and are known not to scale. In the latter category, geographic routing decisions are based on relative locations of sensors and are generally greedy in nature (e.g., nodes forward to neighbors that are geographically closer to the destination).

In position-based routing the next hop is decided by evaluating the coordinates assigned to neighboring nodes relative to the coordinates of the destination and the current node. For example, a node may choose to forward a message to its neighbor that further minimizes the remaining Euclidean distance to the destination or that maximizes the savings in energy. Position-based routing is particularly attractive due to its scalability. In the best case a node need communicate only its position and only store information regarding its immediate one hop neighborhood. Thus the storage, length of communication, and data over which decisions are made is $O(1)$.

However, position-based routing is not without its drawbacks and challenges. When positions are known, for example, messages may get trapped in local minima where no neighboring nodes further optimize the decision criteria. When positions are not known, there exist the additional challenges of establishing and maintaining some coordinate system in order to determine node positions in the network. The localized nature of position-based routing may also render algorithms blind to obstacles that might cause routing decisions to fail.

Until recently, the focus of position-based routing has been on the design of protocols that assume the existence of a globally known coordinate system—for example, through preprogramming or the Global Positioning System (GPS). More recent advances demonstrate that it is possible to establish a network-centric coordinate system, the design of which is often coupled to the routing mechanism. Our attentions concentrate on families of algorithms that are thought to be feasible and that, in the case of references 10 and 13, have implementations in development.

Position-based routing algorithms for static sensor networks may be classified according to the chart in Figure 8.1. This chart matches the categorization of position-based routing protocols used to organize this chapter. Our discussion continues in the next section by introducing the topic that follows the left branch of our

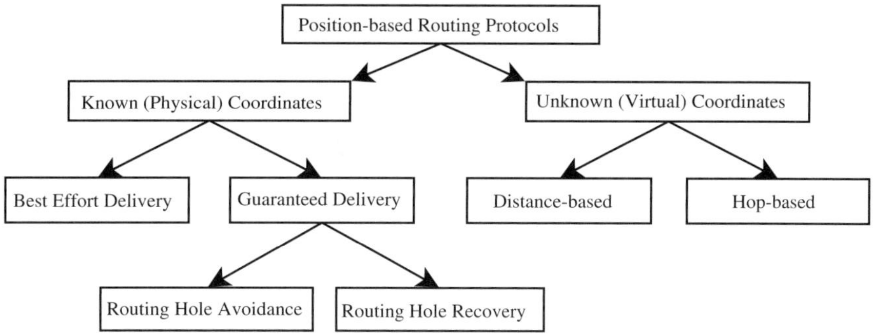

Figure 8.1. A classification of position-based routing protocols.

categorization, those routing protocols that assume that sensors are aware of their physical coordinates.

8.3 ROUTING WITH KNOWN POSITIONS

When investigating position-based routing algorithms for sensor networks, it is natural to assume the a priori knowledge of location information via preprogramming or GPS-like services. Thus, no efforts are made by these protocols to establish location within a network. Current methods make some necessary assumptions. First, links are bidirectional: If node x can receive messages from node y, then y can receive messages from x. In addition, communication models often adhere to the simple unit disk graph (UDG) model where all communication ranges are normalized to some range r. In the UDG neighbors are pairs of nodes separated by a Euclidean distance $\leq r$. The UDG model is relaxed later in our discussion. Feasible schemes must, at a minimum, prove to be loop-free and scalable. We demonstrate some of these subtleties by starting with a discussion of the naive approaches to position-based routing.

8.3.1 Naive Forwarding Mechanisms

Research into scalable forwarding methods for sensor networks has explored a variety of forwarding schemes. Scalability is maintained by keeping knowledge only of the nodes in communication range and choosing the next hop based on this knowledge. All position-based schemes share a common theme: The next hop is determined by maximizing or minimizing some criteria associated with local nodes' positions. We call this the *progress* criteria.

Notions of "Progress". We begin with a formal definition of *progress*. Say a node s holds a message to be forwarded to a destination location d (see Figure 8.2) and has knowledge of all node locations within its communication radius. Then we

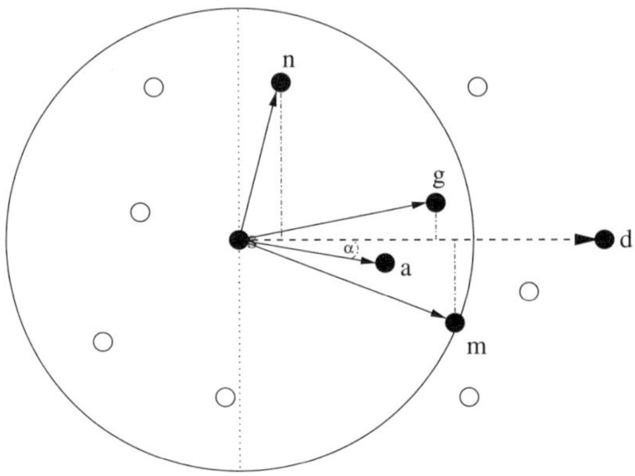

Figure 8.2. Progress.

have the following definition used to better understand the behavior of position-based forwarding schemes.

Definition 8.3.1. The *progress* of a node x en route from sender s to destination d is defined as the orthogonal projection of the location of x onto the line $\overline{sd}$.

We use Figure 8.2 to better demonstrate various notions of progress as discussed below. In this figure, sensor node s holds a message destined for sensor node d. The large circle centered at s represents the communication range of s; thus all nodes inside the circle are neighbors of s. The dotted horizontal line $\overline{sd}$ is the line onto which orthogonal projections are made in order to evaluate and compare progress criteria.

The first forwarding mechanism based on the idea of progress was proposed by Takagi and Kleinrock [14]. In their *Most Forward within Radius* scheme, or MFR, the node with the greatest progress is chosen to be the next hop. Referring to Figure 8.2 we can see that node m projects furthest onto the line that joins s and d. It is important to note that m provides the greatest amount of progress though it is not the node closest to the destination. The motivation behind the use of this myopic routing strategy was to allow for tractable analysis. Recall that progress is defined as a projection onto a line. Clearly, because we are working with projections on a line, the dot product of $dm \cdot ds$ will be minimal over all other neighbors of s when using MFR. This simple notion allowed the authors of reference 14 to determine an optimal neighborhood size for their specific problem.

The converse scheme *Nearest Forward Progress*, or NFP was later proposed in reference 15. In this work the node with the minimum progress, depicted as node n in Figure 8.2, is selected as the next hop. This approach may seem counter-intuitive; but consider that if the broadcast range is variable, then this method has the least probability of collision as well as improved energy savings.

More recently, the direction of nodes was proposed as a criteria for progress in reference 16. The node chosen to be the next hop is the node that is closest to the direction of the destination. Referring to Figure 8.2, we can see that node c, with angle $\angle csd$, is smallest among all of the neighbors of s. Thus, the goal is to minimize the change in direction from the source to the destination.

The DREAM [17] and LAR [18] projects, simultaneously proposed, use an idea similar to compass routing. However, these two approaches are best suited in networks where nodes are mobile. (Despite their application in mobile environments, we provide an overview for completeness.) The node that holds a message m with destination d calculates an angular range where the message must be forwarded. The angular range is calculated using (i) the circle centered at d with radius equal to the maximum movement of d since the last update and (ii) the tangents from the current location to the aforementioned circle. All nodes within this angular range are sent a copy of the message. Clearly, the success of both methods relies on some knowledge of the global network, as well as duplication of messages. Such methods fall outside of the domain of position-based routing.

Each of MFR, NFP, and Compass Routing is myopic. They are localized algorithms that require knowledge only of the immediate neighborhood. Unfortunately, their global behavior is such that none of these approaches can claim to be loop-free. Still, consider that their forwarding decisions attempt to optimize some local criteria, the effect of which is to approximate the shortest Euclidean path between the source and destination. The shortest path may be approximated using another approach that referred to as *greedy forwarding*. Greedy forwarding is a localized forwarding scheme whose express goal is to traverse the Euclidean shortest path. It is this approach on which most research and development, where there is a known and underlying Euclidean coordinate system, has focused.

Greedy Forwarding. Greedy forwarding was first proposed in reference 19 as a routing protocol for wired networks. Referred to as Cartesian routing, the next hop was chosen to be the neighbor that is closest to the destination. Because this work predated GPS and other localization services, knowledge of the global topology was required. The same idea has been reapplied in wireless network settings as the foundation of innumerable routing schemes and algorithms. It is known to behave especially well in dense wireless networks such as those envisioned in many sensor networks.

Greedy forwarding works as follows. Say node s has the neighborhood $N = n_1, \ldots, n_k$ of size k where each n_i is a potential next hop in a traversal that passes through s. Any message that arrives at s has embedded within it the destination d. The greedy approach says that the successor to s will be the neighbor n_i that minimizes the Euclidean distance to d. In Figure 8.2 we can see that node s will select g, the node that most further reduces the distance to the destination among its neighbors.

It can be shown that greedy forwarding is loop-free (using the simple fact that every hop reduces the distance to the destination). The delivery rate of greedy forwarding is known to be quite high in dense networks, but diminishes quickly as the density falls [20]. The problem is that a greedy path may terminate in a routing hole, or void, as shown in the next section.

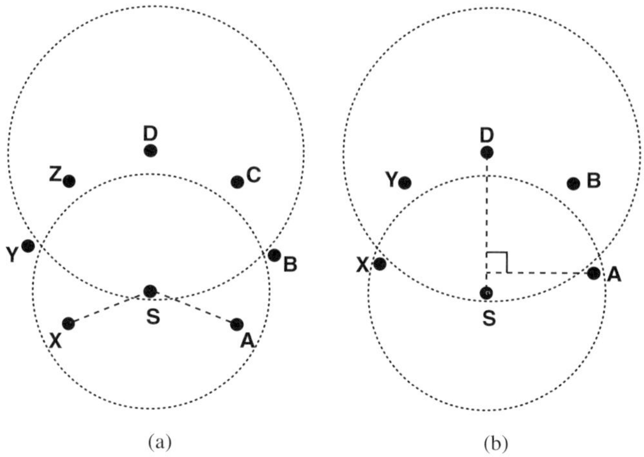

Figure 8.3. Neighbors of stuck nodes may or may not make progress. (a) Convex case. (b) Concave case.

8.3.2 Routing Holes

Greedy forwarding may be loop-free, but delivery is not guaranteed. A consequence of greedy forwarding is that a route may terminate at local minima, where nodes have no neighbors that further reduce the distance to the destination.

We refer our reader to Figure 8.3, which depicts the types of local minima that may occur: The smaller dotted circle represents the communication range of node S, and the larger circle centered at D is used to show that all neighbors of S are further from D than from S itself. Consider that a message destined for node D reaches a minima at node S. There are two cases to consider. The first, as shown in Figure 8.3a, occurs where neighbors of S make no progress toward D according to Definition 8.3.1. This is the obvious case. Less obvious is the case where neighbors of S may actually make progress toward the destination, yet increase the distance from the current location. This example is demonstrated in Figure 8.3b, where S clearly lies within the circle centered at D, while the neighbors of S, x, and A lie outside of the same circle.

These local minima are commonly referred to as voids, holes, or stuck regions. Their occurrence largely determines the performance of greedy forwarding, whose performance varies with network density and distribution. A robust sensor network routing protocol must perform well despite the occurrence of local minima. We proceed in our discussion with proposed solutions to the routing hole problem.

8.3.3 Algorithms for Recovery from Routing Holes

Greedy forwarding is known to perform well in sufficiently dense networks [10], yet there are no delivery guarantees. Many sensor network applications are loss-sensitive and have little to no tolerance for undelivered information. Thus greedy forwarding schemes aiming to guarantee delivery demand that routing voids be circumvented or avoided. Work in reference 20 presents evidence that the frequency and impact of routing holes is manageable. Current solutions to this problem are generally categorized as either broadcasting, planar subgraph methods, or hole mappings, as discussed below.

Broadcasting, BFS, and DFS Approaches. The naive solution to the routing hole problem is simply to broadcast some number of hops from the stuck node until a node is found that is closer to the destination. Two of the first solutions which aimed to be more efficient were the Geographic Routing Algorithm (GRA) in reference 21 and the Optimal Transmission Ranges (OTR) approach in reference 22.

The authors of GRA implement greedy forwarding as the primary means of transport. If, however, a message reaches a stuck region, GRA launches a route discovery packet from the stuck node. The route discovery itself takes the form of a breadth-first search amounting to a limited broadcast, or a depth-first search that eventually produces a single acyclic path. Each node visited during the recovery phase appends its own location to the recovery packet, and all discovered paths are stored in routing tables. In networks of size n, routing tables are found to have a mean size of $O(L \log n)$, where L is the average path length between two nodes. In addition, the authors provide some mechanisms for dealing with location inaccuracy, node failure, and node mobility.

The goal of OTR differs in that it creates Quality of Service (QoS) paths through the network. The necessary characteristics are collected and disseminated during path traversals, which occur from s to d via the depth-first search method. Routing tables are not used, but instead each node stores the next and previous hop for each packet until the status of that packet can be confirmed. Such methods face scalability issues as traffic volume grows, and they are additionally challenged where acknowledgments are not returned.

Hole Recognition and Mappings. A second approach to the recovery phase is to recognize and map the hole boundaries in advance. This is the subject of work in references 23–25. In reference 23, for example, a geometric relationship is described and is labeled as the *tent rule*. The tent-rule is used by each sensor node in the network to determine whether it lies on the boundary of a stuck region. We exemplify the tent rule using the pictorial representation that appears in Figure 8.4. At each node, x neighbors are sorted angularly. For every pair of neighbors u and v where $\overrightarrow{xu}$ is left of $\overrightarrow{xv}$, now consider the perpendicular bisectors of each link (labeled $b1$ and $b2$ in Figure 8.4). If the intersection of the bisectors falls outside of the range of x, then x has no neighbors closer to the region bounded by the communication range and the bisectors (i.e., the shaded region in Figure 8.4). Note that this method locates regions, rather than nodes, where no neighbors are closer. This is particularly suitable in sensor networks where sensing regions are of particular interest.

As stated previously, every node must evaluate its neighborhood using the tent rule. Only at nodes where the tent rule fails may packets get stuck. Upon detecting a failure, stuck nodes are responsible for initiating the following hole-mapping process.

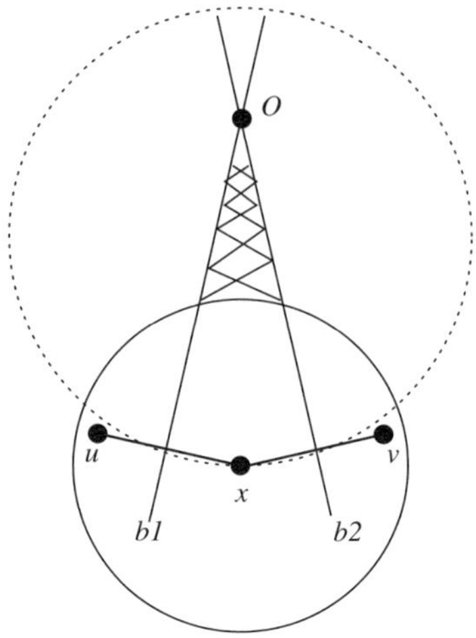

Figure 8.4. Tent rule [23]: Node x has no neighbors closer to shaded region as determined by the intersection of bisectors of adjacent links.

Assume that a packet destined for a node d gets stuck at x. Upon reaching x, the packet is marked for discovery and forwarded to the neighbor whose connecting edge is immediately counterclockwise, or left, from $\overrightarrow{xd}$. Each node that receives the discovery packet appends the newly traversed edge, and then it forwards the packet along the next counterclockwise edge.

During the mapping process, a discovery packet may traverse an edge that intersects with another edge that was previously traversed. Where edge intersections occur, care must be taken. A traversal according to left-hand rule will always return to the source node, yet intersections may cause the traversal to be led away from the region that requires a map as can be seen in Figures 8.5a and 8.6a. If an intersection

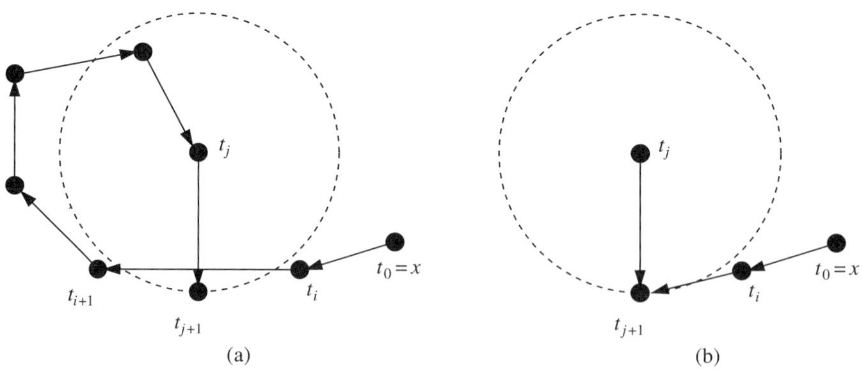

Figure 8.5. "Inside" intersections: (a) Before pruning and (b) after pruning.

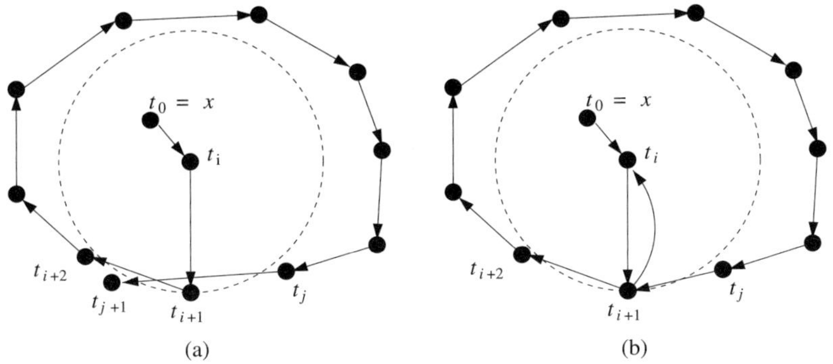

Figure 8.6. "Outside" intersections: (a) Before pruning and (b) After pruning.

appears during a traversal of nodes $t_0 \cdots t_{k-1}$, then we can say that $t_j t_{j+1}$ intersects $t_i t_{i+1}$, with $j > i$. There are only two such cases (the proof of which is available in reference 23).

- *Inside intersections* where node t_j is not visible to nodes t_i and t_{i+1}. Upon detection of this intersection, node t_j replaces segments $t_{i+1}t_{i+2} \cdots t_j t_{j+1}$ with $t_i t_{j+1} tj$, before forwarding to the node that next appears in counterclockwise order from t_{j+1}. The effect of this pruning may be seen in Figure 8.5.
- *Outside intersections* where node t_i is not visible to nodes t_j and t_{j+1}. Upon detection of this intersection, node t_j ignores t_{j+1} and forwards instead to t_{i+1}, thereby ignoring the intersection all together. This example may be seen in Figure 8.6.

The traversal terminates upon its return to x, having recorded the path P that maps the complete stuck region. Once known, a recovery path is shared with all nodes along the path so that resources required in processing and storing discovery packets may be avoided in the future.

As an alternative, the boundhole algorithm may terminate upon reaching some node u such that $|\overline{ud}| < |\overline{xd}|$. If the boundhole is to terminate early, then every recovery packet must have encoded within it the subpath $p \in P$ traversed from x. We call p the escape path. Clearly, the number of hops in $|p|$ is less than that in $|P|$; otherwise x could not be a stuck node. Experiments in reference 20 reveal that the escape path p is significantly smaller than P, perhaps enough to warrant early termination when a stuck region is encountered instead of a complete mapping.

Hole recognition, in general, has thus far received little attention in the research community. Additional algorithms are proposed in references 24 and 25. The former uses a statistical approach to recognize holes in very dense, uniformly distributed networks. The latter study is the first where the hole recognition algorithm does not rely on location information. Neither provides methods for circumventing the stuck regions, yet it is important to recognize that hole recognition is an important problem in and of itself.

Planar Subgraph Methods. The recovery mechanisms so far described all rely on resources that may be taxing on sensing devices given energy and scheduling constraints. DFS, BFS, and mapping methods, for example, require storage either in-device, in-packet, or both. Also, in the worst case, their communication amounts to a limited broadcast. Such communication requires little to no additional memory, but demands additional energy consuming transmissions. Drawbacks such as these often render broadcast-like protocols unsuitable for point-to-point services in sensor networks. To solve this challenge, many projects have investigated the restriction of routing to subgraphs of the original network graph. This class of algorithm is exemplified by one very desirable feature: Such algorithms are stateless; that is, a node requires no knowledge of the network outside of its own neighborhood, yet is able to guarantee delivery. All members of this class all share a simple characteristic: They rely on the construction of the network's planar subgraph. For our purposes a planar subgraph is one that contains all the vertices of G, but where edge intersections occur only at a vertex. (Rather, no edges overlap.)

The most prominent and best-known recovery algorithms route around the routing hole face (or perimeter) in the planar subgraph. This method is equivalently known as *face routing* [9, 26] and *perimeter routing* [10]. Face routing was first proposed by Bose et al. [26] with some theoretical bounds. Karp et al. [10] independently proposed an identical mechanism, but with work on a MAC-compatible implementation.

The inspiration behind these methods comes from the application of a simple rule known to guarantee escape from a maze. In a maze, by keeping one hand against the wall at all times, one is guaranteed to find an exit, or return to the point of origin if no exit exists. This is referred to as the left- or right-hand rule. This works because of a salient feature of maze construction: If we represent the walls of the maze as edges in a graph and represent the intersection of those walls as vertices, then the resulting graph is planar, where no edges intersect.

Wireless network graphs may consist of intersecting edges, so it is necessary for the planar subgraph method to prune edges from the network graph so that it is planar and so that it remains connected. Gabriel graphs (GGs) and relative neighborhood graphs ($RNGs$) are planar graphs whose constructions are localized, a characteristic particularly suitable to sensor environments. Intersecting edges are eliminated by connecting pairs of nodes through *witness* nodes, if such a node exists in a common region.

We refer the reader to Figure 8.7 for examples of each construction where large circles centered at nodes u and v represent the communication ranges of nodes u and v, respectively. The GG construction is depicted in Figure 8.7a. For any pair of neighbors u, v, if a witness node w exists within the circle whose diameter is $|\overline{uv}|$ (shown as the shaded region in Figure 8.7a), then $\overline{uv}$ is removed from the graph. Similarly for the RNG in Figure 8.7b, if w appears within the intersection of the two circles centered at u and v with radii equal to $|\overline{uv}|$ (again shown as the shaded region), then $\overline{uv}$ is removed from the graph. In each, pruning is valid since communication may continue through w. In references 9 and 26 and later in reference 10, it was shown that if the unit disk graph is connected, then the intersections with Gabriel graph $UDG \cap GG$ and with relative neighborhood graph $UDG \cap RNG$ remain connected.

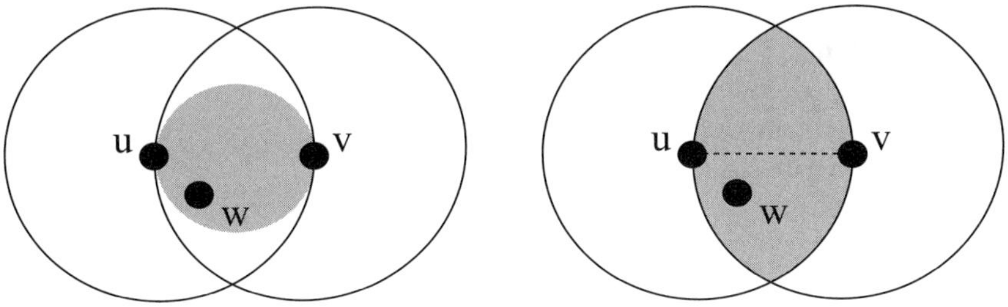

(a) (b)

Figure 8.7. Localized constructions of planar subgraph. (a) Gabriel graph construction. (b) Relative neighborhood graph construction.

Face- and perimeter-routing techniques choose to route greedily whenever possible. The recovery phase is initiated only when a message gets stuck at some node. Upon receipt of message m destined for node t, node s inspects the message to reveal it either in *greedy* or *recovery* mode. The corresponding algorithms, executed at each node, are listed in Algorithms 1 and 2.

ALGORITHM 1. Greedy Mode

let s be the current node
let w be the neighbor closest to t
if $(w, t) < (s, t)$ **then**
    forward m to w
else $\{$no such w exists; m is stuck at $s\}$
    mark packet as *recovery* with location of s
    forward to neighbor that is left of $\overrightarrow{s, t}$
end if

ALGORITHM 2. Face/Recovery Mode

let u be the node from which m was received
let v be the current node
let w be the neighbor closest to t
if $(w, t) < (s, t)$ **then**
    mark packet as *greedy*
    forward to w
else
    forward to neighbor that is left of $\overrightarrow{v, u}$
end if

Algorithms 1 and 2 assume that a stuck node has first constructed the portion of the planar subgraph that occurs within its view, as above. If in greedy mode a packet is forwarded according to Algorithm 1, where a sensor forwards to the neighbor closest to the destination. If no such neighbor exists, then the sensor node forwards according to Algorithm 2. Once stuck, the message m is marked *recovery* and is forwarded to the neighbor that appears first in a counterclockwise direction. While in recovery mode, each sensor that receives the message first checks for a neighbor closer to the destination than the point at which the message was marked *recovery*. Returning to Figure 8.3a, we can see s is a stuck node. In the case of the left-hand rule, $\overline{sx}$ is left of $\overline{sd}$. The recovery path sxy terminates upon finding z since z is closer to d than s, where recovery began.

One special case occurs where an edge $\overline{uv}$ intersects $\overline{sd}$ during recovery. The solution is left as an exercise.

These algorithms are especially suited to sensor networking environments: They guarantee delivery, are localized in their operation, and are stateless in the sense that they require no information outside of the location of their neighbors. However, we see in the next section that subplanar methods are not without their drawbacks.

Drawbacks and Challenges of Planar Methods. Planar subgraph methods, while promising, face three major challenges before their deployment may be considered feasible. First, planarization assumes locations are accurate, an assumption that may be untrue. Second, the localized nature of the planarization process means that one sensor node may be blind to environmental obstacles that are visible to neighboring sensors. Finally, the unit disk model on which these algorithms are constructed is a poor representation of the real world. We use this section to touch on each of these challenges.

Routing protocols that operate over pre-established coordinate systems are generally designed under the assumption that location information is accurate. This assumption is challenged by real-world limitations. GPS, for example, offers high-resolution localization, but is subject to line-of-sight constraints rendering it ill-suited to underground, underwater, and undercoverage applications, to name a few. Furthermore, the added expense incurred by supporting GPS in all nodes is restrictive. Many proposals exist to resolve this issue by supporting some small location-aware infrastructure [27–30], from which all other nodes may learn their locations; yet even these fall prey to poor resolution and estimation errors.

There are cases where localization and estimation errors have little adverse effect. For example, in reference 31 greedy routing was evaluated in simulated networks with localization errors. The results show that the performance of greedy routing is largely unaffected by inaccuracies up to 40% of the radio range. Conversely, there are contexts in which localization errors can be destructive to correct protocol operation. One such example is demonstrated in reference 32, which models and evaluates location error on face-routing techniques; here we find that errors of as little as 20% of the communication range caused high packet drop rates,

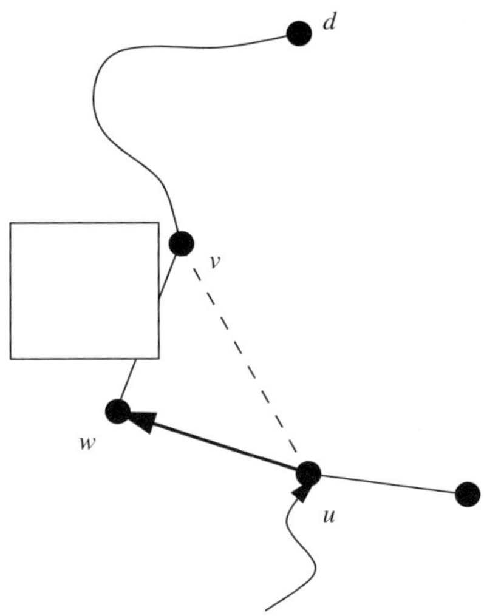

Figure 8.8. Planar methods may fail without inter-neighbor link information.

nonoptimal path selection, and looping. These experiments were reinforced in reference 33, which investigates the types of location errors that lead to performance degradation in face-routing. In doing so, the authors were able to suggest a simple check, consisting of mutual agreement between nodes, to resolve many of the problem cases.

In addition to position errors, a second challenge faced by planarization processes lies in its best feature, namely, that all operations are local and occur without need for negotiation with neighboring nodes. The level of localization inherent in the planarization process leaves open the possibility for incorrect output. Specifically, because a node planarizes its neighborhood using only the locations of its 1-hop neighbors, a node may assume links exist where they do not. Consider the example in Figure 8.8 where a packet destined for node d gets stuck at node u. The planarization of the neighborhood of u removes the link $\overline{uv}$ believing v is reachable via w. Since u is unaware of any obstacles or interference between w and v, u's planarization of its neighborhood is incorrect. This phenomenon was first demonstrated in references 34 and 35.

Finally, we must address the likelihood that the unit disk graph correctly represents wireless network graphs. Experimental evidence in [36, 37] suggests that radio ranges are inconsistent and irregular. Experiments in reference 38 conducted on two sensor networks further demonstrate and enumerate the difficulties that occur because the unit disk assumption is violated. The experimental evidence is used to design a protocol that corrects the failings of the Gabriel and relative neighborhood graph constructions. Protocols that avoid the unit disk assumption are discussed in the next section.

8.3.4 Guaranteed Delivery in the Real World

Face-routing algorithms are attractive because they are localized and efficient. Yet as previously discussed, they are known to be ill-suited to physical environments. There are two reasons for this. First, the Gabriel and relative Neighborhood graph constructions rely on the assumption that all communication ranges in the network are identical and uniform (the attributes associated with the UDG model). Moreover, these distributed constructions are unable to resolve links broken by obstacles or interference. Recent breakthroughs have begun to surmount the impracticalities of face-routing while maintaining delivery guarantees [13, 39].

The first-known protocol to guarantee delivery over a global coordinate system without planarization or the use of left-hand rule is the Greedy Distributed Spanning Tree Protocol (GDSTR) algorithm in reference 39. GDSTR builds on the fact that any message can be successfully delivered via depth-first search if the network is connected via a spanning tree. (This fact alone does not solve the problem: Delivery would be inefficient, needing up to $2n-3$ hops.) Leong et al. [39] describe a new type of spanning tree, the *hull tree*, to route more efficiently. A hull tree is a spanning tree with one added piece of information: Each node records the convex hull that contains all of its descendants in the tree. (The convex hull of a set of points is the smallest polygon that contains all the points.) In GDSTR, forwarding occurs greedily, as with most position-based protocols. If a message reaches a void, a recovery mode is initiated where convex hulls are used to determine the regions of the network that contain unreachable destination. This information is used by GDSTR to route along the spanning tree to forward to the appropriate convex hull. If a node is found en route that is closer to the destination than the node where the message was stuck, then GDSTR returns to greedy forwarding. GDSTR is known to scale well as the neighborhood size grows. Furthermore, the use of multiple hull trees adds fault-tolerance to the network; and if multiple trees are rooted at opposite ends of the network, routing efficiency improves.

The Cross-Link Detection Protocol (CLDP) proposed in reference 13 circumvents voids by face-routing, using left-hand rule over a planar subgraph of the network; its design, however, is motivated by the observation that routing difficulties in planar subgraph methods arise, in part, due to the constructions themselves. (Recall from previous sections that successful local planar subgraph constructions rely on the unit disk graph.) For this reason, CLDP proposes an alternate construction of planar subgraphs that assumes only that links are bidirectional. CLDP operates in a distributed fashion, exchanging some localized operation for accurate information. The idea behind CLDP is that each node is able to probe the vicinity for intersecting links. A probe packet is initialized with the endpoints of the first link to be probed. Figure 8.9 shows the simplest example of a probe traversing a graph using the left-hand rule. Say a probe starts at node d for link (d, a). Sensor node a then forwards to b as determined by the left-hand rule. When the probe reaches node b, the intersection (b, c) is recorded, before the probe packet continues its traversal. (Recall that a traversal according to the left-hand rule will eventually return to its starting point.) Prompted by the return of the probe packet, d proceeds to prune links. Figure 8.9 depicts only

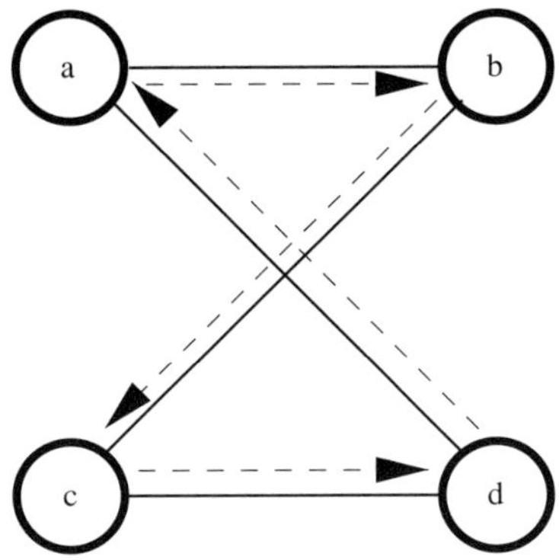

Figure 8.9. CLDP probes links using left-hand rule.

the naive case; the graph remains connected following a removal of either of the two intersecting edges. We leave as an exercise the identification of remaining cases and the care that must taken when pruning links to keep the graph connected. Furthermore, to avoid the slow process of scheduling serial probing by neighboring nodes, a system for concurrent probing is proposed. Concurrent probing is achieved by implementing a mechanism to "lock" links so that no more than one link is removed at a time from any vicinity. CLDP is one of very few protocols to have been implemented on testbeds [13]. The associated communication complexities and storage costs revealed in this process (see reference 38) are motivation to develop alternative approaches to guarantee delivery.

Protocols such as CLDP and GDSTR, in order to be feasible for physical networks, sacrifice efficiency for accuracy. CLDP requires negotiation within each neighborhood in order to prune appropriate links, and GDSTR must broadcast information to construct and maintain its hull trees. It remains an open question whether such tradeoffs are a necessity.

8.4 ROUTING WITH UNKNOWN POSITIONS

By nature of its name, position-based routing protocols may be thought to assume that position location is available for use by the routing protocol. Often this information is not and cannot easily be made available from the outset: For example, the inclusion of GPS in the manufacturing process is cost prohibitive; and despite this, sensor nodes may be deployed in adverse environments where GPS information cannot penetrate. This begs the question, Can position-based routing be applied to contexts where location information is available to only a few nodes? This question is furthered by the notion that many proposed applications do not require any knowledge of geography, but do require advanced point-to-point routing services.

Applications such as data querying [1, 2, 40], reactive-tasking where events are triggered by local events, and data-centric storage [3, 4] all demand robust routing yet ignore physical location. Network applications such as these will identify a node by its identifier or by the data it stores, each of which provides no means by which to route using position. In such contexts a network may implement position-based routing by use of a distributed hash table (DHT), which partitions keys/identifiers from the owner of the key [41]. In a DHT sensor network, position-based routing is an appealing means to bridge this partition.

In this section we explore the potential for position-based routing where location information is generally unavailable. The solution that has received the widest attention is to develop protocols that construct their own maps and create their own coordinate systems. Such protocols are often said to construct and rely upon *virtual coordinate systems*. The routing in virtual coordinate systems is often coupled with the construction of the coordinate system, a coupling that necessitates an understanding of the coordinate construction itself.

One of the first feasible algorithms to construct a virtual coordinate system for the purpose of routing in wireless systems appears in reference 42. This method requires no infrastructure. However, its centralized nature and expensive cost ($O(n^3)$) render it poorly suited for sensor networks.

The establishment of a coordinate system in a distributed fashion, in order to route messages between nodes, is nontrivial. We introduce two methods under investigation. The first is the distance-based approach where the aim is to create some semblance of a Cartesian space (without which the direction between nodes is unclear). The alternative is a hop-based protocol that addresses using the distance between nodes, in hops. Each of these share drawbacks, a discussion we reserve until later.

8.4.1 Distance-Based Coordinates and Routing

One of the earliest protocols promoting a virtual coordinate system construct with a corresponding routing mechanism is the GEM infrastructure [43]. In GEM a labeled graph is constructed and embedded into the network topology in a distributed fashion, where the label encodes a node's position within the graph. The idea is demonstrated by coupled coordinate setup and routing protocols called VPCS (Virtual Polar Coordinate Space) and VPCR (Virtual Polar Coordinate Routing), respectively. Using VCPS, the GEM system is able to embed a ringed tree into the network in order to create a polar coordinate system. The VCPR algorithm designed to route over the polar coordinates is the first to guarantee delivery in a point-to-point fashion with no a priori geographic information. While the initialization scales well, the recovery process initiated on node failure or link degradation is expensive. Recovery may force a large number of nodes to participate in a recomputation process.

To better demonstrate the ideas behind distance-based constructions, we focus our discussion on the NoGeo method proposed in reference 12. In it the authors provide a mechanism to construct a virtual Cartesian space and implement the simplest form of greedy routing. Their work is a derivation from previous work intended to test graph connectivity [44], and it assumes that nodes may accurately measure the distances to their immediate neighbors.

Virtual Coordinate Construction. For clarity, explanation of the NoGeo algorithm is presented in a cumulative fashion where at each additional step we remove some knowledge from the system. Hence we begin by describing the coordinate construction under the assumption that nodes on the network boundary, or perimeter, are aware of their position relative to the network as well as their exact location. The algorithm in reference 12 is based on an iterative relaxation procedure from reference 44 which is used to determine the locations of all remaining nodes in the network. The procedure is such that each link is represented by a force that pulls its adjoining neighbors together. The force in each of the x, y directions is proportional to the difference in the x, y coordinates. At any iteration a node's neighbors are held with fixed position; its *equilibrium* position, where the sum of the forces acting on it is zero, is the average of all its neighbors' x coordinates in the x direction, and similarly for the y direction. This relationship motivates the iterative procedure repeated periodically at each node i using the relaxation equations

$$x_i = \frac{\sum_{k \in neighbor\_set} x_k}{size\_of(neighbor\_set(i))} \tag{8.1}$$

$$y_i = \frac{\sum_{k \in neighbor\_set} y_k}{size\_of(neighbor\_set(i))} \tag{8.2}$$

in each of the x and y directions, respectively.

As an example, consider the wireless sensor network in Figure 8.10a. This network consists of 3200 nodes within a 200×200 unit space where each node has a

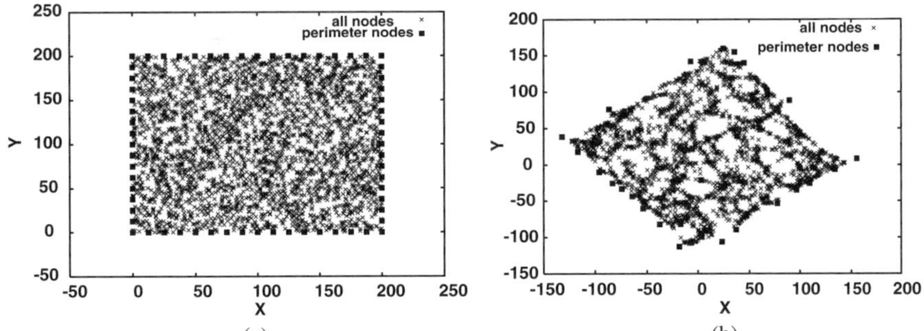

Figure 8.10. The (a) real and (b) virtual coordinates of a 3200-node network. (a) Physical network in 200×200 space. (b) Virtual network where perimeter nodes are initially unknown.

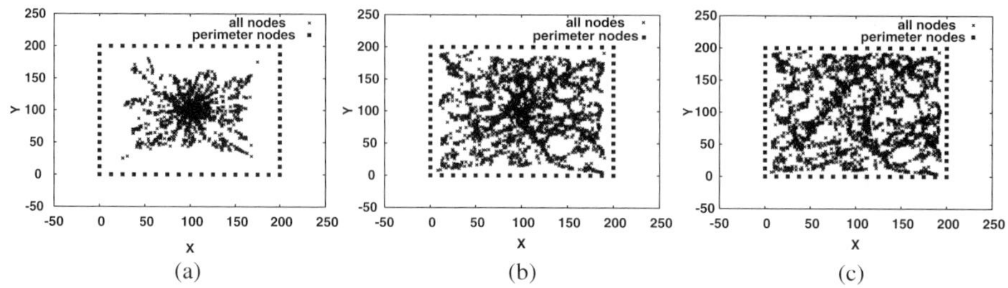

Figure 8.11. Intermittent virtual coordinates of 3200 nodes in a 200 × 200 space where nodes are known in advance to lie on the network's perimeter. (a) After 10 iterations. (b) After 100 iterations. (c) After 1000 iterations.

communication range of 8 units. As the iterative procedure described above is executed over this network, sensor coordinates will gradually shift to match their true coordinates within the network. This "shaping" is depicted following 10, 100, and 1000 iterations in Figures 8.11a – 8.11c, respectively. In this example the initial coordinates of each node have been set to the center of the network at (100, 100). (We later return to compare the virtual and actual topologies.)

This coordinate construction may be prefaced with additional steps if there is no advance knowledge of perimeter nodes, provided there are two beacons somewhere in the network. [Beacons are distinguished from the remaining network because they either (a) hold and disseminate information or (b) play some specific coordination role required for successful setup and communications.] Either beacon may be used to identify nodes on the perimeter, after which point the perimeter nodes exchange messages to determine their locations.

The first step in accomplishing this task is to identify nodes on the perimeter. Recall the assumption that two beacon nodes exist. Either beacon is designated as the primary beacon, which broadcasts a HELLO message. Each node uses receipt of the HELLO message to determine its location from the beacon, according to the *perimeter node criterion*. The *Perimeter Node Criterion* says that if a node is farthest away from the bootstrap beacon among all nodes in its two-hop neighborhood, then this node decides it lays on the perimeter of the network. This is by no means an exact determination, but simulations show that it identifies a sufficient number of perimeter nodes. Once the nodes have identified their perimeter status, they must coordinate to exchange information and calculate their coordinates according to Algorithm 3.

Each perimeter node, following Step 3 in Algorithm 3, has established its own virtual coordinate. One challenge remains. Each perimeter node has established its own virtual coordinate, yet the network lacks a global orientation. The reason is that any set of coordinates satisfying Eq. (8.3) may be rotated, translated, or transposed. The result is that perimeter nodes cannot be guaranteed to share the same sense of direction.

ALGORITHM 3. Coordinate Determination among Perimeter Nodes in NoGeo

1. Each perimeter node broadcasts a HELLO message so that each may calculate its distance to every other perimeter node. The result is stored in a *perimeter vector*.
2. Each perimeter node broadcasts its perimeter vector to the network so that each perimeter node knows the distance between every pair of perimeter nodes in the network.
3. Each perimeter node triangulates to find the coordinates of every perimeter node in the network. Coordinates are chosen so as to minimize

$$\sum_{i,j \in perimeter} (measured\_dist(i, j) - euclid\_dist(i, j))^2 \tag{8.3}$$

where $measured\_dist(i, j)$ represents the distance between nodes i, j as measured in 1, and $dist(i, j)$ is the Euclidean distance between the virtual coordinates of i and j.

The solution is the reason why two beacons must exist. The two beacons share their perimeter vectors and participate in the triangulation process. During this process, each of the perimeter nodes calculates the center of gravity of the network. The combination of the two beacons with the center of gravity provides a set of axes on which all nodes can agree: The center of gravity becomes the origin, while each of the two beacons define the positive x, y axes.

An example topology at equilibrium, following the execution of perimeter identification, coordinate determination, and dissemination, appears in Figure 8.10b. This is the virtual network space that corresponds to the actual network in Figure 8.10a. Using this scheme, the virtual coordinates maintain a structure similar to the actual coordinates, but with some caveats. For example, the network appears to have been rotated about the virtual center of gravity. However, the rotation is uniform and consistent, meaning that the rotation is network-wide. Note that holes in the virtual networks constructed in Figures 8.10b and 8.11c are much larger in size than in the actual network of Figure 8.10a. As discussed below, this salient feature does not reflect itself in the routing mechanism.

Routing in the virtual Cartesian coordinate system takes the form of pure greedy routing as described in Section 8.3.1: Messages are forwarded to the neighbor that most reduces the distance to destination until no such neighbor exists or the destination is found. Despite the inaccuracies of location determination and the increase in hole sizes (as reflected in physical space), greedy routing over virtual coordinate systems performs better than it does over physical coordinate systems. The rate of successful delivery rate is higher when routing over virtual coordinate systems.

The underlying reason for an increase in the rate of delivery is that locations are assigned coordinates relative to the locations of other nodes in the network, not relative to the space that the network occupies. Hence, a virtual coordinate system accurately reflects not network geography, but rather network connectivity. This relationship, when reflected in actual space, is responsible for the apparent growth in hole sizes. (Despite this fact, the physical coverage area is unaffected, a necessary property of sensor networks.) It is also responsible for the improvement in the performance of greedy routing because it reflects network paths and connections more than node location in the physical space.

One drawback to approaches in this class of position-based routing is the expense incurred by the computation and transmission necessary to establish a reasonable coordinate system. Consider in NoGeo, for example, that it is somewhat impractical that initialization requires $O(\sqrt{n})$ nodes to flood the network, and it is also inpractical for at least this number of nodes to store $O(n)$ state (since the distance vectors occupy $O(\sqrt{n} \times \sqrt{n})$ space). We proceed with an approach that reduces this expense in exchange for a loss in topological accuracy.

8.4.2 Hop-Based Coordinates and Routing

One alternative approach to constructing virtual coordinates foregoes Euclidean approximations, altogether. Instead, designated coordinates are comprised of a vector that contains a set of hop-distances to beacons located around the sensor network [45–48]. As is the case with NoGeo, the goal of these methods is to deliver packets in an environment with no a priori knowledge of node locations, in a point-to-point manner. We demonstrate this approach to position-based routing using Beacon Vector Routing (BVR) in reference 47 and reserve discussion for the differences with other projects until later.

BVR is a protocol that assigns routing coordinates and defines a distance function used in forwarding decisions. A node's coordinate is a tuple recording the hop distance to each of a subset of available beacons, information that is disseminated using reverse path tree constructions. (A reverse path tree construction occurs when a beacon broadcasts its existence to the network and all remaining nodes record the shortest hop distance to that beacon.) A distance function is used to route greedily. In the event that greedy routing fails, a correction mechanism exists to guarantee delivery.

Let r denote the total number of beacons, and let q_i denote the distance in hops from a node to beacon i. The position of node q is the tuple $(q_1, q_2, \ldots, q_r)$. By this definition, it is possible for multiple nodes to share the same coordinate so a node identifier is necessary to disambiguate between nodes with identical coordinates. The distance function must favor greedy forwarding to maintain a high level of efficiency. When evaluating the distance function, the BVR metric aims to minimize the difference in coordinates componentwise. This metric is based on the idea that it is better to move toward a beacon close to the destination than it is to move away. Hence, the distance function δ is designed to move a message toward a beacon if the destination is closer to the beacon than the current node; that is, it also moves a message away from a beacon if the destination is further away. (Note that using this intuition, movement toward a beacon always reduces the distance to the destination but moving away is not: The destination may sit on the other side of the beacon from the current node.)

Let the distance function $\delta(p, d)$ measure the goodness of node p as a next hop to d. The above intuition is encapsulated into the distance function using the sums

$$\delta_k^+(p, d) = \sum_{i \in C_k(d)} \max(p_i - d_i, 0) \tag{8.4}$$

$$\delta_k^-(p, d) = \sum_{i \in C_k(d)} \max(d_i - p_i, 0) \tag{8.5}$$

where $C_k(d)$ is the set of k beacons closest to d. The metric is a sum of differences derived from Eqs. (8.4) and (8.5). δ_k^+ is the sum of the differences of beacons closer to the destination than the node p, while δ_k^- is the sum of the differences of beacons further away.

BVR routes greedily as follows. The next hop is chosen to be the node that minimizes δ_k^+; any tie that may occur is broken by minimizing δ_k^-. Note that the k beacons may number fewer than the total number of beacons in the system and that the smallest difference $\delta^{\min}$ encountered during a traversal must be stored in the message header for reference.

ALGORITHM 4. Overview of Beacon Vector Routing (BVR)

1. (Greedy) Where possible, forward to the neighbor that minimizes Eq. (8.4), breaking ties using Eq. (8.5).
2. (Recovery) If no such neighbor exists, record the current distance in the packet as δ^{min} and forward to the beacon closest to the destination.
3. (Recovery) If message has reached a beacon without reverting back to greedy mode, broadcast with a time-to-live equal to hop distance from the destination node.

A global view of the main BVR algorithm is summarized in Algorithm 4. As is the case with other greedy schemes, there are occasions where forwarding may terminate prematurely, failing to find a neighbor that improves on δ^{min}. BVR is able to guarantee delivery using a two-tier recovery mode. First, if a node has no neighbor closer to the destination than itself, it will forward the message to the beacon closest to the destination. The idea is that if a sensor node is unable to find the destination, then it should send the message in the direction of a node that can. Interim nodes that receives the message will forward to the destination as if recovery never occurred, if possible. Second, if a beacon is unable to further minimize $\delta^{\min}$, then it initiates a scope flooding to find the destination. (The scope of the flood is known since the destination coordinates record the hop-distance from each beacon.) While this recovery mechanism is an expensive means of guaranteeing delivery, it is found to occur infrequently in simulations.

Similar ideas have been proposed in references 45 and 46. FT-BVR in reference 46, for example, extends BVR using fault-tolerant techniques to improve on a variety of performance metrics. Logical Coordinate Routing (LCR), in reference 45, is similar to BVR in its coordinate construction and routing mechanisms. Where it differs is in recovery. Where BVR ultimately resorts to a scoped broadcast, LCR backtracks along the path-thus-far until an alternate path is found or the message returns to the originating node where delivery is deemed to have failed. Clearly, LCR avoids the expense of a broadcast in exchange for additional state (either in the message or stored at interim nodes).

8.4.3 Drawbacks and Challenges with Unknown Coordinates

Generally speaking, mechanisms that construct and route over virtual coordinate systems manage to overcome many of the challenges that face routing in physical coordinate systems. To wit, virtual coordinate routing schemes make no assumptions pertaining to the unit disk graph, an assumption that is shown to be violated in practice [36, 37, 49], nor do they require GPS-like services that may be unavailable. This benefit does come at some expense, however.

The first limitation is related to the infrastructure necessary for correct operation. Constructs that create coordinate systems that reflect connectivity require the use of beacons. (Beacons are defined as nodes that have knowledge of their location via global positioning systems that have some preprogramming or that are placed strategically.) Each of these incurs a cost increase on the manufacturing and deployment process that is intended to otherwise be cheap. In addition, many environments well-served by sensor-network applications can be volatile. In such environments a small number of beacons may be lost, destroyed, or otherwise disabled. The use of beacons may be restrictively expensive and complicated when deployed broadly, yet present fault-tolerance and failure issues when deployed narrowly. If beacons are required, does it suffice to select them randomly? If not, what determines the goodness of a beacon? Despite these challenges, beaconing may be an effective solution to a difficult problem.

In addition, virtual coordinate and routing methods also incur additional complexity in communication, and sometimes computation. Coordinates in the network cannot be learned unless information is broadcast so that all nodes share similar knowledge. Energy consumption is a greater concern since transmission is known to be a high-energy operation, and broadcasts increase the chance of collision (though there are proposals to *intelligently* broadcast such as in reference 50. Often the determination of coordinates requires additional computation.

Finally, we address the accuracy of virtual coordinates. Coordinates that record hop counts are likely to be duplicated throughout the network, and so additional care must be taken to deliver messages to the intended recipient. Coordinate systems that measure physical distances reflect network connectivity well, but are subject to limits in resolution. In reference 51, work on Nagpal's algorithm, a set of algorithms to construct and improve coordinates based on distances from three distinct beacons, reveals the smallest resolution to be $\frac{\pi}{4n}r$, where n is the average neighborhood

TABLE 8.1. Summary of Various Protocol Attributes

	MFR, NFP, Compass	Greedy	GRA, DFS-QoS	Mapping	Planar Subgraph Methods	CLDP	GDSTR	NoGeo	BVR
UDG Assumption	No	Yes	No	Yes	Yes	No	No	No	No
Loop-free	No	Yes	NA	NA	Yes[a]	NA	Yes[a]	Yes	Yes
Critical nodes	No	No	No	No	No	No	Yes	Yes	No
Beacon requirements	No	No	No	No	No	No	No	Yes	Yes
Multiple routes	No	No	No	Yes[b]	No	NA	Yes[b]	No	No
Startup coordination	Local	Local	Local	Extended	Local	Extended	Global	Global	Global
Void recovery cost	NA	NA	Medium	Medium	Low	High	Medium	NA	Medium
Guaranteed delivery	No	No	Yes	Yes	Yes	NA	Yes	No	Yes
Obstacle-resilient	Yes	Yes	Yes	Yes	No	Yes	No	Yes	Yes
Coords reflect connectivity	No	No	No	No	No	No	No	Yes	Yes

[a] A loop only occurs when a packet returns to its origin, indicating that the destination is unreachable.

[b] The choice of route occurs only when recovering from a void region.

NA, not available.

size and r is the radius. If achievable, this limit may or may not be acceptable in practice, see Table 8.1 for a summary of routing algorithms.

8.5 SUMMARY

In this chapter we have explored algorithms and mechanisms for position-based routing in sensor networks, whose communication graphs are large and dense relative to traditional wireless ad hoc networks. Position-based routing occurs where forwarding decisions are based on a metric that reflects the position of nodes in reference to the physical space or to reference nodes.

Our discussion began under the assumption that location information is available and accurate. In such environments it is necessary to contend with the appearance of local minima if any delivery guarantees are to be made. This problem has simple solutions that are impractical in physical deployment. It has thus far been necessary to sacrifice simplicity for accuracy in order to resolve the challenges of implementation and deployment.

Furthermore, we presented algorithms that require no a priori location information. Nodes that use such algorithms must cooperate to establish their own coordinates within the network. The construction of the coordinate system is often coupled with routing. Local minima are less of a problem since coordinates emphasize the network topology and reflect connectivity. However, their success relies on the use of beacons. This raises issues related to fault-tolerance, selection, deployment, and additional communication and computational cost in network initialization and maintenance. For this reason we present some open problems and future directions.

8.6 FUTURE WORK AND OPEN PROBLEMS

There are few position-based routing algorithms implemented as protocols, making difficult the task of measurement and evaluation. Many additional open problems remain. Beacon selection remains a challenging task. Current research assumes that beacons may be chosen efficiently at random or that a selection infrastructure exists. A random selection may yield less-than-optimal resolution, while any specialized infrastructure removes from the uniformity of the network and the ease of deployment. Another current challenge that faces the community is to efficiently guarantee point-to-point services without planarity. As described in this chapter, current solutions require broadcasting or locking mechanisms, each of which is costly. Lastly, there is little work which allows us to understand the sacrifices necessary in efficiency and resource conservation that are necessary to provide reasonable accuracy without resorting to broadcasts. Open avenues of exploration exist where recognition and circumvention of local minima are concerned. Also, it is entirely unclear how position-based routing in general will affect (or can be made to interact with) other physical and application layers.

Despite these challenges, the growing research into position-based routing and its associated challenges is reflective of its potential for providing scalable and efficient data transport in sensor networks.

8.7 EXERCISES

1. In mechanisms that map void regions in advance, a path results. It is possible for multiple nodes along this path to identify themselves as stuck nodes, in which case each stuck node initiates the mapping protocol. This is a redundant exercise. How might mapping techniques be augmented to deal with this occurrence?
2. Assume the unit-disk graph (UDG) is connected. Show that the Gabriel (GG) and relative neighborhood (RNG) graphs do not disconnect the unit-disk graph. In other words, show the intersections $GG \cap UDG$ and $RNG \cap UDG$ are connected.
3. In face-routing methods, a packet in recovery mode is routed greedily as soon as possible (i.e. some node is found that is closer to the destination than where recovery began). However, it is conceivable that only an edge sits closer to the destination (as shown in Figure 8.12). Modify face-routing so that it resolves this problem and maintains its delivery guarantees.
4. Nodes in CLDP probe incident links in order to find and remove intersections. However, there are cases where concurrent probing may disconnect the network graph. Enumerate these cases (there are 3) and establish a mechanism to guarantee a disconnection does not occur.
5. In our discussion of BVR and other mechanisms relying on beacons, we left open any mechanism to replace a failed beacon. Assume that a minimum number r of beacons is required for correct operation. There are two operations at issue.

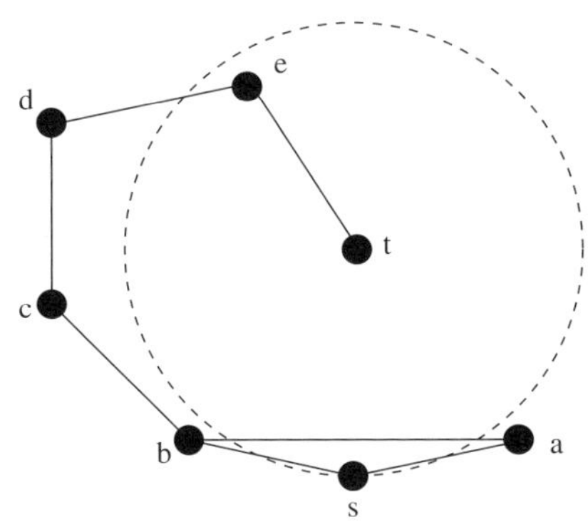

Figure 8.12. A message stuck at s will traverse edge (ba) according to left-hand rule before returning to s.

(a) How might a sensor node in the network detect that less than r beacons are functional?

(b) Once it is detected that too many beacons have failed, how does a node decide to become a beacon itself? What if multiple nodes simultaneously decide to become beacons?

BIBLIOGRAPHY

1. D. Ganesan, D. Estrin, and J. Heidemann. Dimensions: Why do we need a new data handling architecture for sensor networks? *SIGCOMM Computer Communication Review (CCR)*, **33**(1):143–148, 2003.
2. B. Greenstein, D. Estrin, R. Govindan, S. Ratnasamy, and S. Shenker. Difs: A distributed index for features in sensor networks. *Elsevier Journal of Ad Hoc Networks*, **2**:39–46, 2003.
3. S. Ratnasamy, B. Karp, S. Shenker, D. Estrin, R. Govindan, L. Yin, and F. Yu. Data-centric storage in sensornets with ght, a geographic hash table. *Mobile Network Applications*, **8**(4):427–442, 2003.
4. S. Shenker, S. Ratnasamy, B. Karp, R. Govindan, and D. Estrin. Data-centric storage in sensornets. *SIGCOMM Computer Communication Review*, **33**(1):137–142, 2003.
5. D. Estrin, R. Govindan, J. Heidemann, and S. Kumar. Next century challenges: scalable coordination in sensor networks. In *Proceedings of the fifth annual ACM/IEEE International Conference on Mobile Computing and Networking*, Seattle, WA, 1999, 263–270.
6. D. B. Johnson and D. A. Maltz. Dynamic source routing in ad hoc wireless networks. In T. Imielinski and H. Korth, editors, *Mobile Computing*. Kluwer Academic Publishers, Norwell, MA, 1996, pp. 000–000.
7. C. Perkins and P. Bhagwat. Highly dynamic destination-sequenced distance-vector routing (DSDV) for mobile computers. In *ACM SIGCOMM'94 Conference on Communications Architectures, Protocols and Applications*, 1994, pp. 234–244.
8. C. E. Perkins and E. M. Royer. Ad hoc on-demand distance vector routing (aodv). In *Proceedings of the 2nd IEEE Workshop on Mobile Computing Systems and Applications*, February 1999, pp. 90–100.
9. P. Bose, P. Morin, I. Stojmenović, and J. Urrutia. Routing with guaranteed delivery in *ad hoc* wireless networks. *Wireless Networks*, **7**(6):609–616, 2001.
10. B. Karp and H.T. Kung. Gpsr: Greedy perimeter stateless routing for wireless networks. In *Proceedings of ACM MobiCom*, Boston, MA, September 2000.
11. F. Kuhn, R. Wattenhofer, Y. Zhang, and A. Zollinger. Geometric ad-hoc routing: Of theory and practice. In *Proceedings of the Symposium on Principles of Distributed Computing (PODC)*, 2003, pp. 63–72.
12. A. Rao, S. Ratnasamy, C. Papadimitriou, S. Shenker, and I. Stoica. Geographic routing without location information. In *Proceedings of ACM MobiCom*, San Diego, CA, September 2003.
13. Y.-J. Kim, R. Govindan, B. Karp, and S. Shenker. Geographic routing made practical. In *Proceedings of the 2nd USENIX Symposium on Networked Systems Design and Implementation (NSDI)*, Boston, MA, May 2005.

14. H. Takagi and L. Kleinrock. Optimal transmission ranges for randomly distributed packet radio terminals. *IEEE Transactions on Communications*, **32**(3):246–257, 1984.

15. T. C. Hou and V. O. K. Li. Transmission range control in multihop packet radio networks. *IEEE Transactions on Communications*, **34**(1):38–44, 1986.

16. E. Kranakis, H. Singh, and J. Urrutia. Compass Routing on Geometric Networks. In *Proceedings of the* 11th *Canadian Conference on Computational Geometry*, Vancouver, August 1999, pp. 51–54.

17. S. Basagni, I. Chlamtac, V.R. Syrotiuk, and B.A. Woodward. A distance routing effect algorithm for mobility (dream). In *Proceedings of the 4th Annual ACM/IEEE International Conference on Mobile Computing and Networking (MobiCom*, pages 76–84, 1998.

18. Y. B. Ko and N. H. Vaidya. Location-aided routing (lar) in mobile ad hoc networks. In *Proceedings of the 4th Annual ACM/IEEE International Conference on Mobile Computing and Networking (MobiCom)*, 1998, pp. 66–75.

19. G. G. Finn. Routing and addressing problems in large metropolitan-scale internetworks. Technical Report ISI/RR-87-180, Information Sciences Institute (ISI), March 1987.

20. M. Fayed and H. T. Mouftah. Characterizing the Impact of Routing Holes on Geographic Routing. In *Proceedings of Systems Communications 2005 (ICW/ICHSN/ICMCS/SENET 2005)*, August 2005, pp. 401–406.

21. A. Puri, R. Jain, and R. Sengupta. Geographical routing using partial information for wireless ad hoc networks. *IEEE Personal Communications*, **8**(1):48–57, 2001.

22. I. Stojmenovic, M. Russell, and B. Vukojevic. Depth first search and location based localized routing and qos routing in wireless networks. In *Proceedings International Conference on Parallel Processing (ICPP)*, Washington, DC, 2000, IEEE Computer Society, New York, p. 173.

23. Q. Fang, J. Gao, and L. Guibas. Locating and Bypassing Routing Holes in Sensor Networks. In *Proceedings of IEEE/ACM Infocom*, Hong Kong, China, March 2004.

24. S. P. Fekete, A. Kröller, D. Pfisterer, S. Fischer, and C. Buschmann. Neighborhood-based topology recognition in sensor networks. In *Algorithmic Aspects of Wireless Sensor Networks (ALGOSENSORS)*, Vol. 3121 of *Lecture Notes in Computer Science*, Springer, Berlin, 2004, pp. 123–136.

25. S. Funke. Topological hole detection in wireless sensor networks and its applications. In *Proceedings of the 2005 Joint Workshop on Foundations of Mobile Computing (DIALM-POMC '05)*, New York, 2005, pp. 44–53.

26. P. Bose, P. Morin, I. Stojmenović, and J. Urrutia. Routing with guaranteed delivery in *ad hoc* wireless networks. In *Workshop on Discrete Algorithms and Methods for Mobile Computing and Communications (DialM)*, 1999.

27. J. Hightower and G. Borriello. Location systems for ubiquitous computing. *IEEE Computer Magazine*, **August**: 57–66, 2001.

28. L. Hu and D. Evans. Localization for mobile sensor networks. In *Proceedings of the 10th Annual International Conference on Mobile Computing and Networking (MobiCom*, 2004, pp. 45–57.

29. T. Li, A. Ekpenyong, and Y. Huang. A location system using asynchronous distributed sensors. In *Proceedings of IEEE/ACM Infocom*, Hong Kong, China, March 2004.

30. L. M. Ni, Y. Liu, Y. Lau, and A. P. Patil. LANDMARK: Indoor location sensing using active RFID. In *Proceedings of IEEE International Conference on Pervasive Computing (PerCom)*, Fort Worth, TX, March 2003.

31. T. He, C. Huang, B. M. Blum, J. A. Stankovic, and T. Abdelzaher. Range-free localization schemes for large scale sensor networks. In *Proceedings of the 9th Annual International Conference on Mobile Computing and Networking (Mobicom)*, 2003, pp. 81–95.
32. Y. Kim, J.-J. Lee, and A. Helmy. Modeling and analyzing the impact of location inconsistencies on geographic routing in wireless networks. *SIGMOBILE Mobile Computer and Communication Review (MC2R)*, **8**(1):48–60, 2004.
33. K. Seada, A. Helmy, and R. Govindan. On the effect of localization errors on geographic face routing in sensor networks. In *Proceedings of the 3rd International Symposium on Information Processing in Sensor Networks (IPSN)*, 2004, pp. 71–80.
34. B. Karp. Challenges in geographic routing: Sparse networks, obstacles, and traffic provisioning. Presented at DIMACS Workshop on Pervasive Networking, May 2001.
35. C. Lochert, M. Mauve, H. Fler, and H. Hartenstein. Geographic routing in city scenarios. *ACM SIGMOBILE Mobile Computing and Communications Review (MC2R)*, **9**(1):69–72, 2005.
36. A. Woo, T. Tong, and D. Culler. Taming the underlying challenges of reliable multihop routing in sensor networks. In *Proceedings of the 1st International Conference on Embedded Networked Sensor Systems (SenSys)*, 2003, pp. 14–27.
37. J. Zhao and R. Govindan. Understanding packet delivery performance in dense wireless sensor networks. In *Proceedings of the 1st International Conference on Embedded Networked Sensor Systems (SenSys)*, 2003, pp. 1–13.
38. Y.-J. Kim, R. Govindan, B. Karp, and S. Shenker. On the pitfalls of geographic face routing. In *Proceedings of the 2005 Joint Workshop on Foundations of Mobile Computing (DIALM-POMC*, 2005, pp. 34–43.
39. B. Leong, B. Liskov, and R. Morris. Geographic routing without planarization. In *Proceedings of the 3rd USENIX Symposium on Networked Systems Design and Implementation (NSDI)*, San Jose, CA, May 2006.
40. X. Li, Y. J. Kim, R. Govindan, and W. Hong. Multidimensional range queries in sensor networks. In *Proceedings of the 1st International Conference on Embedded Networked Sensor Systems (SenSys)*, 2003, pp. 63–75.
41. H. Balakrishnan, M. F. Kaashoek, D. Karger, R. Morris, and I. Stoica. Looking up data in p2p systems. *Communications of the ACM*, **46**(2):43–48, 2003.
42. Y. Shang, W. Ruml, Y. Zhang, and M. P. J. Fromherz. Localization from mere connectivity. In *Proceedings of the 4th ACM International Symposium on Mobile Ad Hoc Networking & Computing (MobiHoc)*, New York, 2003, pp. 201–212.
43. J. Newsome and D. Song. Gem: Graph embedding for routing and data-centric storage in sensor networks without geographic information. In *Proceedings of the 1st International Conference on Embedded Networked Sensor Systems (SenSys)*, 2003, pp. 76–88.
44. N. Linial, L. Lovasz, and A. Widgerson. Rubber bands, convex embeddings, and graph connectivity. *Combinatorica*, **8**(1):91–102, 1988.
45. Q. Cao and T. Abdelzaher. A scalable logical coordinates framework for routing in wireless sensor networks. In *Proceedings of the 25th IEEE Real-Time Systems Symposium (RTSS)*, 2004, pp. 349–358.
46. L. Demoracski and D. R. Avresky. Performance analysis of fault-tolerant beacon vector routing for wireless sensor networks. In *MSWiM '05: Proceedings of the 8th ACM International Symposium on Modeling, Analysis and Simulation of Wireless and Mobile Systems*, New York, 2005, pp. 40–44.

47. R. Fonseca, S. Ratnasamy, J. Zhao, C. T. Ee, D. Culler, S. Shenker, and I. Stoica. Beacon vector routing: Scalable point-to-point routing in wireless sensornets. In *Proceedings of the 2nd USENIX Symposium on Networked Systems Design and Implementation (NSDI '05)*, Boston, MA, May 2005.

48. Y. Zhao, B. Li, Q. Zhang, Y. Chen, and W. Zhu. Efficient hop id based routing for sparse ad hoc networks. In *Proceedings of the 13TH IEEE International Conference on Network Protocols (ICNP)*, 2005, pp. 179–190.

49. D. S. J. De Couto, D. Aguayo, B. A. Chambers, and R. Morris. Performance of multihop wireless networks: Shortest path is not enough. *SIGCOMM Compututer Communication Review (CCR)*, **33**(1):83–88, 2003.

50. W. Heinzelman, J. Kulik, and H. Balakrishnan. Adaptive protocols for information dissemination in wireless sensor networks. In *Proceedings of the 5th Annual ACM/IEEE International Conference on Mobile Computing and Networking (Mobicom)*, 1999 pp. 174–185.

51. R. Nagpal, H. Shrobe, and J. Bachrach. Organizing a global coordinate system from local information on an ad hoc sensor network. In *the 2nd International Workshop on Information Processing in Sensor Networks (IPSN)*. Springer Verlag Lecture Notes in Computer Science LNCS 2634, April 2003.

CHAPTER 9

Node Positioning for Increased Dependability of Wireless Sensor Networks

MOHAMED YOUNIS

Department of Computer Science and Electrical Engineering, University of Maryland Baltimore County, Baltimore, MD 21250

KEMAL AKKAYA

Department of Computer Science, Southern Illinois University, Carbondale, IL 62901

9.1 INTRODUCTION

Advances in microelectronics have enabled the development of very tiny sensor nodes that have the ability of measuring ambient conditions such as temperature, pressure, humidity, light intensity, vibration, and so on. The sensed data can then be transmitted through an onboard radio transmitter to a single or multiple base stations where it can be further processed. The cost and size advantage of such emerging sensor nodes has encouraged practitioners to explore using them collaboratively in a network formed in ad hoc manner. Such networked sensor systems are not only cost effective but can also provide fast and accurate information gathering in remote and risky areas. Figure 9.1 depicts a typical sensor network architecture. The base station acts as a gateway for linking the sensors to multiple command nodes.

The past few years have witnessed increased interest in the potential use of wireless sensor networks (WSNs) in applications such as disaster management, combat field reconnaissance, border protection, and security surveillance [1, 2]. Sensors in these applications are expected to be remotely deployed and to operate autonomously in unattended environments. While the initial view of the community was that WSNs will play a complementary role that enhances the quality of these applications, recent research results have encouraged practitioners to envision an increased reliance

Algorithms and Protocols for Wireless Sensor Networks, Edited by Azzedine Boukerche

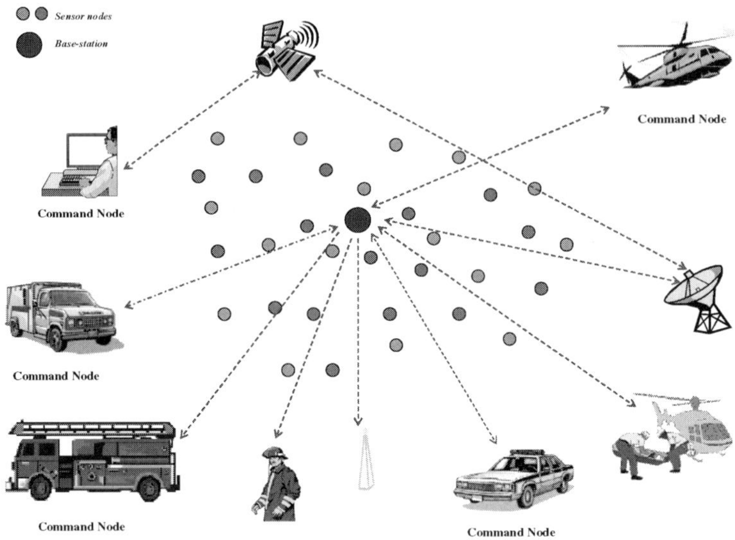

Figure 9.1. A sensor network for a disaster management application.

on WSNs. In order to best realize their potential, dependable design and operation of WSNs have to be ensured.

Dependability is a property that indicates the ability of a system to deliver services to the user subject to a required level of quality. Dependability can be specified in terms of attributes, such as responsiveness, availability, safety, security, and so on. Dependability in WSNs is complicated by many factors including the following: (1) Sensors are significantly constrained in the amount of available resources such as energy, storage, and computation; (2) sensors are expected to be deployed in very large numbers in normal as well as forbidding environments; and (3) WSNs suffer from structural weakness and limited physical protection. Moreover, dependability requirements may vary according to a network's mission, such as field of deployment (e.g., hostile versus friendly), type of application (e.g., monitoring, tracking, data collection), mode of operation (e.g., normal, exception, post-event recovery), and time.

The bulk of the research on WSNs has focused on the effective support of the functional (e.g., data latency) and the nonfunctional (e.g., data integrity) requirements while coping with the resource constraints and on the conservation of available energy in order to prolong the life of the network. Contemporary design schemes for WSNs pursue optimization at the various layers of the communication protocol stack. Popular optimization techniques at the network layer include multihop route setup, in-network data aggregation, and hierarchical network topology [3]. For medium access control, collision avoidance, minimizing idle listening of radio receivers, and output power control are a sample of proposed schemes [1, 4]. At the application layer, examples include adaptive activation of nodes, lightweight data authentication and encryption, load balancing, and query optimization [5, 6].

One of the design optimization strategies is to deterministically place the sensor nodes in order to meet the desired performance goals. In such cases, the coverage can be ensured through careful planning of node densities and field of view, and thus the network topology can be established at setup time. However, in most WSN applications, sensors deployment is random and little control can be exerted in order to ensure coverage and yield uniform node density while achieving strongly connected network topology. Also, the location of the base station can have great influence on the network performance. For example, routing data to a base station that is distant from the source sensor usually involves numerous relaying nodes and thus increases the aggregate delay and energy consumption and risks a packet loss due to link errors. Therefore careful selection of the base-station location may affect various performance metrics such as energy consumption, delay, and throughput. Unlike sensors deployment, positioning of the base stations can be somewhat controlled and is feasible in many application setups.

Optimal node placement is a very challenging problem that has been proven to be NP-complete for base stations [7] and for most of the formulations of sensors' deployment [8–10]. To tackle such complexity, several heuristics have been proposed to find suboptimal solutions [7, 11–14]. However, the context of these optimization strategies is mainly static in the sense that assessing the quality of candidate positions is based on structural quality metric such as distance, network connectivity, and/or basing the analysis on a fixed topology. Therefore, we classify them as static approaches. We note, however, that dynamically adjusting nodes' location can further increase the dependability of WSNs since the optimality of the initial positions may become void during the operation of the network depending on the network state and various external factors. For example, traffic patterns can change based on the monitored events, load may not be balanced among the nodes causing bottlenecks, application-level interest can vary over time, and the available network resources may change due to the depletion of energy of some nodes and/or the addition of more nodes.

In this chapter we opt to categorize the various strategies for statically positioning base stations and sensor nodes in WSNs. We contrast a number of published approaches highlighting their strengths and limitations. Our aim is to help applications designers identify alternative solutions and select appropriate strategies. In addition to surveying static positioning approaches, we show that dynamically repositioning nodes while the network is operational can be a very effective means for boosting the network's dependability attributes. We describe scenarios for which node relocation can be pursued to counter holes in coverage, achieve/maintain strong network connectivity, preserve sensor's energy, increase data timeliness, and boost the node's physical security. We highlight the issues, report on the state of the art, and outline open esearch problem.

The chapter is organized as follows. The next section is dedicated to static strategies for node positioning. We separately cover approaches for sensor placement as well as single and multiple base stations positioning. In Section 9.3 we turn our attention to dynamic positioning schemes. We highlight the technical issues and describe sample techniques that exploit sensors and base-station repositioning to enhance the network dependability. The bulk of the discussion applies to a single base-station setup or to clustered network architectures where multiple base stations operate independently. Section 9.4 highlights the challenges of coordinated repositioning of multiple nodes and describe few attempts to tackle these challenges. Finally, Section 9.5 concludes the chapter.

9.2 STATIC POSITIONING OF NODES

Node placement schemes prior to network startup usually base their choice of the particular nodes positions on metrics that are independent of the network state or assume a fixed network operation pattern that stays unchanged throughout the lifetime of the network. Examples of such static metrics are area coverage, internode distance, and so on. Static network operation models often assume periodic data collection over preset routes. In this section we give an overview of the different categories of static node positioning and discuss sample work from the literature. We cover both sensor and base-station placement in distinct subsections.

9.2.1 Sensor Node Placement

As mentioned before, the position of nodes have dramatic impact on the effectiveness of the WSN and the efficiency of its operation. In this section we discuss contemporary sensor placement strategies in the literature. We classify them according to the deployment schemes, the optimization objective of the placement, and how they relate to the network topology.

Deployment Schemes. Sensors can generally be placed in an area of interest either deterministically or randomly. The choice of the deployment scheme depends highly on the type of sensors, application, and the environment in which the sensors will operate. Controlled node deployment is viable and often necessary when sensors are expensive or when their operation is significantly affected by their position. Examples of such scenarios include when populating an area with highly precise seismic nodes and when placing imaging and video sensors. On the other hand, in some applications, random deployment schemes are the only feasible option. This is particularly true for harsh environments such as a battlefield or a disaster region. Depending on the node distribution and the level of redundancy, random node deployment can achieve the required performance goals.

Deterministic deployment is usually pursued for indoor applications of WSNs. Examples of indoor networks include the Active Sensor Network (ASN) project at the University of Sydney in Australia [15], the Multiple Sensor Indoor Surveillance (MSIS) project at Accenture Technology Labs in Chicago [16], and the Sensor Network Technology projects at Intel [17]. The ASN and MSIS projects gear for serving surveillance applications such as secure installations and enterprise asset management. At Intel, the main focus is on applications in manufacturing plants and engineering facilities—for example, preventative equipment maintenance (Figure 9.2).

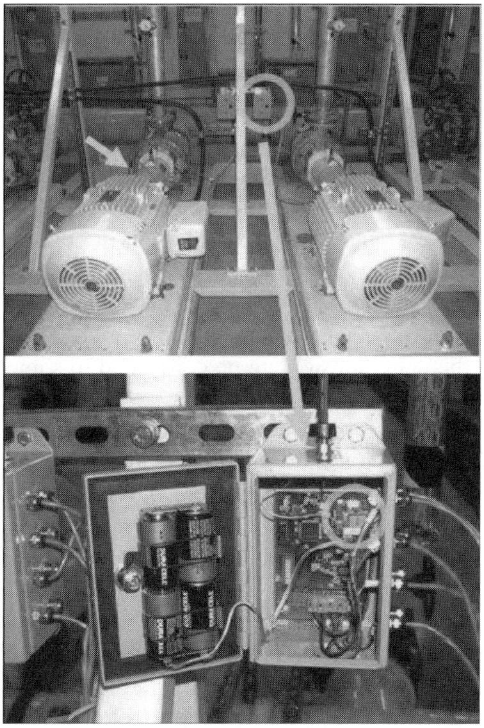

Figure 9.2. Sensors are mounted to analyze the vibration and assess the health of equipment at a semiconductor fabrication plant. (Photographs are from 17.)

Hand-placed sensors are also used to monitor the health of large buildings in order to detect corrosions and overstressed beams that can endanger the structure integrity [18, 19]. Another notable effort is the Sandia Water Initiative at Sandia National Lab which addresses the problem of placing sensors in order to detect and identify the source of contamination in the air or water supplies [20, 21].

Deterministic placement is also very popular in applications of range finders, imaging, and video sensors. In general, these sensors are involved in three-dimensional (3-D) application scenarios, which is much more difficult to analyze compared to two-dimensional deployment regions. Poduri et al. [10] investigated the applicability of contemporary coverage analysis and placement strategies pursued for 2-D space to 3-D setups. They concluded that many of the popular formulation such as art-gallery and sphere-packing problems, which are optimally solvable in 2-D, become NP-hard in 3-D. Most placement approaches for these types of sensor strive to enhance the quality of visual images and/or accuracy of the assessment of the detected objects.

For example, Gonzalez-Banos and Latombe [22] studied the problem of finding the minimum number of range finders, which estimate the proximity of a target, or video sensors and their location in order to cover an area. Unlike other sensors, such as acoustic or temperature, the authors had to account for the restricted capabilities of the range finders, which provide lower and upper bounds only, and for the reliability of detecting objects at grazing angles. The problem is formulated as an Art-Gallery model, for which the fewest guards are to be placed to monitor a gallery. Similarly, the work of Navarro et al. [23] addresses the orientation of video cameras so that high quality images of target objects are captured.

Randomized sensor placement often becomes the only option. For example, in application of WSNs in reconnaissance missions during combat, disaster recovery, and forest fire detection, deterministic deployment of sensors is very risky and/or infeasible. It is widely expected that sensors will be dropped by helicopter, grenade launchers, or clustered bombs. Such means of deployment lead to random spreading of sensors, although the node density can be controlled to some extend. Although it is somewhat unrealistic, many research projects (e.g., reference 24), have assumed uniform node distribution when evaluating the network performance. The rationale is that with the continual decrease in cost and size of microsensors, a large population of nodes is expected and thus a uniform distribution becomes a reasonable approximation.

Ishizuka and Aida [25] have investigated random node distribution functions by trying to capture the fault-tolerance properties of stochastic placement. They compared three deployment patterns (Figure 9.3, taken from reference 25); namely simple diffusion (two-dimensional normal distribution), uniform, and R-random where the nodes are uniformly scattered with respect to the radial and angular directions from the base-station. The R-random node distribution pattern resembles the effect of an exploded shell. The experiments tracked coverage and node reachability as well as data loss in a target tracking application. The simulation results indicated that the initial placement of sensors has a significant effect on network dependability measured in terms of tolerance of node failure that may be caused by damage and battery exhaustion. The results also showed that the R-random deployment is a better placement strategy in terms of fault-tolerance.

While a flat architecture is assumed in reference 25, Xu et al. [26] have considered a two-tier network architecture in which sensors are grouped around relaying nodes that directly communicate with the base station. The goal of the investigation is to identify appropriate node deployment strategies in order to maximize the network lifetime. They first showed that uniform node distribution often does not extend the

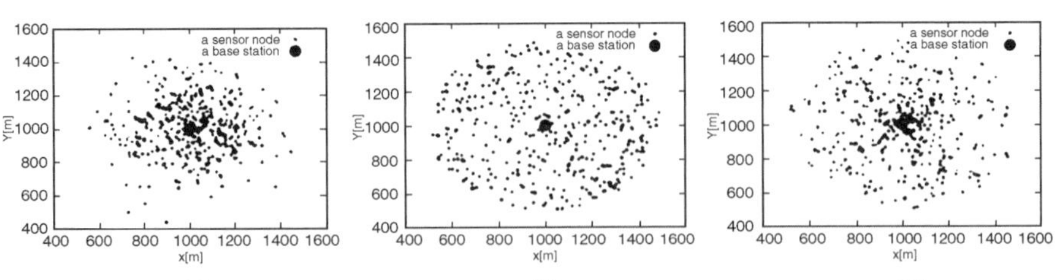

Figure 9.3. (a) Simple diffusion. (b) Constant (uniform) placement. (c) R-random placement.

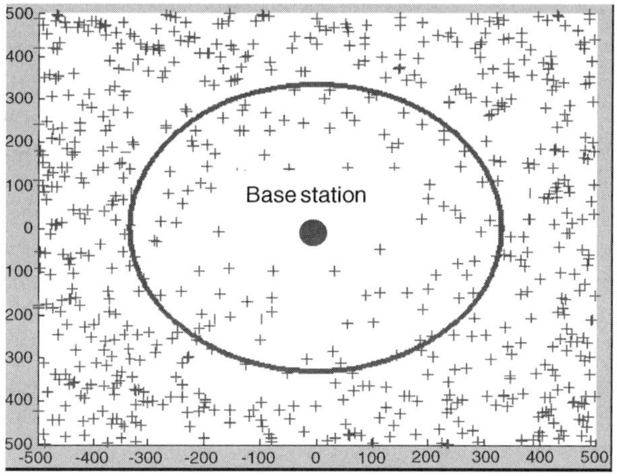

Figure 9.4. An illustration of weighted random deployment (from reference 26). The density of relay nodes inside the circle (close to the base station) is lower and the connectivity is weaker than outside.

network lifetime since relay nodes will consume energy at different rates depending on their proximity to the base station. Basically, the further away the relays are from the base station, the higher the energy they deplete in transmission. To counter such shortcomings, a weighted random node deployment strategy is then proposed to account for the variation in energy consumption rate in the different regions. Such strategy, as illustrated in Figure 9.4, increases the density of relays away from the base station to split the load among more relays and thus extend their average lifetime. Although it has a positive impact on network lifetime, the weighted random distribution may leave some relay nodes disjoint from the base station since some relays may be placed so far that the base station becomes out of their transmission range. Finally, a hybrid deployment strategy is introduced to balance the network lifetime and connectivity goals. The analysis is further extended in reference 27 for the case where relay nodes reach the base station through a multihop communication path. The conclusion regarding the three strategies was found to hold in the multihop case as well.

Primary Objectives for Deployment. Application developers surely like the sensors to be deployed in a way that leverages the overall design goals. Therefore, most of the proposed deployment schemes in the literature focused on increasing the coverage, achieving strong network connectivity, prolonging the network lifetime, and/or boosting the data fidelity. A number of secondary objectives such as tolerance of node failure and load balancing were also considered. Most of the work strives to maximize the deployment objectives using the least amount of resources (e.g., number of nodes). Obviously, meeting the design objectives using random deployment schemes is an utmost challenge. Meanwhile, although intuitively deterministic placement can theoretically meet all primary and secondary objectives, the quest for minimizing the required network resources keeps the problem very hard.

Area Coverage is the deployment objective that has received the most attention in the literature. Assessing the coverage varies based on the underlying model of sensor's field of view and the metric used to measure the collective coverage of deployed sensors. The bulk of the published work (e.g., reference 28) assumed a disk coverage zone centered at the sensor with a radius that equals its sensing range. However, some recent work has started to employ more practical sensor's field of view in the form of irregular polygons [29]. Some of the published papers, especially early ones, use the ratio of the covered area to the overall deployment region as a metric for the quality of coverage [28]. Recent work, however, has focused on the worst-case coverage, usually referred to least exposure, measuring the probability that a target would travel across an area or an event would happen without being detected [30]. The advantage of exposure-based coverage assessment is the inclusion of a practical object detection probability that is based on signal processing formulations (e.g., signal distortion) as applicable to specific sensor types.

As mentioned earlier, optimized sensor placement is not an easy problem, even for deterministic deployment scenarios. Complexity is often introduced by the quest for employing the least number of sensors for meeting the application requirements and by the uncertainty in a sensor's ability to detect an object due to distortion that may be caused by terrain or the sensor's presence in a harsh environment. Dhillon and Chakrabarty [13] considered the placement of sensors on a grid approximation of the deployment region. They formulated a sensing model that factors in the effect of terrain in the sensor's surroundings and inaccuracy in the sensed data (Figure 9.5). The model is then used to identify the grid points on which sensors are to be placed, so that an application-specific minimum confidence level on object detection is met. They proposed a greedy heuristic that strives to achieve the coverage goal through the least number of sensors. The algorithm is iterative. In each iteration, one sensor

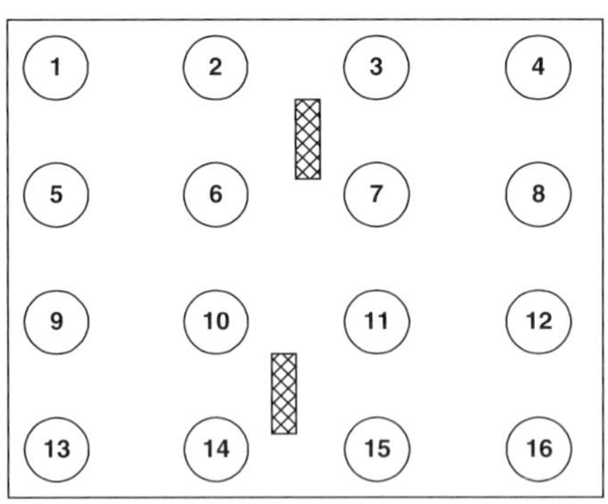

Figure 9.5. Knowing the coordinate of obstacles, the sensor's field of view is adjusted. In the shown example, redrawn from reference 13, the sensor on grid point 14 is not sufficient to cover the area in direction to point 11. The same applies for (2,7), (6,3), and (10,15).

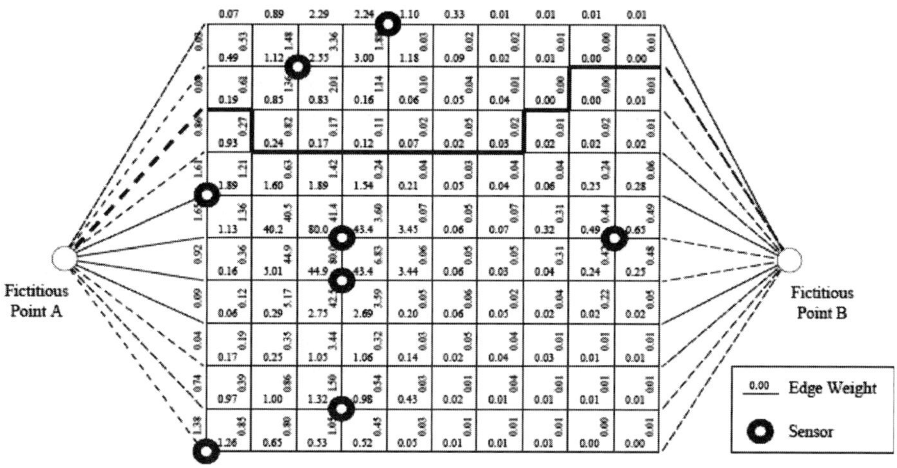

Figure 9.6. The dark circle marks the location of sensors. The weights on the individual line segments are based on the probability that a target is detected by all sensors (combined exposure). The least exposure path, marked in dark line, is found by applying Dijkstra's algorithm.

is placed at the grid point with the least coverage. The algorithm terminates when the coverage goal is met or a bound on the sensor count is reached.

Clouqueur et al. [14] also studied the problem of populating an area of interest with the least number of sensors so that targets can be detected with the highest probability. Unlike reference 13, random deployment is assumed in this work. The authors propose a metric called path exposure to assess the quality of sensor coverage. The idea is to model the sensing range of deployed nodes and establish a collective coverage map of all sensors based on a preset probability of false alarm (detection error). The map is then checked in order to identify the least exposure path on which a target may slip by, with the highest probability of being undetected. Figure 9.6, taken from reference 14, illustrates the idea on a grid structure. Employing such a metric, the authors further introduced a heuristic for incremental node deployment so that every target can be detected with a desired confidence level using the fewest sensors count. The idea is to randomly deploy a subset of the available sensors. Assuming that the sensors can determine and report their positions, the least exposure path is identified and the probability of detection is calculated. If the probability is below a threshold, additional nodes are deployed in order to fill holes in the coverage along the least exposure path. This procedure would be repeated until the required coverage is reached. The paper also tried to answer the question of how many additional nodes are deployed per iteration. On the one hand, it is desirable to use the least number of sensors. On the other hand, the means for sensor deployment may be expensive or risky (e.g., sending a helicopter). The authors derive a formulation that accounts for the cost of deploying nodes and the expected coverage as a function of sensors count. The formulation can be used to guide the designer for the most effective way to populate the area.

The sensor placement problem considered by Biagioni and G. Sasaki [31] is more difficult. They opt to find a placement of nodes that achieves the coverage goals using the least number of sensors and also maintains a strongly connected network topology even if one node fails. The authors review a variety of regular deployment topologies (e.g., hexagonal, ring, star, etc.) and study their coverage and connectivity properties under normal and partial failure conditions. They argue that regular node placement simplifies the analysis due to their symmetry despite the fact that they often do not lead to optimal configuration. The provided analytical formulation can be helpful in crafting a placement as a mix of these regular topologies and in estimating aggregate coverage of nodes.

Network connectivity is also a major concern when deploying sensor nodes. Unlike coverage, which has constantly been an objective or constraint for node placement, connectivity was deemed a non-issue in some of the early work based on the assumption that the transmission range T_r of a node is much longer than its sensing range S_r. The premise is that good sufficient coverage will yield a connected network when T_r is multiple of S_r. However, if the communication range is limited (e.g., $T_r = S_r$), connectivity becomes an issue unless there is a substantial redundancy in coverage. It is worth noting that some work tackled the connectivity concern through deploying relay nodes that have long haul communication capabilities. Such approaches will be discussed in detail later in this section.

Kar and Banerjee [32] considered sensor's placement for complete coverage and connectivity. Assuming that the sensing and radio ranges are equal, the authors first define an *r-strip* as shown in Figure 9.7a. In an *r-stripe*, nodes are placed so that neighbors of a sensor along the x-axis are located on the circumstance of the circle that defines the boundary of its sensing and communication range. Obviously, nodes on an *r-strip* are connected. The authors then tile the entire plane with *r-strips* on lines $y = k(0.5\sqrt{3} + 1)r$ such that the *r-strips* are aligned for even values of the integer k and shifted horizontally $r/2$ for odd values of k, as illustrated in Figure 9.7b. The goal is to fill gaps in coverage with the least overlap among the r-disks that define the boundary of the sensing range. To establish connectivity among nodes in different *r-strips*, additional sensors are placed along the y-axis, which marked by the shaded disks in Figure 10.7b. For every odd value of the integer k, two sensors are placed at $[0, k(0.5\sqrt{3} + 1)r \pm 0.5\sqrt{3}r]$ to establish connectivity between every pair of *r-strips*. For a general convex-shaped finite-size region, connectivity among nodes in horizontal *r-strips* is established by another *r-strip* placed diagonally within the boundary of the region (Figure 9.7c). The authors generalize their scheme for the case where points of interest are to be covered rather than the whole area. However, unless the base station is mobile and can interface with the WSN through any node, having a strongly connected network is not essential in WSNs since data are gathered at the base station. Therefore, ensuring the presence of a data route from a node to the base station would be sufficient, and thus fewer nodes can be employed to achieve network connectivity than the presented approach would use. In addition, vertically placed nodes or diagonal *r-strips* can become a communication bottleneck since they act as gateways among horizontal *r-strips*, which may require the deployment of more sensors to split the traffic.

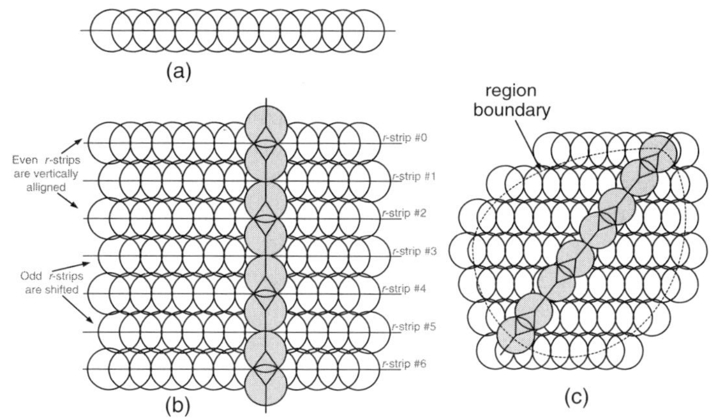

Figure 9.7. Illustration of the placement algorithm in a plane and a finite size region. Figure is redrawn from reference 32.

The focus of reference 33 is on forming *K*-connected WSNs. *K*-connectivity implies that there are *K* independent paths among every pair of nodes. For $K > 1$, the network can tolerate some node and link failures and guarantee certain communication capacity among nodes. The authors study the problem of placing nodes to achieve *K*-connectivity at network setup time or to repair a disconnected network. They formulate the problem as an optimization model that strives to minimize the number of nodes required to maintain *K*-connectivity. They show that the problem is NP-hard and propose two approximation algorithms with varying degree of complexity and closeness to optimality. The algorithms are graph-theory-based. The idea is to compute a weighted complete graph on the same set of vertices (nodes) and then find an approximate minimum-weight *K*-vertex-connected subgraph *g*. Finally, missing links (edges) in *g* are established by deploying the least number of nodes. Again in most WSNs, it is not necessary to achieve *K*-connectivity among sensors unless the base station changes its location frequently.

Network lifetime has been the optimization objective for most of the published communication protocols for WSNs. The positions of nodes significantly impact the network lifetime. For example, variations in node density throughout the area can eventually lead to unbalanced traffic load that causes the rapid drain of the energy reserve of some sensors [26]. In addition, uniform node distribution may lead to the depletion of energy of nodes that are close to the base station at a higher rate than other nodes and thus shorten the network lifetime [34]. Some of the published work—such as reference 27 which we discussed earlier—has focused on prolonging the network lifetime rather than area coverage. The implicit assumption is that there is a sufficient number of nodes or that the sensing range is large enough such that no holes in coverage may result.

Chen et al. [35] studied the effect of node density on network lifetime. Considering the one-dimensional placement scenario, the authors derived an analytical formulation for the network lifetime per unit cost (deployed sensor). They also argued that the network lifetime is not growing proportionally to the increased node population and thus a careful selection of the number of sensors is necessary to balance the cost and lifetime goals. Considering the network to be functional until the first node dies, an optimization problem was defined with the objective of identifying the least number of sensors and their positions so that the network stays operational for the longest time. An approximate two-step solution is proposed. In the first step, the number of sensors is fixed and their placement is optimized for maximum network lifetime. They formulate such optimization as a multivariant nonlinear problem and solve it numerically. In the second step, the number of sensors is minimized in order to achieve the highest network lifetime per unit cost. A closed-form solution is analytically derived for the second step.

Hou et al. [36] considered a two-tier sensor network architecture where sensors are split into groups; each is led by an aggregation-and-forwarding (AFN) node (Figure 9.8). A sensor sends its report directly to the assigned AFN, which aggregates the data from all sensors in its group. The AFNs and the base station form a second tier network in which an AFN sends the aggregated data report to the base station over a multihop path. The authors argue that AFNs can be very critical to the network operation and that their lifetime should be maximized. Two approaches were suggested to prolong the AFNs lifetime. The first is to provision more energy to AFNs. The second is to deploy relay nodes (RNs) in order to reduce the communication energy consumed by an AFN in sending the data to the base station. The RN placement and energy provisioning problem was formulated as a mixed-integer nonlinear programming optimization. For a pool of an E energy budget and M relay nodes, the objective of the optimization is to find the best allocation of the additional energy to

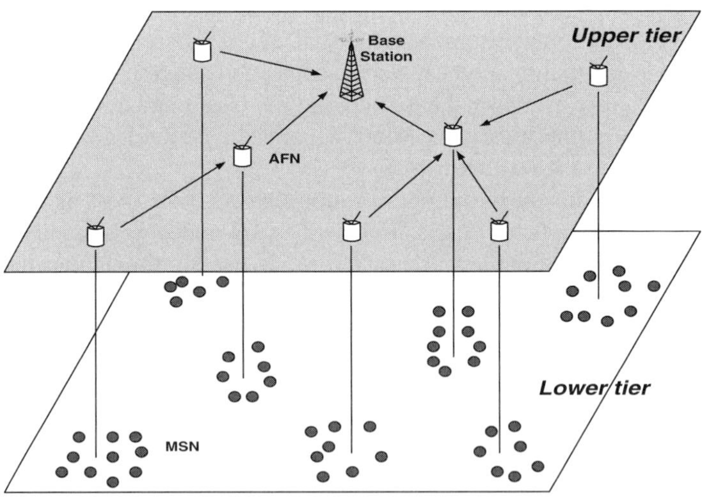

Figure 9.8. Logical view of the assumed two-tier network architecture (redrawn from reference 36). MSN stands for micro-sensor nodes.

existing AFNs and the best positions for placing the M relays. To efficiently solve the optimization problem, the formulation was further simplified through a two-phase procedure. In the first phase a heuristic is proposed for optimized placement of the M relay nodes. Given the known positions of the RNs, in the second phase the energy budget is allocated to the combined AFNs and RNs population, which is a linear programming optimization.

Data fidelity is obviously an important design goal of WSNs. A sensor network basically provides a collective assessment of detected objects by fusing the readings of multiple independent, and sometimes heterogeneous, sensors. Data fusion boosts the creditability of the reported incidents by lowering the probability of false alarms and of missing a detectable object. From a signal processing point of view, data fusion tries to minimize the effect of the distortion by considering the readings from multiple sensors so that a high-fidelity assessment can be made regarding the detected phenomena. Although increasing the number of sensors reporting in a particular region will sure boost the accuracy of the fused data, redundancy in coverage would require an increased node density, which can be undesirable due to cost and other constraints such as the potential of detecting the sensors in a combat field.

Zhang and Wicker [37] looked at the sensor placement problem from a data fusion point of view. The observation that they made is that there is always an estimation distortion associated with a sensor reading which is usually countered by getting many samples. They thus mapped the problem of finding the appropriate sampling points in an area to that of determining the optimal sampling rate for achieving a minimal distortion, which is extensively studied in the signal processing literature. In other words, the problem is transformed from the space to the time domain. The optimal sampling points are actually the best spots where sensors can be placed. The approach is to partition the deployment area into small cells, and then optimal sampling rate per cell is determined for minimal distortion. Assuming that all sensors have the same sampling rate, the number of sensors per cell can be determined.

Similar to Zhang and Wicker [37], Ganesan et al. [38] studied sensors placement to meet some application quality goals. The problem considered is to find nodes positions so that the fused data at the base station meets desired level of fidelity. Unlike Zhang and Wicker [37], a tolerable distortion bound is imposed as a constraint, and minimizing energy consumption in communication is set as an objective of the optimization formulation. Also, in this work the number of sensors is fixed and their position is to be determined. Given the consideration of energy consumption, data paths are modeled in the formulation, making the problem significantly harder. The authors first provided a closed-form solution for the one-dimensional node placement case and used it to propose an approximation algorithm for node placement in a circular region. Extending the approach to handle other regular and irregular structures is noted as future work.

Wang et al. [39] also exploited similar ideas for a WSN that monitors a number of points of interest. Practical sensing models indicate that the ability of detecting target/events diminishes with increased distance. One way to increase the creditability of the fused data is to place the sensors so that a point of interest would be in the high-fidelity sensing range of multiple nodes. Given a fixed number of sensors, there is a tradeoff between deploying a sensor in the vicinity of one point of interest to enhance the probability of event detection and the need to cover other points of interest. The probability of event detection by a sensor is called the utility. The utility per point of interest is thus the collective utility of all sensors that cover that point. The authors formulate a nonlinear optimization model to identify the locations of the sensors so that the average utility per point of interest is maximized. To limit the search space, the area is represented as a grid with only intersection points considered as candidate positions.

Finally we would like to note that the work of Clouqueur et al. [14], which we discussed earlier, can also be classified under data-fidelity-based sensor placement. They estimate the creditability of fused data from multiple sensors and use it to identify the position of sensors for maximizing the probability of target detection. Table 9.1 categorizes the sensor nodes placement mechanisms discussed in this section.

9.2.2 Base-Station Placement

A considerable research has been done on optimal initial (i.e., at network setup time) positioning of single or multiple base stations in WSNs. Published work generally differs based on the assumptions made, the considered network model, the available network state information, and the metrics to be optimized. Popular objectives of base-station positioning include maximizing the overall network lifetime, minimizing the longest data path, and achieving maximal data rate. Given the location, onboard energy supply, and number of sensors, such objectives are optimized using techniques like integer linear programming, network flow, and computational geometry. In this section, we categorize prior work on static single and multiple base-station positioning.

Optimized Positioning of a Single Base Station. The limited energy supply onboard a sensor node has made network longevity a key performance metric. Some published work exploited the flexibility in base station positioning in order to extend the network lifetime. Nonetheless, multiple variants of the base-station positioning were pursued. The difference is due to either the definition of a network lifespan, the network operation model, and the network state parameters that are included in the formulation. While some considered the network to be functional until the first sensor node dies [7], many used the failure of a percentage of the deployed sensors [12, 40, 41] as indicative of the network lifespan. Other work strived to extend the network lifetime through minimizing the total power consumed in collecting the readings of all sensors [11]. All these static base-station positioning approaches have assumed a periodic data collection model for the network. That is, each sensor node transmits a certain amount of data at a fixed rate. Some schemes also factored in the necessary transmission scheme for the sensor nodes [7]. However, no in-network data aggregation has been considered; that is, each node should transmit the packets it received without any concatenation, suppression, or compression. While most of these

TABLE 9.1. A Comparison Between the Various Approaches for Sensors Placement

Reference	Application	Space	Deployment	Primary Objective	Secondary Objective	Constraint
15	Surveillance	3-D	Deterministic	Coverage	Connectivity	—
16	Surveillance	3-D	Deterministic	Coverage	Data fidelity	—
17	Manufacturing	3-D	Deterministic	Data fidelity	Connectivity	
18	Structural health monitoring	3-D	Deterministic	Data fidelity	Connectivity	—
19	Structural health monitoring	3-D	Deterministic	Data fidelity	Connectivity and Fault-tolerance	—
20	Contamination detection	3-D	Deterministic	Coverage	—	Fixed sensors count
21	Contamination detection	3-D	Deterministic	Coverage	Delay	Fixed sensors count
22	Generic	3-D	Deterministic	Data fidelity	Min. sensor count	—
23	Generic	3-D	Deterministic	Data fidelity	—	—
24	Outdoor	2-D	Deterministic	Min. relay count	Fault-tolerance	Connectivity
25	Outdoor	2-D	Random	Coverage and connectivity	Fault-tolerance	—
26	Outdoor	2-D	Random	Network lifetime	—	—
27	Outdoor	2-D	Random	Network lifetime	—	—
13	Surveillance	2-D	Deterministic	Coverage	Minimum sensor count	—
14	Outdoor	2-D	Random	Data fidelity and coverage	Minimum sensor count	—
31	Outdoor	2-D	Deterministic	Coverage and connectivity	Minimum sensor count and fault-tolerance	—
35	Generic	1-D	Deterministic	Network lifetime	—	Coverage
36	Generic	2-D	Deterministic	Network lifetime	—	Fixed relays count
32	Outdoor	2-D	Deterministic	Coverage and Connectivity	Minimum sensor count	—
33	Outdoor	2-D	Deterministic	Connectivity	Fault-tolerance	—
37	Outdoor	2-D	Random	Data fidelity	—	—
38	Generic	1-D 2-D	Deterministic	Data fidelity	Minimal energy consumption in communication	Lower bound on tolerable distortion; fixed sensors count
39	Surveillance	2-D	Deterministic	Data fidelity	—	—

approaches (e.g., see reference 41) rely on the availability of exact nodes locations obtained through GPS or simply by deterministic sensor deployment, a few such as those in references 12 and 42 use internode proximity estimates provided through ranging technology.

The considered network topology and system model are also differentiating factors among published static base-station positioning approaches. In a flat network, topology sensors are homogeneous having the same amount of initial energy, and usually form multihop routes to relay their data to the base station. The base-station positioning problem gets more challenging in case of multihop routing since each sensor node not only transmits its own data but also forwards data from its neighbors when requested [7]. The problem becomes simpler when sensors directly transmit to the base station [42]. However, given the limited transmission range for the sensor nodes, this assumption cannot be applied to WSNs that cover large areas. To support scalability in such setups, hierarchical network topologies are often pursued. The most common hierarchical topology is based on grouping sensors into clusters with a designated cluster head. In such cases, the scope of the base-station positioning problem is reduced to the inter-cluster-head network and thus becomes simpler [12]. It is worth noting that if base stations act as cluster heads, the optimization becomes a multi-base-station positioning problem.

Depending on the assumptions and network models, most of the optimal base-station positioning problem formulations are NP-complete. A common way to counter such complexity is to employ approximation. For instance, in reference 7 the search space is restricted to the sensor locations and the best position "s" among them in terms of network's lifetime is picked. This solution is shown to be a constant approximation of the optimal solution; for example, it achieves a fraction of the optimal network lifespan. The approximation ratio was further improved to $(1 - \varepsilon)$, where $\varepsilon > 0$ is any desired error bound, by factoring in the routes and transmission schedule. However, the improvement came at the cost of increased computation for solving multiple linear programs. To limit such high computation, a technique that explores the potential overlap among the elements of the search space is proposed [43]. The idea is to replace an infinite search space for each variable by a finite-element search space with guaranteed bound on possible loss in performance. Specifically, the search space increases exponentially with the increasing number of variables, and such an increase can be reduced by exploring the potential overlap among the elements of the search space. In order to determine the potential overlap, the variables are expressed in the form of a geometric progression and a common factor among theses geometric progression is identified.

The approaches in references 12 and 42 do not consider data relaying, and thus the problem becomes solvable in a polynomial time. In order to minimize the total communication power, the base station is to be located such that the maximum distance to a sensor node is minimized [42]. A computational geometry-based algorithm, whose complexity is linear in the number of nodes n, is proposed. This algorithm tries to determine the circle with the least diameter that encloses the nodes. Such circle can be formed with at most three points picked among the locations of the sensors. The base station will then be positioned at the center of the circle [42]. The same

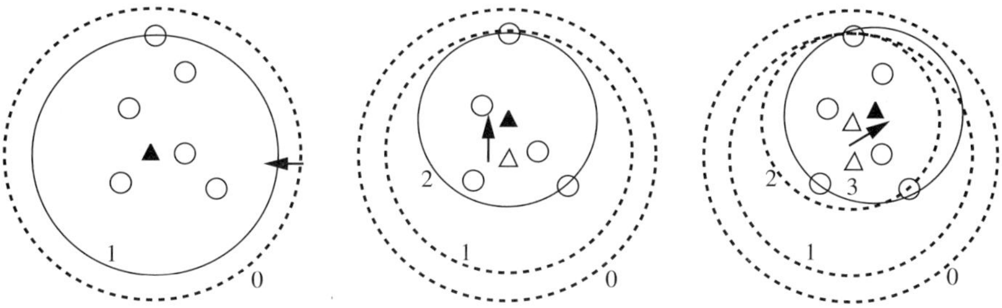

Figure 9.9. Finding the minimum enclosing circle for six application nodes [12]. The base station marked as a triangle is placed at the center of the smallest disk that contains all application nodes (the small circles).

algorithm is further extended for a two-tiered WSN where special "application" nodes are designated as cluster heads [12]. The application node interfaces the cluster with the base station. The minimum enclosing circle is thus found for the application nodes as shown in Figure 9.9, redrawn from reference 12. Although the time complexity for these approaches are quite promising, they imply a centralized network management strategy, including routing and/or MAC, which may not be desirable particularly in large-scale settings.

Positioning Multiple Base-Stations. Multi-base-station positioning is even more challenging given the higher scale and the fact that individual sensors can select among multiple destinations to which they send the data. The positioning problem is typically defined as the optimal layout for a known number of base stations in order to maximize some performance metric such as total communication energy and throughput [11] or area coverage [44]. In some cases, the number of base stations may not be known in advance, and thus the optimal number and location of base stations are to be found [40]. In the context of wireless sensor and actor networks, the base stations (actors) need to be positioned for maximized area coverage and reduced data delivery latency [44].

In general, the complexity of the multi-base-station positioning problem varies based on the planned network architecture. When a flat network topology is pursued, the problem stays NP-complete [7]. The same applies to clustered networks in which some sensors are designated as cluster heads forming a two-tier topology. However, when the nodes are grouped into clusters that are led by the individual base stations, the complexity depends on the order of the network clustering and base-station positioning procedures. If the sensors are assigned to base stations prior to placing or finalizing the positions of base stations, the scope of the problem becomes local to the individual clusters and concerns only each base station independently from the others [40]. In other words, the problem becomes similar to a single base-station positioning. However, if the base-station positioning precedes the network clustering, the complexity remains NP-complete [11].

To counter the high complexity, most published approaches try to limit the search space (i.e., the cardinality of the set of candidate locations) in the hope of converging to some locations that achieve near-optimal performance. For example, in reference 41, candidate locations are determined a priori by the application designers. Meanwhile, in reference 11 the feasible set is restricted to where sensors are. The latter case also applies when the base-station positioning is combined with the network clustering scheme. For example, it may be desired for the base stations to be placed so that each sensor would reach a base station in at most K hops. By deciding to place base stations next to some sensors, the positioning/clustering problem in that case becomes finding the K-dominating set that has some polynomial time solution [45]. It is worth noting that if the number of base stations is fixed, the problem stays NP-complete.

Possible approaches for the multi-base-station positioning include approximation algorithms [11] and integer programming [41]. To maximize the achievable rate of collecting sensor's data, the base-station positioning problem is solved by using two approximation strategies, namely, greedy and local search [11]. In the greedy algorithm the base-station position is restricted to the location of sensors and then the base stations are individually placed in an arbitrary order for increased data rate. The local search starts with a random configuration of base stations and is then followed by a search for a better layout that boosts the data rate. An example for local search for four base stations on a 10×10 grid with a sensor transmission radius of 2.2 units is shown in Figure 9.10, which is redrawn from [11]. On the other hand, the Integer Linear Programming (ILP) formulation proposed in reference 41 is geared for splitting the data routing load among the sensors as evenly as possible. The objective function is to minimize the maximum energy consumption at the individual sensors while minimizing the total communication energy. The constraints of the ILP formulation include a bound on the total energy consumed by a node in a data collection round and a restriction on the candidate base-station location to be picked from a set of predetermined positions. Other constraints are also specified to ensure a balanced

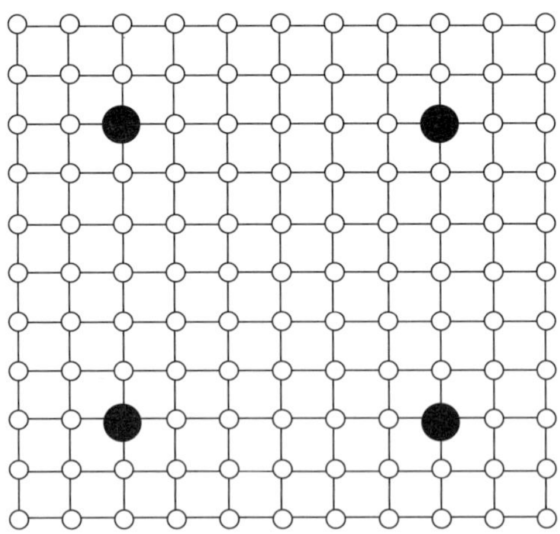

Figure 9.10. Optimal solution with local search.

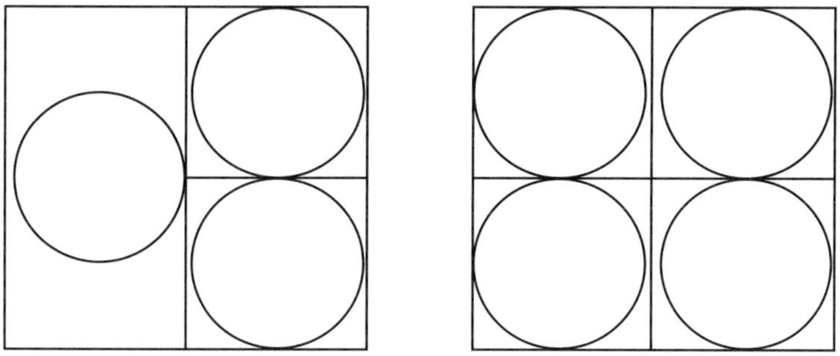

Figure 9.11. Initially, base stations (actors) are uniformly placed in the area of interest. Circles define the acting range of an actor.

flow through the individual nodes and to allow transmission of messages to a feasible site only if a base station exists at that site. The base-station positions are recomputed periodically to cope with changes in the network state.

In addition to energy metric considered in references 11 and 41, area coverage can also be optimized through careful positioning of multiple base stations (actors) [44]. In some applications, scenarios like disaster management and combat field reconnaissance base stations not only collect and process the data but also can do some reactive actions such as extinguishing a fire and de-mining a travel path. In this case, minimizing the delivery latency of sensors' data and the time for a base station to reach the spot that needs an action would be design objectives. Initially the actors are positioned uniformly in order to maximize the coverage (i.e., minimize the overlap among the action ranges) of the area as shown in Figure 9.11. Sensors are then grouped into clusters; each is led by an actor. After clustering, each actor considers the positions of its assigned sensors as vertices and computes the vertex 1-center [7]. Relocating the actor at the vertex 1-center location ensures minimum delay from the farthest sensor node; that is, it minimizes the maximum latency for data delivery.

However, when relocating an actor to its 1-center location, it may lose connection with the other actors in the network. In order to also ensure network connectivity, the approach in reference 44 is further extended in reference 46. Connectivity is maintained by moving an actor close to the vertex 1-center of its cluster as much as possible without breaking the links with its neighbors. The relocations are strived to be performed in a global order based on the IDs of actors so as not to disconnect the network with simultaneous relocations of neighboring actors.

Table 9.2 compares the characteristics of the approaches discussed in this section. It is worth noting that the approach of reference 41 reevaluates the optimality of the picked base-station positions unlike the previously discussed static approaches. However, it still falls short from being a dynamic approach since it simply pauses the network operation between successive base-stations placements. In the next section, we argue that if this repositioning can be done dynamically without disrupting the network operation, it not only can improve certain performance metrics but also adjust the network topology based on the changes in the event area.

TABLE 9.2. Comparison of Approaches for Static Base-Station Positioning

	Reference 7	Reference 12	Reference 11	Reference 40	Reference 41	Reference 44
Data rate of sensors	Fixed	Fixed	Fixed	NA	Fixed	NA
Network lifetime definition(s)	First sensor node to die	K-of-N sensor nodes to die	Total power	Percentage of unreachable nodes	Percentage fo dead sensors	NA
Number of BS	1	1	Fixed and known	Fixed (known or unknown)	Fixed and known	Fixed and known
Initial candidate BS Locations	NA	NA	The set of sensor locations	Depends on the clustering algorithm	Predefined random set of locations	Uniformly distributed
Consider routing and MAC schemes	Yes	No	Yes	No	Yes	No
Clusters	No	Yes	No	Yes	No	Yes
Network topology	No restriction	Two-tier (sensors, cluster head) with random graphs	Regular grid, random graph, preferential attachment graph	Uniform random graph	Random graph	Two-tier uniform
Sensor location	Available	Available Not available	Available	Available	Available	Available
Multihop routes	Yes	No	Yes	Yes	Yes	Yes
Solution	Integer programming	Computational Geometry	Local search and greedy heuristics	Computational geometry	Integer programming	Vertex 1-center heuristics

NA, not available.

9.3 DYNAMIC REPOSITIONING OF NODES

Most of the protocols described above initially compute the optimal location for the nodes and do not consider moving them once they have been deployed. Moreover, the context of the pursued optimization strategies is mainly static in the sense that assessing the quality of candidate positions are based on performance metrics like the data rate, sensing range, path length in terms of the number of hops from a sensor node to the base station, and so on. In addition, the placement decision is made at the time of network setup and does not consider dynamic changes during the network operation. For example, traffic patterns can change based on the monitored events, load may not be balanced among the nodes causing bottlenecks, application-level interest can vary over time, and the available network resources may change due to the depletion of energy of some nodes and/or the addition of more nodes.

Therefore, dynamically repositioning the nodes while the network is operational is necessary to further improve the performance of the network. For instance, when many of the sensors in the vicinity of the base-station become dysfunctional due to the exhaustion of their batteries, it is better for the base station to reposition itself in order to become easily and reliably reachable to data sources. Such repositioning can boost the network longevity and reduce the effect of packet drops caused by link and node failures. Similarly, instead of repositioning the base station, some redundant sensors from other parts of the monitored region can be identified and relocated to replace the dead sensors in the vicinity of the base-station to improve the network lifetime. Such dynamic relocation can also be very beneficial in a target tracking application where the target is mobile. For instance, some of the sensors can be relocated close to the target to increase the fidelity of the sensor's data. Moreover, in some applications it may be wise for the base station to keep a distance from harmful targets (e.g., an enemy tank) by relocating to safer areas in order to ensure its availability.

Relocating the nodes during regular network operation is very challenging. Unlike initial placement, such relocation is pursued in response to a network- or environment-based stimulus and thus requires continual monitoring of the network state and performance as well as analysis of events happening in the vicinity of the node. In addition, the relocation process would need careful handling since it could potentially cause disruption in the data delivery. The basic issues can be enumerated as follows: when it would make sense for a node to relocate, where it should go, and how the data will be routed while the node is moving. In this section we discuss these issues in detail and discuss sample published approaches on dynamic sensor and base-station repositioning.

9.3.1 Relocation Issues

Relocation. The decision for a node movement has to be motivated by either (a) an unacceptable dependability measure despite setting up the most efficient network topology or (b) a desire to boost such measures beyond what is achievable at the present node position. Motives vary based on the targeted dependability attributes. Examples include the observation of bottlenecks in data relaying, a decrease in node coverage in an area, an increase in packet latency, excessive energy consumption per packet delivery, an increased risk to important nodes such as the base station from an approaching target, and so on. A weighted average may also be pursued to combine multiple dependability metrics based on the application in hand.

Once a node has the motive, it will consider moving to a new position. Such consideration does not necessarily lead to an actual relocation. The node would need to qualify the impact of repositioning at the new location on the network dependability. Therefore the "when" and "where" issues of the node movement are very closely interrelated. In addition, the node would have to assess the relocation overhead. Such overhead can be incurred by the node and the network. For example, if the base station or the sensor node is a robot, the energy consumed by the mechanical parts in the movement is a significant overhead to the lifetime of the robot's battery and thus should be minimized. Moreover, when energy and timeliness metrics are of utmost concern, the impact on the lifetime of individual sensors and on route maintenance has to be considered, respectively.

Where to Relocate. When having a motive to relocate, the node needs to identify a new position that would boost the network dependability. Again, the qualification of the new position and possibly the search criteria may vary based on the dependability attribute. Finding an optimal location for the node in a multihop network is a very complex problem. The complexity is mainly resulting from two factors. The first is the potentially infinite number of possible positions to which a node can be moved. The second factor is the overhead of keeping track of the network and the node state information for determining the new location. In addition, if the base station is to be relocated, for every interim solution considered during the search for an optimal position, a new multihop network topology may need to be established in order to qualify that interim solution in comparison to the current or previously picked positions.

A mathematical formulation of the base-station relocation problem may involve a huge number of parameters including the positions of all deployed nodes, their state information such as energy reserve, transmission range, and so on, and the data sources in the network. Similar observations can be made for sensor repositioning problems. In that case, a sensor may need to know the boundaries of the monitored region, the current coverage ratio of the network, the location of dead sensor nodes, and so on, in order to determine its new location. Given the large number of nodes typically involved in applications of WSNs, the pursuance of exhaustive search will be impractical. In addition, the dynamic nature of the network makes the sensor state and sources of data variant, and thus the optimization process may have to be repeated frequently. Moreover, it may be undesirable to involve the nodes in complex computation in order to ensure sufficient computation capacity for application level processing (e.g., data fusion) or save enough energy for movement of the node. Therefore, approximate and local solutions or search heuristics are more attractive in the context of WSNs [7, 47, 48].

relocation is exploited: (1) post-deployment and (2) on-demand relocation. We discuss these two categories of relocation in details in the following subsections.

Post-Deployment Sensor Relocation. This type of relocation is pursued at the conclusion of the sensor deployment phase when the sensor nodes are being positioned in the area. As we discussed earlier, in most of the WSN applications, sensor deployment is performed randomly due to the inaccessibility of the monitored areas. However, this random configuration usually does not provide an adequate coverage of the area without deploying an excessive number of nodes. Alternatively, the coverage quality can be improved by moving the sensor nodes if they are able to do so. In that case, the sensor nodes can be relocated to the regions that do not have the desired level of coverage or even are not covered at all. Given the energy cost of mechanical movement and the communication messages involved in directing the motion, the relocation process should be lightweight and should conclude in a reasonable time.

Wang et al. [48] utilizes sensor's ability to move to distribute the sensor nodes as evenly as possible in the region. The goal is to maximize the area coverage within the least time duration and with minimal overhead in terms of travel distances and inter-sensor message traffic. The main idea is that each sensor assesses the coverage in its vicinity after deployment and decides on whether it should move to boost the coverage. To assess the coverage, a sensor node creates a Voronoi polygon with respect to neighboring sensors, as illustrated in Figure 9.12. Every point in a Voronoi polygon is closer to the sensor of that polygon (i.e., S_i in Figure 9.12) than any other sensor. The intersection of the disk that defines the sensing range and the Voronoi polygon would identify any uncovered area, which would motivate a sensor to move.

In order to decide where to reposition a sensor, three methods were proposed: vector-based (VEC), Voronoi-based (VOR), and minimax. The main idea of the VEC method is borrowed from electromagnetics where close particles are subject to an

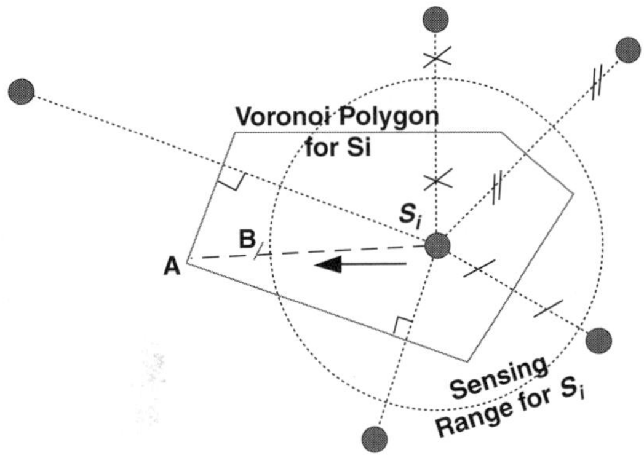

Figure 9.12. Every sensor S_i forms Voronoi polygon with respect to the position of its neighboring sensor. The part of the polygon that lies outside the sensing range is not covered by S_i.

Managing and Justifying the Move. Once the new location of the been picked and confirmed to enhance some desired dependability attrit node should identify a travel path to this new location. The main contributir to the path selection are the total distance to be traveled, suitability of th the path safety, and the risk of disrupting the network operation. Minim travel distance for both the base station and the sensor nodes is very crucial energy consumed by the mechanical parts in such a movement is much n the communication and computation energy. Therefore, the shortest poss should be identified to reach to the new location. However, the node also has path that is physically feasible to travel over. The node may need to consult map or rely on the skills of a carrier (e.g., a robot with cameras or a human avoid obstacles and dead ends. The other concern is protecting the node di move. Since a WSN is usually deployed in harsh environments to detect a dangerous targets/events, the node should avoid exposure to harm or getting For example, the node should not go through a fire to reach the new locatio

The node should also minimize any negative impact on the network o While the node is on the move, it must ensure that data continue to flow. For e if the base station is relocating, it has to arrange for sensor transmission to c planned travel path in order to make sure that packets will continue to reach tinual data delivery prevents the relocation from causing an application leve by missing important reports. Such application level robustness is a deper attribute in itself. Therefore, it is desirable to restrict changes to the networ ogy. Avoiding radical changes to the data routes limits the disruption of ongc traffic and also curtails the overhead that the relocation introduces. Again, t performs a tradeoff analysis between the gain achieved by going to a new and the overhead in terms of additional energy consumption that the motion on sensors and the base station. If the motion is justified, the node can ph relocate.

The last issue is whether there are constraints on the time duration that t budgets for the move. These constraints may arise in very dynamic environr which the traffic pattern changes frequently. In such a case the gains achi going to a location may be lost or degraded very quickly and the node would that it has to move yet to a third location or even get back to the old positior worst case, the node keeps switching back and forth. Therefore, a gradual ap to the new location may be advisable in order to prevent this scenario.

9.3.2 Sensor Repositioning Schemes

While the bulk of published work envisioned sensors to be stationary, some gated the possibility of attaching sensors to moveable entities such as robots [4 Sensor's mobility has been exploited to boost the dependability of WSNs. I ample, sensors can re-spread in the area to ensure uniform coverage, move to loaded nodes in order to prevent bottlenecks, increase bandwidth by ca data to the base station, and so on [51–55]. Proposed schemes for dynam sors positioning in the literature can be categorized into two groups based or expelling force to keep them apart. In the context of WSNs, virtual forces are applied to a sensor node by its neighbors and by the boundaries of its Voronoi polygon in order to change its location. While in VEC the nodes are pushed away from the densely populated areas, VOR pulls the sensors to the sparsely populated areas. In VOR, the sensor node is pulled toward the farthest Voronoi vertex to fix the coverage hole in the polygon, point “A” in Figure 9.12. However, the sensor will be allowed to travel only a distance that equals half of its communication range, point “B” in Figure 9.12, in order to avoid stepping into the area handled by another sensor that was out of reach prior to the move (i.e., is not a current neighbor of S_i,), which can lead to an unnecessary move backward later on. In the minimax method, a sensor also gets closer to its farthest Voronoi vertex. However, unlike VOR, the minimax approach strives to keep most of the other vertices of the Voronoi polygon within the sensing range. It thus relocates the sensor to a point inside the Voronoi polygon whose distance to the farthest Voronoi vertex is minimized. The minimax scheme is more conservative in the sense that it avoids creating coverage holes by going far from the closest vertices, leading to more regularly shaped Voronoi polygon.

The conserved departure from current sensor location leads to a gradual relocation, round by round as shown in Figure 9.13. This usually causes zigzag movement of each sensor rather than directly going to the final destination. In order to shorten the total travel distance, a proxy-based approach is proposed in reference 54. In this approach, the sensor nodes do not move physically unless their final destination is computed. The authors consider a network with stationary and mobile sensors. Mobile sensors are used to fill coverage holes identified in a distributed way by stationary nodes. Thus, mobile sensors only move logically and designate the stationary sensor nodes as their proxies. With this approach, significant improvements can be made to the total and average distance traveled by mobile nodes while at the same time achieving exactly the same level of coverage reported in reference 48. The approach only increases the message complexity. However, given that movement is more costly in terms

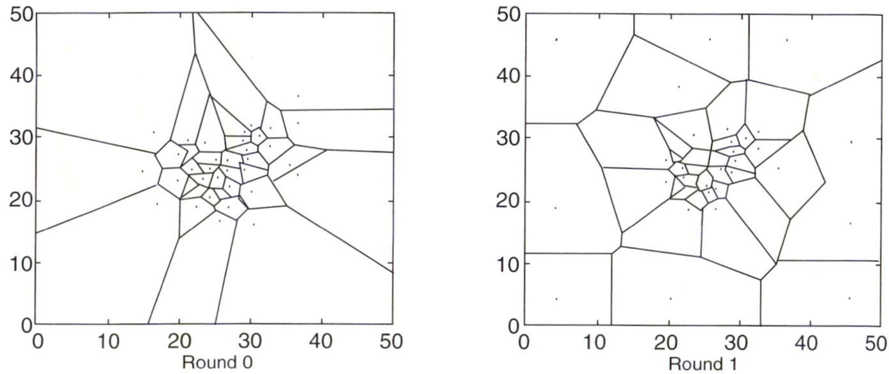

Figure 9.13. Sensors pursue relocation iteratively until no improvement in coverage can be achieved. (Taken from reference 51.)

1	2	3	2	1	3
2	22	3	7	56	18
4	12	2	30	1	5
3	7	2	1	2	3
1	5	19	3	21	77
6	7	10	10	4	5

(a)

2	2	2	2	2	2
18	18	18	18	18	18
9	9	9	9	9	9
3	3	3	3	3	3
21	21	21	21	21	21
7	7	7	7	7	7

(b)

10	10	10	10	10	10
10	10	10	10	10	10
10	10	10	10	10	10
10	10	10	10	10	10
10	10	10	10	10	10
10	10	10	10	10	10

(c)

Figure 9.14. Steps for SMART: (a) Initial 2D mesh. (b) Row scan. (c) Column scan.

of energy, such increase can be justified. Nonetheless, this process still can be very slow and hence prolong the deployment time.

With the objective of reducing the overall deployment time, Wu and Yang [53] proposed another solution to the same problem based on two-dimensional scan of a clustered network, called SMART. The approach adopts a popular scheme for balancing load among nodes in parallel processing architectures by assigning an equal number of tasks to each processor. This idea is applied to a multicluster WSN where each cluster is represented with a square cell forming a 2D mesh, as seen in Figure 9.14, which is redrawn from reference 53. The number of sensors annotated on every cell represents the load of that cluster. Each cluster head knows only its location within the mesh (i.e., row and column indices) and the number of sensors in its cluster. It is assumed that a cluster head can only communicate with its counterparts in neighboring cells. Achieving uniform coverage is then mapped to the load-balancing problem with a goal of evening the distribution of sensors among the clusters. To achieve this goal, each cluster head performs both a row-based and a column-based scan to exchange the load information. In a row-based scan, the leftmost cluster head forwards its load (i.e., number of sensors) to its right neighbor. Each neighbor on the row adds the received load to its own load and forwards it until the rightmost cluster head is reached. This cluster head computes the average load for its row and sends a message back until the leftmost cluster head gets such average (Figure 9.14b). After the scan process, the sensors are relocated to match the desired node count per cluster. That is, the overloaded clusters give sensors while the underloaded clusters take sensors. The same procedure is applied for each column (Figure 9.14c). The approach also handles possible holes in the network when there are clusters with no sensors. The simulation results compared VOR (discussed above) and SMART with respect to the number of moves made by the sensors and the number of rounds until termination. SMART was shown to provide the minimum number of moves. Although it was also shown that SMART converges in a fewer number of rounds for densely populated WSNs, VOR was found to be superior for sparsely populated networks.

Another similar post-deployment relocation work for improving the initial coverage and providing uniform distribution of sensors is presented in reference 56. Although the idea is similar to the VEC mechanism reported in reference 48, this time it is inspired by the equilibrium of particles in Physics. The particles follow the law of Coulomb and push themselves to reach equilibrium in an environment. Therefore, the authors define forces for each sensor node in the network based on the internode distances and the local density of nodes. The partial force on a node i from node j at time t is expressed as follows:

$$f_t^{i,j} = d(R - |p_t^i - p_t^j|)\frac{p_t^j - p_t^i}{|p_t^j - p_t^i|}$$

where d is the density, R is the transmission range, and p_t is the location at time t. Each sensor's movement is decided by the combined force applied to that sensor by all neighboring nodes. A sensor keeps on moving until it travels a distance below a certain threshold at a given time. In some cases, the node can move back and forth between two locations leading to an oscillation. If such oscillation is noticed to occur more than a preset limit, the node stays at the center of gravity of the oscillation points.

To validate the performance, the approach is implemented and compared to a simulated annealing-based solution, which provides an optimal coverage. The validation results indicated that the proposed self-spreading approach performs very close to optimal in terms of coverage and delivers a superior performance in terms of the total distance traveled and the time to converge.

On-Demand Repositioning of Sensors. Instead of relocating the nodes at the deployment phase, sensor relocation can be used on demand to improve certain performance metrics such as coverage, network lifetime, and so on. This can be decided during the network operation based on the changes in either application-level needs or the network state. For instance, the application can be tracking a fast-moving target that may require repositioning of some sensor nodes based on the new location of the target. Furthermore, in some applications, there can be an increasing number of dysfunctional nodes in a particular part of the area necessitating the redistribution of available sensors. In addition to improving coverage, the energy consumption can be reduced through on-demand relocation of sensors in order to reach the best efficient topology.

The approach presented in reference 47 performs sensor relocation to counter holes in coverage caused by sensors failure. The idea is simply to identify some spare sensors from different parts of the network that can be repositioned in the vicinity of the faulty nodes. The selection of the most appropriate choice among multiple candidate spare nodes is based on the recovery time and overhead imposed. Both criteria would favor close-by spares over distant ones. Minimizing the recovery time can be particularly crucial for delay sensitive applications. The overhead can be in the form of energy consumption due to the node's travel and due to the message exchange, especially if spares are picked in a distributed manner. In order to detect the closest redundant sensor with low message complexity, a grid-based approach is proposed. The region is divided into cells with a designated head for each cell. Each cell head advertises/requests redundant nodes for its cell. A quorum-based solution is proposed

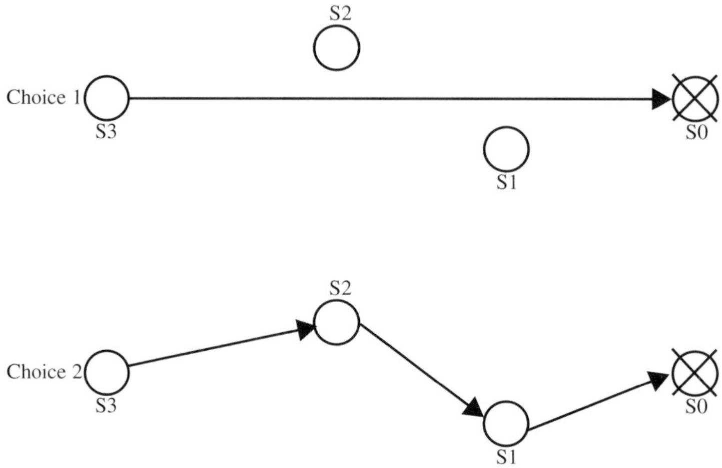

Figure 9.15. Cascaded Movement of Sensors; S3 replaces S2, S2 settles in S1's position and S1 move to where S0 is.

to detect the intersection of advertisements and requests within the grid. Once the redundant sensor is located, it is relocated to the desired cell without disrupting the data traffic and affecting the network topology.

Since directly moving the node can drain significant amount of energy, a cascaded movement is proposed. The idea is to determine intermediate sensor nodes on the path and replace those nodes gradually. That is, the redundant sensor will replace the first sensor node on the path. That node will also move and replace the second sensor node, and so on. For the example shown in Figure 9.15 (which is redrawn from reference 47), rather than directly moving S3 to the location of S0, in choice 2 all sensors S_3, S_2, and S_1 move at the same time and replace S_2, $S_{1,}$ and $S_{0,}$ respectively, in order to minimize the relocation time. The path is selected such that it will minimize the total mechanical movement energy and at the same time maximize the remaining energy of sensor nodes. In order to determine such a path, Dijkstra's least-cost path algorithm is used. The overall solution is also revisited to provide a distributed approach for determining the best cascading schedule.

When validated, the approach outperformed VOR of reference 48 with respect to the number of sensors involved in the relocation, the total consumed energy, and the total remaining energy. In addition, cascaded movement delivered much better performance than direct movement in terms of relocation time, energy cost, and remaining energy. However, obviously the cost of maintaining a grid, selection of the cell head and redundant nodes will grow dramatically with the increasing number of nodes. For scalability, a hierarchical solution might be needed to restrict the size of the region and the cost of movements.

Coverage improvement was the objective of relocating imaging sensors in reference 57. Stationary cameras may not provide the desired coverage when there

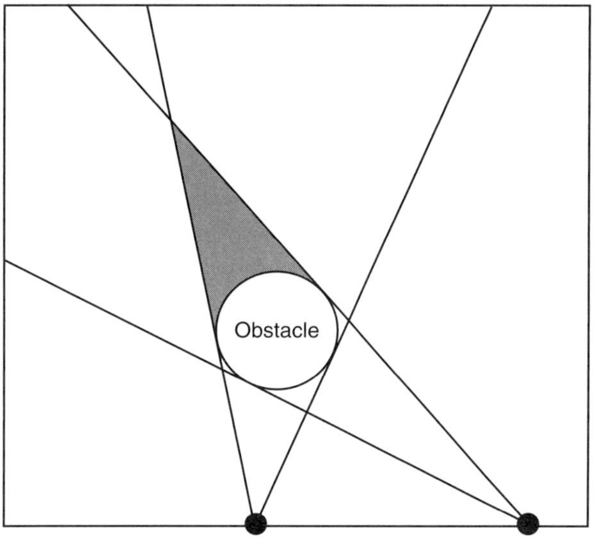

Figure 9.16. An obstacle reducing the coverage of two cameras directed at different areas.

are environmental peculiarities (e.g., moving obstacles) in the event area as seen in Figure 9.16. Thus, moving the cameras in order to avoid obstacles that block their vision would increase the coverage. The mobility for the camera nodes is made possible through providing a traction mechanism under each camera that will enable motion in one dimension, as is also implemented in Robomote [49]. This mobility feature on cameras is actuated when the coverage of the monitored areas falls below a certain ratio. The experiments performed with real-life target tracking applications verified that the mobile cameras can increase the coverage of the monitored area and thus would decrease the target miss ratio significantly when compared to stationary-node based setups.

Dasgupta et al. [58] exploited sensor relocation to improve the lifetime of the network rather than coverage. The maximum lifetime sensor deployment problem with coverage constraints has been investigated. The authors assumed a network operation model in which every sensor periodically sends its data report to the base station. The network is required to cover a number of points of interest for the longest time. The average energy consumption per data collection round is used as a metric for measuring the sensor's lifetime. The problem is then transformed to minimizing the average energy consumption by a sensor per round by balancing the load among sensors. The idea is to spread the responsibility of probing the points of interest among the most number of sensors and to carefully assign relays so that the data are disseminated using the least amount of energy. A heuristic was proposed that tries to relocate sensors in order to form the most efficient topology. First, sensors are sorted decently according to the point of interest that they cover. Starting from the top of the sorted list, the algorithm iterates on all sensors. In each iteration, the sensor is checked for whether it can move to another location to serve as a relay. The new location is picked based on the traffic flow and the data path that this node is part of or will be joining. Basically the relocating node should reduce its energy consumption by getting close to its downstream neighbor. A sensor repositioning is allowed only if it is not risking a loss in coverage.

9.3.3 Base-Station Relocation for Increased Dependability

In this section, we report on some of our investigation of the base-station relocation problem. Three dependability attributes are considered: network longevity, data delivery timeliness, and base-station physical security (safety). For each of these attributes, we explain how we address the three questions; when to relocate, where to place the base station, and how the move is managed.

Relocation for Increased Network Longevity. Although energy-aware multihop routing does dynamically adapt to changes in sensor's energy and traffic pattern, sensors nearby the base station die quickly because they are the most utilized nodes in the network. Consequently, nodes that are further away from the base station are picked as substitute relays as depicted in Fig. 9.17, and the node's energy utilized for communication with the base station would be considerably higher. Such effects can spread in a spiral manner, thus draining the sensors energy and hence shortening the lifetime of the network. To stop such pattern of energy depletion, the base station is repositioned [59, 60].

The main idea is to move the base station toward the sources of largest traffic. The traffic density (*PT*) times the transmission power (E_{TR}) is used as a metric for monitoring the network operation and searching for the best base-station location. The idea is to track changes in (a) the nodes that act as the closest hop to the base station and (b) the traffic density going through these hops. If the distance between the base station and some of the nodes that are in direct communication is smaller than a

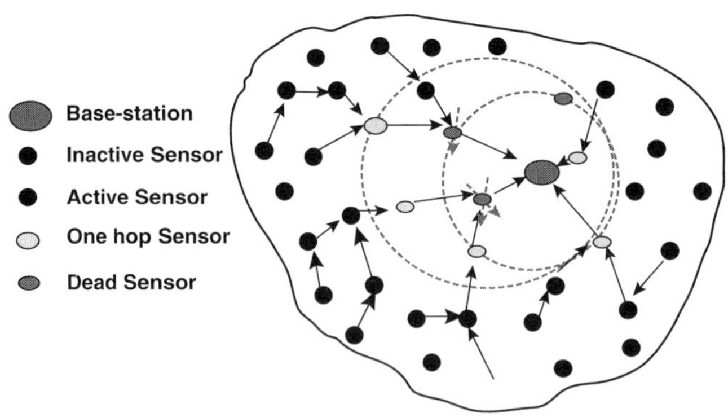

Figure 9.17. Nodes close to the base station die rather quickly due to overload forcing the more distant nodes to relay the data to the base station.

threshold value, the base station will qualify the impact of these nodes on the overall network lifetime by considering the number of packets routed through them. If the $PT * E_{TR}$ is greater than a certain threshold, the base station will consider relocating to a new position.

Such an approach has multiple advantages. First, the base station will be close to the area where more nodes are collecting data and thus communication-related energy consumption will be reduced. It is thus expected that the average energy per packet gets reduced. In addition, nodes collecting the most data will be closer to the base station and most probably fewer hops will be involved lowering the overall latency time for data collection. Moreover, the packet throughput will be higher since it is expected that most messages pass through fewer hops and travel shorter distances, making them less likely to be dropped. In summary, such an approach for relocating the base station not only can increase network longevity but also can enhance other performance metrics like latency and throughput.

While such positioning will be ideal for high-traffic paths, it can worsen the performance on paths with lower traffic density or which are topologically opposite to the direction of the base-station motion. Therefore, before confirming the move, the base station validates the overall impact on transmission energy by factoring in the possible extension of some data paths and the cost of signaling overhead to those sensor nodes affected by the move. In summary, the cost of base-station relocation should be justified before being pursued. When the base station starts to move, the data gathering process still continues and thus the routes should be adjusted before the base station gets out of range of some sensors (if any). Unlike static approaches discussed in Section 9.3, routes adjustment is an issue in dynamic base-station positioning. This issue can be handled by either increasing the transmission power or designating additional forwarder sensors. The change in base-station position may also introduce shadowing or multipath fading to some links. Slow or gradual advance toward the new position can be effective in avoiding unexpected link failures that may cause negative performance impacts, there by allowing the base station to rethink the suitability of the newly selected position and/or the decision to move further.

Enhancing Timeliness of Delay-Constrained Traffic. In addition to boosting network longevity, the repositioning of the base station also becomes influential when real-time traffic with certain end-to-end delay requirements is involved. For instance, when routes to the base station get congested, most requests for establishing paths for real-time data may be denied or the deadline miss rate of real-time packets may increase significantly. Traffic congestion can be caused by an increase in the number of real-time data packets coming from nodes close to a recent event. In such circumstances, it may be infeasible to meet the requirements for real-time data delivery. Repositioning the base station can then be beneficial in order to spread the traffic on additional hops and increase the feasibility for meeting the timeliness requirements.

A trigger for such relocation can be the unacceptable increase in the miss rate of real-time packets or just a desire to increase timeliness even if the miss rate is at a level that is tolerable by the application. To boost timeliness, the base station can move to the location of, or close to, the most heavily loaded node and try to split the

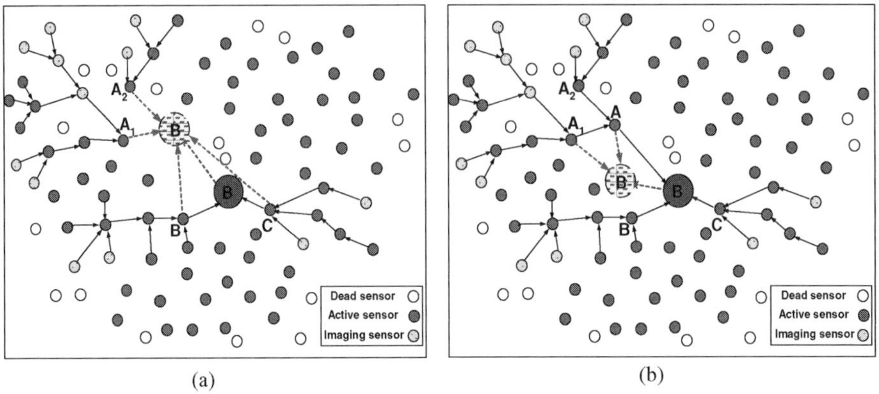

Figure 9.18. (a) The base station (denoted as B) is relocated to the location of *A* if the delay is not extended on the path from *C* to *B*. (b) If real-time traffic through *C* is affected, B is relocated close to *C* while still splitting traffic at A.

incoming traffic passing through that node without extending the delay experienced by real-time packets over other routes as shown in Figure 9.18 [60, 61]. Such loaded nodes are picked based on the real-time traffic service rate, often determined during the route setup in order to allocate bandwidth to both real-time and non-real-time traffic. Again, the pros of the relocation have to be qualified to make sure that the overhead is justified. It is also worth noting that in this approach the impact on link quality is not a major concern when the base station moves close to a heavily loaded node since it is unlikely that the node is experiencing a disruptive level of interference while being able to relay a high volume of real-time packets.

Handling the base-station motion is similar to the approach of the network longevity relocation, described above. As long as the base station remains within the transmission range of all the last-hop nodes, the current routes can be maintained by only adjusting the transmission power. If it is expected that the new location will put the base station out of the transmission range of some of the last hop nodes in the current routes, new forwarder nodes that are not involved in any routing activity are selected. Such unused nodes will introduce very little queuing delay, which is desirable for on-time delivery of all real-time packets that use these nodes as relays. Note that designating new forwarder nodes is still more costly than just adjusting the transmission power. Therefore, if the new location imposes excessive topology changes, alternative positions that will cause no or minimal topology changes are to be considered.

Protecting the Base Station. As discussed earlier, moving the base station toward highly loaded nodes would have the potential for enhancing network performance in terms of energy consumption, throughput, and delay. However, not all locations would be safe for the base station even if a substantial gain in performance is perceived. In many situations, sensors are placed in hostile environments where there is always a lurking danger to the base station as it gets closer to the data sources. For example, in a disaster management application, sensors may be reporting about fires, collapsing building, gas leaks, and so on. Similarly, in combat field surveillance, the sensors may be reporting tanks or enemy troops. Moving too close to the reported events in these scenarios would be very risky. Thus, handling such scenarios would be subject to performance and safety tradeoff. In such cases, the base station would have to consider the potential performance degradation while getting away from danger.

To assess the safety implication of repositioning the base station, a stochastic or a cognitive formulation can be pursued [62]. The idea is to track the base-station safety levels at different locations and use them to define the parameters of the base-station safety model. Then the threat implication is estimated as a function of the proximity to reported events and the severity of these events. If the location of the event is not accurately known, the data volume related to that event and the location of the reporting sensors are factored in. An objective function is then formed to balance safety and performance goals and used to guide the search for the new location of the base station. Figure 9.19 shows some performance improvements through the base-station repositioning and its safety aware version named SAFER. In these experiments, sensor nodes are randomly placed in a 500 $\times$ 500-m^2 area while the network is tasked with a target tracking mission. The base station is dynamically repositioned during the network operation based on the location of high volume traffic in the event area.

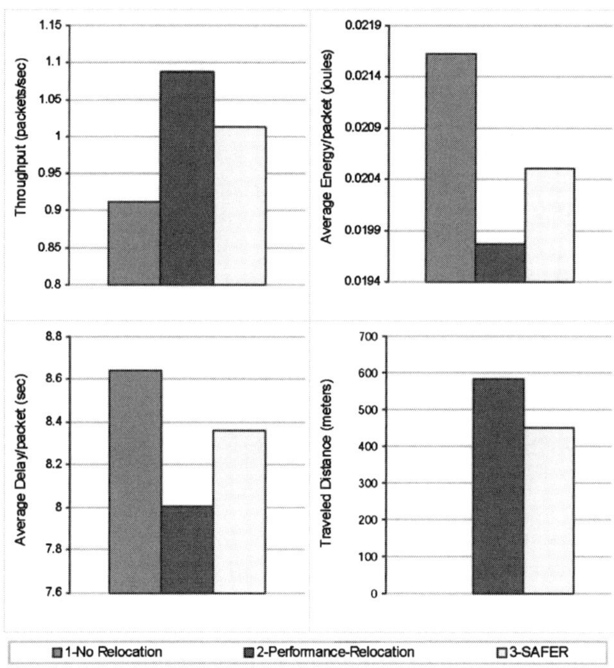

Figure 9.19. Performance improvements in terms of throughput, energy and delay by repositioning of the base-station and its safety aware version.

The same experiments are performed by taking into consideration the existing risks in the event region while the base station is on the move.

Finally we would like to note that all computation related to the presented heuristics is performed at the base station, which usually is not resource-constrained when compared to sensors. Therefore, the base station should be able to track/collect the necessary state information of the network in order to trigger relocation and determine the new position. The base station would identify the necessary network information based on the goal of the relocation (e.g., energy, timeliness, safety, etc.). We argue that this information can be obtained even in distributed route setups by keeping track of packets arriving from the sensor nodes over time. For example, the base station can monitor the number of received packets from a particular node, estimate the remaining energy, determine the length of the path, and so on.

9.4 COORDINATED MULTINODE RELOCATION

In many application setups, nodes coordinate among themselves in order to efficiently and effectively handle application-level requirements. Examples of such applications include robotic-based land-mine detection and deactivation, multi-rover exploration of distance planets, employing a sensor fleet for oceanic studies, and so on. However, unlike the case discussed in the previous section, such application-level coordination requires the relocation to care for internode networking issues. For instance, consider the scenario depicted in Figure 9.20a. A set of sensors are deployed in an area of interest. Four base stations are employed to serve an application based on data collected from these sensors. The sensors have been partitioned into nonoverlapping clusters; each is managed by a distinct base station.

Given the commonly uneven distribution of sensor nodes, some events would be hard to monitor or would overburden the scarce network resources in the proximity of the event. In Figure 9.20a, two sample events articulate such an issue. The depicted events, may be targets or fires, are reported by very few sensors for which the events happen to be within their detection range. Clearly, the intracluster network topology for both clusters 1 and 4 are not efficient since the data are routed over many hops and may not be arriving at B_1 and B_4 reliably and/or on time. In addition, some of the nodes that are involved in relaying data may not have abundant energy reserve to keep the path stable for an extended duration. The following are some of the challenges that base stations may face when trying to boost the efficiency of the data collection and dependability of the intracluster network:

1. B_1 may decide to relocate to better serve the event tracked by sensors in its cluster—for example, monitor the event in a timely manner or to minimize the energy consumption at relaying sensor nodes. However, such repositioning may get B_1 out of the communication range of B_4. Such a move can only be acceptable if B_1 does not need to interact with B_4, which is unlikely, or when an alternative path can be established with an acceptable delay bound. Referring to Figure 9.20a, relocating B_1 would be feasible if the communication path (B_1, B_3, B_4) meets the timeliness requirements.

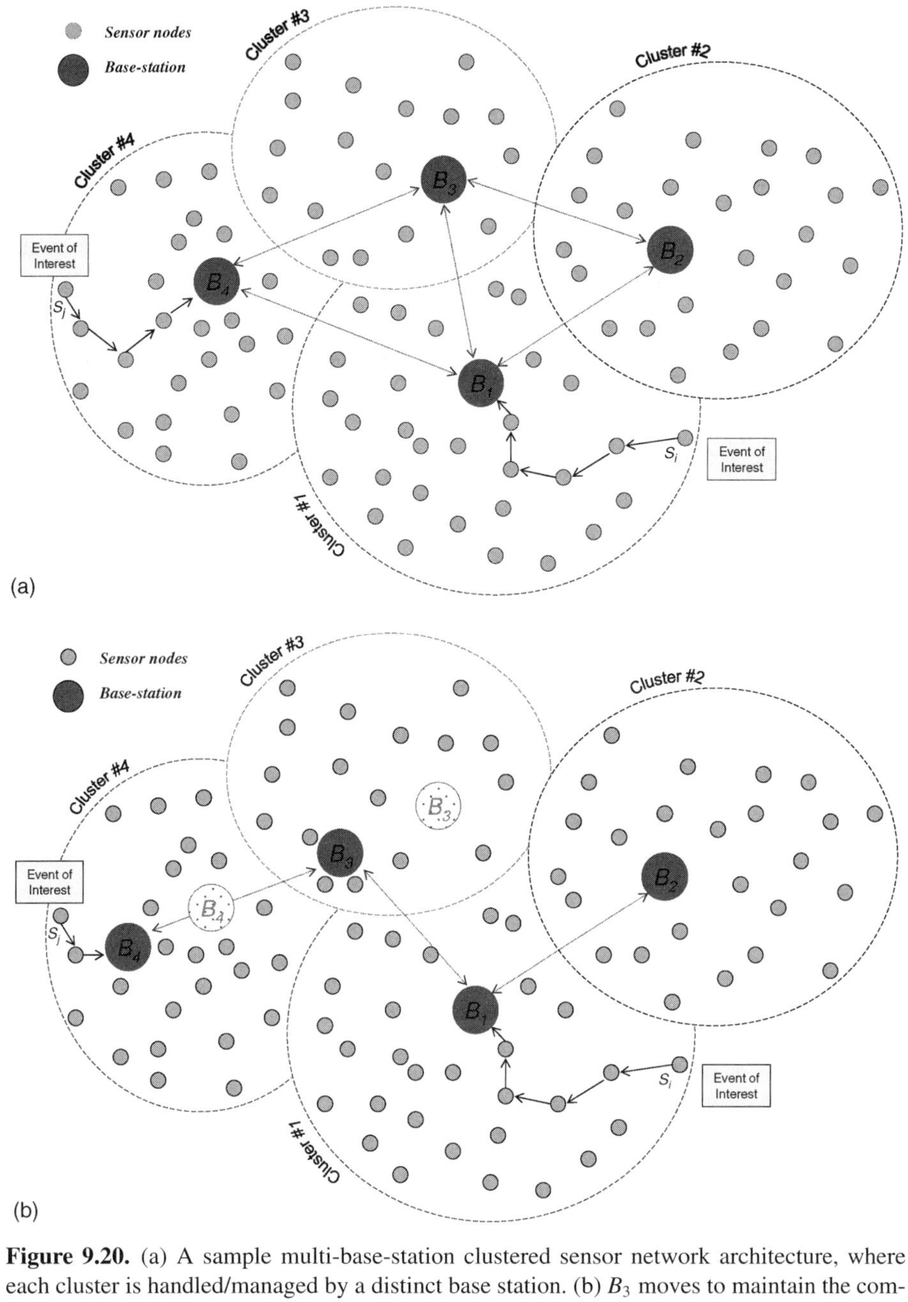

Figure 9.20. (a) A sample multi-base-station clustered sensor network architecture, where each cluster is handled/managed by a distinct base station. (b) B_3 moves to maintain the communication link with B_4, probably losing connectivity to B_2. However, base station still form a connected graph.

2. The possibility of handling the event by a different cluster is another issue of consideration. For example, while B_1 can relocate to better serve the event, the data can also be routed from S_i to B_2 over a shorter and more energy-efficient path. That would require changing the association of S_i from cluster 1 to cluster 2, at least temporarily. Such modification of cluster membership would sure impose an overhead. Such a tradeoff will be unavoidable.
3. When relocation of a base station is the only option for boosting the dependability of the operation within a cluster, a ripple effect may be caused throughout the network. Consider, for example, the relocation of B_4 close to S_j. Although such a close proximity to the event would have positive impact on the operation in cluster 4, moving B_4 that far makes it unreachable to other base stations. Such a scenario is not acceptable in a collaborative computing environment. A more complex alternative is to relocate multiple base stations in order to maintain connectivity among base stations. Figure 9.20b illustrates a possible multi-base-station relocation that better serves the event reported by S_j. Basically, B_3 gets closer to the new location of B_4 in order to prevent their communication link from getting broken, possibly at the expense of losing connectivity to B_2. This solution could have been more involved if B_1 does not have a link to B_2 forcing it to move as well to be in the range of B_3 and B_2.

Obviously, a multinode relocation can be significantly more complex and can introduce lots of overhead. We envision coordinated multinode relocation to be a promising research direction. Only little attention has been paid to tackle the multinode relocation challenges. One of the few attempts was reported in reference 63, where a dynamic multi-base-station positioning is proposed in order to improve the network longevity. The motion of base stations is further restricted to maintain the connectivity of the inter-base-station network as seen in Figure 9.21. When a BS is to be relocated, its links with the neighbors are checked first before the relocation is performed. If changing the position of a base station is to cause partitioning in the network, its

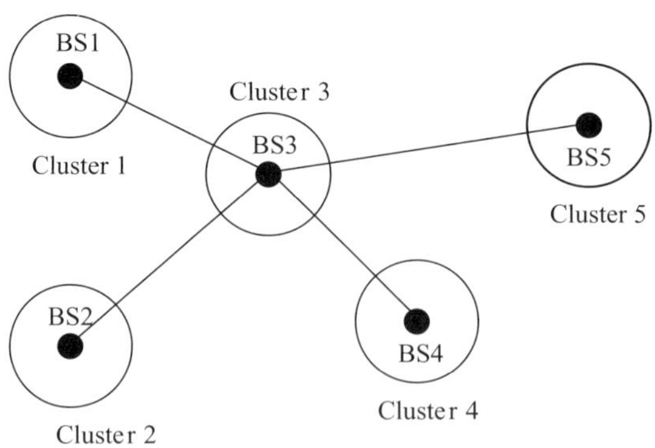

Figure 9.21. Connected inter-base-station network.

neighbors are to move to restore broken links. For example, in Figure 9.21, if BS_3 is to move toward BS_2, BS_5 would follow through to avoid being disconnected from the network. In order to prevent simultaneous relocations, a mutual-exclusion-based mechanism is used. The idea is to access to a global token exclusively in order to perform the relocation. A base station will cease any motion until a token is granted.

9.5 CONCLUSION

Wireless sensor networks (WSNs) have attracted lots of attention in recent years due to their potential in many applications such as border protection and combat field surveillance. Given the criticality of such applications, maintaining a dependable operation of the network is a fundamental objective. However, the resource-constrained nature of sensor nodes and the ad hoc formation of the network, often coupled with an unattended deployment, pose nonconventional challenges and motivate the need for special techniques for dependable design and management of WSNs. In this chapter, we have discussed the effect node placement strategies on the dependability of WSNs. We categorized the various approaches in the literature on sensor and base-station positioning for enhanced dependability. We have further highlighted the potential of dynamic repositioning of the base-station and sensor nodes, as a viable means for increasing the dependability of WSNs. Unlike the initial careful placement, node repositioning can assist in dealing with dynamic variations in the network resources and surrounding environment. We have identified the technical issues pertaining to relocating the nodes—namely, when to reposition a node, where to move it, and how to manage the network while the node is in motion. We have discussed sample techniques that employ sensor and base-station relocation for boosting the coverage, tolerating node failure, extending network lifetime, increasing responsiveness in data delivery, and protecting the network assets. We have further identified the coordinated multinode repositioning problem as an open area for research.

9.6 EXERCISES

1. List some examples of sensor network applications and enumerate the dependability requirements that these applications are subject to. Discuss how node placement affects the network dependability attributes in the context of these applications.
2. What are the most important metrics to be considered in performing sensor relocation? Discuss possible tradeoffs among these metrics.
3. When would dynamic node placement be most appropriate?
4. How would the communication range of relay and sensor nodes affect the complexity and placement strategy of relay nodes?
5. Describe some scenarios for which dynamic placement of nodes can hurt the network performance rather than enhancing it.

6. Based on the analysis and deployment techniques discussed in this chapter, how would you design a sensor network for the following applications? (i) Underwater surveillance, (ii) Border protection, and (iii) Lunar exploration.
7. Pick a randomly deployed sensor network with five nodes. Show the final configuration of the sensor network after applying the Voronoi-based sensor relocation (VOR) technique of reference 48.
8. Discuss the relation between network clustering and the multiple base-station placement problems. How would the order of applying the clustering and placement techniques affect the network design/performance?

BIBLIOGRAPHY

1. I. F. Akyildiz, W. Su, Y. Sankarasubramaniam, and E. Cayirci. Wireless sensor networks: A survey. *Computer Networks*, **38**:393–422, 2002.
2. C.-Y. Chong and S. P. Kumar. Sensor networks: Evolution, opportunities, and challenges. *Proceedings of the IEEE*, **91**(8):1247–1256, 2003.
3. K. Akkaya and M. Younis. A survey on routing protocols for wireless sensor networks. *Elsevier Journal of Ad Hoc Networks*, **3**(3):325–349, 2005.
4. P. Naik and K. M. Sivalingam. A survey of MAC protocols for sensor networks. In *Wireless Sensor Networks*, C. S. Raghavendra, K. M. Sivalingam, and T. Znati, editors. Kluwer Academic Publishers, Norwell, MA, 93–107.
5. N. Sadagopan, B. Krishnamachari, and A. Helmy. Active query forwarding in sensor networks (ACQUIRE). *Elsevier Journal of Ad Hoc Networks*, **3**(1):91–113, 2005.
6. A. Cerpa and D. Estrin, ASCENT: Adaptive self-configuring sensor networks topologies. *IEEE Transactions on Mobile Computing*, **3**(3):272–285, July-August 2004.
7. A. Efrat, S. Har-Peled, and J. S. B. Mitchell. Approximation algorithms for two optimal location problems in sensor networks. In *Proceedings of the 3rd International Conference on Broadband Communications, Networks and Systems (Broadnets 2005)*, pp. 714–723, Boston, MA, October 2005.
8. X. Cheng, DZ Du, L. Wang, and B. Xu. Relay sensor placement in wireless sensor networks. *ACM/Springer Journal of Wireless Networks*, **14**(3):347–355, June 2008.
9. E. L. Lloyd and G. Xue. Relay node placement in wireless sensor networks. *IEEE Transactions on Computers*, **56**(1):134–138, January 2007.
10. S. Poduri, S. Pattem, B. Krishnamachari, and G. S. Sukhatme. Sensor network configuration and the curse of dimensionality. In *Proceedings of the 3rd IEEE Workshop on Embedded Networked Sensors*, Cambridge, MA, May 2006.
11. A. Bogdanov, E. Maneva, and S. Riesenfeld, Power-aware base station positioning for sensor networks. In *Proceedings of the 23rd International Annual Joint Conference of the IEEE Computer and Communications Societies (INFOCOM 2004)*, pp. 575–585, Hong Kong, March 2004.
12. J. Pan, L. Cai, Y. T. Hou, Y. Shi, and S. X. Shen. Optimal base-station locations in two-tiered wireless sensor networks. *IEEE Transactions on Mobile Computing*, **4**(5):458–473, September/October 2005.

13. S. S. Dhillon and K. Chakrabarty. Sensor placement for effective coverage and surveillance in distributed sensor networks. In *Proceedings of the IEEE Wireless Communications and Networking Conference (WCNC '03)*, New Orleans, LA, March 2003.

14. T. Clouqueur, V. Phipatanasuphorn, P. Ramanathan, and K. K. Saluja. Sensor deployment strategy for target detection. In *Proceedings of the 1st ACM international Workshop on Wireless Sensor Networks and Applications (WSNA '02)*, pp. 42–48, Atlanta, GA, September 2002.

15. A. Brooks, A. Makarenko, T. Kaupp, S. Williams, and H. Durrant-Whyte. Implementation of an indoor active sensor network. In *Proceedings of the 9th International Symposium on Experimental Robotics*, Singapore, June 2004.

16. V. A. Petrushin, G. Wei, O. Shakil, D. Roqueiro, and V. Gershman. Multiple-sensor indoor surveillance system. In *Proceedings of the 3rd Canadian Conference on Computer and Robot Vision (CRV'06)*, pp. 40–47, Québec City, June 2006.

17. L. Krishnamurthy et al. Design and deployment of industrial sensor networks: Experiences from a semiconductor plant and the North Sea. In *Proceedings of the 3rd ACM Conference on Embedded Networked Sensor Systems (SenSys '05)*, pp. 64–75, San Diego, CA, November 2005.

18. J. Paek, K. Chintalapudi, J. Cafferey, R. Govindan, and S. Masri. A wireless sensor network for structural health monitoring: Performance and experience. In *Proceedings of the Second IEEE Workshop on Embedded Networked Sensors (EmNetS-II)*, pp. 63–74, Syndney, Australia, May 2005.

19. K. Mechitov, W. Kim, G. Agha, and T. Nagayama. High-frequency distributed sensing for structure monitoring. In *Proceedings of the First International Conference on Networked Sensing Systems*, Tokyo, Japan, June 2004.

20. J. Berry, L. Fleischer, W. E. Hart, and C. A. Phillips. Sensor placement in municipal water networks. In *Proceedings of the World Water and Environmental Resources Conference*, Philadelphia, PA, June 2003.

21. J.-P. Watson, H. Greenberg, and W. E. Hart. A multiple-objective analysis of sensor placement optimization in water networks. In *Proceedings of the World Water and Environment Resources Congress*, Salt Lake City, UT, June 2004.

22. H. H. Gonzalez-Banos and J. C. Latombe. A randomized art-gallery algorithm for sensor placement. In *Proceedings of the 17th ACM Symposium on Computational Geometry (SoCG'01)*, pp. 232–240, Boston, MA 2001.

23. L. Navarro, J. Dolan, and P. Khosla. Optimal sensor placement for cooperative distributed vision. In *Proceedings of the International Conference on Robotics and Automation (ICRA)*, pp. 939–944, New Orleans, LA, April 2004.

24. J. Tang, B. Hao, and A. Sen. Relay node placement in large scale wireless sensor networks. *Computer Communications (special issue on wireless sensor networks)*, **29**:490–501, 2006.

25. M. Ishizuka, and M. Aida. Performance study of node placement in sensor networks. In *Proceedings of the 24th International Conference on Distributed Computing Systems Workshops, W7:EC (Icdcsw'04)*, Vol. 7, pp. 598–603, Hachioji, Tokyo, Japan, March 2004.

26. K. Xu, H. Hassanein, G. Takahara, and W. Wang. Relay node deployment strategies in heterogeneous wireless sensor networks: Single-hop communication case. In *Proceedings of the IEEE Global Telecommunication Conference (Globecom '05)*, pp. 21–25, St. Louis, MO, November 2005.

27. K. Xu, H. Hassanein, G. Takahara, and Q. Wang. Relay node deployment strategies in heterogeneous wireless sensor networks: Multiple-hop communication case. In *Proceedings of 2nd IEEE Conference on Sensor and Ad Hoc Communications and Networks (SECON '05)*, pp. 238–241, Santa Clara, CA, September 2005.
28. C.-F. Huang and Y.-C. Tseng. The coverage problem in a wireless sensor network. In *Proceedings of the ACM 9th Annual International Conference on Mobile Computing and Networking (MobiCom 2003)*, pp. 115–121, San Diego, CA, September 2003.
29. A. Boukerche, X. Fei, and R. B. Araujo. A coverage preserving and fault tolerant based scheme for irregular sensing range in wireless sensor networks. In *Proceedings of the 49th Annual IEEE Global Communication Conference (Globecom'06)*, San Francisco, CA, November 2006.
30. S. Meguerdichian, F. Koushanfar, M. Potkonjak, and M. B. Srivastava. Coverage problems in wireless ad-hoc sensor networks. In *Proceedings of the IEEE Annual Joint Conference of the IEEE Computer and Communications Societies (INFOCOM'01)*, pp. 1380–1387, Anchorage, AL, April 2001.
31. E. S. Biagioni and G. Sasaki. Wireless sensor placement for reliable and efficient data collection. In *Proceedings of the 36th Annual Hawaii international Conference on System Sciences (HICSS'03), Track 5*, Vol. 5, Big Island, HI, January 2003.
32. K. Kar and S. Banerjee. Node placement for connected coverage in sensor networks. In *Proceedings of the Workshop on Modeling and Optimization in Mobile, Ad Hoc and Wireless Networks (WiOpt'03)*, Sophia Antipolis, France, 2003.
33. J. Bredin, E. Demaine, M. Taghi Hajiaghayi, and D. Rus. Deploying sensor networks with guaranteed capacity and fault tolerance. In *Proceedings of the 6th ACM International Symposium on Mobile Ad Hoc Networking and Computing*, pp. 309–319, Urbana-Champaign, IL, May 2005.
34. M. Younis, M. Youssef, and K. Arisha. Energy-aware management in cluster-based sensor networks. *Computer Networks*, **43**(5):649–668, 2003.
35. Y. Chen, C. Chuan, and Q. Zhao. Sensor placement for maximizing lifetime per unit cost in wireless sensor networks. In *Proceedings of the IEEE Military Communication Conference (MILCOM '05)*, Vol. 2, pp. 1097–1102, Atlantic City, NJ, October 2005.
36. Y. T. Hou, Y. Shi, and H. D. Sherali. On energy provisioning and relay node placement for wireless sensor networks. *IEEE Transactions on Wireless Communications*, **4**(5):2579–2590, 2005.
37. X. Zhang and S. B. Wicker. How to distribute sensors in a random field? In *Proceedings of the 3rd International Symposium on Information Processing in Sensor Networks (IPSN '04)*, pp. 243–250, Berkeley, CA, April 2004.
38. D. Ganesan, R. Cristescu, and B. Beferull-Lozano, Power-efficient sensor placement and transmission structure for data gathering under distortion constraints. In *Proceedings of the 3rd International Symposium on Information Processing in Sensor Networks (IPSN '04)*, pp. 142–150, Berkeley, CA, April 2004.
39. Q. Wang, K. Xu, G. Takahara, and H. Hassanein. Deployment for information oriented sensing coverage in wireless sensor networks. In *Proceedings of the 49th IEEE Global Telecommunication Conference (Globecom'06)*, San Francisco, CA, November 2006.
40. E. I. Oyman and C. Ersoy. Multiple sink network design problem in large scale wireless sensor networks. In *Proceedings of the IEEE International Conference on Communications (ICC 2004)*, pp. 3663–3667, Paris, June 2004.

41. S. R. Gandham, M. Dawande, R. Prakash, and S. Venkatesan. Energy efficient schemes for wireless sensor networks with multiple mobile base stations. In *Proceedings of the 46th IEEE Global Telecommunication Conference (Globecom'03)*, pp. 377–381, San Francisco, CA, December 2003.

42. J. Pan et al. Locating base-stations for video sensor networks. In *Proceedings of the IEEE Vehicular Technology Conference (VTC-Fall'03)*, Vol. 5, pp. 3000–3004, Orlando, FL, October 2003.

43. Y. Shi, Y. T. Hou, and A. Efrat. Algorithm design for base station placement problems in sensor networks. In *Proceedings of IEEE International Conferenfsfce on Quality of Service in Heterogeneous Wired/Wireless Networks*, Waterloo, Ontario, Canada, August 7–9, 2006.

44. K. Akkaya and M. Younis. COLA: A coverage and latency aware actor placement for wireless sensor and actor networks. In *Proceedings of IEEE Vehicular Technology Conference (VTC-Spring '06)*, Montreal, Canada, September 2006.

45. T. W. Haynes, S. T. Hedetniemi, and P. J. Slater. *Domination in Graphs: Advanced Topics*. Marcel Dekker, New York, 1998.

46. K. Akkaya and M. Younis. Coverage and latency aware actor placement for wireless sensor and actor networks. Technical Report, Department of Computer Science and Electrical Engineering, University of Maryland Baltimore County, April 2006.

47. K. Akkaya, M. Younis and M. Bangad. Sink repositioning for enhanced performance in wireless sensor networks. *Computer Networks*, **49**(1):512–534, 2005.

48. G. Wang, G. Cao, T. La Porta, and W. Zhang. Sensor relocation in mobile sensor networks. In *Proceedings of the IEEE Annual Joint Conference of the IEEE Computer and Communications Societies (INFOCOM'05)*, pp. 2302–2312, Miami, FL, March 2005.

49. http://www-robotics.usc.edu/~robomote/

50. http://www.symbio.jst.go.jp/symbio/morph.html

51. G. Wang, G. Cao, and T. La Porta. Movement-assisted sensor deployment. In *Proceedings of IEEE INFOCOM*, Hong Kong, March 2004.

52. A. Kansal, E. Yuen, W. J. Kaiser, G. J. Pottie, and M. B. Srivastava. Sensing uncertainty reduction using low complexity actuation. In *Proceedings of Information Processing in Sensor Networks (IPSN 2004)*, Berkeley, CA, April 2004.

53. J. Wu and S. Yang. SMART: A scan-based movement assisted sensor deployment method in wireless sensor networks. In *Proceedings of the IEEE Annual Joint Conference of the IEEE Computer and Communications Societies (INFOCOM'05)*, Miami, FL, March 2005.

54. G. Wang, G. Cao, and T. La Porta. Proxy-based sensor deployment for mobile sensor networks. In *Proceedings of the 1st IEEE International Conference on Mobile Ad-hoc and Sensor Systems (MASS)*, Fort Lauderdale, FL, October 2004.

55. A. Kansal et al. Controlled mobility for sustainable wireless sensor networks. In *Proceedings of IEEE Sensor and Ad Hoc Communications and Networks (SECON)*, Santa Clara, CA, October 2004.

56. N. Heo and P. K. Varshney. A distributed self spreading algorithm for mobile wireless sensor networks. In *Proceedings of IEEE Wireless Communications and Networking Conference, WCNC 2003*, New Orleans, LA, March 2003.

57. A. Kansal, E. Yuen, W. J. Kaiser, G. J. Pottie, and M. B. Srivastava. Sensing Uncertainty Reduction Using Low Complexity Actuation. In *Proceedings of Information Processing in Sensor Networks (IPSN 2004)*, Berkeley, CA, April 2004.

58. K. Dasgupta, M. Kukreja, and K. Kalpakis. Topology-aware placement and role assignment for energy-efficient information gathering in sensor networks. In *Proceedings of 8th IEEE Symposium. on Computers and Communication (ISCC '03)*, Kemer-Antalya, Turkey, July 2003.
59. M. Younis, M. Bangad, and K. Akkaya, Base-station repositioning for optimized performance of sensor networks. In *Proceedings of the IEEE VTC 2003—Wireless Ad hoc, Sensor, and Wearable Networks*, Orlando, FL, October 2003.
60. K. Akkaya, M. Younis, and M. Bangad. Sink repositioning for enhanced performance in wireless sensor networks. *Computer Networks*, **49**(1):512–534, 2005.
61. K. Akkaya and M. Younis. Relocation of gateway for enhanced timeliness in wireless sensor networks. In *Proceedings of the IEEE Workshop on Energy-Efficient Wireless Communications and Networks (EWCN 2004), in conjunction with the 23rd IEEE International Performance Computing and Communications Conference (IPCCC'04)*, Phoenix, AZ, April 2004.
62. W. Youssef, M. Younis, and K. Akkaya. An intelligent safety-aware gateway relocation scheme for wireless sensor networks. In *Proceedings of the IEEE International Conference on Communications (ICC 2006)*, Istanbul, Turkey, June 2006.
63. J. English, M. Wiacek, and M. Younis. CORE: Coordinated relocation of sink nodes in wireless sensor networks. In *Proceedings of the 23rd IEEE Biennial Symposium on Communications (QBSC 2006)*, Kingston, Ontario, Canada, May 2006.

CHAPTER 10

Mobility in Wireless Sensor Networks

STEFANO BASAGNI

ECE Department, Northeastern University, Boston, MA 02115

ALESSIO CAROSI and CHIARA PETRIOLI

Dipartimento di Informatica, Università di Roma "La Sapienza," Roma 00198, Italy

10.1 WIRELESS SENSOR NETWORKS: GENERALITIES AND APPLICATIONS

Technological advances as well as the advent of 4G communications and of pervasive and ubiquitous computing have fostered a renewed interest in multihop (ad hoc) communications [1]. In particular, the interest is in self-organizing wireless multihop networks composed of a possibly very large number of nodes. These nodes can be either static or mobile and are usually constrained as for the most critical resources, such as power and computation capabilities.

Wireless sensor networks (WSNs) are a typical example of this kind of networks [2, 3]. In this case, the well-known paradigm of ad hoc networking specializes to consider the following characteristics.

Mobility. Whereas mobility is a fundamental aspect of all nodes in an ad hoc networks, mobility in WSNs pertains mostly to a subset of the network elements, and it is more specifically application-dependent.

Volume of Nodes. WSNs are usually comprised of a higher number of nodes (in the thousands) rather than the few hundreds that are typical of ad hoc networks.

Resource Availability. Even if an ad hoc node is portable, power and computational resources are not usually crucial elements. Laptops and PDAs have rechargeable batteries, have plenty of volatile and durable memory, and run the same software available for bigger, static computers. WSN nodes are heavily resource-constrained. Sensor nodes are usually unreplaceable, and they become unusable after failure or energy depletion. Available memory is in the order of

---

Algorithms and Protocols for Wireless Sensor Networks, Edited by Azzedine Boukerche

KBs, and computational capabilities are limited. This imposes custom operative systems, reduced protocol complexity, and highly specialized software.

Data Communications. Ad hoc data communication is typically peer-to-peer. One node that wishes to communicate with one or more nodes does so using one of the many available protocols. In WSNs the emphasis is on data transport from the sensor to specific data collection nodes (*sinks*). Nodes are informed about *events of interest* to the users. When such an event is detected by the sensor, a corresponding data packet is created and sent to the sinks possibly via a multihop route (data routing).

Given these very specific characteristics of WSN, it is crucial to devise WSN protocols for topology organization, interest dissemination, and data routing that are energy-conserving, scalable, and able to prolong the overall network longevity, especially in networks with a large number of devices. Differently from what is commonly understood and found in the WSNs literature, such protocols often have to deal with the mobility of some of the network components. This is the case of applications recently been proposed, which include:

- Underwater monitoring, such as submerging a network of sensors in an ocean bed to detect debris from plane crashes for recovery and identification purposes. In this case, beyond (partially) mobile sensors, the network comprises *unmanned* or *autonomous underwater vehicles* (AUV) that are sent roaming through the network for data collection [4].
- Networks in support of mobile small-scale robot squads that coordinate for performing a common task. While performing their operations, the robots exchange information among each other and/or transmit collected measures to remote centers by using the network [5, 6].
- Sensor networks for collecting information about the location of a user/piece of equipment, and so on, like tracking objects as well as humans in a hospital, items in a warehouse, and so on [7].
- Enabling autonomous albeit controlled living of elderly people (independent assisted living) through *e-health systems* made up of wearable sensors as well as telemedicine equipment, home networking (*domotics*) techniques, and sensor-based *smart spaces* [8].

All these applications involve mobility: Sensor nodes can move (e.g., assisted living through wearable sensors, tracking, etc.) as well as the data collectors (robots or AUVs).

In this chapter we motivate and illustrate the use of mobility in WSNs. Furthermore, we demonstrate how mobility is actually advantageous for improving the performance of these networks.

10.1.1 Motivating the Use of Mobility in WSNs

Most of the research in WSNs concerns networks whose nodes do not move and cannot be replaced. Nodes sense events of interest, and some energy-efficient routing protocol is used for delivering the sensed data to static sinks. In this scenario, it has been observed that the nodes that more than all the others have their energy drained from data communication are those closer to the sinks. These nodes relay data for all the other nodes in the network as well as packets from the sinks to the sensors. As a consequence, nodes that are closer to the sinks soon "die" from energy depletion resulting in the disconnection of the sinks from the rest of the network. The problem of energy drainage at the sink neighbors is referred to as the "sink neighbors problem." The consequences of this problem are illustrated in Figure 10.1, which shows the average nodal residual energy when the nodes closest to the sink die. The figure refers to networks with 400 homogeneous nodes, each with a short transmission range (25 m), placed on a 20×20 grid. The static sink is (optimally) located at the center of the deployment area. Packets are periodically sent to the sink by using a shortest path-like multihop routing protocol [9, 10]. The picture shows the quite uneven node average residual energy (percentage of the nodal initial energy) at the time the four sensor nodes that relay packets to the sink die, leaving the sink unable to receive any more data from the network. The remarkably high variance among the residual energies is due to the different distance of each node from the sink and, in general, to the different number of sensor-to-sink routes to which a node belongs, which implies different number of packets to relay. Nodes along the "cross" centered at the sink tend to be the preferred data relays. The closer these nodes are to the sink, the higher the number of packets they receive and transmit, and consequently the higher their energy consumption. These are the nodes with the lowest residual energy in the figure. In particular, when the nodes around the sink die, the energy at the nodes along the cross arms averages at 71.07% of their initial energy, while 42.75% of the

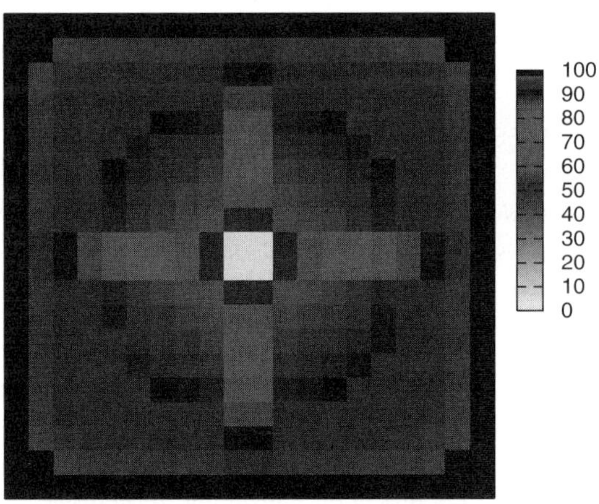

Figure 10.1. Node residual energy (%).

network nodes have more than 95% of their initial energy available! This incapacity of balancing node energy consumption results in short *network lifetime* (the time until packets are no longer deliverable to the sink) and inefficient use of available resources. The sink is soon disconnected from the network, while a large number of the deployed nodes are still fully operational!

This chapter explores solutions for mitigating the sink neighbors problem illustrated above, by exploiting the mobility of some of the network components, thus improving network performance.

Given the typical mobile nature of general ad hoc networks, the impact of mobility on their performance has been extensively explored [11–19]. The primary objective of these works is that of exploiting the mobility of some of the nodes for message delivery in disconnected ad hoc networks, for improving network throughput, and for studying mobility-assisted routing in general.

This chapter concerns approaches designed specifically for WSNs, and in particular those solutions where the mobility of some network elements brings considerable improvement in key performance metrics that are typical of sensor networking. In the rest of the chapter we describe research on mobility in WSNs proposed recently. In particular, we will survey works where (a) the *sensor nodes can move*, so that batches of "fresher" nodes are kept close to the sink for providing uninterrupted data forwarding; and (b) *relays* are sent throughout the network to collect data and bringing them to the sink. Finally, (c) we will describe works where the *sink* itself moves to collect sensed data. The last two approaches appear to be the more promising for energy efficiency and longer network lifetime, since sink and relays are usually considered resource-rich. Therefore, energy consumption and network lifetime are not impacted by the energy needed to move them. In the case where the sensor nodes move, a great deal of the nodes' energy is spent on the movement itself, thus having a detrimental impact on the lifetime of the nodes. The final part of the chapter concerns a quite thorough discussion on how to compare different approaches to mobility in WSNs, focusing on sink and on relay mobility. In particular, we show how, by using simulations, one can effectively assess pros and cons of different solutions and gain useful insights for defining and testing new ones.

10.2 SENSOR NODES MOBILITY

Mobility of the sensor nodes has been exploited for improving, or enabling altogether, sensing and communication coverage [20–22]. The idea presented in reference 20 is to have the sensors move into positions that minimize the energy cost of reporting streams of data to the sink, which is statically placed. The protocols proposed by Wang et al. [21] aim at moving mobile sensors from densely deployed areas to areas with coverage holes, where for some reasons a limited number of sensors have been deployed. The three protocols are proven by simulations to be effective in terms of coverage, deployment time, and moving distance. Minimization of the energy consumption of moving nodes has been subsequently addressed by the authors in reference 22 by letting the node "move logically"—that is, only after they can decide whether their moving maximizes the coverage or not. The idea of using mobile sensors has also been explored within the robotic community, where mobile robots are (also) equipped with sensors. Typical examples are the works by Howard et al. [23, 24]. In reference 23 an algorithm for the deployment of the members of a robotic team into an unknown environment is given. The aim of this algorithm is the maximization of the coverage area, while maintaining line-of-sight contact among the robots. In reference 24 the same authors draw from the theory of potential fields to distribute the mobile sensors throughout a given area. The fields are constructed in such a way that each node is repelled by obstacles and other nodes, thereby forcing the node to spread throughout the area.

Finally, distributed algorithms for the mobility of sensor nodes have been investigated in reference 25. In this work, mobility algorithms are proposed that move the nodes to positions that reduce the transmission power needed to send the data to the (static) sink. The positions for the moving sensors are determined via distributed simulated annealing, as opposed to a greedy strategy that could lead to a suboptimal placement. By using distributed simulated annealing, a node based only on information on its current neighbors accepts a "bad move" with a positive probability. This move could be locally nonoptimal, but could benefit the network globally in the longer run.

Works that consider mobile sensors and robots are mostly concerned with sensor deployment time and with sensing coverage. The costs associated with sensor movements as well as the cost of transmitting sensed data are often not considered, and network lifetime is rarely a metric of interest.

10.3 MOBILE RELAYS

Chatzigiannakis and Nikoletseas [26] explore the possibility of using the coordinated motion of a small number of users in the network for efficient communication between any pair of other mobile nodes. Some of the nodes act as *mobile relays* in that they carry packets for other nodes. Packets are exchanged when the source node and the relay are neighbors (namely, they are in the radio vicinity of each other), and they are then delivered to the intended destination when the relay passes by it.

This is basically what has been introduced to WSNs by Shah et al. in their works on *data MULEs* [15, 27]. The MULEs are mobile nodes roaming among the sensors to act as forwarding agents. Energy conservation is possible because of single-hop communications (from a sensor to the MULE that is passing by) rather than in a multihop way (from the sensor to the sink). The packet now reaches the sink when the MULE eventually passes by it and transfers all collected sensed data to it. MULEs are effective for energy conservation in the so-called *delay tolerant networks* [28]. Energy is traded off for latency; that is, the energy needed to communicate a packet to the sink is decreased at the cost of waiting for a MULE to pass nearby and at the cost of waiting for the MULE to move to the vicinity of a sink. Other problems, such as scheduling sensor-to-sink transmissions within this model, have been studied in reference 29.

Unmanned vehicles for data collection have been further investigated in reference 30. Nodes send their packets to nearby cluster heads via multihop routing. The moving relays then pass by the cluster heads to collect the data according to some predefined schedule. Therefore, the collector has to visit only the cluster heads and not all the nodes. Furthermore, multihop routing is reduced to a smaller number of hops since the data are sent from a sensor node to a clusterhead who is nearby. Data delivery to the static sink occurs periodically, when the collectors return to the sink to drop their packets and recharge. This concept and architecture has been further explored in reference 31. In this case the authors consider different classes of nodes, where the collectors roam (controllably) among the nodes, grouped into clusters, and can be (uncontrollably) mobile. The goal here is that of determining the schedules of the collectors visits to the nodes that minimize transmission energy consumption, data latency, and nodal buffer requirements.

A first discussion on how to include controllable mobile *relays* into the network has been presented in reference 32. In the paper, an implementation of a sensor network with an autonomous mobile relay (a robot) is presented. The robot visits the (static) sensors, collects their data (single-hop exchange), and delivers them to the sink, similarly to what happens for the MULEs. However, in this case the movements of the robot adapt to the network application priorities, which dictate data collection performance parameters. In other words, the robot is part of the system, and it is the system that *controls* its mobility. The testbed-based experimental results concern the evaluation of methods for controlling the speed of the robot for optimizing data collection. The robot traverses networks with different densities following a straight trail and collects the data. The data are then delivered to the sink. Methods are defined for routing the sensed data to nodes that are one hop from the robot route when the robot itself does not pass sufficiently close to some sensors. Further development of this work with multiple controlled mobile elements (here called explicitly data MULEs) has been presented in reference 33. This work considers the two cases where nodes are deployed uniformly and randomly in a given geographic area and when, more realistically, they are distributed differently. In the first case, criteria are given for the choice of the number of MULEs and for dealing with nodes that can be served by multiple mules. In the case of nonuniform nodal distribution, a load balancing algorithm is introduced for distributing the number of sensor nodes to the various MULEs so that each MULE serves approximately the same number of sensors. MULEs roam through the network in straight lines and gather information about the nodes they can reach. Then the MULEs, which can wirelessly talk to each other directly, elect a leader and send the information they gather to it. Based on this information, the leader executes the load balancing algorithm and associate nodes to MULEs. Data collection is finally performed by the MULEs that travel through the network (in straight lines) and explicitly poll the assigned sensor nodes for collecting their data. The problem of scheduling the visit to the sensor nodes of a single relay (called ME, for Mobile Element) so that there is no data loss (due to buffer overflow) is tackled in reference 34. The corresponding Mobile Element Scheduling (MES) problem is proven to be NP-complete, and centralized analytical model (ILP) and algorithms are given for solving the problem. In particular, given as input the data generation patterns (sensing rates) at the sensors, the corresponding buffer overflow times, and a matrix cost for the ME movements, the model and algorithms determine the sequence of visit to the nodes so that none of the buffers overflow. General discussions on the advantages and challenges of controlled mobility are conducted in references 35 and 36. In the first paper a new approach to exploiting mobility, termed *Morph*, is defined, and the benefits to sensor networks sustainability are shown and discussed. In reference 36, considerations about the use of controllable mobility are more general and are followed by a detailed description of how mobility induces improvements in WSNs performance.

The problem of controlled data MULE-like mobility has been also recently addressed in reference 37. The authors first propose an algorithm for avoiding sensor nodes buffer overflow while minimizing the speed of the mobile relay. They also extend this algorithm to the case where some of the packets have delay constraints (i.e., they are "urgent messages" that have to be delivered to the sink within a given time since their generation). Finally, an investigation of the controlled use of relay nodes for data collection and subsequent report to the sink has also been proposed in reference 38. Although the authors recognize that moving the sink directly yields better resource utilization and hence longer lifetime, they argue that for certain applications, moving the sink is infeasible. Therefore, having one or more resource-rich mobile relay nodes is remarkably helpful. Given that the sensor nodes know about the current location of the relay node, routing protocols are presented for delivering the data from the sensors to the relay and from the relay to the sink and, finally, for determining the route of the relay. Improvements on network lifetimes are fourfold with respect to the case of a static sink.

10.4 MOBILE SINKS

A data dissemination protocol, termed SEAD (Scalable Energy-Efficient Asynchronous Dissemination), is defined by Kim et al. [39] where a tree-like communication structure is built and maintained. Differently from previous data centric solutions à la *directed diffusion* [40], which also use tree structure for data and interest gathering and dissemination, according to SEAD the *sink moves* (randomly) to certain sensor nodes in the tree (access points) from which it collects the data. Communication between the sink and the access points can be multihop. This happens when the sink moves away from the access points. The tradeoffs that SEAD attempts to obtain concern data latency and the energy needed for tree reconfiguration so that the access points are closer to the current position of the sinks. In this way, SEAD is shown to outperform other solutions for data dissemination in WSNs where the sink does not move, such as directed diffusion [40], TTDD [41], and ADMR [42].

Building and maintaining routes to a mobile sink is the topic of references [43–45]. The aim is that of minimizing these operations overhead. In the first paper, local update techniques are described for detecting disconnections and perform route repair in "sink-oriented trees." The ERUP protocol is proposed in reference 44 for conducting route rediscovery only in the vicinity of the damaged route. In reference

45, initial routes are constructed from the nodes to the sink according to any viable WSN routing. Given that the sink is moving, if the routes are no longer valid, forwarder nodes are designated to extend the current routes.

All these works have in common that the mobility of the sink is unpredictable and references *uncontrollable*. For example, in references 39 and 46, sinks move according to the random waypoint model.

The use of mobile sinks with *predictable* mobility has been more recently presented in 47–49 and 50. In these works the sinks (airplanes) fly over the sensor field and gather the sensed data periodically. While the movement of the sink is fully controllable, it is external to the network infrastructure; that is, the trajectories are not determined by network components and activity. The main contribution of these papers is the energy-efficient transmission to the passing sink [47, 48, 51]. In reference 49 the authors consider heterogeneous sensor networks made up of two types of nodes. Type 0 nodes do the basic sensing, perform short-range communications, and are partitioned into clusters whose cluster head is a type 1 node. The cluster heads take care of receiving data from type 0 nodes (possibly through multihop routes), do some sensing, aggregate data, and perform long-range transmissions to the aircraft. Each passing of the aircraft triggers a new data sensing and data collection cycle. The aim of the paper is to determine the optimum node deployment (i.e., the densities of each type of nodes) and the nodal energy needed to achieve a given network lifetime (numbers of data gathering cycles) while ensuring sensing coverage and radio connectivity with high probability.

Inherent patterns of the sink movement are exploited in reference 52 for the design of robust and energy-efficient routing. This work assumes that there is a certain degree of predictability in the sink movement, such as the routine route of a ranger patrolling a forest. Sensor nodes learn about the sink whereabouts at given times via statistics techniques as well as methods from distributed reinforcement learning.

In reference 50 a model for sink movements is presented where the sinks, called "observers," move along the same route repeatedly and collect data from the sensors. The sensed data are collected while the observer traverses the network. In particular, when passing by sensor nodes, the observer wakes them up and receives their data (if any). The authors describe a prototype system developed at Rice University where the observers are carried by shuttles, and the sensors are spread out throughout the campus. In particular, the authors determine the transmission range needed to collect data from a predefined percentage of the sensor nodes, given the observer speed, the time required to transmit a packet, and different traffic patterns. The correlation among the various system parameters is investigated analytically.

Network-controlled sink mobility for reducing energy consumption and for maximizing the lifetime of a sensor network has been considered initially in reference 53 and then in references 54–56, 9 and 10. In these works the sink moves among the (static) sensor nodes and, while sojourning at given locations, collects data that are sent to it via multihop routes. The first work concerns minimizing energy consumption. An ILP model is presented that determines the locations of multiple sinks as well as the routes from the sensors to the sinks. Time is divided into rounds. At the beginning of each round, information on the nodes' residual energy is centrally gathered and the ILP problem is solved for finding the new locations for the sink that minimize the maximum energy consumption spent at the nodes during that round. Minimizing the energy consumption results in increased network longevity. No constraints are enforced on the sink movements, and there is no relation between the number of the sinks and their position in subsequent rounds: The model is solved from scratch every time.

The problem of network lifetime maximization through controlled sink mobility is explicitly addressed in reference 54 for networks with a single sink. Sink locations and sink sojourn times at those locations are determined that maximize the network lifetime via a new LP formulation of the problem. Maximizing the network lifetime is maximizing the sum of sojourn times of the sink at the visited locations (in this case they coincide with the sensor locations). The experiments in reference 54 refer to scenarios where $n = L^2$ nodes are arranged in a $L \times L$ grid. The sink has no limitation on the time $t_k \geq 0$ it can spend at sensor k and can move from any location to any other location in the network. The network lifetime is improved by a factor of 5. This happens when the sink sojourns at the nodes located at the four corner areas and in the central area of the grid.

Papadimitriou and Georgiadis [55] present another centralized solution for the problem of maximizing network lifetime by combining the model presented in reference 54 and the LP formulation for maximum lifetime routing described in reference 57. A constant of the model in reference 54 is here turned into a variable so that the resulting model presented solves both (a) the problem of determining the sink sojourn times at the given sites and (b) the routing of the packets to the current position of the sink. This routing-dependent solution achieves improvements with respect to the lifetime values of reference 54 that are twofold.

Lifetime maximization as a min–max problem is the idea explored by Luo and Hubaux [56]. The authors consider together sink mobility and data routing. A load balancing solution is presented that, while keeping the sink moving along the external perimeter of the network, achieves lifetimes 500% higher than when the sink stays still in the center of the network.

These centralized solutions concern how to drive the sink to places in the network where data collection has the least impact on the sensor nodes' residual energy. However, several aspects of sensor networking (such as the possibility of collision and corresponding energy cost) are not considered. Similarly, the cost of building, maintaining, and releasing routes to the current position of the sink and its impact on network lifetime and other metrics is never taken into account. A Mixed Integer Linear Programming (MILP) model has been presented in references 58 and 59 and in more detail in references 9 and 10, where more realistic data communication costs and constraints of sensor networking and on sink mobility are explicitly considered. Moreover, those papers introduce and evaluate the first distributed and localized heuristic for controlled sink mobility. The MILP model and the distributed solution are described thoroughly in the next section.

10.5 COMPARING DIFFERENT APPROACHES TO MOBILITY

The most promising use of mobility for enhancing the performance of WSNs has been shown to be obtained when resource-rich network components move—that is, when either the sink itself or mobile relays are left free to roam through the network for collecting data.

This part of the chapter is dedicated to describe and discuss how to compare two very different approaches to mobility in WSNs. Our aim is twofold. First of all, we want to gain an understanding of which among the many proposed solutions are the most promising ones in terms of usage and performance enhancement. More than specific mobility protocols, we are interested in determining which network architecture is the best for supporting efficiently the application listed above. Then, we want to establish clear metrics and tools for comparing mobility solutions and obtain from this performance comparison new insights for protocol design and effective use of mobility.

By the study of what proposed in the literature, we could observe that when considering mobility of sinks and relays, the major difference in performance is made by whether routing between the sensor nodes and a sink/relay is single-hop or multihop. This choice for the routing greatly affects essential metrics such as nodal energy consumption, packet latency, and probability of successfully delivering data to the sink/relay. It has also immediate consequences on the design and cost of the wireless sensor node: When a multihop data routing protocol needs to be implemented, a node has to provide resources (storage, computation, etc.) for enabling the forwarding of packets to the data collector; that is, a protocol stack has to be available. When routing is single-hop, instead, a simpler nodal architecture suffices for storing the packets and for transmitting them to a relay when it passes by.

In the following we describe and investigate two network architectures that clearly show the impact of choosing multihop versus single-hop routing when sinks or relays roam through the network. We consider single-hop routing according to the data MULE approach. We will consider the original model [27], which exploits the *uncontrolled* mobility of multiple relays, and more recent works where it is shown that the relay mobility can be controlled by the network current conditions [33]. We then show the pros and cons of multihop routing by describing an analytical model and distributed heuristics for sink mobility. As in the previous case, we consider and compare solutions where the mobility of the sink is controlled by the network itself with those where sink mobility does not depend on what is going on in the network [9, 10].

10.5.1 The Data MULE Approach to Mobility in WSNs

The idea of a *Mobile Ubiquitous LAN Extension* (MULE) [15, 27] stems from that of deploying mobile agents for carrying information [60], especially among nodes in possibly disconnected networks. The concept is quite simple: MULEs are resource-rich nodes that roam freely throughout the network. When as a result of its motion a MULE gets to be in the radio proximity of a wireless sensor node, it receives all the packets generated by that sensor so far (if any). Upon getting close to one of the network *access points* (i.e., the sinks), the MULE transfers to it all the collected packets. In this way, whether there is a route from a sensor to one of the sinks or not, (i.e., independently of network connectivity), a packet is eventually delivered to the sink. The key concept here is that sensor-relay routing as well as relay-sink routing is single-hop.

The data MULE architecture is made up of three layers. The highest layer comprises one or more access points—that is, the sinks. These are the resource-rich, static components to whom the MULEs deliver the data collected throughout the network. The middle layer is the set of the MULEs. These are nodes characterized by mobility, large storage, and renewable power and by the ability to communicate with the sensor nodes as well as with the sinks wirelessly. Most importantly, MULEs movements cannot be predicted in advance and are considered completely random (uncontrolled mobility). Examples of MULE nodes are (a) animals roaming in a given area (as in an environmental monitoring application [11]) and (b) vehicles traveling in a city or university campus. Finally, the bottom tier is comprised of a possibly large number or resource-constrained static sensor nodes that can communicate with a MULE that is passing by.

Communication in the MULE architecture is single-hop: A node stores the sensed data until a MULE passes by. Once the MULE arrives, the node transmits all the stored packets to the MULE. When the MULE passes by the sink, it transmits to it all the packets it has collected.

Clear advantages of the MULEs approach to data gathering and delivery in wireless sensor networking include the following.

- *Robustness*. No node depends on any single MULE. Any of the roaming MULEs would do. If one of the MULEs fails, a node can still count on another MULE to pick up its packets.
- *Lower Complexity*. The one-hop nature of the sensor-to-MULE communication allows very simple nodal protocol stack. Unsophisticated MAC is often enough to guarantee safe transfer of data packets from their source to the passing MULE.
- *Decreased Energy Consumption*. The typical overhead associated with multihop ad hoc routing is not present in the MULE system. This overhead is oftentimes the major culprit of energy consumption in WSNs.

However, the flip side of this quite simple and effective system presents non-negligible drawbacks. The first that comes to mind is the average latency incurred by the packets from the time they are created to when they are delivered to the sink. Given the serendipitous nature of the MULEs movements, there is no guarantee on deterministic delay bounds. It is not possible to surely state when and where a MULE will show up close to a sensor or to the sink. Therefore, a packet can sit in a sensor queue, or in the queue of a MULE for quite a long time, which imposes very high overall sensor-to-sink latency. The uncontrolled MULE mobility has also a detrimental effect on the packet delivery ratio; that is, the ratio between all packets that are created in the network over those that are successfully delivered to the sink. Sensor nodes, having limited storage, can buffer only a limited amount of packets. If the inter-arrival time of a MULE in the radio vicinity of a sensor is too large, chances are that the sensor is forced to discard all the packets in excess of its buffer size. This forces down the amount and type of applications that could use the MULE architecture for data gathering and dissemination. Only applications that are delay-tolerant can endure the possibly very high latencies imposed by the MULE system.

Recent studies have shown that it is possible to mitigate the drawbacks of the MULE architecture by having the relay movement guided by the network itself—that is, by *controlling* the mobility of the MULEs. One of the first attempts is described by Kansal et al. [32]. As mentioned earlier, this work has quite an experimental flavor and is concerned with a network where one MULE (a mobile robot) travels along a straight trail up and down the sensor node deployment area to collect data. Variation in the speed of the MULE is determined by the network load: The robot will go slower if more data are to be transmitted by the nearby sensors, but will go faster otherwise. Since the MULE's route is always the same straight line, it is highly likely that some sensors are not able to transmit their data to the MULE directly. For this reason the solution proposed in reference 32 organizes the network operation into two parts. The first, named the *network algorithm*, takes care of how the sensor nodes interact with each other so that packets from those nodes that are not sufficiently close to the passing MULE are delivered to it by nodes that are close. The second part specifies how the MULE moves, and it is called the *motion control algorithm*. Here is a description of the two parts.

The **Network algorithm** defines the way in which those nodes that cannot directly transmit to the sink find data routes to send their packets via intermediate nodes. The network algorithm is performed in three steps: *initialization*, *local multihop*, and *data collection*.

1. *Initialization*. During the first run through the network deployment area, along its only route, the MULE beacons a simple "hello"-like packet. All nodes that receive this packet are neighbors to the passing MULE, set a hop-count variable to 1, and broadcast the hello packet further, updating the hop-count field to its own. A node that receives multiple hello messages from neighboring nodes updates its own hop-count variable to the least values received plus 1, and it keeps forwarding the message (with the updated hop-count). At the same time, it records the node closer to the MULE path that sent the hello packet with the least values (possible ties are easily broken). In this way each node become aware of (a) its hop distance from a node that can directly transmit to the passing MULE and (b) a route to that node. Overall, a forest (set of disjoint trees) is built as a first step of the network algorithm, with the nodes closer to the MULE's route as the roots.
2. *Local Multihops*. Data packets are sent at the root nodes from any sensor node in the corresponding tree. Any suitable routing algorithm (e.g., *directed diffusion* [40] or a simpler "Convergecasting," such as the one defined in reference 61) can be used for implementing this step.

3. *Data Collection by the MULE*. After its first passage through the network, the MULE keeps moving up and down its route collecting data from the nodes along its path. Nodes polled by the passing MULE transmit their packets along with those of the nodes in their trees.

Motion Control Algorithm. Since the route of the MULE is fixed, motion control reduces to speed regulation. This is done by fixing the round trip time (RTT) of the MULE. Then, three possible strategies are described in reference 32 that correspond to different strategies of MULE movements.

1. The path is traversed at a certain fixed speed. Let us assume that at this speed the MULE takes T < RTT to travel its route (and back). The remaining "spare time" RTT-T is divided equally among the nodes along the MULE's route, and the MULE stops at each node for this time. The MULE might not know about the node location. Therefore, the MULE stops as soon as it hears from the node.
2. The MULE travels the route at the constant speed of (length of route)/RTT and does not stop at the nodes.
3. The speed of the MULE adapts to the current situation of (the queue of) the nodes. This can be realized by having the MULE traveling at the speed 2∗(length of route)/RTT. The extra amount of time is divided among those nodes from which the MULE has collected a limited amount of data (determined by a given threshold) in the previous round. (The set of nodes the MULE stops at changes at each round.)

Further details about the described algorithms as well as its implementation on a real-hardware testbed can be found in reference 32.

Here we describe the natural evolution of this *single*-MULE solution to controlled mobility in the "mostly single-hop" routing realm: *multiple* controlled mobile elements (data MULEs) for data collection [33]. The motivations are clear: The single MULE approach does not scale well. If the number of nodes in the network increases, there are more nodes from which data must be collected in the same amount of time. Therefore, packets might be dropped (buffer overflow) because the MULE cannot get to visit certain nodes on time. Deploying multiple MULEs solves this scalability problem. If the nodes are scattered randomly and uniformly throughout the deployment area, the obvious solution consists in dividing the area into same-size parts and having a MULE in each of these parts. Given the deployment distribution, each MULE would have roughly the same amount of nodes to serve. In this case, there are two main problems to deal with: deciding the number of data MULEs and deciding what to do with nodes that are shared by two MULEs.

- *Number of Data MULEs*. The number of data MULEs needed to collect data while avoiding packet loss is computed based on the RTT of a MULE (defined as the time it takes to the MULE for traversing the assigned area along a straight line) and on the *buffer_fill_time*, which is the time needed to a node to fill its buffer. The RTT is defined as follows:

$$(l/s) + n \times st + (l/s)$$

where l is the side of the deployment area (a square), s is the MULE speed, n is the number of nodes in the network, and st is the service time—that is, the time needed to transfer packets from a node's buffer to the servicing data MULE. It is assumed that the MULE stops at each node while traversing the networks for st time. This happens only one way. Once the MULE has reached the end of the line, it goes back without collecting data. In the case where there are nodes that are not in the transmission range of any of the MULEs, multihop routing is used to send the packet to a node (root of the data distribution tree, as in the single-MULE case described above) close to the MULE route. In order not to lose packets because of buffer overflow, the number of data MULEs required will be

$$\left\lceil \frac{\text{RTT}}{\text{buffer\_fill\_time}} \right\rceil$$

(One MULE suffices when RTT $\leq$ buffer_fill_time.)

- *Sharing Nodes*. Whenever a node can be serviced by two MULEs, most likely because it is equidistant (in hops) from the MULEs routes, ties are broken randomly or by simple techniques. For instance, the node might decide to send one packet to one MULE and the next one to the other MULE, or it may just flip a coin every time it has a packet to send, and decide which MULE is going to get the packet accordingly.

Deployment of multiple MULEs is also described in case nodes are not scattered randomly and uniformly. This can happen, for instance, when nodes are manually placed in specific places, or when the terrain induces a nonuniform nodal distribution. The MULEs move along straight lines that are not necessarily equally spaced throughout the area. Assuming that each node is at most one hop from a MULE's route (i.e., that every sensor is directly served by at least a MULE), nodes are divided into two classes: *nonshareable (NS) nodes* and *shareable (SH) nodes*. The first set comprises those nodes that can be served only by one of the MULEs. The set SH instead contains nodes that are in the transmission range of multiple MULEs (at least two). For NS nodes there is no option other than to be served by the only MULE that passes close to them. Shareable nodes should instead be assigned to only one MULE. The challenge here is to define a way to assign nodes in SH to the MULEs so that the number of nodes served by each MULE is roughly the same. The algorithm for data gathering in this case is made up of five phases: *initialization, leader election, load balancing, node assignment*, and *data collection*. The following is a brief description of the algorithm (details can be found in reference 33).

- *Initialization*. The MULEs go on their first trip beaconing around their presence. Nodes that receive this beacon reply back with their ID. By the end of this phase, the MULEs are back at their starting points and know of all the nodes they can serve.
- *Leader Election*. All the MULEs are assumed to be able to communicate with each other (i.e., at least when at their initial position, the topology of the "MULE network," is a clique). One of the MULEs is elected as the leader of the bunch (for instance, the MULE with the smaller ID). At this point, every MULE sends to the leader the information gathered in the first phase.
- *Load Balancing*. The leader divides the network nodes into shareable and nonshareable. The shareable nodes are further classified according to the MULEs that can serve them. This is the most delicate part of the algorithm, all performed locally by the leader MULE. It is divided into two main parts: The first part takes care of calculating the average number of nodes that could be assigned to each MULE. The second takes care of the actual load balancing [33], determining the nodes that could be assigned to each MULE.
- *Node Assignment*. As a result of the previous phase, some of the shareable nodes can be served by two MULEs. In the node assignment, phase nodes that are shared by two MULEs are ordered according to their unique ID. At this point the first part of the set is assigned to the MULE of the two with the smallest ID, and the remaining nodes are assigned to the other MULE. After the assignment has been calculated, the leader MULE communicates to all the other MULEs the nodes they have been given to serve.
- *Data Collection.* In the data collection phase the MULEs start their journey along their routes, looking for data. At this time the nodes do not know the MULE to which they are assigned; and hence when they hear the MULE beacon, they respond with their packet. They will receive an acknowledgment only from the MULE that is supposed to collect their packets. A node records this information, and in the future it will respond only to its assigned MULE.

Simulation results are shown in reference 33 that demonstrate how effective this algorithm is in distributing nodes to MULEs.

10.5.2 Multihop Routing and Controlled Sink Mobility

We consider here the case of a mobile sink traveling through the network and receiving data at its current location via multihop routing. We start by describing in detail a solution where the sink moves according to the current network condition (controlled mobility), with the aim of balancing energy consumption among the network nodes and thus prolonging the network lifetime.

This solution, presented in references 9 and 10, illustrates two fundamental research steps: mathematical modeling (for the centralized determination of an optimal solution) and the definition of distributed heuristics. A mathematical model is defined as follows: Taking as input realistic parameters such as the cost of route setup, maintenance, and release, the cost of transmitting and relaying packets to the sink, and limitations in sink movement will determine where the sink should go and how much it should stay there so that network lifetime is maximized. The solution, clearly centralized, is based on a Mixed Integer Linear Programming (MILP) modeling. Although MILP models are computationally hard to solve, the model defined in reference 10 is solvable for realistically large instance of the problem. The optimum sink route determined by solving the MILP model is used for comparison with the route determined by a distributed heuristic which is more suited for use in WSNs. The distributed heuristic, termed Greedy Maximum Residual Energy (GMRE), determines sink movements and sojourn times by greedily sending the sink from one point to the one surrounded by the nodes with the highest residual energy.

Here is the typical scenario. A large number $|N|$ of resource constrained, static nodes with sensing and wireless communication capabilities are scattered in a given geographic area. Data packets are generated periodically at the sensor nodes: At a given data rate r_i, node $i \in N$ transmits packets that are delivered to the sink for processing.

While the nodes are static, the sink can be mobile. More specifically, a set $S = \{1, \ldots, q\}$ of q sink *sites* is considered which are the points within the geographic area the sink can visit. For instance, Figure 10.2 shows a typical scenario where 36 nodes (represented by circles) are placed on a 6×6 grid and 25 sink sites (squares)

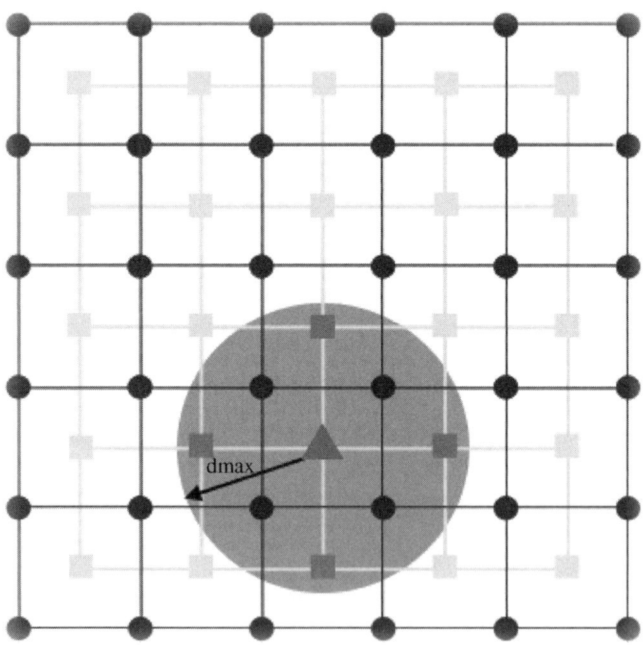

Figure 10.2. Sensor nodes, sink sites, and sink movements.

are organized according to a 5 × 5 grid. A link between two nodes indicates that those two nodes are neighbors (i.e., they can hear each other's transmissions). A lighter-colored link between two sites indicates that the sink can move from one site to the other and vice versa. Because of the sink neighbors problem, the sink moves throughout the network in an attempt to balance the energy consumption among the nodes. A problem that arises is that of how to let the nodes know what is the new position of the sink. To this aim, when the sink arrives at a new site, it broadcasts a packet f to all the network nodes, making them aware of its current site. When a node has a packet to send, it now sends it toward the new site of the sink. Every routing scheme that works with the topological information provided by f, such as geographic or shortest paths-based routing, is a viable routing for data delivery to the sink. The main advantages of routing independence are the following. It guarantees the longest possible network lifetime given the specific routing. It allows the network users to design or choose the routing algorithm that best meets the WSN application requirements, which go beyond improving network lifetime and may consider other metrics. Every time the sink leaves a site, it again broadcasts a packet to all nodes to communicate that it has moved. Upon receiving this packet, a node stops forwarding data (remaining packets are buffered) and waits to receive a new packet f from the sink, carrying its new location.

While the sink is traveling the senser do not transmit and they buffer the packets. The farther the sink travels, the more it takes to get to the new site and the higher the delay suffered by the data. For containing this delay, a new parameter called d_{MAX} is introduced that represents an upper bound on the distance that the sink can travel from a site to the following one. The pair (S, d_{MAX}) uniquely defines a graph of sink sites where there is a link between two sites if and only if their (Euclidean) distance is $\leq d_{MAX}$. Figure 10.2 shows the four sites (darker squares) the sink (the triangle) can reach from its current position. The lighter lines between the sites of Figure 10.2 indicate that the sink can only move horizontally or vertically in the 5 × 5 grid.

In the case of high rate sink mobility and low data traffic, the energy cost for route construction and release can be significant. Therefore, this cost should be explicitly taken into account. Finally, in order to evaluate the impact of different sink mobility rates, the parameter t_{min} is introduced to represent a mandatory minimum time the sink has to sojourn at a site. (High t_{min} values slow the sink down, while low t_{min} values allow it to move faster.) The problem we set out to solve here is the following: *Determine the starting site and the route for the mobile sink over the graph* (S, d_{max})*, together with the sojourn times* $t_k \geq t_{min}$ *of the sink at each visited site* $k \in S$ *so that network lifetime is maximized.*

MILP formulation. Here are the sets, the parameters, and the variables used for formalizing the problem.

- Let S be the set of sink sites—that is, the locations at which the sink may sojourn: $S = \{1, \ldots, q\}$.

- N is the set of the network nodes: $N = \{1, \ldots, n\}$.
- e_0: Initial energy (Joules) of each node.
- b_{ik}: Energy consumption (Joules) at node $i \in N$ for setting up/releasing routes when the sink moves to site $k \in S$.
- c_{ik}: Power consumption (Watts) for receiving and transmitting packets at node $i \in N$ when the sink sojourns at site $k \in S$.
- t_{min}: Mandatory minimum time (seconds) for which the sink is required to stay at site $k \in S$.
- d_{jk}: Euclidean or shortest path distance (meters) between any two sink sites $j, k \in S$.
- d_{max}: Maximum distance (meters) the sink is allowed to travel each time it moves.
- A: The set of directed edges joining sink sites whose distance is less than or equal to d_{max}, that is, $A = \{(j, k) \in S \times S : j \neq k, d_{jk} \leq d_{MAX}\}$.
- O: The set of directed edges $(0, k)$, $k \in S$, joining a fictitious site 0 (origin) with the sites in S.
- D: The set of directed edges $(k, q+1)$, $k \in S$, joining the sites in S with a fictitious site $q+1$ (final destination).
- X: The union of A, O, and D.

The following variables are also considered:

- t_k: Sojourning time (seconds) of the sink at site $k \in S$.
- z_k: Binary variable taking the value 1 if the sink sojourns at site $k \in S$ ($t_k > 0$); 0 otherwise ($t_k = 0$).
- x_{jk}: Binary variable indicating the status of $(j, k) \in X$. $x_{jk} = 1$ if and only if arc (j, k) is on the sink movement route; $x_{jk} = 0$ otherwise.
- v_k: Auxiliary variable used to enforce a unique sink route.

The MILP formulation is defined by the following objective function and constraints.

$$\text{Max} \sum_{k \in S} t_k \tag{10.1}$$

$$\text{subject to:} \quad \sum_{k \in S} c_{ik} t_k + \sum_{k \in S} b_{ik} z_k \leq e_0 \qquad (i \in N) \tag{10.2}$$

$$t_{min} z_k \leq t_k \leq M z_k \qquad (k \in S) \tag{10.3}$$

$$\sum_{k \in S} x_{0k} = 1 \tag{10.4}$$

$$\sum_{k \in S} x_{k,q+1} = 1 \tag{10.5}$$

$$\sum_{\substack{j \in S \cup \{0\} \\ (j,k) \in O \cup A}} x_{jk} = \sum_{\substack{j \in S \cup \{q+1\} \\ (k,j) \in A \cup D}} x_{kj} \qquad (k \in S) \tag{10.6}$$

$$\sum_{\substack{j \in S \cup \{0\} \\ (j,k) \in O \cup A}} x_{jk} = z_k \qquad (k \in S) \tag{10.7}$$

$$v_j - v_k + q x_{jk} \leq q - 1 \qquad ((j,k) \in A) \tag{10.8}$$

$$v_k \geq 0 \qquad (k \in S) \tag{10.9}$$

$$z_k \in \{0, 1\} \qquad (k \in S) \tag{10.10}$$

$$x_{jk} \in \{0, 1\} \qquad ((j,k) \in X) \tag{10.11}$$

The objective function (10.1) maximizes the sink total time at sojourning sites, $\sum_k t_k$, which is the effective network lifetime. Constraint (10.2) states that the combined energy spent at node i for data delivery ($\sum_{k \in S} c_{ik} t_k$) and data route construction and release ($\sum_{k \in S} b_{ik} z_k$) during $\sum_k t_k$ (the time before the death of the first node) should not exceed the node's initial energy e_0. The right part of double inequality (10.3) forces z_k to take the value 1 if the sink sojourns at site k ($t_k > 0$), thus linking the binary variable z_k [constraint (10.10)] with the continuous variable t_k. M is a significantly large number. The left part of double inequality (10.3) restricts the sojourn time t_k to be at least equal to the mandatory minimum sojourn time t_{min} if the sink sojourns at site k ($z_k = 1$) and at the same time forces z_k to take the value 0 if the sink does not sojourn at site k ($t_k = 0$). The first sojourning site in the sink movement route is allowed to be any site in S. To implement this, a fictitious fixed initial site 0 (origin) is introduced. At the beginning of the sensor network's lifetime, the sink moves in zero time (and cost) from the origin to some site $\alpha \in S$, determined by the model. This is that particular site such that $x_{0\alpha} = 1$ [Eq. (10.4)]; that is, it is the optimum starting point of the sink journey. Then, the sink sojourns at that first site and at subsequent other sites in S to be determined by the model. Finally, from the last sojourning site ω the sink moves to a second fictitious site "$q + 1$" (destination), again in zero time [Eq. (10.5)]. The site ω completes the sink route started at site α. This is the last site at which the sink sojourns, and it marks the end of the sensor network lifetime. The arcs $(j, k) \in X$ on the sink route are associated with binary variables x_{jk} equal to 1. The variable x_{jk} is equal to 0 for all the $(j, k) \in X$ that do not belong to the route. Equivalently, one can think of a unit of flow moving from the origin to the destination. Constraint (10.4) induces a unit of flow from the origin to some node $\alpha \in S$, while constraints (10.5) cause the destination to absorb a unit of flow coming from some node $k \in S$. Constraint (10.6) forces flow conservation at all sites $k \in S$, thus ensuring the generation of a route. Constraint (10.7) ensures that the sites $k \in S$ on the generated route are sites at which the sink sojourns ($k | z_k = 1$). To elaborate, if z_k in constraint (10.7) equals 1, then the sink sojourns at site k, and therefore there must be one and only one arc on the sink movement route reaching site k. On the other hand, if z_k equals 0, then there will not be any incoming arc to that site. Finally, constraint (10.5) induces a unit of flow from the last node in the sink route (ω) to the fictitious final node $q + 1$.

A possible optimum sink route is depicted in Figure 10.3a. This route goes from the initial site $\alpha = a$ to the final site $\omega = g$. Constraints (10.4) and (10.5) ensure that there is only one initial site and one final site for the sink route ($x_{0\alpha} = x_{\omega q+1} = 1$), independently of any α and ω. The two sets of constraints (10.6) and (10.7) generate the route from α to ω that passes through all the sites where the sink has to sojourn in order to maximize the network lifetime. More precisely, constraints (10.6) enforce that for every site $k \in S$ the global number of outgoing visited arcs x_{kl} equal the incoming visited ones x_{jk} (with the exception of the two fictitious sites 0 and $q + 1$). For instance, this is the case of site c in Figure 10.3a: In the route from α to ω, there is only one incoming arc and one outgoing visited arc ($x_{bc} = 1$ and $x_{cd} = 1$). According to constraints (10.7), for every site k on the sink route, there must be a way to get there; that is, there must be exactly one site j (which includes the fictitious site 0) from which k is reachable directly ($x_{jk} = 1$ and $z_k = 1$). At the same time, the other sites not belonging to the sink route are not visited by the sink. For instance, sites $k \in \{a, b, c, d, e, f, g\}$ in Figure 10.3a are all and only those for which $z_k = 1$. These are all and only the sites in the sink route. All other sites h are such that $z_h = 0$.

Constraints (10.6) and (10.7) ensure flow conservation. However, they do not prevent the formation of cycles disjoint from the sink route from α to ω. The (disjoint, non-simple) route depicted in Figure 10.3b comprising nodes in $\{a, b, c, d, e, f, g\} \subseteq S$ and nodes $\{h, i, j, k, l, m\} \subseteq S$ (cycle) is possible according to this model up to constraints (10.7), since none of the constraints from (10.2) to (10.7) is violated by having $z_k = 1$ for $k \in \{h, i, j, k, l, m\}$ as well as $x_{h,i} = x_{i,j} = x_{j,k} = x_{k,l} = x_{l,m} = 1$. This situation is not realistic and is undesirable. It is practically impossible, for instance, to have the sink moving from site d (a site in the connected route from site a to site g) to site i (a site in the cycle), since the distance between these two sites is greater than d_{max}. Constraints (10.8) ensure that no such cycles are formed. (Similar constraints have been used in the integer programming formulation of the Traveling Salesman Problem (TSP) to avoid subtours [62].) According to constraints (10.9), a site k is associated with a "weight" $v_k \geq 0$. Constraints (10.8) introduce a site ordering in the sink sojourns. The sites visited by the sink are traversed in increasing order; that is, if $x_{jk} = 1$, then $v_j < v_k$. In this way it is clearly impossible to return to the same node and, hence, to generate sink routes with disjoint cycles like the one in Figure 10.3b.

Some comments are in order. Parameter t_{min} has been introduced for investigating the effect of higher or lower sink mobility rates on the network performance. The model solution produces sink route and sojourn times $t_k \geq t_{min}$ at site k that maximizes network lifetime. Varying t_{min} enables us to explore a number of tradeoffs. For example, higher t_{min} values should induce lower overhead for route construction and release. Lower t_{min} values instead should enable a finer tuning of sojourn times at different sites (which may be useful in achieving a longer network lifetime), at the price of increasing overhead.

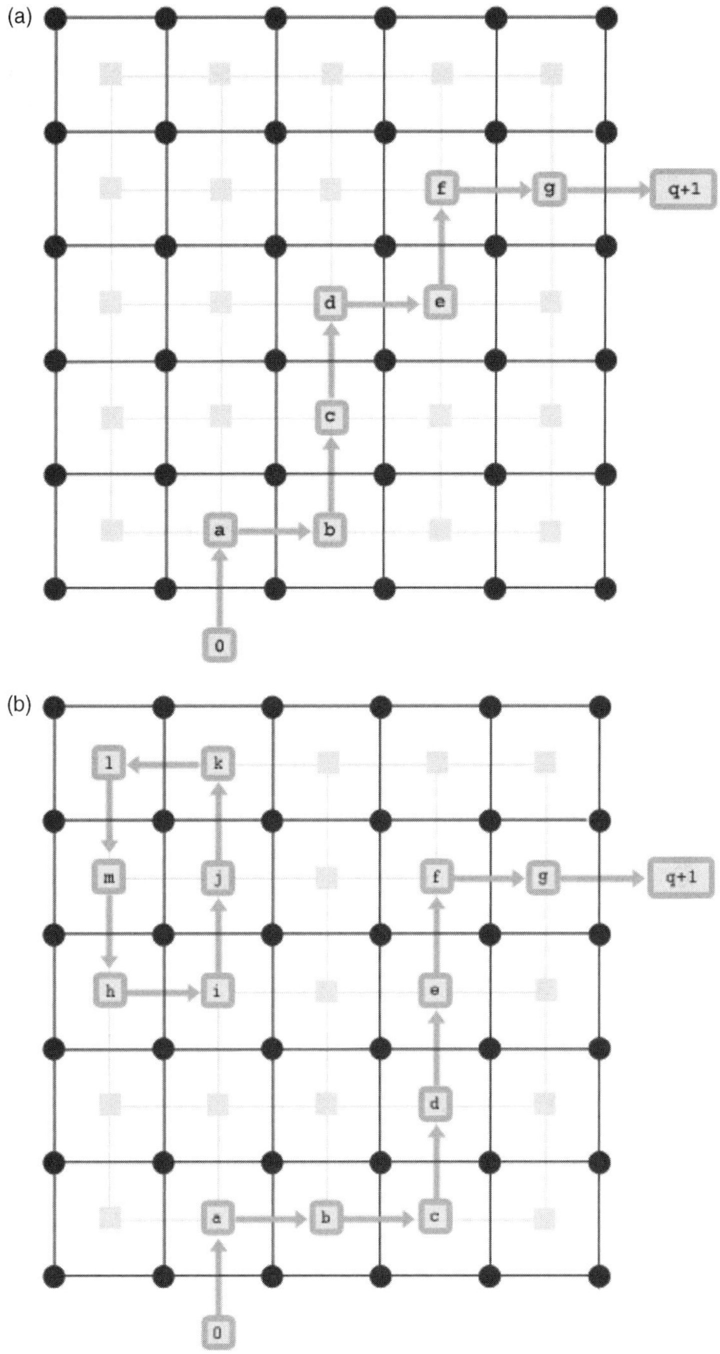

Figure 10.3. Sink optimum routes produced by constraints (10.4) to (10.7). (a) A sink route. (b) A disjoint cycle in the sink route.

Figure 10.4. Two adjacent physical sites, $n + m$ logical sites and their interconnections.

The power consumption rate c_{ik} of each node $i \in N$ when the sink sojourns at site $k \in S$ can be computed in a different way (e.g., analytically [54]), or provided as input to the model from simulations or from real-data traffic traces [9, 10]. In other words, the model can be customized to find the optimum lifetime for different routing protocols (by computing the corresponding values of b_{ik} and c_{ik}).

The model does not allow the sink to pass twice for the same site. However, this can be made possible by having a single "physical" represented by h "logical" sites, where h is the number of times we want the sink to be able to pass through that site. The logical sites have no arcs between them and are connected to all the (logical) sites of adjacent (physical) sites. Figure 10.4 concerns the case of two adjacent physical sites and their logical sites. With this simple modification, optimal lifetime is obtained where the sink is allowed to visit each site at most h times.

The MILP formulation is an improvement over previously proposed models. The model is independent of a number of factors such as sensor node deployment and sensor density, the sink site topology, the size and shape of the geographic area of deployment, and the sensor node technical features such as transmission radius, energy model, and so on. The formulation also includes a number of realistic constraints, such as the noninstantaneous movement of sink between sites potentially far apart from each other. Most importantly, and differently from all previously proposed LP solutions, this formulation explicitly considers costs for changing location.

The Greedy Maximum Residual Energy (GMRE) Protocol. The solution presented above determines the movements and the sojourn times of the mobile sink at different sites so that network lifetime is maximized. Movements and times are determined by solving the described analytical model by providing as input a host of information concerning the *whole* networks. In other words, this solution is *centralized*: Information is collected at a "solver" and the resulting output is the best route for the sink.

Collecting information about network condition can be overwhelmingly expensive in terms of energy and time, and most of the times it is unfeasible in resource constrained networks like WSNs. Therefore, protocols for controlled mobility have to be designed and deployed that can realistically be deployed in WSNs. The optimality is traded off for feasibility, as often happens. This motivates the definition of the following heuristic.

In the Greedy Maximum Residual Energy (GMRE) protocol [9, 10] the sink periodically moves to a new site. More specifically, every t_{min}, it decides whether to move or to stay at the current site. If it moves, the sink selects the site within d_{max} from its current position surrounded by nodes that have the most energy left. The choice is performed greedily—that is, moving toward the site that at the current time appears the best to move to. In time, this should highly likely result in balanced energy consumption at the network nodes and therefore result in longer network lifetime. For deciding whether to move or not, the sink collects information about the residual energy at the nodes around each of the potential future sites (we call this energy value the residual energy at the site) and compares it with the residual energy at the current site. If there are adjacent sites with a residual energy higher than that at the current site, the sink moves to the site with the highest residual energy (selecting randomly among sites with the same residual energy in case of ties). Otherwise the sink stays at the current location. The communication to the sink of the energy level at other network locations proceeds in two phases.

First Phase. For each of the adjacent sites the sink identifies one *sentinel* sensor node that measures and reports the residual energy at the site when requested by the sink. The second phase concerns the sink inquiring the sentinels. This happens every time the sink has to decide to move. To implement the first phase, we take advantage of the flooding performed by the sink when it makes the nodes aware of its new location. For this heuristic protocol we assume that a node that is in the "transmission vicinity" of a site (i.e., whose Euclidean distance from a site is less than or equal to the nodes transmission range) is aware of that. This can be obtained by endowing the nodes with a suitable localization mechanism (such as one of those described in references 63 and 64). The flooding message contains the coordinates of the current location of the sink. Upon receiving this packet, a node knows if it is in the vicinity of a possible future sink site. If this is the case, it sends to the sink a (small) packet to candidate itself as sentinel. When the sink receives these packets, it decides which is the sentinel for a given site. An example of the sentinel identification mechanism is depicted in Figure 10.5. The sink current site E is indicated by a triangle. The squares A, B, C, D, F, G, H, and I identify the adjacent sink sites. The figure shows what happens upon flooding the route construction message. Potential sentinels for sink site G are marked as black circles. Their distance from G is less than or equal to the node transmission range. When such potential sentinels receive the flooded message, they answer back to the sink, identifying themselves as candidates. It is the sink that selects the (single) sentinel for a given possible future site.

Second Phase. At the time the sink has to decide whether to move or not, it interrogates the selected sentinels about the residual energy at their sites by sending them a (small) packet. At this time the sentinels query their neighboring sensor nodes about their residual energy and communicate back to the sink the minimum of the obtained values or any suitable function that can express how critical (for the network lifetime) it is to place the sink in that area. This information enables the sink to select the next site, depending on the residual energy of its area.

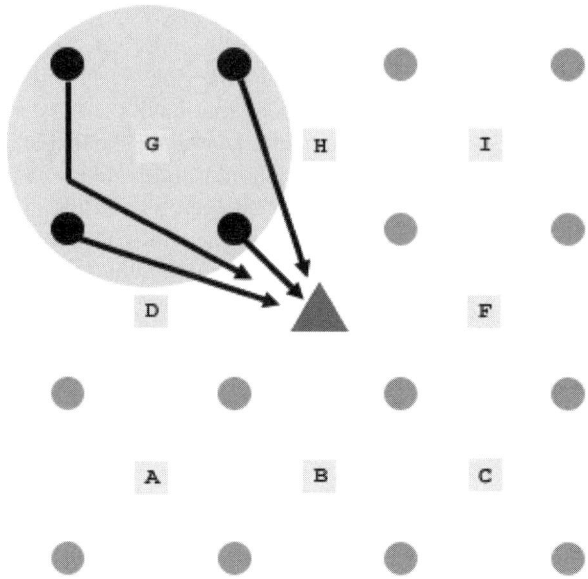

Figure 10.5. Sentinels at adjacent sites.

10.5.3 Multihop Routing and Uncontrolled Mobility

The mathematical model and the distributed heuristic (GMRE) are clear examples of controlled mobility: The sink moves, or attempts to move, toward network locations where nodal energy is available.

For the sake of benchmarking, it is interesting also to compare controlled mobility—and especially the more realistic, distributed GMRE—with a simple protocol for sink mobility that captures the case of uncontrolled, *random mobility* (RM) of the sink à la data MULEs. The difference here is that routing from the sensors to the sink is multihop. The heuristic is extremely simple: Every t_{min} the sink selects randomly and uniformly the new location among all the sites within distance d_{max} from the current. In case a site different from the current is selected, the sink moves to that site. We observe that even in this case, where the sink does not necessarily move to areas with high residual energy, improvements in network lifetime are obtained with respect to the case with an optimally placed static sink.

There are cases where the mobility, although neither random nor controlled by the network, can be exploited to improve network lifetime. Luo and Hubaux [56] successfully demonstrate this by simply moving the sink around the deployment area. Specifically, by searching the space of all possible sink mobility strategies that are periodic (recurrent sink movements within a constant period), the authors claim that a good mobility strategy is obtained by having the sink traveling at the periphery of the network. While controlled, the sink movements are not guided by the network situation/activities.

Inspired by the solution by Luo and Hubaux, we devise a mobility strategy, termed Perimeter Mobility (PM). According to PM, the sink visits cyclically the sites along the perimeter of the (square) deployment area. It stays at each site for t_{min} and then moves onto the next site in a counterclockwise fashion.

It makes sense to compare this different cases of controlled and uncontrolled mobility, evaluating which solution performs more vigorously with respect to selected metrics. This is the natural next step: After having discussed different types of mobility and the corresponding algorithmic solutions (models and protocols), we proceed at selecting scenarios and metrics for comparing them.

10.6 COMPARATIVE EVALUATION

A fundamental part of the creation and investigation of solutions for networking consists in testing the new solutions (protocols, methods, techniques, algorithms) and comparing them. We describe here the way this is usually done, giving some hints and insights on the process of evaluating the performance of a WSN.

Since experiments on real testbeds are costly and usually scale to a few tens of nodes, it is common to (start to) evaluate a protocol for WSNs via simulations. In the attempt to consider the different layers of a WSN architecture, we use the VINT project network simulator ns2 [65]. This widely used software tool allows us to consider some physical layer properties and a MAC protocol CSMA/CS-like which are typical of wireless sensor networking.

Concerning the protocols for mobility presented here, the first question one is lead to investigate is about the properties of GMRE and of data MULEs mobility in a realistic, typical WSN scenario (static sensor nodes). The comparison should be about properties corresponding to relevant metrics. These include node energy consumption, network lifetime, end-to-end packet latency, and probability of successfully delivering the monitored information to the sink.

The next natural step is that of investigating in deeper details the data MULE architecture and the case of mobility with multihop routing (GMRE, RM, and PM) with the aim of identifying the impact of varying scenarios and protocol parameters on their performance.

Overall, the purpose of the following sections is that of introducing the reader to a deeper understanding of the performance trends when considering different application scenarios.

10.6.1 The Choice of a Suitable Scenario

It is meaningful to choose a scenario that resembles a realistic deployment of a WSN with static nodes. Therefore, parameters like the number or network nodes, the size of the deployment area, the nodal transmission radius and initial energy, and so on, should be chosen so that what is depicted is substantially similar to a scenario one could find in real life. Here is a suitable example: 400 sensor nodes are deployed on a 20×20 regular grid covering a squared area of side $L = 400$ m. Sensor nodes have a transmission radius equal to 30 m, which imposes a maximum of 8 neighbors per node. Nodes periodically generate packets at a rate of 0.5 bps (low traffic scenario, typical of sensor networking). Packet size is equal to 512 B. The buffer size of the sensor nodes varies among 10, 50, and 100 packets. Energy consumption follows the TR1000 specifications [66]. An ideal awake/asleep schedule is assumed in which nodes consume energy only when transmitting or receiving packets. The channel is ideal (no errors occurs when transmitting) and it has a data rate equal to 250 Kbps. In the case of the data MULE architecture, the number of available MULEs is an important parameter: 1, 2, and 4 data MULEs can be a good starting point. In the case of data MULEs the deployment area is divided into 81 cells. The MULEs move from the center of one cell (source) to the center of another one (destination) according to the random waypoint model [27]. The MULE travels at a speed of 1 m/s. While traveling from the source to the destination cell, a MULE stops at intermediate cells for data collection, gathering packets from all (and only) the sensors in a cell. The sojourn time spent in each cell is 1 s. MULE queues are considered unbounded (in fact, they can contain 1000 packets, which ensure no overflow, i.e., data loss). The sink to which the MULEs deliver the collected packets is (statically) placed at the center of the deployment area. Once at the sink, a MULE stays in its proximity for a time necessary to empty its queue.

Meaningful parameters for GMRE and the other heuristics are as follows. The sink moves among 8×8 sink sites (grid deployment) at the speed of 1 m/s. A shortest path-like routing is used to deliver data in a multihop fashion from the sensor to the current site of the sink. The parameter t_{min} is set to 50 K (best-performing value among the many tested). The parameter d_{max} is equal to 190 m.

In all the experiments, every node generates over 3000 packets during the simulation time. Finally, in order to obtain statistically meaningful results, the same experiment (simulation run) should be performed on different networks for an enough number of times. That is why results are obtained by averaging over experimental outcomes from 100 different network topologies.

Table 10.1 gives a bird's-eye view of the results of the comparison between GMRE and the data MULE solutions.

When the network traffic is low and the network deployment area size is limited (as considered here), we observed that both GMRE and the data MULEs deliver all

TABLE 10.1. GMRE and Data MULEs, General Comparison

	GMRE	Data MULEs
Data latency	•	•••••
Energy consumption	•••	•
Packet delivery ratio	•••••	•••
Computational needs	•••	•

Low •
High •••••

generated packets to the sink. As a rule of thumb, successful packet delivery for low/medium network traffic is more challenging for the MULE solution than for the multihop approach (GMRE). The packet delivery ratio degrades as the deployment area grows in size, since the inter-arrival time of a MULE at a cell grows and overflow of sensor node queues can. On the other hand, GMRE is always able to deliver packets to the sink.

The advantage of using solution data MULEs-like is clear when energy consumption and sensor node computational capability are considered. Given the single-hop nature of data exchange between the sensors and the MULEs, nodes are not required to implement a full protocol stack. Basic physical and MAC layer functions are sufficient for all data communications. This implies lower nodal energy consumption and lower node complexity and cost. In the scenario described above, sensor nodes in a data MULE setting consume one order of magnitude less energy than in the multihop scenario. However, the gain in energy conservation is heavily paid in terms of end-to-end data latency. The difference in this case is as high as four orders of magnitude! This is due to (a) the extremely long time it takes for a MULE to visit the same cell twice and (b) the time needed to go back to the sink to deliver the packets. These two factors are of course dependent on the number of MULEs, their speed, and the size of the deployment area. We observed that varying MULEs speed from pedestrian (as shown here) to slow vehicular speed would not lead to improvements of more than one order of magnitude. Only introducing a high number of data MULEs would satisfy the latency requirements of many WSN applications. This is sometimes impossible and is certainly costly.

10.6.2 Experiments for Data MULEs

The kind of experiments that are meaningful to perform in the data MULE architecture mainly concern the average number of packet discarded, and hence lost, because of buffer overflow and the sensor-to-sink packet data latency. It is interesting to notice that in order to interpret correctly the experimental results on these fundamental metrics, it is sometimes necessary to investigate other metrics and patterns, such as (a) the inter-arrival times of the MULEs at the different cells and (b) the average sojourn time of a MULE in a cell.

Here is an example of experimental investigation in the basic scenario depicted above. We observed that the average number of packets discarded by the sensor nodes in the case of one MULE roaming among sensors with buffer size set to 10 (packets) is not high at all, especially for nodes that are toward the center of the deployment area. There are a few nodes, however, that, given the presence of a single MULE and small buffers, have to drop some packets. In this case the problem is the movement of the MULE itself, given that the random waypoint mobility leads the MULE toward the center of the deployment area with high probability. The longer at the center, the longer the MULE inter-arrival times at the cells along the perimeter. This is the reason for the observed 4.5% packet loss at the nodes in the corners.

This behavior is confirmed by an investigation of the inter-arrival times of 1, 2, and 4 MULEs at the same cell. In all the considered scenarios the inter-arrival times values are extremely high (in the thousands of seconds or more), explaining why the data MULEs approach results in high packet latency. As the number of MULEs increases, it takes less and less time before a MULE arrives at a given cell, as expected. At the same time, given that the MULEs tend more to visit central cells than peripheral ones, the inter-arrival times are higher at the area borders.

MULEs visit each cell for the enforced 1 s. The only cell that experiences a longer sojourn time is the one where the sink is located. The higher the traffic and the area size and the lower the number of MULEs, the larger the amount of data stored in a MULE queue, which imposes longer times to unload the data to the sink. Therefore, the time needed for data transfer may exceed 1 s.

The study of these last two metrics makes us understand the results about what it is perhaps the most important metric for assessing whether or not the data MULE architecture is suitable for certain applications: data latency. Measures on sensor-to-sink latency should take into account the latency incurred by data packets from the sensor to the sink, in a MULE queue and in a (source) node queue. While it has an impact on the number of packets that are dropped, the size of a node queue has little influence on the packet latency. Packets are transferred from the source queue (node or MULE) to the destination queue (MULE or sink) all at once, and the transfer time is quite small. As a consequence, the end-to-end delay is given almost entirely by the time from when the packet is generated and a MULE arrives at the cell, plus the time it takes to the MULE to deliver it to the sink. The time a packet sits on its node queue depends on the number of MULEs, as expected. When the number of MULEs doubles, this time halves. We observe that this component is the most significant one. Given that the sink is placed at the center, the inter-arrival time of a MULE at the sink is on average shorter than that at a (noncentral) cell. The time spent by a packet on a MULE queue is basically independent of the number of MULEs. As a consequence, the end-to-end packet latency *almost* halves with doubling the MULEs, since only one of the delay component halves. In any case, and independently of the number of MULEs, the average latency of a packet is in the *thousands* of seconds. As expected, this is quite high.

The trends observed in the basic scenario are also confirmed in a setting where while the nodes are as dense as in the case with $L = 400$ m, the deployment area is larger ($L = 572$ m).

10.6.3 Experiments on GMRE

As outlined in Section 10.6.1, controlled sink mobility and multihop data communication lead to higher energy consumption than in the data MULE case. However, this is the solution for satisfying more stringent latency constraints. What we show here is one possible, meaningful choice for evaluating a distributed heuristic like GMRE. In particular, we explore GMRE to explain the impact of different protocol parameters on its performance and to assess the effectiveness of an energy-aware sink mobility heuristic over other sink mobility schemes. It makes sense to compare GMRE to the following schemes.

- Static sink placement (termed STATIC in the following). In this case the sink is optimally placed at the center of the deployment area.
- The sink mobility as generated by solving the MILP model presented in Section 10.5.2 (OPT).
- RM, the mobility heuristic outlined in Section 10.5.3, which is similar to the scenario with a randomly moving MULE, but with multihop routing.
- The scheme that we term PM (for Perimeter Mobility), according to which the sink moves cyclically along the perimeter of the deployment area (Section 10.5.3).

Experiments should be performed in different settings. For instance, one should vary the transmission radius (from 25 m to 30 m), the data routing protocol (shortest path-like versus geographic), and the way nodes are deployed (grid positioning versus random scattering). The results obtained are consistent with what we describe here (which refers to the basic scenario with transmission ranges of 30 m and 25 m). The conclusion is that GMRE is a simple but effective solution, achieving a network lifetime close to OPT's—that is, the best that can be obtained by exploiting sink mobility in a multihop WSN. RM and PM have similar performance in terms of network lifetime with PM outperforming RM at higher t_{min} values. Both schemes improve considerably over STATIC, which advocates for the use of mobility in WSNs. However, the lack of energy awareness does not allow us to match the results seen with GMRE (see references 10 and 59 for further details).

Table 10.2 shows the average network lifetimes achieved by STATIC, RM, PM, GMRE, and OPT in networks whose nodes have transmission ranges of 25 m and 30 m. (The network lifetime has been defined as the time until the first node dies because of energy depletion.)

Improvements with respect to the static case can be as high as 350% when the sink moves according to GMRE in scenarios with 64 sink sites. In this case the GMRE lifetime is never more than 22% (25%) shorter than the OPT lifetime when the transmission range is 25 m (30 m) and t_{min} is kept below 250 K. Improvements on network lifetime are obtained even when the sink moves randomly (RM heuristic) or

TABLE 10.2. Average Network Lifetime for Different Transmission Ranges (seconds $\times 10^6$)

	Radius 25 m				Radius 30 m			
t_{min}	50 K	250 K	500 K	1 M	50 K	250 K	500 K	1 M
OPT	36.2	36	35.9	33.43	41.2	41	40.8	39.3
GMRE	30.9	28.79	24.4	21.9	31.9	31	29	27.9
RM	23.18	20.44	17.6	13.08	28.9	25.7	21.9	16.4
PM	21.9	20.9	19.9	16.17	28.7	26.1	24	19.67
STATIC	7	7	7	7	6.7	6.7	6.7	6.7

when the sink moves along the area perimeter (PM heuristic). The latter experiences performances only slightly better than RM when the transmission range is 25 m. RM leads to a network lifetime halfway between the STATIC lifetime and that obtained by OPT mobility.

In general, for GMRE, RM, and PM we notice that the lower the t_{min} value, the higher the network lifetime. At higher t_{min} values, the sink cannot finely decide the time to spend at each site, being stuck at a site for a longer time. This implies a less balanced energy consumption at the nodes. When t_{min} is very high, it is not even possible for the sink to sojourn at all network sites, since the lifetime is reached before the sink can visit them all.

Even in the case of OPT mobility, lower t_{min} values correspond to longer lifetimes. In this case, however, the decrease in the network lifetime corresponding to higher t_{min} values is not as evident as for GMRE and RM. At this point, one becomes curious about why this happens. First of all, in the OPT case, t_{min} is simply a lower bound on the sojourn time: The sink has to stay at a site for that time, but does not have to move after it. It can stay an (optimum) amount of time after t_{min} and then move. In case of GMRE, RM, and PM sink mobility, the decision about whether to move or not is due every t_{min} (after which the sink often moves). This has a twofold consequence. On one side, increasing t_{min} is more restrictive for GMRE than for OPT in terms of fine-tuning the sojourn time at a site. Furthermore, being able to decide optimum sojourn times implies much lower sink mobility (OPT case), which corresponds to lower overhead for route management and hence to lower energy consumption with respect to that incurred by GMRE, RM, and PM. Secondly, GMRE, RM, and PM do not have a global view of the network topology and do not know the network traffic—that is, how the nodes energy consumption evolves over time. This induces decreased performance with respect to OPT. The RM and PM heuristics do not enforce any energy-based criterion for sink movement, resulting in the worst performance among all the schemes with mobile sink we investigated. Even if GMRE takes into account nodes residual energy, the decision about whether to move and where is based on the current status of the network and on a local view of the residual energy. According to the best "greedy" tradition, this could lead to a bad move with respect to global network lifetime maximization. The impact of such a bad move is clearly higher for high t_{min} values: The wrong toll is paid for a longer time.

The OPT performance also degrades for higher values of t_{min}. In this case it converges to values that are typical of when the sink is kept static. For instance, for values of t_{min} approaching 7.013.801s the MILP model positions the sink at one of the sites in the center of the deployment area and leaves it there (static). However, for what explained above, increasing values of t_{min} are less critical in the case of OPT mobility than in the case of GMRE, RM, and PM. OPT network lifetime values start to decrease steeply at very high t_{min}values. Considering that OPT needs global information for deriving optimum sink mobility and sojourn times, and considering also the more "philosophical" algorithmic differences between OPT and GMRE, the fact that the heuristic never obtains network lifetimes more than 25% lower than OPTs, for relatively low sink mobility rate (low t_{min} values), can really be considered an excellent result.

Now, onto the flip side of this mobile story. Increases in network lifetime due to the sink mobility are paid in term of increased data packet latency. The reasons are quite clear. First of all, packets that are generated while the sink is moving, as well as those in transit toward the sink, have to wait until routes to the new position of the sink are established. Furthermore, in order to balance energy depletion, the sink will spend time not only at the center of the deployment area but also along the borders. This imposes longer average route length and hence a higher packet latency than the latency experienced when the sink is statically placed at the center. (The latter is actually the dominant reason for increased latency in low sink mobility scenarios.) In any case, the average packet latency is *always* (i.e., for all the solution investigated) below half a second! This is far beyond the measures of latency observed in the data MULEs case, which were in the thousands of seconds (independently of the number of MULEs). The improvement is four/five orders of magnitude!

Table 10.3 reports the average data latency incurred by packets in OPT, GMRE, RM, PM, and STATIC for t_{min} =50,000 s and 1,000,000 s. (Intermediate values are consistent with these ones.) Although small and always below .5 s, there are differences among the data latencies of the four schemes that we have investigated and compared. It is interesting to see why. When the sink stays on the perimeter or at the corner sites (as in GMRE, OPT, and PM), we know that lifetime increases. However, these are also the cases when the average length of the routes to the sink increases, which implies, in turn, a (slightly) higher packet latency. It is thus reasonable to expect that a packet experiences lower latency when the sink is statically placed at the center of the sensor deployment area. The RM heuristic, which tends to move the sink to sites located centrally, is the second best mobility scheme in terms of latency. As t_{min} increases, the sink tends to stay less at central sites, leading to higher average latencies experienced by RM packets. The opposite trend is observed for GMRE. For low t_{min} values, the sink stays at sites on the corners and on the perimeter, which leads to and average packet latency up to 30% higher than those experienced in the RM case. When t_{min} increases, the sink sojourns less at corner sites, which implies a decrease of the average latency.

Although the above discussion allows us to understand the subtle differences between the performance of the investigated schemes, there is very little variance among

TABLE 10.3. Average Packet Latency (seconds)

	8 × 8 Sink Sites	
t_{min}	50 K	1 M
OPT	.315	.315
GMRE	.32	.29
RM	.25	.26
PM	.30	.30
STATIC	.19	.19

the measured latencies. All values range between .19 s (STATIC) and .32 s (GMRE and OPT).

As a result of this simple investigation, which the reader is encouraged to reproduce and vary, we obtain that GMRE shows a good tradeoff between network lifetime gains and latency increases, promising to be a suitable solution for those application scenarios that cannot be covered by the data MULEs mechanism.

10.7 SOME OPEN PROBLEMS AND RESEARCH DIRECTIONS

Research on *mobile* WSNs is quite interesting and recent. Mobility in these networks can be tackled from several points of view. Many research directions could be taken. For instance, as demonstrated in the case of OPT and GMRE, having mathematical models is quite an important benchmark. Although centralized, and also often computationally tough, an optimal solution is quite useful for giving an idea of what can be obtained by designing and deploying a more realistic distributed heuristic. While progresses are being done in the multihop routing/mobile sink realm, this is quite uncharted territory in the case of the data MULEs approach.

In the initial approach to controlling the mobility of the MULEs, as described here (Section 10.5.1), it is pointed out that when not enough MULEs are available, some nodes might not fall close enough to a MULE route. A proposed solution is that of introducing multihop routes (organized as trees) to nodes that can directly communicate with a MULE when it passes. This approach could be studied in more detail and improvement can be obtained by determining, for instance, the depth of the tree that best keeps energy consumption low while allowing packets to get to the MULE—that is, without discarding them. One can also think of dividing the data packets into priority classes, so that the MULE trajectory is controlled by the network in such a way that the MULE goes first where high-priority data are to be collected.

10.8 CONCLUSIONS

This chapter explores ways for using the mobility of network components in large networks of resource constrained devices, like wireless sensor networks (WSNs). The aim is that of exploiting mobility for improving the performance of these networks in terms of network lifetime, throughput, and connectivity without significantly affecting data routing and end-to-end latency. Whereas energy-efficient solutions have been proposed for WSNs with statically placed sensors and one or more static data collection points (the *sinks*), our investigation delves into the various ways the sensor nodes, some mobile relays, or the sinks can be made mobile for improving network performance. We review the most recent proposals for mobility use and management in WSNs, considering the pros and cons, as well as the costs and tradeoffs, of using mobile sensors, mobile relays, and moving sinks. In the final part of the chapter we give examples of how, once we have selected typical solutions for dealing with the mobility of sinks and relays, these solutions are compared. The aim is to (a) introduce the reader to the selection of suitable scenarios and performance metric and (b) illustrate a general approach to "reading" experimental results. In particular, we illustrate controlled and uncontrolled mobility both in the case of networks with single-hop routing of the packets to a passing mobile relay (data MULEs architecture) and in the case of controlled and uncontrolled mobility of a sink roaming through the network and having the data delivered to it in a multihop fashion. The two different network architectures, with single-hop versus multihop data routing, are described in details and compared. Our comparison concerns key metrics of interest for WSNs, which include network lifetime and sensor-to-sink data latency. Besides providing evidence of the advantages of using mobility, the chapter also provides insights on different techniques for protocol design and performance evaluation and comparison.

10.9 EXERCISES

1. **Review questions**. (a) Why is it less convenient to use resource-constrained mobile sensor nodes rather than resource-rich mobile elements like relays and sinks? (b) Give an example that clearly motivates the use of mobility in WSNs. (c) What metric is most affected by the use of mobile relays rather than mobile sinks? (d) For what kind of application it is clearly more advisable to use the data MULE approach instead of a multihop solution with mobile sink? (e) List five metrics of interest that should be evaluated when dealing with mobile WSNs. (f) Describe pros and cons of developing a mathematical formulation of a problem for WSNs like the one described in Section 10.5.2. (g) What are the major differences, in terms of mobility, when comparing GMRE, RM, and PM? (h) If the MULEs move according to the random waypoint mobility model, where are they more likely to be found? What does this imply in terms of network performance? (That is, which metric is the most affected in this case?)
2. Consider the MILP model presented in Section 10.5.2. What are the problems that would occur if constraints (10.8) are removed? Why is this removal a problem?
3. Consider the MILP model presented in Section 10.5.2. Imagine that instead of knowing the sink sites, several different *routes* for the sink are available. Rewrite the model in such a way that the output is the sequence of routes that maximize network lifetime.
4. Consider again the MILP model presented in Section 10.5.2. Rewrite the model so that the sink can pass for the same site at most five times.
5. Consider the pattern of energy consumption depicted in Figure 10.6 (which refers to the basic scenario described in Section 10.6.1 with 8×8 sink sites). The darker the area, the higher the energy consumption. Given this pattern, describe where an heuristic will more likely send the sink (i.e., to which sites). Justify your answer.
6. What is the main difference between the data MULE approach and the PM scheme described in Section 10.5.3?

(a) (b) (c)

Figure 10.6. Average nodal energy consumption when the sink sojourns at different sites. (a) Corner site. (b) Perimeter site. (c) Central site.

ACKNOWLEDGMENTS

This work was supported in part by the European FP6 027227 IP Project "E-Sense (Capturing Ambient Intelligence for Mobile Communications through Wireless Sensor Networks) and by NSF grant CNS-0738720."

BIBLIOGRAPHY

1. S. Basagni, M. Conti, S. Giordano, and I. Stojmenovic, editors. *Mobile Ad Hoc Networking*. IEEE Press and John Wiley & Sons, Piscataway, NJ, and Hoboken, NJ, 2004.
2. I. F. Akyildiz, W. Su, Y. Sankarasubramaniam, and E. Cayirci. Wireless sensor networks: A survey. *Computer Networks*, vol. **38**(4):393–422, 2002.
3. E. H. Callaway, Jr. *Wireless Sensor Networks: Architectures and Protocols*. Auerbach Publications, Boca Raton, FL, 2003.
4. I. F. Akyildiz, D. Pompili, and T. Melodia. Underwater acoustic sensor networks: Research challenges. *Elsevier's Journal of Ad Hoc Networks*, **3**(3):257–279, 2005.
5. G. Kantor, S. Singh, R. Peterson, D. Rus, A. Das, V. Kumar, G. Pereira, and J. Spletzer. Distributed search and rescue with robot and sensor team. In *Proceedings of the 4th International Conference on Field and Service Robotics, FSR 2003*, Lake Yamanaka, Japan, July 2003, pp. 327–332.
6. K. Kotay, R. Peterson, and D. Rus. Experiments with robots and sensor networks for mapping and navigation. In *Proceedings of the International Conference on Field and Service Robotics, FSR 2005*, Port Douglas, Australia, July 29–31, 2005.
7. R. Brooks, P. Ramanathan, and A. Sayeed. Distributed target classification and tracking in sensor networks. *Proceedings of the IEEE*, **91**(8):1163–1171, 2003.
8. G. Virone, A. Wood, L. Selavo, Q. Cao, L. Fang, T. Doan, Z. He, R. Stoleru, S. Lin, and J. A. Stankovic. An assisted living oriented information system based on a residential wireless sensor network. In *Proceedings of the 1st Distributed Diagnosis and Home Healthcare (D2H2) Conference*, Arlington, VA, April 2–4, 2006, pp. 95–100.

9. S. Basagni, A. Carosi, E. Melachrinoudis, C. Petrioli, and M. Z. Wang. A new MILP formulation and distributed protocols for wireless sensor networks lifetime maximization. In *Proceedings of the IEEE International Conference on Communications, ICC 2006*, Vol. 8, Istanbul, Turkey, June 11–15 2006, pp. 3517–3524.
10. S. Basagni, A. Carosi, E. Melachrinoudis, C. Petrioli, and M. Z. Wang. Controlled sink mobility for prolonging wireless sensor networks lifetime. *ACM/Elsevier Wireless Networks*, 2008.
11. P. Juang, H. Oki, Y. Wang, M. Martonosi, L.-S. Peh, and D. Rubenstein. Energy-efficient computing for wildlife tracking: Design tradeoffs and early experiences with zebranet. In *Proceedings of the 10th International Conference on Architectural Support for Programming Languages and Operating Systems, ASPLOS-X*, San Jose, CA, October 5–9, 2002, pp. 96–107.
12. Q. Li and D. Rus. Sending messages to mobile users in disconnected ad-hoc wireless networks. In *Proceedings of 6th ACM Annual International Conference on Mobile Computing and Networking, MobiCom 2000*, Boston, MA, August 6–11 2000, pp. 44–55.
13. M. Grossglauser and D. N. C. Tse. Mobility increases the capacity of ad-hoc wireless networks. *IEEE/ACM Transactions on Networking*, **10**(4):477–486, 2002.
14. W. Zhao, M. H. Ammar, and E. W. Zegura. A message ferrying approach for data delivery in sparse mobile ad hoc networks. In *Proceedings of the 5th ACM International Symposium on Mobile Ad Hoc Networking and Computing, MobiHoc 2004*, Roppongi Hills, Tokyo, Japan, May 24–26, 2004, pp. 187–198.
15. S. Jain, R. C. Shah, W. Brunette, G. Borriello, and S. Roy. Exploiting mobility for energy-efficient data collection in sensor networks. In *Proceedings of the IEEE Workshop on Modeling and Optimization in Mobile, Ad Hoc and Wireless Networks, WiOpt 2004*, Cambridge, UK, March 24–26, 2004.
16. W. Zhao, M. H. Ammar, and E. W. Zegura. Controlling the mobility of multiple data transport ferries in a delay-tolerant network. In *Proceedings of IEEE Infocom 2005*, Vol. 2, Miami, FL, March 13–17, 2005, pp. 1407–1418.
17. H. Jun, W. Zhao, M. H. Ammar, E. W. Zegura, and C. Lee. Trading latency for energy in wireless ad hoc networks using message ferrying. In *Proceedings of the 3rd IEEE International Conference on Pervasive Computing and Communications, PerCom 2005*, Kawai, HI, March 8–12, 2005, pp. 220–225.
18. M. Mukarram Bin Tariq, M. H. Ammar, and E. W. Zegura. Message ferry route design for sparse ad hoc networks with mobile nodes. In *Proceedings of the 7th ACM International Symposium on Mobile Ad Hoc Networking and Computing, MobiHoc 2006*, Firenze, Italy, May 22–25, 2006, pp. 37–48.
19. T. Spyropoulos, K. Psounis, and C. S. Raghavendra. Performance analysis of mobility-assisted routing. In *Proceedings of the 7th ACM International Symposium on Mobile Ad Hoc Networking and Computing, MobiHoc 2006*, Firenze, Italy, May 22–25, 2006, pp. 49–60.
20. D. K. Goldenberg, J. Lin, A. S. Morse, B. E. Rosen, and Y. R. Yang. Towards mobility as a network control primitive. In *Proceedings of the 5th ACM International Symposium on Mobile Ad Hoc Networking and Computing, MobiHoc 2004*, Roppongi Hills, Tokyo, Japan, May 24–26, 2004, pp. 163–174.
21. G. Wang, G. Cao, and T. F. La Porta. Movement-assisted sensor deployment. In *Proceedings of IEEE INFOCOM 2004*, Vol. 4, Hong Kong, March 7–11, 2004, pp. 2469–2479.

22. G. Wang, G. Cao, and T. F. La Porta. Proxy-based sensor deployment for mobile sensor networks. In *Proceedings of the First IEEE International Conference on Mobile Ad Hoc and Sensor Systems, MASS 2004*, Fort Lauderdale, FL, October 25–27, 2004, pp. 493–502.
23. A. Howard, M. J. Matarić, and G. S. Sukhatme. An incremental deployment algorithm for mobile sensor networks. *Kluwer Autonomous Robots, Special Issue on Intelligent Embedded Systems*, G. S. Sukhatme, editor, Vol. 13, no. 2, pp. 113–126, September 2002.
24. A. Howard, M. J. Matarić, and G. S. Sukhatme. Mobile sensor networks deployment using potential fields: A distributed, scalable solution to the area coverage problem. In *Proceedings of the 6th International Symposium on Distributed Autonomous Robotic Systems, DARS 2002*, Fukuoka, Japan, June 25–27, 2002.
25. R. Rao and G. Kesidis. Purposeful mobility for relaying and surveillance in mobile ad hoc sensor networks. *IEEE Transactions on Mobile Computing*, **3**(3):225–232, 2004.
26. I. Chatzigiannakis and S. Nikoletseas. An adaptive compulsory protocol for basic communications in highly changing ad-hoc mobile networks. In *Proceedings of the International Parallel and Distributed Processing Symposium, IPDPS 2002*, Fort Lauderdale, FL, April 15–19 2002, pp. 193–202.
27. R. C. Shah, S. Roy, S. Jain, and W. Brunette. Data MULEs: Modeling a three-tier architecture for sparse sensor networks. In *Proceedings of the First IEEE International Workshop on Sensor Network Protocols and Applications, SNPA 2003*, Anchorage, AK, May 11, 2003, pp. 30–41.
28. T. Small and Z. J. Haas. Resource and performance tradeoffs in delay-tolerant wireless networks. In *Proceedings of the ACM SIGCOMM 2005 Workshop on Delay-Tolerant Networking and Related Networks, WDTN 2005*, Philadelphia, PA, August 22–26, 2005.
29. L. Song and D. Hatzinakos. Dense wireless sensor networks with mobile sinks. In *Proceedings of the IEEE International Conference on Acoustic Speech, and Signal Processing, ICASSP 2005*, Vol. 3, Philadelphia, PA, March 18–23, 2005, pp. 677–680.
30. Y. Tirta, Z. Li, Y.-H. Lu, and S. Bagchi. Efficient collection of sensor data in remote fields using mobile collectors. In *Proceedings of the 13th International Conference on Computer Communications and Networks, ICCCN 2004*, Chicago, IL, October 11–13, 2004, pp. 515–519.
31. Y. Tirta, B. Lau, N. Malhotra, S. Bagchi, Z. Li, and Y.-H. Lu. Controlled mobility for efficient data gathering in sensor networks with passively mobile nodes. In *IEEE Monograph on Sensor Network Operations*, S. Phoha, T. F. La Porta, and C. Griffin, editors, IEEE Press & John Wiley & Sons, Piscataway, NJ, and New York, 2005, 49–58.
32. A. Kansal, A. A. Somasundara, D. D. Jea, M. B. Srivastava, and D. Estrin. Intelligent fluid infrastructure for embedded networks. In *Proceedings of the 2nd ACM/SIGMOBILE International Conference on Mobile Systems, Applications, and Services, MobySys 2004*, Boston, MA, June 6–9, 2004, pp. 111–124.
33. D. D. Jea, A. A. Somasundara, and M. B. Srivastava. Multiple controlled mobile elements (data mules) for data collection in sensor networks. In *Proceedings of the 1st IEEE International Conference on Distributed Computing in Sensor Systems, DCOSS 2005*, LNCS series, V. K. Prasanna, S. Iyengar, P. G. Spirakis, and M. Welsh, editors, Vol. 3560, Marina del Rey, CA, June 30–July 1 2005, pp. 244–257.
34. A. A. Somasundara, A. Ramamoorthy, and M. B. Srivastava. Mobile element scheduling for efficient data collection in wireless sensor networks with dynamic deadlines. In

Proceedings of the 25th IEEE International Real-Time Systems Symposium, RTSS 2004, Lisbon, Portugal, December 5–8 2004, pp. 296–305.

35. A. Kansal, M. Rahimi, D. Estrin, W. J. Kaiser, G. J. Pottie, and M. B. Srivastava. Controlled mobility for sustainable wireless sensor networks. In *Proceedings of the First IEEE Annual Conference on Sensor and Ad Hoc Communications and Networks, SECON 2004*, Santa Clara, CA, October 4–7, 2004, pp. 1–6.
36. A. A. Somasundara, A. Kansal, D. D. Jea, D. Estrin, and M. B. Srivastava. Controllably mobile infrastructure for low energy embedded networks. *IEEE Transactions on Mobile Computing*, **5**(8):958–973, 2006.
37. E. Ekici, Y. Gu, and D. Bozdag. Mobility-based communications in wireless sensor networks. *IEEE Communications Magazine*, vol. **44**(7):56–62, 2006.
38. W. Wang, V. Srinivasan, and K.-C. Chua. Using mobile relays to prolong the lifetime of wireless sensor networks. In *Proceedings of the 11th Annual ACM/SIGMOBILE International Conference on Mobile Computing and Networking, MobiCom 2005*, Cologne, Germany, August 28–September 2, 2005, pp. 270–283.
39. H. S. Kim, T. F. Abdelzaher, and W. H. Kwon. Minimum energy asynchronous dissemination to mobile sinks in wireless sensor networks. In *Proceedings of the First International Conference on Embedded Networked Sensor Systems, SenSys 2003*, Los Angeles, CA, November 5–7, 2003, pp. 193–204.
40. C. Intanagonwiwat, R. Govindan, D. Estrin, J. Heidemann, and F. Silva. Directed diffusion for wireless sensor networking. *IEEE/ACM Transactions on Networking*, **11**(1):2–16, 2003.
41. F. Ye, H. Luo, J. Cheng, S. Lu, and L. Zhang. A two-tier data dissemination model for large scale wireless sensor networks. In *Proceedings of the 8th ACM Annual International Conference on Mobile Computing and Networking, MobiCom 2002*, Atlanta, GA, September 23–28, 2002, pp. 148–159.
42. J. G. Jetcheva and D. B. Johnson. Adaptive demand-driven multicast routing in multi-hop wireless ad hoc networks. In *Proceedings of the 2nd ACM International Symposium on Mobile Ad Hoc Networking & Computing, MobiHoc 2001*, Long Beach, CA, October 4–5, 2001, pp. 33–44.
43. K. Hwang, J. In, Y. Yun, and D. Eom. Dynamic sink oriented tree algorithm for efficient target traking of multiple mobile sink users in wide sensor fields. In *Proceedings of the 60th IEEE Vehicular Technology Conference, VTC 2004-Fall*, Vol. 7, Los Angeles, CA, September 26–29, 2004, pp. 4607–4610.
44. X. Hu, Y. Liu, M. J. Lee, and T. N. Saadawi. A novel route update design for wireless sensor networks. *ACM Mobile Computing and Communications Review*, **8**(1):18–26, 2004.
45. K. Akkaya and M. Younis. Energy-aware to mobile gateway in wireless sensor networks. In *Proceedings of the IEEE Globecom 2004 Workshops*, Dallas, TX, November 29–December 3, 2004, pp. 16–21.
46. H. L. Xuan and S. Lee. A coordination-based data dissemination protocol for wireless sensor networks. In *Proceedings of the 2004 Intelligent Sensors, Sensor Networks and Information Processing Conference, ISS-NIP 2004*, Melbourne, Australia, December 14–17, 2004, pp. 13–18.
47. L. Tong, Q. Zhao, and S. Adireddy. Sensor networks with mobile agents. In *Proceedings of the IEEE Military Communication Conference, MILCOM 2003*, Vol. 1, Boston, MA, October 13–16, 2003, pp. 705–710.

48. P. Venkitasubramaniam, S. Adireddy, and L. Tong. Sensor networks with mobile agents: Optimal random access and coding. *IEEE Journal on Selected Areas in Communications*, **22**(6): 1058–1068, 2004.

49. V. P. Mhatre, C. Rosenberg, D. Kofman, R. Mazumdar, and N. Shroff. A minimum cost heterogeneous sensor network with a lifetime contraint. *IEEE Transactions on Mobile Computing*, **4**(1), 4–15, 2005.

50. A. Chakrabarti, A. Sabharwal, and B. Aazhang. Using predictable observer mobility for power efficient design of sensor networks. In *Proceedings of the Second International Workshop on Information Processing in Sensor Networks, IPSN 2003*, LNCS series, Vol. 2634, F. Zhao and L. Guibas, editors, Palo Alto, CA, April 22–23, 2003, pp. 129–145.

51. P. Venkitasubramaniam, Q. Zhao, and L. Tong, Sensor networks with multiple mobile access points. In *Proceedings of the 38th Annual Conference on Information Sciences and Systems, CISS 2004*, Princeton, NJ, March 17–19, 2002.

52. P. Baruah, R. Urgaonkar, and B. Krishnamachari, Learning-enforced time domain routing to mobile sinks in wireless sensor fields. In *Proceedings of the 29th Annual IEEE International Conference on Local Computer Networks, LCN 2004*, Tampa, FL, November 16–18, 2004, pp. 525–532.

53. S. R. Gandham, M. Dawande, R. Prakash, and S. Venkatesan. Energy efficient schemes for wireless sensor networks with multiple mobile base stations. In *Proceedings of IEEE Globecom 2003*, Vol. 1, San Francisco, CA, December 1–5, 2003, pp. 377–381.

54. Z. M. Wang, S. Basagni, E. Melachrinoudis, and C. Petrioli. Exploiting sink mobility for maximizing sensor networks lifetime. In *Proceedings of the 38th Hawaii International Conference on System Sciences*, Big Island, HI, January 3–6, 2005.

55. I. Papadimitriou and L. Georgiadis. Maximum lifetime routing to mobile sink in wireless sensor networks. In *Proceedings of the 2005 International Conference on Software, Telecommunications and Computer Networks, SoftCOM 2005*, Split, Croatia, September 15–17, 2005.

56. J. Luo and J.-P. Hubaux. Joint mobility and routing for lifetime elongation in wireless sensor networks. In *Proceedings of IEEE Infocom 2005*, Vol. 3, Miami, FL, March 13–17, 2005, pp. 1735–1746.

57. J.-H. Chang and L. Tassiulas. Maximum lifetime routing in wireless sensor networks. *IEEE Transactions on Networking*, **12**(4): 609–619, 2004.

58. S. Basagni, A. Carosi, E. Melachrinoudis, C. Petrioli, and Z. M. Wang. Controlling sink mobility in wireless sensor networks: A new model and protocols. *Poster at ACM/SIGMOBILE MobiCom 2005*, Cologne, Germany, August 28–September 2, 2005.

59. S. Basagni, A. Carosi, E. Melachrinoudis, C. Petrioli, and M. Z. Wang. Protocols and model for sink mobility in wireless sensor networks. *ACM Mobile Computing and Communication Review, (MC^2R)*, **10**(4): 28–30, 2006.

60. A. Vahdat and D. Becker. Epidemic routing for partially connected ad hoc networks. Duke University, Durham, NC, Technical Report CS-200006, April 2000.

61. S. Basagni, M. Elia, and R. Ghosh. ViBES: Virtual backbone for energy saving in wireless sensor networks. In *Proceedings of the IEEE Military Communication Conference, MILCOM 2004*, Monterey, CA, October 31–November 3, 2004.

62. C. E. Miller, A. W. Tucker, and R. A. Zemlin. Integer programming formulation of traveling salesman problems. *Journal of the ACM*, **7**(4): 311–325, 1960.

63. A. Savvides and M. B. Srivastava. Location discovery. In *Mobile Ad Hoc Networking*, S. Basagni, M. Conti, S. Giordano, and I. Stojmenovic, editors, IEEE Press and John Wiley & Sons, Piscataway, NJ, and Hoboken, NJ, 2004, chapter 8, pp. 231–254.

64. M. Battelli and S. Basagni. Localization for wireless sensor networks: Protocols and perspectives. In *Proceedings of IEEE CCECE 2007*, Vancouver, Canada, April 22–26, 2007, pp. 1074–1077.

65. The VINT Project, *The ns Manual*. http://www.isi.edu/nsnam/ns/, 2002.

66. ASH transceiver designer's guide. www.rfm.com, May 19, 2004.

CHAPTER 11

Localization Systems for Wireless Sensor Networks

AZZEDINE BOUKERCHE

School of Information Technology and Engineering, University of Ottawa, Ottawa, Ontario K1N 6N5, Canada

HORACIO A. B. F. OLIVEIRA

School of Information Technology and Engineering, University of Ottawa, Ottawa, Ontario K1N 6N5, Canada; Federal University of Minas Gerais, Brazil; and Federal University of Amazonas, Brazil

EDUARDO F. NAKAMURA

Federal University of Minas Gerais, Brazil; and FUCAPI—Analysis, Research, and Technology Innovation Center, Brazil

ANTONIO A. F. LOUREIRO

Department of Computer Sciences, Federal University of Minas Gerais, Belo Horizonte, Brazil

11.1 INTRODUCTION

Wireless sensor networks (WSNs) [1–4] are composed of a large number of sensor nodes used to monitor an area of interest. This type of network has become popular due to its applicability, which includes several areas such as environmental, health, industrial, domestic, agricultural, meteorological, spatial, and military. Despite the fact that the main goal of a WSN is to monitor an area of interest, several secondary objectives, or prerequisites, have to be achieved in order to reach the main objective (Figure 11.1).

The definition of a localization system [5–16] among the sensor nodes is one of these prerequisites needed in order to make viable many of the WSNs applications. The localization problem consists in identifying the physical location (e.g., latitude, longitude, and altitude) of a determined object. Such a problem is very ample and extensive, relating areas such as robotics, ad hoc networks, wireless sensor networks,

Algorithms and Protocols for Wireless Sensor Networks, Edited by Azzedine Boukerche

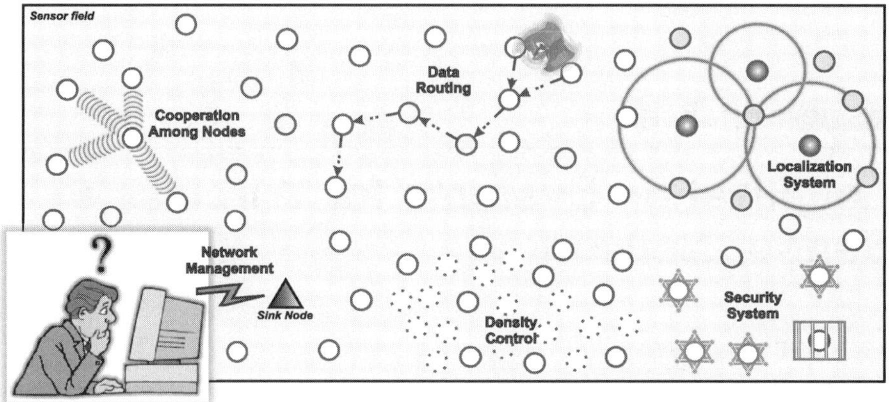

Figure 11.1. Several areas of the WSNs that work together in order to achieve one common goal: to monitor an area of interest.

cellular telephony, military, aviation, and astronomy. In this chapter the localization problem will be addressed under the viewpoint of the WSNs.

The localization systems have been identified as a key technology to the development and operation of the WSN [2]. Although its use is not exclusive to these networks, the localization systems, especially the ones with multihop, differentiates from the other areas because its applicability had been greatly extended with the advent of the WSN.

11.1.1 Problem Statement

In this work, we consider a WSN as being composed of n nodes, with a communication range of r, and distributed in a two-dimensional squared sensor field $Q = [0, s] \times [0, s]$. For the sake of simplification, we consider symmetric communication links; that is, for any two nodes u and v, u reaches v if, and only if, v reaches u and with the same signal strength w. Thus, we represent the network by the Euclidean graph[1]:

- $V = \{v_1, v_2, \ldots, v_n\}$ is the set of sensor nodes;
- $\langle i, j \rangle \in E$ iff v_i reaches v_j; that is, the distance between v_i and v_j is less than r;
- $w(e) \leq r$ is the weight of edge $e = \langle i, j \rangle$, that is, the distance between v_i and v_j.

In an Euclidean graph, each node has a coordinate $(x_i, y_i) \in \mathcal{R}^2$ in two-dimensional space, which represents the location of the node i in Q. For the sake of simplicity, we will only consider two dimensions in this chapter, but the methods explained here can be easily extended to provide position information in three dimensions.

Some terms can be used to designate the current state of a node:

[1] A Euclidean graph is a weighted graph where the weights are equal to the Euclidean lengths of the edges.

Definition 1 (Unknown Nodes, $\mathcal{U}$). Also known as free or dumb nodes, $\mathcal{U}$ refers to the nodes of the network that do not know their localization information. To allow these nodes to estimate their position is the main goal of the localization systems.

Definition 2 (Settled Nodes, $\mathcal{S}$). These nodes were initially unknown nodes, but managed to estimate their positions using the localization system. The number of settled nodes and the estimated position error of these nodes are the main parameters of quality of a localization system.

Definition 3 (Beacon Nodes, $\mathcal{B}$). Also known as landmarks or anchors, these are the nodes that do not need the localization system in order to estimate their physical positions. Their localization is obtained by manual placement or by external means such as the GPS. These nodes form the base of the majority localization systems for WSN.

Definition 4 (Reference Nodes, $\mathcal{R}$). These are the nodes in which the localization information will be used by an unknown node to estimate its location. A reference node must also be a beacon or a settled node.

The localization problem can then be simplified and defined by the following definition.

Definition 5 (Localization Problem). Given a multihop network $G = (V, E)$ and a set of beacon nodes $\mathcal{B}$ and their positions (x_b, y_b), for all $b \in \mathcal{B}$, we want to find the position x_u of almost all unknown node $u \in \mathcal{U}$, transforming these unknown nodes into settled nodes $\mathcal{S}$.

11.1.2 Importance of the Localization Systems

The position of sensor nodes need not to be engineered or predetermined. This allows random deployment in inaccessible terrains or disaster relief operations [1]. Thus, a localization system is required in order to provide the position information to the nodes. The importance of the localization information arise from several factors, many of them related only to the WSNs. Some of these factors include:

- *Gathered Data Identification:* This consists in mapping the data/events to their location of gather/occurrence. One of the main goals of a WSN is to monitor an area of interest. However, once the data are collected, it becomes important to identify the region to which these data belong.
- *Gathered Data Correlation:* This allows the intermediate nodes to correlate and perform information fusion of the data gathered on the same region while these data are forwarded through the network [17–19].
- *Nodes Addressing:* This refers to the possibility of using the physical location of the nodes that have their unique identification in the network [20, 21].

- *Network Management:* This allows the management and query of nodes localized in a determined region [22], evaluation of the nodes coverage [23], and energy maps generation [24].
- *Geographic Algorithms:* These are algorithms that use the localization information of the nodes to optimize the use of the network resources. Some of these algorithms include routing [22, 25–27], density control [23], and object tracking [28] algorithms.

11.1.3 Requirements of a Localization System

Due to its limitations and applications, the WSNs impose a number of prerequisites that must be taken into consideration by a localization system. Some of these prerequisites include:

- *Auto-organization,* Which is independent of any infrastructure.
- *Scalability,* Where the algorithms can be applicable to large-scale and/or dense sensor networks.
- *Robustness,* which consists of tolerance to communication problems and also to inaccurate distance and position information.
- *Efficiency* on using the network resources, because even being indispensable to the most WSNs, the localization system is not the main goal of these networks.

11.1.4 The Components of the Localization Systems

The localization systems can be divided into three distinct components:

1. *Distance/Angle Estimation:* This component is responsible for estimating information of distances and/or angles between two nodes. Such information will be used by the other components of the localization system.
2. *Position Computation:* This component is responsible for computing a node's position based on the available information of distances/angles and positions of the reference nodes.
3. *Localization Algorithm:* This is the main component of a localization system. It determines how the available information will be manipulated in order to allow most or all the nodes of the WSN to estimate their positions.

Figure 11.2 depicts this component division. Besides being a didactic viewpoint, the importance of such a division into components comes, as we will see, from the need to recognize that the final performance of the localization systems depends directly on each one of these components. Also, each component has its own goal and methods of solution. They can be seen as subareas of the localization problem that need to be separately analyzed and studied.

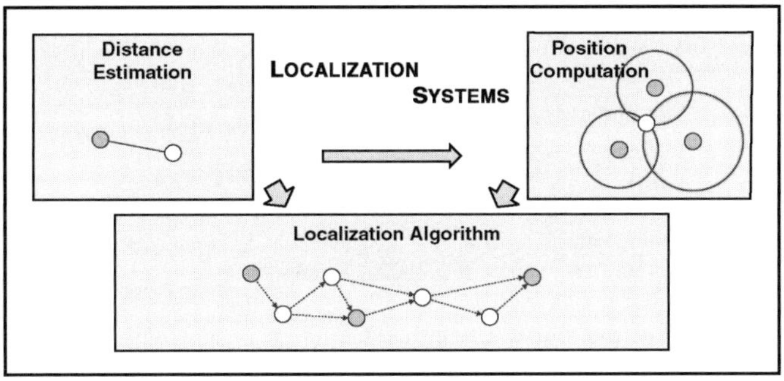

Figure 11.2. The division of the localization systems into three distinct components: distance/angle estimation, position computation, and localization algorithm. The arrows indicate the dependency relations—that is, the information flow from one component to another.

The rest of this chapter defines and discusses each one of these components by showing, analyzing, and comparing the main methods used by each one of them. This way, in Section 11.2 the first component, the Distance/Angle Estimation, and its techniques are studied. Section 11.3 presents the second component, the Position Computation, also with some of its techniques. The third component, the Localization Algorithm, is discussed in Section 11.4 along with some known localization algorithms proposed in the literature. Finally, Section 11.5 presents a summary along with some concluding remarks.

11.2 DISTANCE/ANGLE ESTIMATION

The distance/angle estimation is obtained by identifying the distance or angle between two nodes. Such estimates constitute an important component of the localization systems, because they are used by the position computation and also by the localization algorithm.

Different methods can be used to estimate such information. Some of them are very accurate, but with higher costs (in terms of hardware, energy, and processor resources), while others are not accurate, but already available on most sensor nodes.

In the following sections, some of the main methods used by the localization systems to estimate distances/angles will be studied. These methods include RSSI, ToA/TDoA, AoA, and the communication range. Finally, some general comments will be made regarding this component and its methods of estimation.

11.2.1 Received Signal Strength Indicator (RSSI)

RSSI derives the distance between two nodes based on the strength of the signal received by one of the nodes. As depicted in Figure 11.3, a sender node sends a signal with a determined strength that gets reduced as this signal is propagated. The greater

Figure 11.3. Decrease in the signal strength. The signal is sent with a determined strength that decreases, theoretically, proportionally with the square of the distance. In the real world, a number of factors collaborate in the faster and varied decrease in this signal strength.

the distance of the receiver node, the lower the signal strength when it arrives this node.

The signal strength is commonly measured in dBm (decibel in reference to one milliwatt) or in watts. Theoretically, this signal strength decreases as the inverse of the squared distance, and a known radio propagation model (Table 11.1) can be used to convert the signal strength into distance. However, in real-world environments, this indication is highly influenced by noises, by obstacles, and by the type of the antenna, which makes it hard to be modeled by a mathematical formula. In these cases, it is normal to make a system calibration [29], where values of RSSI and distances are previously evaluated in a controlled environment.

This method, like the others, has some advantages and some disadvantages. The main advantage is its low cost, because most receivers are capable of estimating the received signal strength. The disadvantage of this method is that it is very subject to noises and interferences, which results in higher inaccuracies on the distance estimations. Some experiments [6] show errors from 2 to 3 m in scenarios where all nodes are placed in a plane field, 1.5 m from the ground, and with a communication range of 10 m.

Although the RSSI show some plausible results in simulations and controlled experiments, its use in real-world applications is still questionable [32]. But, considering

TABLE 11.1. Two Known Radio Propagation Models

Model	Description	Formula
Free Space [30]	Consider the ideal propagation condition without interferences or obstacles.	$P_r(d) = \frac{P_t G_t G_r \lambda^2}{(4\pi)^2 d^2 L}$
Two-Ray Ground [31]	Like the Free Space, but consider the possibility of signal reflection in the ground.	$P_r(d) = \frac{P_t G_t G_r h_t^2 h_r^2}{d^4 L}$

P_t and P_r are the transmitted and received signal power (strength), G_t and G_r are the antenna gains of the transmitter and receiver, λ is the wavelength, d is the distance between the nodes, L is the system loss, and h_t and h_r are the heights of the transmit and receive antennas.

its low cost, it is possible that a more sophisticated and precise use of the RSSI (e.g., with better transmitters) could become the most used technology of distance estimation by the cost/precision viewpoint [33]. But this technology is not yet available.

11.2.2 Time (Difference) of Arrival (ToA/TDoA)

Different methods try to estimate the distance between two nodes using time-based measures. The most simple and intuitive is the ToA (Time of Arrival) [34]. In this case, the distance between two nodes is directly proportional to the time that the signal takes to propagate from one point to another. This way, if a signal was sent in time t_1 and reached the receiver node in time t_2, the distance between the sender and the receiver is

$$d = s_r(t_2 - t_1) \tag{11.1}$$

where s_r is the propagation speed of the radio signal (speed of light), and t_1 and t_2 are the times when the signal was sent and received (Figure 11.4a). This type of estimation requires precisely synchronized nodes [35], and the time when the signal leaves the node must be in the packet that is sent.

The TDoA (Time Difference of Arrival) is based on (a) the difference of the times that a single signal from a single node arrives in three or more nodes or (b) the difference of the times that multiple signals from a single node arrive in another node. The first case, more common in cellular networks, requires precisely synchronized receiver nodes (in this case, base stations). In the second case, more common and suitable for WSNs, the nodes must be equipped with an extra hardware capable of sending two types of signals simultaneously. These signals must have different propagation speeds, like radio/ultrasound [36] or radio/acoustic [29]. Usually, the first signal is the packet itself, which propagates in the speed of light ($\sim$300, 000

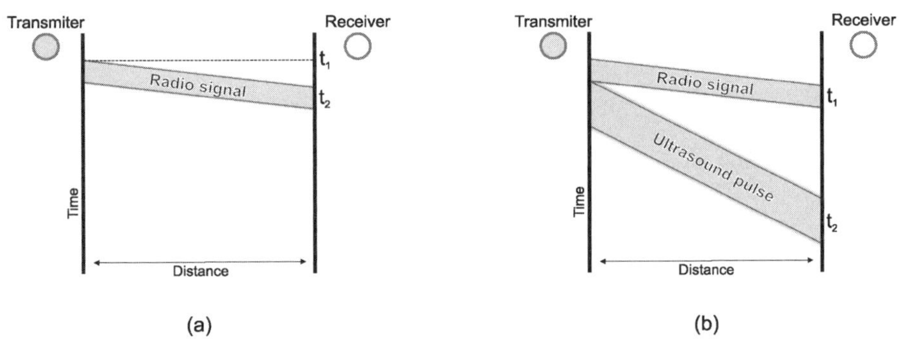

Figure 11.4. Methods to derive the distance from the signal arrival time: (a) ToA (Time of Arrival), where the time that the signal takes to leave the transmitter and arrive in the receiver is computed. (b) TDoA (Time Difference of Arrival), where the difference of the arrival times of two different signals sent simultaneously is used.

km/s), and the second signal is some kind of sound, because of its slower propagation (≈340 m/s), which is six orders of magnitude slower then the first signal.

An example of TDoA that is suitable for WSN is used by reference 6 and depicted in Figure 11.4b, where an ultrasound pulse is sent simultaneously with a radio signal. In this case, nodes compute the difference in the time of arrivals of the two signals. The distance can now be computed by the following formula:

$$d = (s_r - s_s) * (t_2 - t_1) \qquad (11.2)$$

where s_r and s_s are the propagation speed of the radio and ultrasound signals, and t_1 and t_2 are the arrival times of the radio and ultrasound signals, respectively.

The errors in the distance estimations obtained by TDoA are measured in centimeters. Experiments with ultrasound made in reference 6 indicate errors of about 2 or 3 cm (smaller than the sensor node) in scenarios where the nodes were in a distance of 3 m. In reference 37 acoustic sound was studied and the results showed errors of abouts 23 cm, with nodes in a distance of 2 m.

Despite the lower errors, these systems have some disadvantages. The first is the need of extra hardware to send the second signal, which increases the node cost. The second disadvantage is the range of the second signal, which is normally low between 3 m and 10 m with more powerful transmitters.

11.2.3 Angle/Direction of Arrival (AoA/DoA)

The angle of arrival of the signal [36, 38] can also be used by the localization systems. This angle can be in relation to the node itself, in relation to a electronic compass, or in relation to a second signal received by the node.

The estimation of the angle of arrival is done by using directive antennas or using an array of receivers—usually three or more—uniformly separated. In the last case, based on the times of arrival of the signal to each one of the receivers, it becomes possible to estimate the angle of arrival of this signal (Figure 11.5).

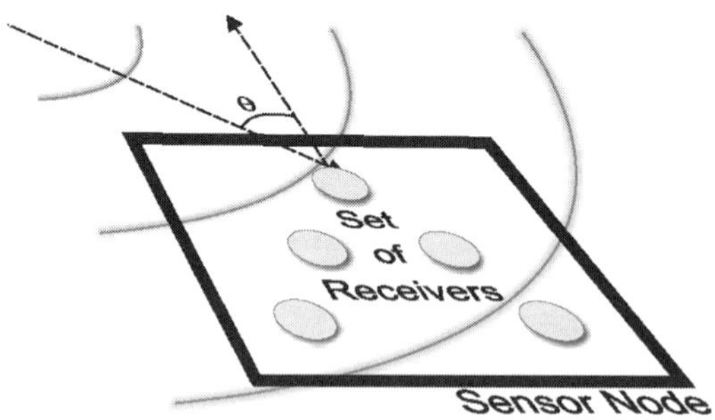

Figure 11.5. Angle of arrival of the signal. The difference in the times of arrival of the signal to each one of the receivers, as well as the difference in the position of these receivers, allows the node to estimate the angle of arrival of the signal.

Experiments show that this method has an inaccuracy of a few degrees (about 5° in reference 36. The need of extra hardware and the need of a minimal distance among the receivers result in some disadvantages in terms of cost and size of the sensor nodes.

11.2.4 Communication Range

In some cases, the only information available to estimate a distance is the communication range of the sensor nodes. If a node receives a data packet from another node, then the distance between these nodes is between zero and the maximum communication range.

Usually, techniques that use this method of distance estimation do not need an accurate distance, but only an interval. To get only one distance (and not an interval), we can choose one point from the interval, like the middle point, for example. In this last case, the maximum error of this estimation will be one-half the communication range.

This method of distance estimation has the advantage of being the most simple and with the lowest cost. No extra hardware is required, nor is extra computation needed to estimate a distance. On the other hand, an error of half the communication range for each distance estimation is not viable for the most localization systems. Consider, for example, a communication range of 100 m. In this case, the error of this method can be about 50 m for each distance estimation.

11.2.5 Comments About the Distance/Angle Estimation

The choice of what method to use to estimate distance between nodes in a localization system is an important factor that influences the final performance of the system. Usually, as will be shown in the next section, to estimate a position, a node uses at least three distance estimations, each of them with an associated error. On the other hand, if only the accuracy of such methods were important, we could just use a TDoA that has lower errors. But factors such as the size and cost (in terms of hardware, processor, and energy) of the nodes must also be taken into consideration. Thus, the chosen method used to estimate distances will depend on the application requirements and also on the available resources. Table 11.2 compares each one of the methods described in this section.

11.3 POSITION COMPUTATION

When a node has enough information of distances and/or angles and positions, it can compute its own position using one of the methods that will be studied in this section.

Several methods can be used to compute the position of a node. Such methods includes trilateration, multilateration, triangulation, probabilistic approaches, bounding box, and the central position. The choice of which method to use also impacts the final performance of the localization system. Such a choice depends on the available

TABLE 11.2. Comparison of the Methods Used to Estimate Distances/Angles Between Two Nodes[a]

Method	Precision	Maximum Distance	Extra Hardware	Challenges
RSSI	Meters (2–4 m)	Communication range	None	Variation of the RSSI, interferences
ToA	Centimeters (2–3 cm)	Communication range	None	Nodes synchronization
TDoA	Centimeters (2–3 cm)	few meters (2–10 m)	Ultrasound transmitter	Maximum distance of work
AoA	A few degrees (5°)	Communication range	Set of receivers	Work on small sensor nodes
Communication Range	Half the communication range	Communication range	None	—

[a]The chosen method depends on the application, scenario, and available resources.

information and on the processor limitations. In the subsequent sections, the cited methods will be studied. After that, some general comments will be made regarding this component and its methods.

11.3.1 Trilateration and Multilateration

Trilateration is the most basic and intuitive method. This method computes a node position by the intersection of three circles, as depicted in Figure 11.6. To estimate

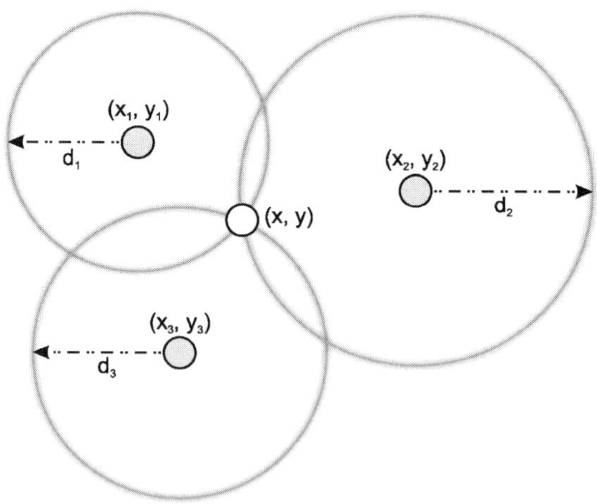

Figure 11.6. Theoretical model of the trilateration: The position of the unknown node corresponds to the intersection of the three circles formed by the positions and distances to the reference nodes.

its position using trilateration, a node needs to know the positions of three reference nodes and the distance to each of these nodes. The distances can be estimated using one of the methods explained in the previous section.

The circles formed by the position and distance to each one of the references can be represented by the formulas:

$$(\hat{x} - x_1)^2 + (\hat{y} - y_1)^2 = d_1^2$$
$$(\hat{x} - x_2)^2 + (\hat{y} - y_2)^2 = d_2^2$$
$$(\hat{x} - x_3)^2 + (\hat{y} - y_3)^2 = d_3^2$$

where $(\hat{x},\hat{y})$ is the position we want to compute, (x_i,y_i) is the position of the ith reference, and d_i is the distance of the ith reference node to the unknown node. In this case, we have three quadratic equation with two unknowns, which can be solved, theoretically, into one solution.

In real-world applications, the distance estimation inaccuracies as well as the inaccurate position information of the reference nodes make it difficult to compute the position. As depicted in Figure 11.7a, the circles do not intersect into only one point, resulting in an infinite set of possible solutions.

Furthermore, when a larger number of reference points are available, we can use the multilateration to compute the nodes position. In this case, an overdetermined system of equations, where we have more equations than unknowns, must be solved. Figure 11.8 depicts this case. Usually, overdetermined systems do not have a unique

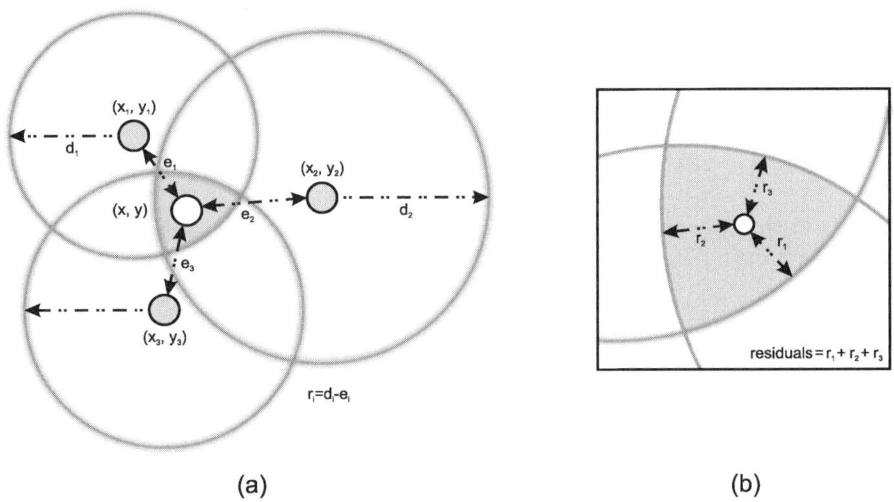

Figure 11.7. A more realistic model of the trilateration: (a) The inconsistencies of the positions and distances generate a system with infinite solutions. (b) The residual value, as the sum of the squared differences between the estimated and computed distances.

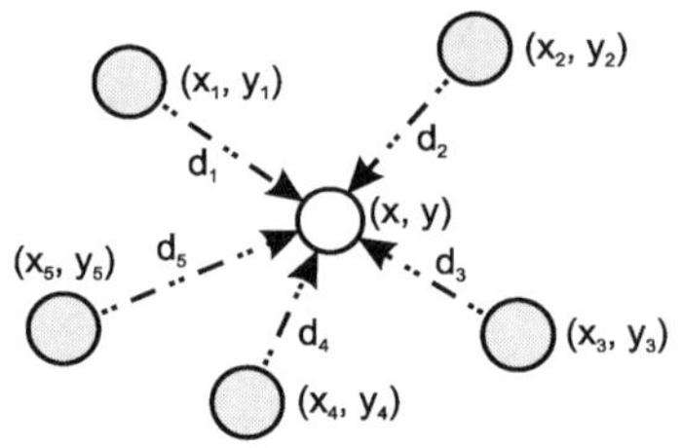

Figure 11.8. Multilateration: similar to the trilateration, but more than three references can be used and an overdetermined system is solved. A method to solve this system includes the least squares method.

solution. When considering n reference points and also the error of the distance estimations, which makes $d_i = \hat{d}_i - \epsilon$, the system of equations becomes

$$(\hat{x} - x_1)^2 + (\hat{y} - y_1)^2 = \hat{d}_1^2 - \epsilon$$

$$\vdots$$

$$(\hat{x} - x_n)^2 + (\hat{y} - y_n)^2 = \hat{d}_n^2 - \epsilon$$

where ϵ is normally considered to be an independent normal random variable with zero mean. This system can be linearized, by subtracting the last equation, into $Ax \approx b$, or

$$\begin{bmatrix} 2(x_1 - x_n) & 2(y_1 - y_n) \\ \vdots & \vdots \\ 2(x_{n-1} - x_n) & 2(y_{n-1} - y_n) \end{bmatrix} \begin{bmatrix} \hat{x} \\ \hat{y} \end{bmatrix} \approx \begin{bmatrix} x_1^2 - x_n^2 + y_1^2 - y_n^2 + d_n^2 - d_1^2 \\ \vdots \\ x_{n-1}^2 - x_n^2 + y_{n-1}^2 - y_n^2 - d_{n-1}^2 + d_n^2 \end{bmatrix} \tag{11.3}$$

This linear system can be easily solved using standard methods like the least squares approach [39, 40]. This can be done by the following parameter estimation:

$$x = (A^T A)^{-1}(A^T b) \tag{11.4}$$

The main idea of this method is to minimize the sum of the squares of the differences between the estimated (e.g., using the RSSI) and the computed distances (using the estimated position). This sum of the differences is known as the residuals, as depicted in Figure 11.7b. In mathematical terms, the computed position is

$$(\hat{x}, \hat{y}) = \min\left(\sum_{i=1}^{n}\left(\sqrt{(\hat{x} - x_i)^2 + (\hat{y} - y_i)^2} - d_i\right)^2\right) \tag{11.5}$$

where (x_i, y_i) is the position of the ith reference node, and d_i is the estimated distance. $\sqrt{(\hat{x} - x_i)^2 + (\hat{y} - y_i)^2}$ is the distance between the computed position and the position of the ith reference node, which is the computed distance.

The number of floating point operations needed to compute a position depends on the method used to solve the system of equations. In the case of the method of least squares, $(m + n/3)n^2$ floating point operations (where m is the number of unknowns and n is the number of equations) are required to estimate a position [39].

11.3.2 Triangulation

In triangulation [36, 38], information of angles are used instead of distances. The position computation can be made remotely or by the node itself. In both cases, the position is computed using the trigonometry laws of sines and cosines.

In the first case—remote positioning, depicted in Figure 11.9a—at least two reference nodes estimate the angle of arrival and remotely compute the position of the unknown node as the point where the lines along the angles from each reference node intersect. This type of triangulation is mostly used in cellular networks.

But for sensor networks, the most important is that the node itself computes its own position. In this case, depicted in Figure 11.9b, at least three reference nodes are required. The unknown node estimates its angle to each one of the three reference nodes and, based on these angles and on the positions of the reference nodes (which form a triangle), computes its own position using simple trigonometrical relationships. This technique is similar to trilateration. In fact, based on the angles of arrival, it is possible to derive the distances to the reference nodes [38].

11.3.3 Probabilistic Approaches

The uncertainty in the distance estimations motivated the appearance of probabilistic approaches to compute a node position. In the probabilistic approach, the position computation does not result in one single point, like in the other cases, but in a set of points and their probabilities of being the real position of the unknown node.

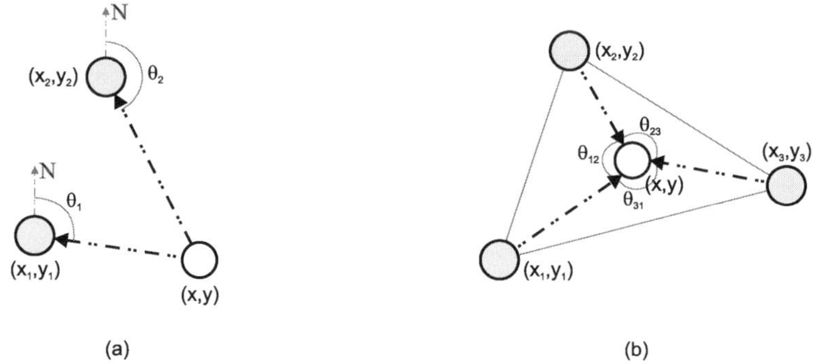

Figure 11.9. Triangulation. (a) Remote positioning: The reference nodes exchange information between them to compute the position of the unknown node using the angle of arrival. (b) self-positioning: Where the unknown node estimates its angle to at least three reference nodes, the unknown node itself is able to compute its own position.

A example of a probabilistic approach is proposed in reference 41. In this work, the error of the distance estimation is modeled as a normal random variable. When an unknown node receives a packet from a reference node, it can be in any place around the reference node with probabilities as depicted in Figure 11.10a. When

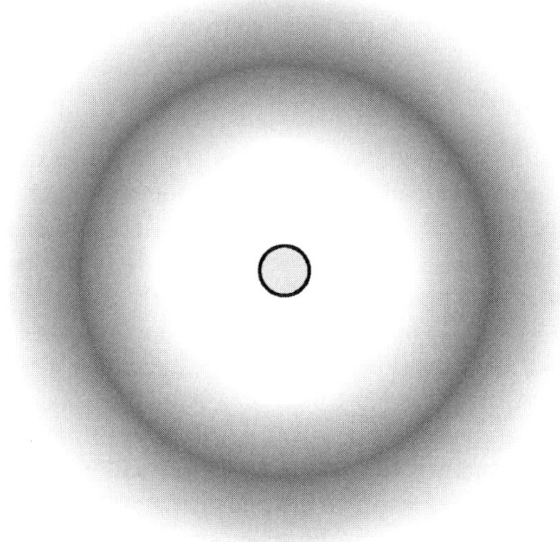

(a)

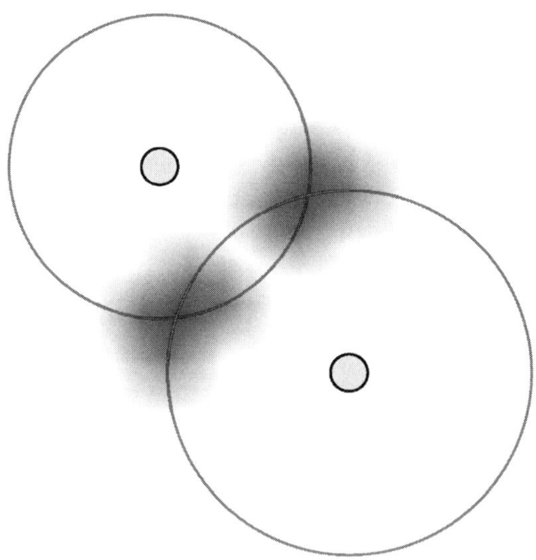

(b)

Figure 11.10. Probabilistic approach. (a) Probability of a node's position using one reference node: The unknown node has equal probability of being on around the reference node. (b) Probability when using two reference nodes: A pair of points with greater probabilities results from the system. (c) probabilities when using more than two references: A unique point with greater probability results from the system.

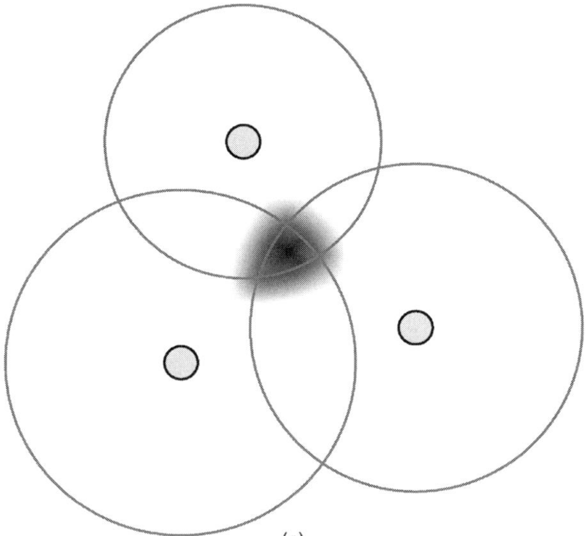
(c)

Figure 11.10. (*Continued*)

another packet is received from another reference node, the unknown node computes its position again as depicted in Figure 11.10b. When new position information of other nodes is received, it becomes possible to identify the probable location of the unknown node, as depicted in Figure 11.10c .

When an application requires a single position, the point with greater probability can be computed. The main problem of this approach is the high computational cost and the space required to store the information. In reference 42 it is shown that if we consider the sample size as a grid of $d \times d$, the complexity of this method would be $O(3d^2)$. One possible application of this method consists in sending the gathered information to a central and more powerful node in order to compute the positions.

11.3.4 Bounding Box

Bounding box, proposed in reference 43, uses squares—instead of circles as in trilateration—to bound the possible positions of a node. An example of this method is depicted in Figure 11.11.

For each reference node i, a bounding box is defined as a square with center in the position of this node (x_i, y_i), with sides of size $2d_i$ (where d is the estimated distance) and with coordinates:

$$(x_i - d_i,\, y_i - d_i) \quad \text{and} \quad (x_i + d_i,\, y_i + d_i) \tag{11.6}$$

The intersection of all bounding boxes can be easily computed without the need for floating point operations by taking the maximum of the low coordinates and the

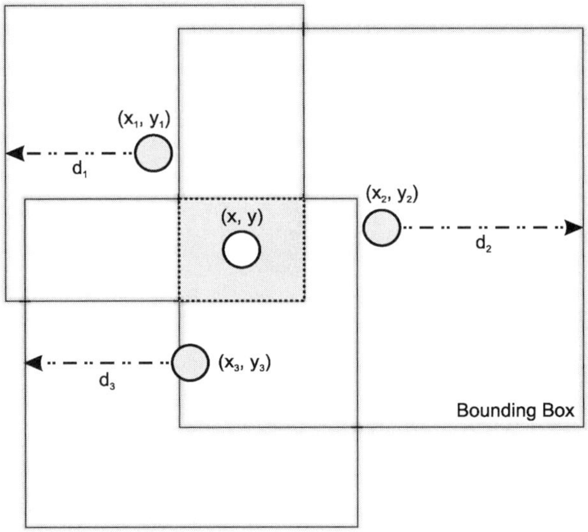

Figure 11.11. Bounding box illustration. The intersection of all bounding boxes (shaded) is computed without the need for floating point operations. The position of the node is computed as the center of this rectangle.

minimum of the high coordinates of all bounding boxes:

$$(\max(x_i - d_i), \max(y_i - d_i)) \quad \text{and} \quad (\min(x_i + d_i), \min(y_i + d_i)) \tag{11.7}$$

This is the shaded rectangle in Figure 11.11. The final position of the unknown node is then computed as the center of the intersection of all bounding boxes:

$$(\hat{x}, \hat{y}) = \left(\frac{\max(x_i - d_i) + \min(x_i + d_i)}{2}, \frac{\max(y_i - d_i) + \min(y_i + d_i)}{2} \right) \tag{11.8}$$

Despite the final error of this method, which is greater than that of trilateration, a smaller amount of processor resources is used compute the intersection of squares than to compute the intersection of circles.

11.3.5 Central Position in Relation to the Reference Nodes

By the assumption that the most probable position of a node is the central point among all the reference nodes, we can compute the position of an unknown node without the need of estimating distances or angles, but only by using the communication range, as explained in Section 11.2.4.

In this case, the position of a node is computed by using the following equation [32]:

$$(\hat{x}, \hat{y}) = \left(\frac{\sum_{i=1}^{n} x_i}{n}, \frac{\sum_{i=1}^{n} y_i}{n} \right) \tag{11.9}$$

where n is the number of reference nodes.

This method is the most simple in terms of computational resources and required information. Only $2n + 2$ float point operations (where n is the number of reference nodes) are required to compute a position. On the other hand, the obtained solutions are not accurate, mainly when the number of reference nodes is low.

11.3.6 Comments About the Position Computation

In previous subsections, we showed a number of methods that can be used to compute the position of an unknown node based on the gathered information of positions and distances to the reference nodes.

A number of other methods exist that aim to compute the position of a node. Location Fingerprinting is a method where the signal characteristics obtained from a set of locations are cataloged, and then the position computation of a node now consists in comparing its signal characteristics with the ones that were previously cataloged. This technique is used by Bahl and Padmanabhan [44] and others indoors localization systems, but the need for generating a signal signature database makes this technique unfeasible for the most scenarios of the WSNs. Some works use a variant of the trilateration that uses only two reference nodes. In this case, two possible solutions, which is the intersection of two circles, result from the system. The choice between these two possible positions is made by using extra multihop information or even the direction of the localization recursion, as done by Oliveira and co-workers [11, 12]. In the APIT algorithm, He et al. [32] use triangles formed by three beacon nodes, and a node decides if it is inside or outside these triangles by comparing its signal strength measurements with the measurements of its neighbors. The position of the node is computed by finding the centroid of the intersection of the beacon triangles that the node is inside. Other works concentrate all information of distances among the nodes into a central node and uses mathematical optimization techniques to compute the positions of the nodes. As examples, we have (a) the work of Doherty et al. [5], which formulated the localization problem as a convex optimization problem based only in connectivity-induced constraints and used semidefinite program (SDP) to solve the problem, and (b) the work of Shang et al. [7, 45] that used multidimensional scale (MDS).

The choice of what method to use can also impact the final performance of the localization system. In the next section, some localization algorithms will be studied. Depending on the used localization algorithm, this error in the position computation can harm in greater or minor degree the localization system as a whole. In some algorithms, for example, the newly computed positions are used to assist the other unknown nodes to compute their positions. In this case, a small error in the position computation can result in a localization system with high errors.

TABLE 11.3. Comparison of the Methods Used to Compute Positions[a]

Method	# Refs	Dist?	Angle?	Comp.	Challenges
Trilateration	3	Yes	No	$O(1)$	Susceptible to inaccurate distances
Multilateration	$n \geq 3$	Yes	No	$O(n^3)$	Computational complexity
Triangulation	3	No	Yes	$O(1)$	Require extra hardware
Probabilistic	$n \geq 3$	Yes	No	$O(3d^2)$ (d=*grid*)	Computational and space complexity
Bounding Box	$n \geq 2$	Yes	No	$O(n)$	Final Position Error
Central Position	$n \geq 1$	No	No	$O(n)$	Final position error

[a] Again, there is no ideal solution that works in all scenarios. The choice of what method to use will depend on the gathered information and on the available processor resources.

The information of positions and distances gathered by a node and the available processor resources also restrict the choice of what method to used. Table 11.3 summarizes and compares the main characteristics of the position computation methods explained in this section.

11.4 LOCALIZATION ALGORITHM

The localization algorithm is the main component of a localization system. This component determines how the information of distances and positions will be manipulated in order to allow most or all the nodes of the WSN to estimate their positions.

The localization algorithms can be classified into some categories:

- *Distributed or Centralized Position Computation.* The positions of the nodes can be computed in a distributed way by the network nodes (self-positioning) [8, 9, 34] or by a single central node (e.g., a more powerful node, the sink node—remote positioning) [5, 46].
- *With or Without an Infrastructure.* If there is no need for infrastructure [8, 9] or if there is the need to redesign the previous infrastructure in order to allow the functioning of the localization algorithm (e.g., manual placement of the beacon nodes) [34, 36, 47],
- *Relative or Absolute Positioning.* The computed positions can be related to global coordinates (e.g., latitude, longitude) [9, 34] or related to a node or point of the network [48, 49].
- *Indoor or Outdoor Scenarios.* If the system is more appropriated for indoors [36, 47] or outdoors [34] scenarios.
- *One Hop or Multi Hop.* If all unknown nodes have direct communication with the beacon nodes [34, 47] or if a multihop communication is needed [8, 9].

Many times in this chapter, the word "performance" was used to make reference to aspects of quality of the localization systems. The following aspects can be used to evaluate this performance:

- *Mean Error and Consistence.* This identifies is the mean error of the position estimates and also shows if this mean error is repeated in similar but different scenarios (consistence of the mean). This mean error limits the usage of the localization system to the applications where this level of inaccuracies is acceptable.
- *Communication Cost.* This identifies the algorithm complexity in terms of packets exchanged. It also identifies the cost of the localization system to the sensor network.
- *Number of Settled Nodes.* This defines the percentage of network nodes that were able to compute their positions at the end of the localization algorithm. The ideal is that all nodes should be able to compute their position, but in many cases it is not possible.
- *Number of Beacon Nodes.* This identifies the number of beacon nodes required to make the algorithm work. Beacon nodes are usually more expensive than the normal nodes, and their usage should be minimized.

Some network characteristics can affect the performance of the localization systems. It is important to make experiments for each proposed localization system to evaluate their behaviors when varying these characteristics, which includes:

- *Network Density.* In high dense networks, we have lower distances among the nodes, which results in lower errors in distance estimations and also in the error of the localization system. Besides, the higher number of neighbors results in more information that can be used by an unknown node to better compute its position.
- *Network Scale.* Increasing the number of nodes (and keeping the network density, which increases the area of the sensor field) results in a higher number of hops. Usually, a higher number of hops results in more inaccurate positions computation, increasing the mean error of the localization system.
- *Number of Beacon Nodes.* When deploying a higher number of beacon nodes in the network, the mean error of the localization system tends to decrease and the number of settled nodes tends to increase.
- *GPS Accuracy.* Although considered by many works, the GPS does not provide perfect localization, especially in sensor networks. Because most of the beacon nodes use the GPS to get their position, the GPS accuracy will impact the final position error of the localization systems that depend on this service.

In the next sections, some proposed localization algorithms will be studied and analyzed. These algorithms are: Ad Hoc Positioning System, Recursive Position Estimation, Localization with a Mobile Beacon, Global Positioning System, and the

Cricket Location Support System. After that, some general comments will be made regarding this component and its methods.

11.4.1 Ad Hoc Positioning System (APS)

In the APS [8], a reduced number of beacon nodes (e.g., three or more) are deployed with the unknown nodes. The author proposes, then, a localization algorithm where each node estimates its distance to the beacon nodes in a multihop way. Once these distances are estimated, the nodes can compute their positions using trilateration. Three methods of hop-by-hop distance propagation are proposed: DV-Hop, DV-Distance, and Euclidean.

In the DV-Hop, the beacon nodes start the propagation of their position information (Figure 11.12a). Working as an extension of the distance vector algorithm, all nodes receive the position information of all beacon nodes as well as the number of hops to these beacons. When a beacon node receives a position information from the other beacon nodes, it has enough information to compute the average size of one hop based on its own position, on the position of the other beacon nodes, and also on the number of hops among them (Figure 11.12b). This last value is then flooded in a controlled way into the network as a correction factor. When an unknown node receives the correction, it is able to convert its distance to the beacon nodes from number of hops to meters (Figure 11.12c). The complete DV-Hop algorithm is shown in Algorithm 1. The complexity of message exchanging of this algorithm is driven by the total number of beacon and normal nodes, which is $O(n * (m + 1))$, where n is the number of nodes and m is the number of beacon nodes.

DV-Distance works like DV-Hop. But, instead of propagating the number of hops, it propagates the estimated distances (e.g., using RSSI); and each node, before forwarding the position information of the beacon nodes, adds its estimated distance to the one contained in the packet and then forwards this packet. In this case, there is

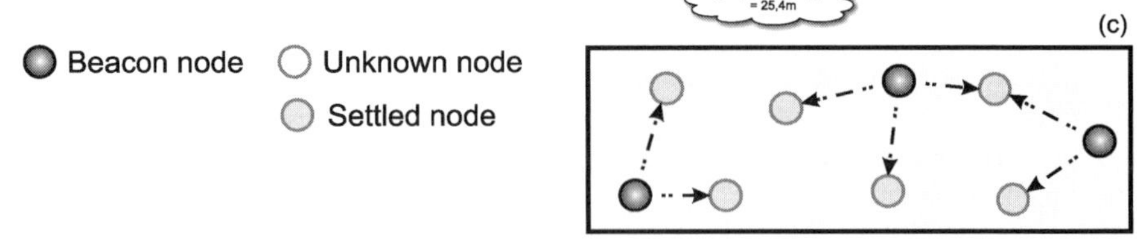

Figure 11.12. Example and phases of the APS: DV-Hop. (a) Initially, the beacon nodes broadcast their positions and (b) Compute the average size of hop. (c) This last value is sent to the network, and the nodes receiving it compute their positions.

ALGORITHM 1. APS DV-HoP Localization Algorithm

▷ **Variables:**

$positions_i = \emptyset$; {Set of beacon positions.}
$correction_i = -1$; {Correction computed/received by this node.}

▷ **Input:**

$msg_i = nil$.

Action:

if $n_i \in \mathcal{B}$ **then** {If this node is a beacon node.}
  $(x_i, y_i) := getGpsPosition()$;
  Send $beaconPos(i, x_i, y_i, 0)$ to all $n_j \in Neig_i$.
end if

▷ **Input:**

$msg_i = beaconPos(k, x_k, y_k, h_k)$.

Action:

if $k \notin \mathcal{R}_i$ **then** {If this node did not already receive this packet.}
  $\mathcal{R}_i := \mathcal{R}_i \cup \{k\}$;
  $positions_i := positions_i \cup \{(x_k, y_k, h_k)\}$;
  if $n_i \in \mathcal{B}$ **then** {If this is a beacon node, wait for more position packets}
    Start $waitTimer$.
  end if
  Send $beaconPos(k, x_k, y_k, h_k + 1)$ to all $n_j \in Neig_i$.
end if

▷ **Input:**

$waitTimer$ timeout. {Executed only by the beacon nodes.}

Action:

$correction_i = \frac{\sum \sqrt{(x_i - x_j)^2 + (y_i - y_j)^2}}{\sum h_j}$ for all $(x_j, y_j, h_j) \in positions_i$;
Send $correction_i$ to all $n_j \in Neig_i$.

▷ **Input:**

$msg_i = correction_k$.

Action:

if $n_i \in \mathcal{U}$ **then** {If this node is an unknown node.}
  $correction_i := correction_k$;
  $h_j := h_j * correction_i$ for all $(x_j, y_j, h_j) \in positions_i$;
  $(x_i, y_i) := positionComputation(positions_i)$; {Becomes a settled node.}
  Send $correction_i$ to all $n_j \in Neig_i$.
end if

---

no need for a correction factor, because the distances to the beacon nodes are already in meters.

The Euclidean method works by propagating the true Euclidean distance of the unknown nodes to the beacon nodes. To allow this, a node needs at least another two nodes that already estimated their distances to a beacon node. It also needs its own distance to these last two nodes, and also the distance between these two nodes. The distance estimation of the unknown node to the beacon node is made using the Pythagorean theorem on the triangles generated by the lines that is formed by the distances.

An advantage of the APS is that its localization algorithm requires a low number of beacon nodes in order to work. However, the way the distances are propagated, (especially in the DV-Hop and DV-Distance), as well as the way the distances are converted from hops to meters in DV-Hop, results in an erroneous position computation, which increases the final localization error of the system.

11.4.2 Recursive Position Estimation (RPE)

In the algorithm of the RPE [9], the nodes estimate their positions based on a set of initial beacon nodes (e.g., 5% of the nodes) using only local information. The localization information iteratively increases as the newly settled nodes become reference nodes.

The RPE algorithm can be divided into four phases as depicted in Figure 11.13. In the first phase, a node determines its reference nodes. In the second phase, the node estimates its distance to the reference nodes using, for example, the RSSI. In the third phase, the node computes its position using trilateration (becoming a settled node). In the final phase, the node becomes a reference node by broadcasting its newly estimated position to its neighbors.

When a node becomes a reference, it can assist the other nodes to also compute their positions. Figure 11.13 depicts this behavior of the RPE. In Figure 11.13a, node 14 has only two reference nodes (nodes 9 and 13), which does not allow the node to compute its position. On the other hand, node 8 also does not know its position, but can compute it (Figure 11.13b), becoming the third reference (Figure 11.13c) that node 14 needs to compute its own position. The complete RPE algorithm is shown in Algorithm 2, and its communication complexity is the same as that of flooding.

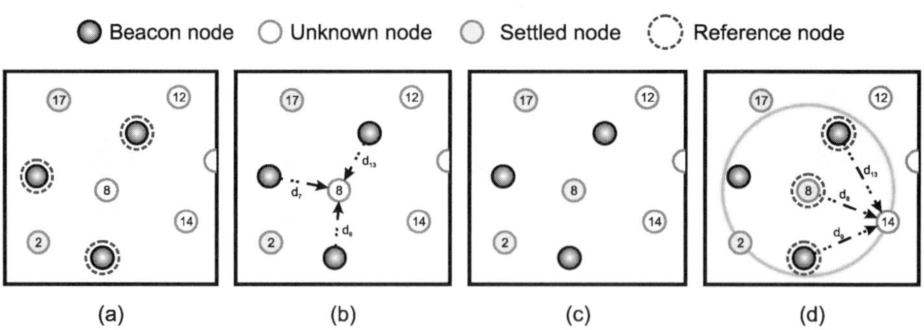

Figure 11.13. Example and phases of the RPE. (a) Initially, the node chooses its reference nodes; then (b) this node estimates its distance to each one of the reference nodes, (c) Computes its position using trilateration, and (d) broadcasts its newly estimated position to assist the other nodes.

ALGORITHM 2. RPE Localization Algorithm

▷ **Variables:**

$positions_i = \emptyset$ {Set of received positions.}
$references_i = \emptyset$ {Set of reference nodes.}

▷ **Input:**

$msg_i = nil.$

Action:

if $n_i \in \mathcal{B}$ **then** {If this node is a beacon node.}
    $(x_i, y_i) := getGpsPosition();$
    Send $position(x_i, y_i, 0)$ to all $n_j \in Neig_i$.
end if

▷ **Input:**

$msg_i = position(x_k, y_k, r_k)$ such that $dist_k = distanceEstimation(msg_i)$.

Action:

if $n_i \in \mathcal{U}$ **then** {If this node is an unknown node.}
    $positions_i := positions_i \cup \{(x_k, y_k, r_k, dist_k)\};$
    Start $waitTimer$.
end if

▷ **Input:**

$waitTimer$ timeout.

Action:

if $size(positions_i) >= 3$ **then** {If there is enough positions.}
    $references_i := chooseThreeBestPositions(positions_i)$
    $(x_i, y_i, r_i) := positionComputation(references_i);$ {Becomes a settled node.}
    Send $position(x_i, y_i, r_i)$ to all $n_j \in Neig_i$. {Becomes a reference node.}
end if

An advantage of this algorithm is that the number of reference nodes quickly increases, in such a way that the majority of the nodes can compute their position. But this technique has the disadvantage of propagating the localization error. This means that the low inaccurate position estimation of a node can be used by other nodes to estimate their positions, increasing this inaccuracy. One example of this problem is that node 14 in Figure 11.13 will get a localization error greater than that of node 8. Furthermore, a node must have at least three reference neighbors to compute its position.

11.4.3 Directed Position Estimation (DPE)

By adding some restriction to a recursive localization system like that in the RPE, we can make the localization recursion go from a single point (recursion origin) and follow a determined and known direction (Figure 11.14a). Once this behavior is guaranteed, it is possible to estimate a node position using only two reference neighbors.

When a node has a position information of only two reference neighbors, a pair of possible points results from the system: One is the right position of the unknown node and the other is a wrong estimate (Figure 11.14b). Once the direction of the

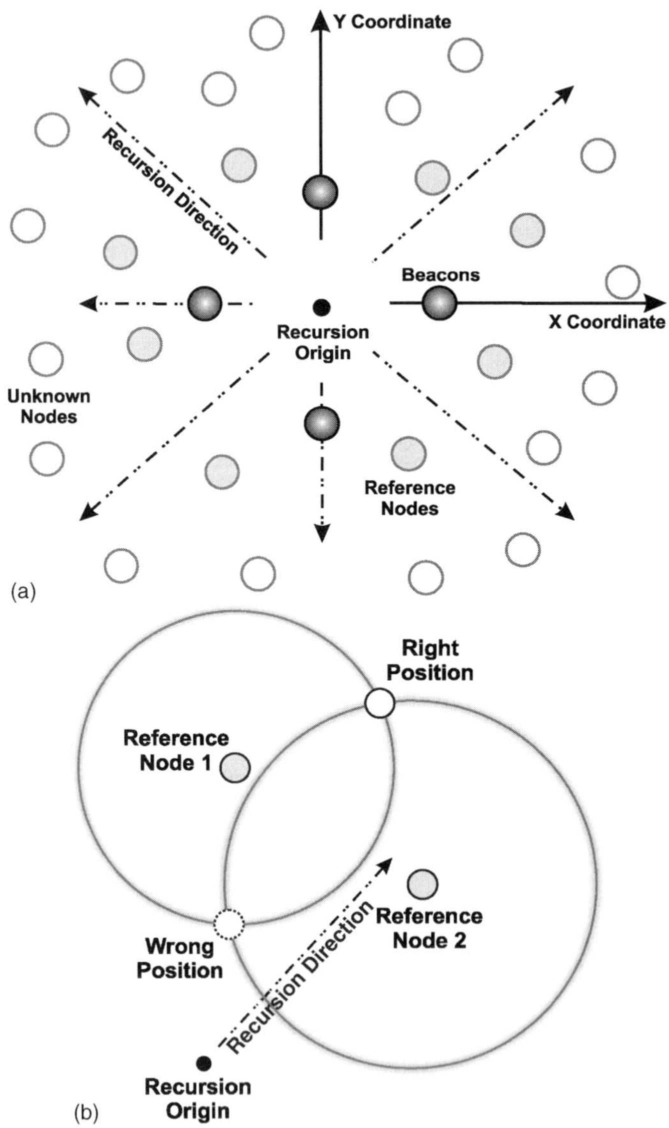

Figure 11.14. (a) The DPE doing a directed localization recursion. and (b) A position estimate using only two reference neighbors. A pair of possible solutions result from the system. The right position of the node is the most distant point from the recursion origin.

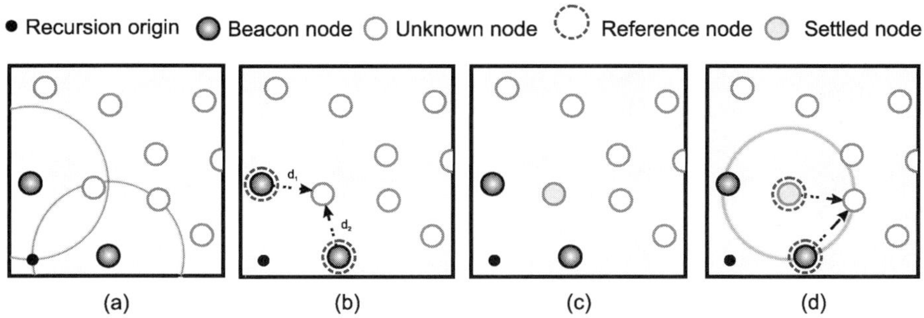

Figure 11.15. Example and phases of the DPE. (a) First, the beacon nodes starts the recursion. (b) Then a node determines its (two) reference nodes, (c) estimates its position, and (d) then becomes a reference by broadcasting this information.

localization recursion is kept, it is easy to choose between the two possible solutions: The most distant point from the recursion origin is the right position of the unknown node. This is the base of the DPE [11, 12].

The DPE algorithm is divided into four phases (Figure 11.15). In the first phase, the recursion of such a system is started from a single point by the beacon structure (Figure 11.15a). In the second phase, a node determines its (two) reference nodes and estimate its distances to these nodes (Figure 11.15b). In the third phase, the node computes its position (Figure 11.15c), and then becomes a reference by sending this information to its neighbors (fourth phase, Figure 11.15d). This way, the recursion of such a system goes from the center to the edge of the WSN. The complete DPE algorithm is shown in Algorithm 3, and its communication complexity is the same as that of flooding.

This approach leads to a localization system that can work in a low dense sensor network. Besides, the controlled way in which the recursion is made will also lead to a system with lower and predictable errors. But like the RPE, this technique has the disadvantage of propagating the localization error.

11.4.4 Localization with a Mobile Beacon (LMB)

Some recent works [42, 46] have proposed the use of mobile beacons to assist the nodes of the WSN in estimating their positions. A mobile beacon is a node that is aware of its position (e.g., equipped with a GPS receiver) and that has the ability to move among the sensor field. This beacon can be a human operator, an unmanned vehicle, an aircraft, or a robot. The localization algorithm proposed in reference 42 uses mobile beacons to allow the nodes to compute their positions, while in reference 46 the mobile beacon itself computes the position of the nodes.

The system operation in reference 42 is quite simple. Once the nodes are deployed, the mobile beacon travels through the sensor field broadcasting messages that contain its current coordinates. When a free node receives more than three messages from the mobile beacon, it computes its position, using a probabilistic approach, based on the received coordinates and on the RSSI distance estimations. Figure 11.16 illustrates

ALGORITHM 3. DPE Localization Algorithm

▷ **Variables:**

$positions_i = \emptyset$ {Set of received positions.}
$references_i = \emptyset$ {Set of reference nodes.}

▷ **Input:**

$msg_i = nil.$

Action:

if $n_i \in \mathcal{B}$ **then** {If this node is a beacon node.}
    $(x_i, y_i) := getGpsPosition();$
    Send $position(x_i, y_i)$ to all $n_j \in Neig_i$.
end if

▷ **Input:**

$msg_i = position(x_k, y_k)$ such that $dist_k = distanceEstimation(msg_i)$.

Action:

if $n_i \in \mathcal{U}$ **then** {If this node is an unknown node.}
    $positions_i := positions_i \cup \{(x_k, y_k, dist_k)\};$
    Start *waitTimer*.
end if

▷ **Input:**

waitTimer timeout.

Action:

if $size(positions_i) >= 2$ **then** {If there is enough positions.}
    $references_i := chooseTwoBestPositions(positions_i)$
    $(x_i, y_i) := mostDistantFromOrigin(intersectCircles(references_i));$ {Becomes a settled node.}
    Send $position(x_i, y_i)$ to all $n_j \in Neig_i$. {Becomes a reference node.}
end if

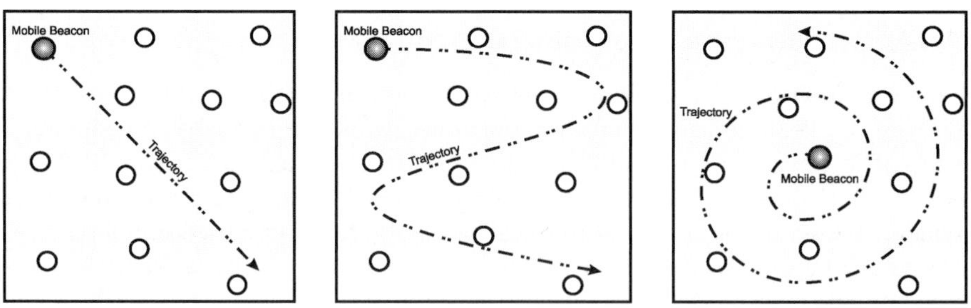

(a) (b) (c)

Figure 11.16. Operation and possible trajectories for the localization with a mobile beacon. (a) The mobile beacon moving along the sensor field in a straight line. (b) A less rectilinear trajectory. (c) A trajectory in spiral form.

ALGORITHM 4. MBL Localization Algorithm

▷ **Variables:**
$positions_i = \emptyset$ {Set of position information.}

▷ **Input:**
$msg_i = nil.$

Action:
if $n_i \in \mathcal{U}$ **then** {If this node is a beacon node.}
    StartWalking();
    Start *posTimer*.
end if

▷ **Input:**
posTimer timeout.

Action:
$(x_i, y_i) := getGpsPosition()$;
Send $position(x_i, y_i)$ to all $n_j \in Neig_i$.
Restart *posTimer*.

▷ **Input:**
$msg_i = position(x_k, y_k)$ such that $dist_k = distanceEstimation(msg_i)$.

Action:
$positions_i := positions_i \cup \{(x_k, y_k, dist_k)\}$;
if $size(positions_i) >= 3$ **then** {If there is enough references.}
    $(x_i, y_i) := positionComputation(positions_i)$;
end if

---

this scenario and three possible trajectories for the mobile beacon. Algorithm 4 details the functioning of the system. The communication cost for the WSN is null, since the nodes (except the mobile beacon) do not need to send any packets.

An advantage of LMB is that the position estimates are computed based on the same node (mobile beacon) keeping the mean localization error low and preventing the propagation of this error. In addition, LMB avoids the use of nodes equipped with GPS, except for the mobile beacon. On the other hand, in this technique a sensor node can estimate its position only when the mobile beacon passes near this node, which may take a long time depending on factors as the size of the sensor field, the beacon mobility capacity, and the node trajectory. Yet, the mobile beacon may never pass nearby some nodes, either because of the trajectory or because of a problem with the mobile beacon.

An important aspect that directly influences the position estimates is the trajectory of the mobile beacon. The less rectilinear the trajectory, the better the estimates. The reason is that the lower the collinearity among the reference points, the lower the estimate error [3]. Thus, rectilinear trajectories such as Figure 11.16a must be avoided.

11.4.5 The Global Positioning System (GPS)

The GPS [34, 50] is a system composed of 24 satellites in operation that orbit around the earth. Each satellite circles the earth at a height of 20.200 km and makes two complete rotations every day. The orbits were defined in a way that in each region of the earth we could "see" at least four satellites in the sky.

A GPS receiver is able to receive the information constantly sent by the satellites, estimate its distance to at least four known satellites using ToA, and, finally, compute its position using trilateration. Once these procedures are executed, the receiver is able to inform its latitude, longitude, and altitude.

One of the solutions for the localization problem in WSN is to equip each one of the sensor nodes with a GPS receiver. One of the main advantages would be the relatively small (2–15 m, depending on the receiver) and precise localization error, because all nodes would have a similar error. However, this solution has many disadvantages [5, 6, 8, 48]: (a) The cost and size of the sensor nodes, are increased, (b) it cannot be used when there is no satellite visible (e.g., indoor scenarios, under water, under climatic conditions, mars exploration), and (c) the extra hardware consumes energy. Due to these disadvantages, the usage of the GPS is usually limited to a small fraction of the nodes (e.g., beacon nodes).

11.4.6 The Cricket Location Support System

Cricket [47] combines active beacons and passive ultrasonic receivers to provide a localization system. Cricket is designed for mobile nodes in an indoor environment, but it may be used with static nodes. The active beacons broadcast their location information over an RF (radio frequency) channel together with an ultrasonic pulse. The other nodes use the TDoA method to estimate their distances to the beacons. When the node has enough information of positions and distances, it can compute its position through multilateration.

To make it possible for the algorithm to work, a grid of beacon nodes must be previously created so that all nodes have at least three beacon nodes in their communication range. This infrastructure can be considered as a disadvantage of the system, but Cricket is very accurate and able to work in indoor scenarios.

11.4.7 Comments About the Localization Algorithm

The localization algorithm is the main component of a localization system. This component defines how the available information that is provided by the beacon nodes, by the distances estimations, and by the position computations will be manipulated in order to allow the localization information to expand from the beacon nodes to the nodes of the sensor network.

As previously mentioned, the localization systems, especially the ones with multihop, had been extensively studied with the advent of the WSN. This way, a number of other localization algorithms have recently been proposed that focus on different aspects like errors, number of beacons, number of settled nodes, or GPS usage,

TABLE 11.4. Localization Algorithms Comparison[a]

Algorithm	Number of Beacons	Position Computation	Infrastructured?	Positioning	Scenarios	Multihop?
APS	≥ 3	Distributive	No	Absolute	Outdoors	Yes
RPE	5% nodes	Distributive	No	Absolute	Outdoors	Yes
DPE	4	Distributive	No	Relative	Outdoors	Yes
LMB	1 mobile	Distributive	No	Absolute	Outdoors	No
GPS	—	Distributive	Yes	Absolute	Outdoors	No
Cricket	Grid	Distributive	Yes	Relative	Indoors	No

[a] Some characteristics of the localization algorithms identify the possible scenarios that can be applicable. The choice of what algorithm to use depends on the application requirements and on the available resources.

among other things. A number of works try to reduce or completely remove the need of GPS receivers on beacon nodes. Bulusu et al. [48] manually placed multiple beacon nodes with overlapping coverage regions so the nodes could directly use the beacons position to compute their positions, but the manual placement of the beacon nodes is not feasible for the most WSNs scenarios. Capkun et al. [49] proposes a distributed, infrastructure-free localization algorithm that does not rely on GPS. The key point of the algorithm is to show that it is possible to build a relative coordinate system without centralized knowledge of the network topology. In APIT proposed by He et al. [32], beacon nodes equipped with high-powered transmitters in a heterogeneous network broadcast their positions generating a series of triangles formed by the positions of three beacon nodes. A node can estimate the triangles in which it is inside and compute its position using the intersection of these triangles.

The choice of which algorithm to use depends on the available resources, on the scenario, on the requirements of the application, and also on the mean localization error acceptable by the nodes. Table 11.4 compares the main characteristics of each one of these algorithms.

11.5 FINAL REMARKS

In this chapter we have discussed the localization problem for wireless sensor networks. We have divided the design of the localization systems for WSNs into the following three components:

1. Distance/Angle Estimation
   - Received Signal Strength Indicator (RSSI)
   - Time [Difference] of Arrival (ToA/TDoA)
   - Angle/Direction of Arrival (AoA/DoA)
   - Communication Range

2. Position Computation
   - Trilateration
   - Multilateration
   - Triangulation
   - Probabilistic Approaches
   - Bounding Box
   - Central Position
3. Localization Algorithm
   - Ad Hoc Positioning System (APS)
   - Recursive Position Estimation (RPE)
   - Directed Position Estimation (DPE)
   - Localization with a Mobile Beacon (LMB)
   - The Global Positioning System (GPS)
   - The Cricket Location Support System

The importance of dividing the localization systems into components, as mentioned before, comes from the necessity of recognizing that the final performance of the localization systems depend directly on each one of these components. For example, a localization system should achieve better results if the TDoA method is used instead of the RSSI to estimate distances. The same principle applies to the other components. These components can be seen as subareas of the localization problem that need to be separately studied.

A general rule in WSN is that there is no perfect solution to a problem that performs best in any situation or application. The same rule applies to the localization problem. This chapter showed a number of proposed localization systems, each of them with emphasis in a specific scenario and/or application. The necessity of different solutions for different applications and also the high number of possible applications of the WSN have greatly motivated the study and proposals of new solutions to the localization problem shown in this chapter. However, there are a number of other localization systems for WSNs which the reader can find in the provided bibliography and in the current literature. For instance, in military applications, WSNs can be deployed in remote—possibly hostile—environments in order to perform tasks such as battlefield surveillance, enemy tracking, and security monitoring of military facilities. In these cases, security techniques must be implemented in order to provide a secure localization system. These techniques, as shown in Chapter 19, are used to propose a whole new set of localization systems for WSNs.

11.6 EXERCISES

1. What is the importance of localization systems in wireless sensor networks? What are the main applications for WSNs that differ from the applications in Ad Hoc Networks?

2. List and discuss the requirements of a localization system to be implemented in wireless sensor networks.
3. What is RSSI? What are that main advantages and disadvantages of using this technique?
4. An unknown node received the positions and estimated the distances from three different beacon nodes. The received positions were (6,3), (3,4), and (5,6) while the estimated distances were 2.3, 1.5, and 2, respectively. Show the system of equations obtained from the received data when computing the unknown node's position using trilateration. Solve the system using MATLAB.
5. Why is GPS not a good solution for the localization problem in WSNs?
6. Explain the key ideas of the APS algorithm and identify the main disadvantages and disadvantages using this APS localization algorithm?

BIBLIOGRAPHY

1. I. F. Akyildiz, W. Su, Y. Sankarasubramaniam, and E. Cyirci. Wireless sensor networks: A survey. *Computer Networks*, **38**(4): 393–422, 2002.
2. D. Estrin, L. Girod, G. Pottie, and M. Srivastava. Instrumenting the world with wireless sensor networks. In *International Conference on Acoustics, Speech, and Signal Processing (ICASSP 2001)*, Salt Lake City, Utah, June 2001, pp. 2033–2036.
3. M. Ilyas and I. Mahgoub. *Handbook of Sensor Networks: Compact Wireless and Wired Sensing Systems*, CRC Press LLC, Boca Raton, FL, 2004, Chapter 20.
4. G. J. Pottie and W. J. Kaiser. Wireless integrated network sensors. *Communications of the ACM*, **43**(5):51–58, 2000.
5. L. Doherty, K. S. J. Pister, and Laurent E. Ghaoui. Convex position estimation in wireless sensor networks. In *IEEE Conference on Computer Communications 2001*, Vol. 3, Anchorage, AK, April 2001, pp. 1655–1663.
6. A. Savvides, C.-C. Han, and M. B. Strivastava. Dynamic fine-grained localization in ad-hoc networks of sensors. In *7th ACM/IEEE International Conference on Mobile Computing and Networking*, Rome, Italy, 2001, pp. 166–179.
7. Y. Shang, W. Ruml, Y. Zhang, and M. P. J. Fromherz. Localization from mere connectivity. In *Proceedings of the 4th ACM International Symposium on Mobile Ad Hoc Networking and Computing*, Annapolis, MD, 2003, pp. 201–212.
8. D. Niculescu and B. Nath, Ad hoc positioning system (aps). In *IEEE Global Communications Conference (GlobeCom '01)*, San Antonio, TX, USA, November 2001, pp. 2926–2931.
9. J. Albowicz, A. Chen, and L. Zhang, Recursive position estimation in sensor networks. In *The 9th International Conference on Network Protocols*, November 2001, pp. 35–41.
10. N. Bulusu, J. Heidemann, D. Estrin, and T. Tran. Self-configuring localization systems: Design and experimental evaluation. *Transactions on Embedded Computing Systems*, **3**(1):24–60, 2004.
11. H. A. B. F. Oliveira, E. F. Nakamura, A. A. F. Loureiro, and A. Boukerche. Directed position estimation: A recursive localization approach for wireless sensor networks. In *Proceedings of the 14th IEEE International Conference on Computer Communications and Networks*

(IC3N '05), S. R. Thuel, Y. Yang, and E. K. Park, editors, San Diego, CA, October 2005, IEEE, New York, pp. 557–562.

12. H. A. B. F. Oliveira, E. F. Nakamura, A. A. F. Loureiro, and A. Boukerche. Error analysis of localization systems in sensor networks. In *Proceedings of the 13th ACM International Symposium on Geographic Information Systems (GIS '05)*, C. Shahabi and O. Boulcema, editors, Bremen, Germany, November 2005, ACM Press, New York, pp. 71–78.
13. A. Boukerche, H. A. B. F. Oliveira, E. F. Nakamura, and A. A. F. Loureiro. A Voronoi approach for scalable and robust dv-hop localization system for sensor networks. In Proceedings of 16th International Conference on, *Computer Communications and Networks, 2007 (ICCCN 2007)*, 2007, pp. 497–502.
14. A. Boukerche, H. A. B. F. Oliveira, E. F. Nakamura, and A. A. F. Loureiro. Localization systems for wireless sensor networks. *IEEE Wireless Communications—Special Issue on Wireless Sensor Networks*, **14**:6–12, 2007.
15. H. A. B. F. Oliveira, E. F. Nakamura, A. A. F. Loureiro, and A. Boukerche. Localization in time and space for sensor networks. In *AINA '07: 21st IEEE International Conference on Advanced Information Networking and Applications*, Niagara Falls, Canada, May 2007, pp. 539–546.
16. A. Boukerche, H. A. B. F. Oliveira, E. F. Nakamura, and A. A. F. Loureiro. Vehicular ad hoc networks: A new challenge for localization-based systems. *Elsevier Computer Communications*, **31**:2838–2849, 2008.
17. R. R. Brooks and S. S. Iyengar. *Multi-Sensor Fusion: Fundamentals and Applications*, Prentice Hall, Englewood Cliffs, NJ, 1998.
18. C. Intanagonwiwat, R. Govindan, and D. Estrin. Directed diffusion: A scalable and robust communication paradigm for sensor networks. In *Proceedings of the 6th ACM International Conference on Mobile Computing and Networking (MobiCom '00)*, Boston, MA, August 2000, , ACM Press, New York, pp. 56–67.
19. E. F. Nakamura, C. M. S. Figueiredo, and A. A. F. Loureiro. Information fusion for data dissemination in self-organizing wireless sensor networks. In *Proceedings of the 4th International Conference on Networking (ICN 2005)*, Reunion Island, April 2005.
20. J. C. Navas and T. Imielinski, GeoCast—geographic addressing and routing. In *Mobile Computing and Networking*, 1997, pp. 66–76.
21. J. Heidemann, F. Silva, C. Intanagonwiwat, R. Govindan, D. Estrin, and D. Ganesan. Building efficient wireless sensor networks with low-level naming. In *SOSP '01: Proceedings of the 18th ACM Symposium on Operating Systems Principles*, New York, 2001, ACM Press, New York, pp. 146–159.
22. Y. Yu, R. Govindan, and D. Estrin. Geographical and energy aware routing: A recursive data dissemination protocol for wireless sensor networks. Technical Report CSD-TR-01-0023, UCLA Computer Science Department, 2001.
23. Y. Xu, J. S. Heidemann, and D. Estrin. Geography-informed energy conservation for ad hoc routing. In *Mobile Computing and Networking*, 2001, pp. 70–84.
24. M. Do Val Machado, O. Goussevskaia, R. Aparecida de Freitas Mini, A. A. F. Loureiro, G. R. Mateus, and J. M. S. Nogueira. Data dissemination using the energy map. In *The Second Annual Conference on Wireless on demand Network Systems and Services (WONS 2005)*, St. Moritz, Switzerland, 2005. CD-ROM.
25. A. Boukerche, H. A. B. F. Oliveira, E. F. Nakamura, and A. A. F. Loureiro. A novel location-free greedy forward algorithm for wireless sensor networks. In *Proceedings of*

the 2008 IEEE International Conference on Communications (ICC 2008), Beijing, China, May 2008.

26. A. Boukerche, H. A. B. F. Oliveira, E. F. Nakamura, and A. A. F. Loureiro. Towards an integrated solution for node localization and data routing in sensor networks. In *ISCC '07: 12th IEEE Symposium on Computers and Communications*, Aveiro, Portugal, July 2007, pp. 449–454.
27. Brad Karp and H. T. Kung. Gpsr: greedy perimeter stateless routing for wireless networks. In *Proceedings of the 6th Annual International Conference on Mobile Computing and Networking*, Boston, MA, 2000, pp. 243–254.
28. S. Kumar, C. Alaettinglu, and D. Estrin. Scalable object-tracking through unattended techniques (scout). In *ICNP '00: Proceedings of the 2000 International Conference on Network Protocols*, Washington, DC, 2000, IEEE Computer Society, New York, p. 253.
29. K. Whitehouse and D. Culler. Calibration as parameter estimation in sensor networks. In *WSNA '02: Proceedings of the 1st ACM International Workshop on Wireless Sensor Networks and Applications*, New York, 2002, ACM Press, New York, pp. 59–67.
30. H. T. Friis. A note on a simple transmission formula. In *Proceedings IRE*, 1946, p. 34.
31. T. S. Rappaport. *Wireless Communications, Principles and Practice*, Prentice Hall, Englewood Cliffs, NJ, 1996.
32. T. He, C. Huang, B. M. Blum, J. A. Stankovic, and T. Abdelzaher. Range-free localization schemes for large scale sensor networks. In *MobiCom '03: Proceedings of the 9th Annual International Conference on Mobile Computing and Networking*, New York, 2003, ACM Press, New York, pp. 81–95.
33. J. Bachrach and C. Taylor. Localization in sensor networks. In *Handbook of Sensor Networks: Algorithms and Architectures*, I. Stojmenovic, editors, John Wiley & Sons, Hoboken, NJ, 2005.
34. B. Hofmann-Wellenho, H. Lichtenegger, and J. Collins. *Global Positioning System: Theory and Practice*, 4th edition, Springer-Verlag, Berlin, 1997.
35. J. Elson. Time synchronization in wireless sensor networks. Department Computer Sciences, University of California, Ph.D. dissertation, Los Angeles, 2003.
36. N. B. Priyantha, A. K. Miu, H. Balakrishnan, and S. Teller. The cricket compass for context-aware mobile applications. In *7th ACM International Conference on Mobile Computing and Networking*. Rome, Italy, July 2001. CD-ROM.
37. K. Whitehouse. The design of calamari: An ad hoc localization system for sensor networks. M.S. thesis, University of California at Berkeley, 2002.
38. D. Niculescu and B. Nath. Ad hoc positioning system (aps) using aoa. In *Proceedings of INFOCOM 2003*, San Francisco, CA, 2003.
39. G. H. Golub and C. F. Van Loan. *Matrix Computations*, 3rd edition, Johns Hopkins University Press, Baltimore, MD, 1996.
40. R. Jain. *The Art of Computer Systems Performance Analysis: Techniques for Experimental Design, Measurement, Simulation, and Modeling*, John Wiley & Sons, New York, 1991.
41. V. Ramadurai and M. L. Sichitiu. Localization in wireless sensor networks: A probabilistic approach. In *Proceedings of the 2003 International Conference on Wireless Networks (ICWN 2003)*, Las Vegas, NV, June 2003, pp. 275–281.
42. M. L. Sichitiu and V. Ramadurai. Localization of wireless sensor networks with a mobile beacon. In *Proceedings of the 1st IEEE International Conference on Mobile Ad Hoc and Sensor Systems (MASS 2004)*, FL, October 2004, pp. 174–183.

43. S. Simic and S. Sastry. Distributed localization in wireless ad hoc networks. Technical Report UCB/ERL M02/26, UC Berkeley, 2002.
44. P. Bahl and V. N. Padmanabhan. Radar: an in-building rf-based user location and tracking system. In *Proceedings of the IEEE Infocom 2000*, Vol. 2, Tel Aviv, Israel, March 2000, IEEE, New York, pp. 775–784.
45. Y. Shang and W. Ruml. Improved mds-based localization. In *IEEE Conference on Computer Communications 2004*, Vol. 4, March 2004, pp. 2640–2651.
46. P. Pathirana, N. Bulusu, S. Jha, and A. Savkin. Node localization using mobile robots in delay-tolerant sensor networks. *IEEE Transactions on Mobile Computing*, **4**(3):285–296, 2005.
47. N. B. Priyantha, A. Chakraborty, and H. Balakrishnan. The cricket location-support system. In *Mobile Computing and Networking*, Boston, MA, August 2000, pp. 32–43.
48. N. Bulusu, J. Heidemann, and D. Estrin. Gps-less low cost outdoor localization for very small devices. *IEEE Personal Communications Magazine*, **7**(5):28–34, 2000.
49. S. Capkun, M. Hamdi, and J.-P. Hubaux. Gps-free positioning in mobile ad hoc networks. *Cluster Computing*, **5**(2):157–167, 2002.
50. E. D. Kaplan, editors. *Understanding GPS: Principles and Applications*, Artech House, Norwood, MA, 1996.

CHAPTER 12

Location Discovery in Sensor Networks

ASIS NASIPURI

Department of Electrical and Computer Engineering,
The University of North Carolina at Charlotte, Charlotte, NC 28223

12.1 INTRODUCTION

Location estimation is a fairly mature topic that has interested researchers for many years for applications such as maritime and aircraft navigation, transportation, geological explorations, robotics, tactical missions, emergency systems, and more. It led to the development of a number of different location discovery systems that use a variety of techniques. Location discovery techniques range from using the direction and elevation of stars to sophisticated distance ranging techniques applied to RF, acoustic, and ultrasonic signals. The most popular of these location estimation systems is perhaps the geographical positioning system (GPS), in which a device uses RF signals received from a system of satellites to calculate its absolute geographical location. Currently, GPS receivers cost less than \$100 and can provide position estimates with errors within a few tens of meters. Despite these advances, location discovery has attracted renewed interest in recent times for a special area of application—that of wireless sensor networks. This is driven by special requirements and constraints in these networks that prohibit the use of existing technology such as GPS. Consequently, a number of novel approaches for location discovery have emerged.

Wireless sensor networks is an exciting new concept that emerged from advancements in embedded systems, wireless, and sensor technologies. A wireless sensor is an embedded system that may have an array of sensors (such as temperature, sound, light, magnetic field, chemical, etc.) as well as on-board microprocessor and memory that can be used for signal processing, storage, and on-the-fly networking. Such devices may be manufactured at low cost and very small form factor, making it possible for disposable use. A large number of such wireless sensors can form a self-configuring wireless sensor network for sensing multiple types of signals over wide regions for

Algorithms and Protocols for Wireless Sensor Networks, Edited by Azzedine Boukerche

extended periods of time without direct human intervention. Such networks have tremendous prospects for distributed monitoring applications at low installation and operating costs. It is envisioned that in the future, dust-sized smart sensors will be embedded in the physical world (buildings, roadways, automobiles, etc.) to perform a variety of tasks that include sensing, computation, communication, and actuation [1]. Examples of applications that have already received attention include environmental monitoring, agriculture, transportation systems, security systems, military applications, industrial monitoring and control, and many others.

The field of sensor networks opened an array of challenging research problems that include design issues on sensor technology, RF communications, networking protocols, information processing, database, algorithm development, power harvesting, and many more. Many of these issues are focused on developing techniques that enable maximum utilization of the limited resources in individual sensors—namely, processing, memory, battery, and so on—to optimize the sensing tasks of the network as a whole. A review of design issues and research directions on sensor networks may be found in reference 2.

In this chapter we focus on the issue of location discovery, which is a very important aspect in sensor networks. For instance, when a number of networked sensors are used for environmental monitoring, it is important to know the locations of the signals obtained (temperature, humidity, etc.). In addition, many of the networking protocols for sensor networking also require location information. Examples include geographic query forwarding [3], location aware routing [4], and multisensor information fusion [5]. Since sensor networks consist of a large number of nodes, it is not practical to deploy the nodes in preestablished locations. Moreover, due to limitations in hardware, cost, and size, many of the preexisting location estimation techniques cannot be applied to sensor nodes. Hence, there has been tremendous interest for developing new low-cost and low-complexity techniques that would enable wireless sensor nodes to determine their locations after they have been deployed in an area of interest. This problem is commonly referred to as *localization* [6] within the scientific community involved with research on wireless sensor networks. The specific design challenges, state of the art, and research issues on localization in sensor networks are presented in this chapter.

12.2 LOCATION DISCOVERY CONCEPTS

Existing localization schemes rely mostly on the principle of *triangulation*. This refers to the process of using *distances* or *angles* from three reference points to an unknown point to compute its position on two-dimensional space. For three-dimensional space, an additional reference point is needed.

Lateration: If the distances from an unknown location to three reference points are known, the location of the unknown point can be obtained using geometrical calculations known as multilateration as shown in Figure 12.1a. Here, the unknown location is obtained as the point of intersection of three circles characteristics in free space is modeled as $1/d^2$, whereas in terrestrial conditions it is determined as $1/d^\alpha$, where $2.5 < \alpha < 4.5$ depending on the terrain and other environmental factors [7]. The path loss factor α may be obtained experimentally for the specific environment where it is to be used. This is often a cost-effective method for distance estimation.

- *Time of Flight:* A useful technique for determining the distance between an RF transmitter and a receiver is to measure the time taken for the RF signal to cover the distance. Here, a receiver measures the time of arrival (ToA) of a signal transmitted from reference point to compute the distance using the velocity of propagation of the signal. This requires the receiver to have accurate time synchronization with the transmitter, which is often hard to implement. An alternative is use an additional transmitter to act as a frame of reference, where the receiver estimates the time differences of arrivals (TDoA) between signals from multiple transmitters to determine distances. These distances can be used for localization using multilateration. Such a technique is fairly accurate when the distances involved are long, such as between a ground-based receiver and a system of satellites orbiting the earth, as used in GPS. However, distance ranging from time of flight is hard to implement for short distances because RF signals propagate in the speed of light and it requires extremely accurate hardware to detect small time differences.

12.2.2 Mechanisms for Estimating Angles

Most of the earlier methods for angle estimation were based on optical methods, such as determining the direction of stars for air and sea navigation. Some of the more technologically advanced methods for angle estimation are as follows:

- *Use of Directional Antennas.* Large phased-array antennas have gain patterns that can be sufficiently directional to allow determination of the angle of arrival of an RF signal. This technique was proposed and evaluated in reference 8 for designing a system for tracking the locations of transportation vehicles. It used rotating directional antennas in each vehicle to determine the angles of arrival of RF signals from fixed base stations for localization. However, even the most expensive phased array antennas can produce errors in angle estimation, which is the primary drawback of the system. It also involves complex and bulky instrumentation (rotating directional antenna arrays) at the receiver.
- *Use of Angular Signatures.* Another example of localization using angulation is the VHS omnidirectional ranging or VOR systems that are used for aircraft navigation. Here, specific RF signals are transmitted from ground-based VOR stations that allow receivers on flying aircrafts to determine the angles from the stations [9]. A simple implementation of this idea involves a combination of periodic transmissions of bursts of an omnidirectional signal along with a continuously rotating beam emitted from the VOR station. A receiving aircraft can determine the angular orientation with respect to the VOR station from the

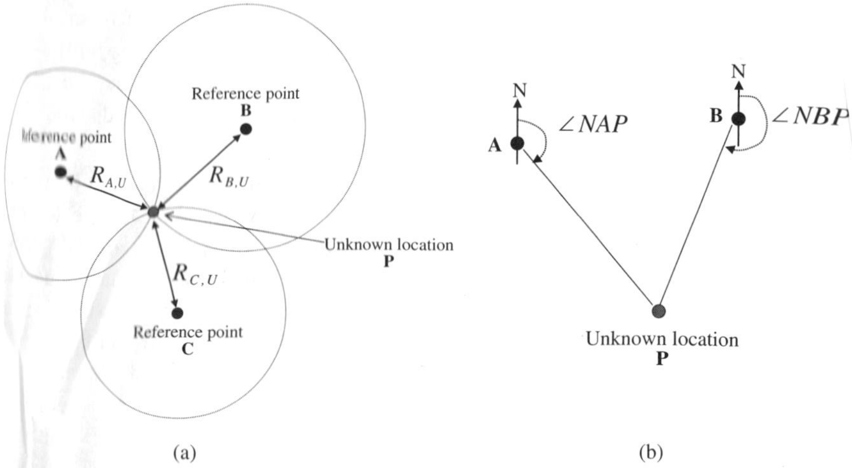

Figure 12.1. Triangulation using (a) lateration, which shows that the unknown location is obtained as the intersection of three circles, and (b) angulation, where the unknown location is obtained from the intersection of lines with appropriate angles.

centered at the reference points having radii equal to the distances from these reference points.

Angulation: Alternatively, the angles from two known reference points may be used to determine the location of an unknown point, as shown in Figure 12.1b. This mechanism is known as angulation. Here, the unknown location is obtained by the point of intersection of the lines drawn from the reference point with appropriate angles.

The key issue for implementing a system that employs triangulation is to measure either the distances or angles from a set of fixed stations acting as reference points. We next describe the methods that can be applied for measuring these.

12.2.1 Mechanisms for Estimating Distances

Some of the known methods for estimating the distance between two points are as follows:

- *Received Signal Strength Indicator (RSSI).* This utilizes the predicted propagation characteristics of a wireless signal to determine the distance from a transmitter. If the power of the transmitted signal is known, a receiver can estimate the distance from the transmitter by measuring the power of the signal that it receives from the transmitter. Although simple and cost effective, this technique suffers from significant errors as the path loss characteristics of wireless signals is highly dependent on the physical environment. For instance, the path los observed time difference between the flash and the rotating beam. The combination of angles estimated from two VOR stations and their locations can be used to determine the position of the aircraft.

12.3 ISSUES ON LOCALIZATION IN SENSOR NETWORKS

With these ideas on location discovery methods, we now turn our attention to the special requirements and issues for location discovery in sensor networks. In most cases, sensor network applications require that the wireless sensor nodes are deployed randomly in an area of interest. Applications may also require sensor nodes to be deployed by dropping them from flying aircrafts and requiring the nodes to self-configure into a wireless network. Because of such deployment methods and the large number of nodes involved, special techniques are required for the nodes to determine their locations after deployment.

Why Not GPS?. Although GPS receivers have evolved to become smaller in size, cheaper, and accurate, several reasons prohibit the use of GPS in most sensor network applications. Firstly, GPS requires the receiver to receive data from four or more satellites for functioning. This is not possible in indoor environments or in locations where foliage and other cover obstructs satellite signals. Secondly, limitations of size, battery, and hardware resources of sensor nodes is prohibitive for using GPS hardware in every node. This is particularly true for wireless sensor nodes of the future, which are envisioned to be of the size of a few cubic millimeters [10]. Thirdly, the addition of GPS hardware would make sensor nodes costly, which clashes with one of the design objectives of sensor nodes. Finally, in most applications of wireless sensor networks, only relative locations of the nodes are needed and not their absolute geographical positions. Hence, a GPS device is considered as an overkill in terms of cost and hardware requirements for applications involving wireless sensor nodes.

12.3.1 Requirements for Localization in Sensor Networks

The special requirements for localization schemes for sensor networks generally depend on the nature of the applications and the constraints imposed by hardware and network infrastructure. Based on these, some of the specific issues concerning the design of the localization scheme are as follows:

1. *Absolute versus Relative Locations.* Systems such as the GPS determine the absolute location in terms of the latitude, longitude, and altitude with respect to the earth's coordinates. Alternatively, locations may be obtained with respect to a given frame of reference, such as the location of a base station that acts as the coordinator of the sensor network. It could also be an arbitrary reference point at the scene of application, such as a location inside a building. In almost all applications of sensor networks, only *relative locations* are required for

performing the monitoring, processing, and communication tasks. Note that a relative location can always be transformed to an absolute location if the absolute location of the reference point is known.

2. *Accuracy and Precision of Localization.* Accuracy implies the closeness of the location estimate to the ground truth. This may depend on a number of factors. Each system has its own *granularity* of measurements, which describes the smallest measurable distance. Depending on the technology and mechanism used, localization schemes may have a granularity of measurements within a few inches or within a room (tens of meters) or even larger. The required granularity of localization in sensor networks depends on the application. In most cases, it depends on the relative distances between nodes in the network. A second issue is the *precision*, which describes the consistency of the estimates. As a frame of reference, some expensive GPS receivers can estimate locations with errors smaller than 10 m with 95% precision.
3. *Scale of Measurements.* A different, but perhaps related issue is the *scale* of measurements, which describes the largest distance over which the localization system can work. GPS can obtain location estimates over the whole world. The required scale of localization in sensor network is the total area of the network, which makes the implementation simpler.
4. *Dynamics of the Nodes.* If the nodes of a network are mobile, the localization system has to face the challenge of taking repeated measurements to continuously track its location. The repetition rate will depend on the dynamics of the network. Most sensor network applications require the nodes to be static after deployment, although dynamic scenarios also exist (such as using wireless sensors on robotic platforms). Hence, it is generally required that the localization system for a sensor network should be able to track displacements of the sensor nodes when necessary.
5. *Communication Requirements.* In addition to using passive measurements of some reference signals for localization, such as those used for schemes employing pure triangulation, localization schemes may benefit from information exchanged between the nodes in the network and/or some reference nodes, such as beacons. Existing systems vary in the communications required for position estimation. For instance, a GPS receiver does not transmit but performs localization by only receiving the RF signals from satellites. On the other hand, most cellular telephone localization schemes rely on two-way communication between the cell phone and the base stations. Communication between a sensor node and a reference node or other sensor nodes can provide important benefits such as time synchronization and improvements in accuracy and precision. However, a central issue in wireless sensor networks is the minimization of communication requirements in the nodes to conserve battery. This introduces special considerations for designing the localization scheme as well.
6. *Self-Localization versus Remote Localization.* The most desirable scheme for localization in this respect is one in which the sensor nodes can self-localize—that is, determine their own positions from signals received from external

sources. Alternatively, localization schemes can allow a remote station, such as a base station, to determine the locations of all sensor nodes in the network from signals received from them. Such remote localization schemes suffer from scalability and communication cost problems. It should be noted that the resulting difference between self- and remote positioning is where the computations take place and the resultant need for communication between the sensor nodes and the remote station.

7. *Cost.* This is an important issue, since the requirements for designing large-scale sensor networks are to (a) keep the cost (and complexity) of each node low and (b) benefit from the collective sensing and computation capabilities of a large number of nodes in the network. Hence, it is highly desirable that the localization system does not require expensive hardware at the sensor nodes. The cost of building external infrastructure for enabling localization also plays a role. However, that is usually considered less critical because it does not increase with the size of the network.
8. *Form Factor.* Wireless sensor nodes must have a small form factor, which eliminates localization schemes that require large components such as antenna arrays. The size of the sensor node thus plays a critical role in determining the mechanism that can be effectively used for localization in sensor networks.
9. *Passive versus Active Localization.* Some location estimation schemes require the unknown node to play an active role for position estimation, while others can work even without any action from the target node requiring localization. In sensor networks, the sensor nodes can allow active localization within the limitations of its hardware and cost constraints. For instance, a sensor node may act as a transponder for a transmitted RF signal to estimate round-trip transmission delay between a remote transmitter and the node. However, it may be difficult to implement a system that requires the sensor node to estimate the direction of arrival of the received signal and send that information back because that usually requires complex antenna arrays.

12.3.2 Challenges

From the above discussion, it can be concluded that the main challenges for designing a localization scheme for wireless sensor nodes arise from the need to deal with the low hardware complexity and cost of implementation, small form factor of the nodes, and their arbitrary locations (indoor, outdoor, and in uncharacterized regions). Typically, the natural choice for location estimation is to use triangulation, which requires estimation of distances or angles from fixed reference points. In this section, we discuss the challenges involved with obtaining these estimates from the perspective of sensor networks.

Ranging Issues. Of the two primary techniques for ranging, measuring time of flight using RF signals is difficult for applications in sensor networks. This is because RF signals travel at the speed of light and measuring the extremely short travel times within the spacial domain of a sensor network (that can be room sized or of equivalent size) is a technical hurdle. It would require extremely accurate clock synchronization between the sensor node and the transmitter that remains to be a technological challenge. Consequently, time-of-flight measurements have been explored using acoustic and ultrasonic signals that have a much slower propagation speed than RF. Acoustic ranging has been used in many different distance measuring devices due to their low cost and accuracy for indoor use. These devices can provide accuracies within 10 cm. The comparatively higher frequency utrasonic signals (frequencies typically within 24–40 kHz) can be used to provide accuracies within 2–5 cm. However, they have a smaller range as compared to acoustic signals. Usage of acoustic or ultrasonic signals for ranging in sensor networks also faces other challenges. Firstly, acoustic sources and detectors are generally larger in size as compared to RF sources, due to their larger wavelength. This poses a problem for small sensor nodes. Secondly, again due to their larger wavelength, acoustic signals cannot propagate through physical obstructions. They also suffer from severe multipath effects, making it hard to design a reliable distance estimation system for arbitrarily placed sensor nodes in unknown environments. Systems such as Cricket [11] and BAT [12] use the time-difference-of-arrivals of ultrasound and RF signals from the same source to perform indoor localization. Assuming that the RF signal is received instantaneously, the corresponding delay of the much slower ultrasound signal provides the needed distance estimate.

Usage of RSSI for ranging requires the knowledge of the corresponding signal propagation model. However, even with extensive channel estimation and modeling, ranging using RSSI faces inaccuracies caused by shadowing, multipath reflections, refractions, and scattering effects. Nevetheless, because of the ease of implementation, localization schemes using RSSI have been researched extensively [6]. It has been applied in a number of localization schemes such as RADAR [13], AHLoS [14], and APS [15]. The RADAR indoor location system applies a wall attenuation factor (WAF) and signal strength maps that are obtained from extensive offline measurements of indoor signal propagation characteristics. AHLoS assumes that a limited number of nodes know their locations, either from using GPS or from manual configurations. Other nodes use a combination of RSSI and ToA ranging techniques to determine their positions with respect to the beacons. AHLoS utilizes collaborative multilateration, a mechanism that is explained in the next section. A decentralized approach to RF-based localization in indoor environments was presented in reference 16.

Angle of Arrival (AOA) Estimation. Estimating the direction of arrival of a wireless signal requires an antenna with extremely small beamwidth. This can be achieved with an antenna array, which can be prohibitvely large for use in wireless sensor nodes. Hence alternative techniques need to be used for AOA estimation for applications in sensor networks. A possible approach to reduce the size of the antenna is to use ultrasound signals, which is used in the Cricket Compass project [17] to determine angles from phase difference and time difference of arrivals of an ultrasound pulse on multiple detectors that are placed in a specific pattern within a space of few centimeters. Such a device can determine the angle of arrival with accuracy of 5° within an angle of $\pm 40^\circ$. A key requirement for the success of this mechanism is having line of sight from the source. Since such AOA estimation heavily relies on phase differences, multipath and scattering can also cause problems. AOA estimation has been used in references 15, 18 and 19. In reference 18, rotating optical beacons are used, and the angle is measured at the sensors by the times at which the optical signal is detected. The concept used in reference 19 is similar, except that it used RF signals from an 802.11 transmitter using a directional antenna. To reduce errors in AOA estimation caused by the nonzero beamwidth of the RF signal, the center of the beam was detected from the angle where the signal strength is maximum.

12.3.3 Other Technical Challenges

The errors in distance or angle measurements need to be minimized either by multiple measurements at the same node or by combining measurements from other nodes in the vicinity, a method that is known is collaboration. In either case, a common requirement is the need for optimization operations that requires extensive computations and data handling. This must be done within the limitations of processor capabilities, memory, and other hardware of the wireless sensor nodes. In addition, collaborative computation also taxes the battery of the nodes from communication. Additional complications arise when collaboration requires routing, which is dependent on location information, and consequently, nodes need to have localization done *before* they can collaborate.

12.4 LOCALIZATION USING RANGING IN SENSOR NETWORKS

A significant amount of work has been reported on the development new techniques that are applicable for location discovery of small, low-powered wireless nodes in a sensor network. Several of these ideas have also been demonstrated in proof-of-concept implementations and commercially available products. In this section, we describe a few of these developments that use ranging as the primary mechanism for localization. Other localization schemes using AOA and alternative schemes are presented in the following sections.

12.4.1 Localization Using Ranging

Localization schemes that use ranging generally address the multilateration problem where the *distance estimates have errors* that are caused by problems with accurate ranging as described in the previous sections. With ranging errors, the three circles centered at the reference points with radii equal to the corresponding distance estimates (shown in Figure 12.1a) do not intersect at a common point. The solution to the localization problem then shifts from a geometric solution to one of *estimation*, where one has to determine the optimum location coordinates that minimizes an objective function involving the estimated distances and the unknown node location. This is explained below with reference to the scenario depicted in Figure 12.2.

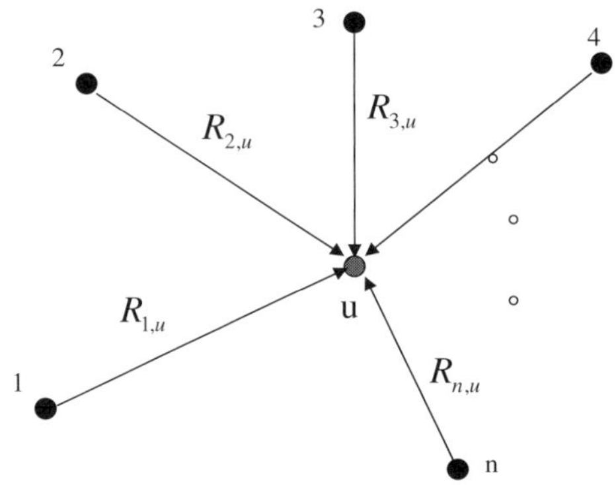

Figure 12.2. Illustration of atomic multilateration for estimation of a location from erroneous distance measurements.

Let (x_u, y_u) be the unknown location of a node **u** that is to be determined, and let (x_i, y_i) be the location of the ith reference node, $i = 1, 2, \cdots, n$. Most often the reference nodes are called *beacons* as they generate RF, acoustic, or ultrasound signals to allow range estimation from the sensor node [14]. If $R_{i,u}$ be the estimated distance between the unknown sensor node and the ith beacon, then the error in this estimate is given by

$$f_i(x_u, y_u) = R_{i,u} - \sqrt{(x_i - x_u)^2 + (y_i - y_u)^2} \qquad (12.1)$$

and consequently, the objective function to minimize for estimating the unknown location (x_u, y_u) can be expressed as

$$F(x_u, y_u) = \sum_{i=1}^{n} f_i^2(x_u, y_u) \qquad (12.2)$$

The solution has been addressed in detail in reference 14, where it is termed as *atomic multilateration*. A solution to linearizing the equations for obtaining the minimum mean square estimate (MMSE) of the unknown location is also proposed under the assumption that the ranging errors in Eq. (12.1) has a Gaussian distribution. The following are some important issues in this regard:

- The solution requires at least three reference nodes or beacons from which distance estimates have been obtained at the sensor node. The locations of these beacons must be known. Also, these beacon nodes should not be located on a straight line.
- If the speed of propagation of the beacon signal that is used for ranging is unknown, it can be estimated by using mesurements of time-of flight of the beacon signal from at least four beacon nodes.

- In a typical scenario of a wireless sensor network, it is likely that a few of the nodes may know their locations using GPS or some other mechanism (perhaps even from manual deployment). These nodes may act as beacons for other nodes to localize in the network. This introduces the issue of solving the problem in a multihop network environment.

12.4.2 Localization over Multiple Hops

A more general scenario for a localization problem in sensor networks is depicted in Figure 12.3. Here, the dark nodes are assumed to know their locations and the rest (unfilled) nodes are not aware of their locations. The problem is to determine the locations of as many of the unfilled nodes as possble from the information at the dark nodes. Each node may transmit beacon signals for other nodes in its neighborhood to estimate its distance from it. Two solutions to this muiltihop localization problem have been explored in reference 14, referred to as the *iterative multilateration* and *collaborative multilateration* algorithms.

Iterative Multilateration. This algorithm uses the method of determining the locations of unknown nodes from those of the known nodes using atomic multilateration and then propagating this knowledge to other unknown nodes. An unknown node acts as a beacon after its location has been determined. The concept is depicted in Figure 12.4. With sufficient number of beacon nodes to start the process, this algorithm can allow all nodes to self-localize iteratively. Moreover, the problem can be solved in a totally distributed manner. A concern with this algorithm is that errors tend to accumulate from one iteration to the next as more unknown nodes start acting as beacons. An alternative where all distance measurements are passed to a central node for *centralized* processing for localization of all nodes in the network can also be considered. This has the advantage that the algorithm can start from the node that has the maximum number of beacons in its neighborhood, thereby reducing the error at the starting point.

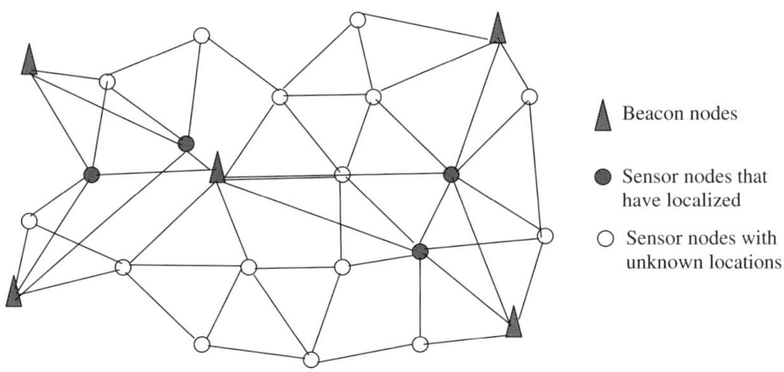

Figure 12.3. Illustration of a multihop network with beacons, nodes with known locations and nodes with unknown locations.

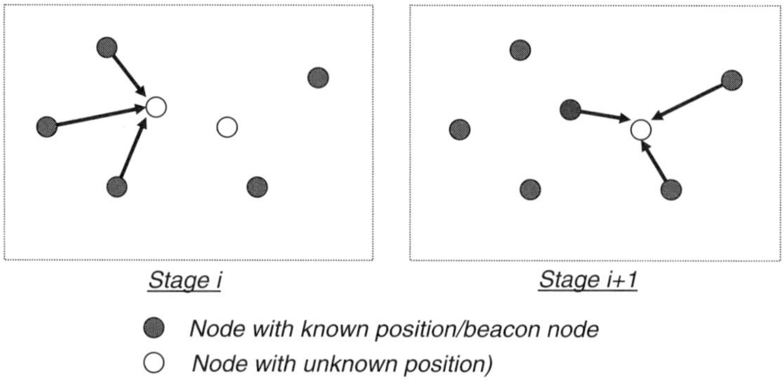

Figure 12.4. Illustration of the iterative multilateration principle.

Collaborative Multilateration. The collaborative multilateration algorithm addresses the problem that one or more unknown nodes may not have a minimum of three beacon nodes in its neighborhood to perform localization using atomic multilateration. An example is shown in Figure 12.5. The proposed solution uses all distance estimates over multiple hops and solves for the unknown locations simultaneously. For the scenario depicted in Figure 12.5, where $R_{i,j}$ represents the estimated distance between nodes i and j, and (x_i, y_i) is the location of node i, the collaborative multilateration algorithm attempts to solve the following problem:

$$\min_{x_3, y_3, x_4, y_4} \left[f_{2,3}^2 + f_{1,3}^2 + f_{4,3}^2 + f_{4,5}^2 + f_{4,6}^2 \right] \tag{12.3}$$

where $f_{i,j}$ represents the expression for error in the estimated distance between nodes i and j, that is,

$$f_{i,j} = R_{i,j} - \sqrt{(x_i - x_j)^2 + (y_i - y_j)^2} \tag{12.4}$$

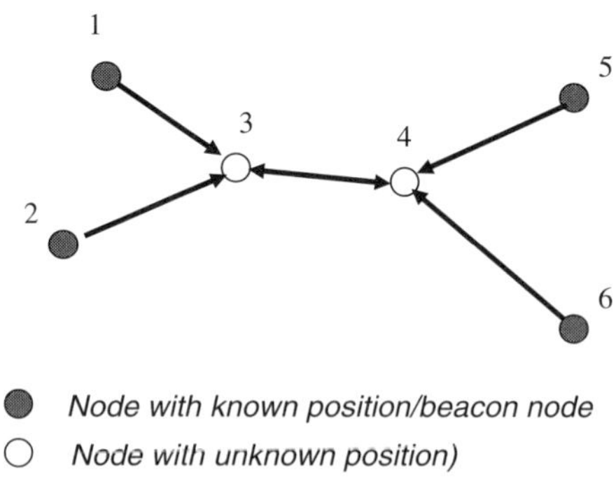

Figure 12.5. Illustration of the collaborative multilateration principle.

Obviously, this problem is more complex to solve. A solution using initial estimates and an iterative least squares method for refinement has been described in reference 14. The solution can be achieved under the assumptions that there is sufficient connectivity, a sufficient number of beacons nodes exist in the network, and the ranging errors have a Gaussian distribution. The authors also present ideas for optimizations using clusters or subtrees as well as distributed computations.

12.5 LOCALIZATION USING ANGLE ESTIMATION IN SENSOR NETWORKS

Since estimation of distances is prone to errors, efforts have also been directed toward localization using angle estimation in sensor networks. The basic idea of angulation is described in Figure 12.1b, where it is assumed that the angles $\angle NAP$ and $\angle NBP$ as well as the locations of reference nodes **A** and **B** are known to the sensor node **P**.

Note that determination of the angles with respect to an absolute angular reference (i.e., north) also requires the knowledge of the orientation of the node. To avoid that, angulation may be performed by determining *relative angles*, that is, the angular spans between nodes **A**, **B**, and **C**, as shown in Figure 12.6. Here, a sensor node located at an unknown location **P** determines the angles α and β, sustained between reference points **A** and **B**, and **B** and **C**, respectively. The sensor node then computes its location (X_p, Y_p) as follows:

$$X_p = x_2 + C\cos(\gamma - \eta), \qquad Y_p = y_2 - C\sin(\gamma - \eta) \tag{12.5}$$

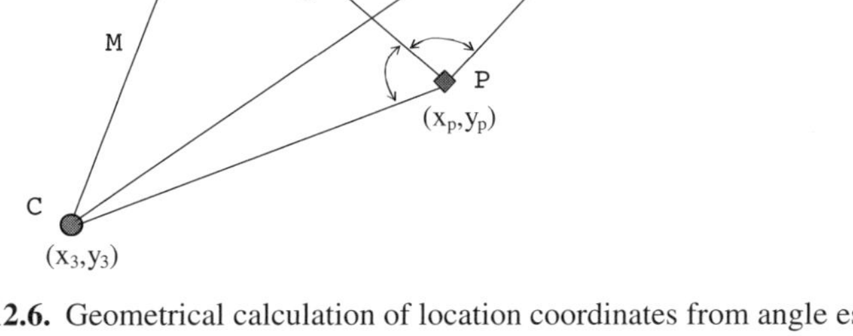

Figure 12.6. Geometrical calculation of location coordinates from angle estimation.

where

$$
\begin{aligned}
\eta &= \tan^{-1}\left(\frac{y_1 - y_2}{x_1 - x_2}\right) \\
\gamma &= \tan^{-1}\left(\frac{R\sin(\alpha) - \sin(\beta)\cos(A) - \cos(\beta)\sin(A)}{\sin(\beta)\sin(A) - \cos(\beta)\cos(A) - R\cos(\alpha)}\right) \\
C &= \frac{L\sin(\alpha + \gamma)}{\sin(\alpha)} \\
R &= \frac{L\sin(\beta)}{M\sin(\alpha)} \\
A &= \cos^{-1}\left(\frac{S^2 - M^2 - L^2}{2ML}\right) \\
S &= \sqrt{(x_1 - x_3)^2 + (y_1 - y_3)^2} \\
M &= \sqrt{(x_1 - x_3)^2 + (y_2 - y_3)^2} \\
L &= \sqrt{(x_1 - x_2)^2 + (y_1 - y_2)^2}
\end{aligned}
\tag{12.6}
$$

Hence, with a mechanism that allows each node to determine the angles sustained between reference nodes **A** and **B**, and **B** and **C**, resepctively, the network can self-localize without the need for collaboration or centralized computations. A method for doing this was proposed in reference 20 and a prototype implementation of the scheme using the Crossbow *mica2* wireless sensor nodes was presented in reference 18. The principle of operation of this AOA-based localization scheme is presented next.

12.5.1 AOA Based Localization Using Rotating Directional Beacons

The principle of operation is somewhat similar to VOR stations that are used in aircraft navigation. The system uses three specially constructed beacon generators to facilitate angle determination at the sensor nodes. Each beacon generator transmits a *directional* beacon signal that *rotates* with a fixed angular velocity. The sensor nodes calculate the angles α and β by determining the times of arrival of the directional beams from different beacon generators. It is assumed that the directional beacon signals have very narrow beamwidths on the horizontal plane so as to enable accurate estimation of times when the beams are pointed toward a sensor. They should have sufficiently large vertical beam angles, so that the beams cover the entire sensor network area; that is, they cover sensors located at different vertical angles with respect to the transmitter. The beacon signal can be RF or optical, as long as it is possible to transmit it at a narrow beam. A schematic illustration of the proposed implementation is depicted in Figure 12.7a, which assumes the special case where the beacon generators (reference points) are located at the vertices of a rectangle.

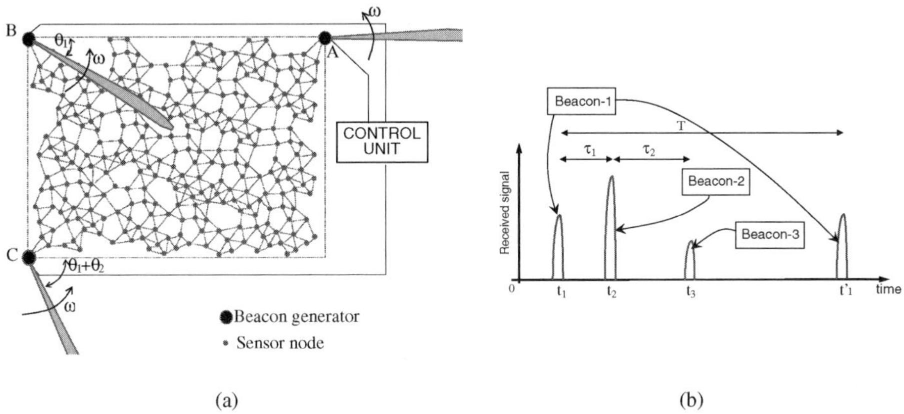

Figure 12.7. Illustration of (a) the system model for AOA-based localization and (b) the times of arrival of beacon signals at a sensor node.

The main considerations required for successful implementation of this scheme are as follows:

- *Identical Angular Velocities.* All three beacon signals must be rotating with identical angular velocities. However, any arbitrary angular velocity may be used, which can be measured at the sensor nodes by noting the time period of rotations.
- *Beacon Identification.* The sensor nodes should be capable of identifying the beacons from their signals. This can be implemented by various mechanisms, such as using coding or modulation of the beacon signals.
- *Phase Reference.* The phase differences between the rotating beacon signals (θ_1 between **A** and **B** and θ_2 between **B** and **C** in Figure 12.7) must be known at the sensors. To remove this problem, θ_1, θ_2, and θ_3 may all be made zero by appropriate initial configurations.

Each sensor will then receive the beacon signals periodically as shown in Figure 12.7b. The required angle estimates α and β can then be obtained by measuring the times t_1, t_2, and t_3 as follows:

$$\alpha = \omega(t_2 - t_1) - \theta_1$$
$$\beta = \omega(t_3 - t_2) - \theta_2 \tag{12.7}$$

Possible Sources of Error. The primary source of error in the above technique is due to inaccurate angle estimation that can be caused by nonzero beamwidth of the beacon signals. It is difficult to get very narrow beams of RF signals unless large antenna arrays are used at the beacon generators. To avoid such expensive infrastructure in case RF signals are used, the sensor nodes may be allowed to detect the center of an incoming RF beam by detecting the time at which the received signal strenght is maximum [19]. Alternatively, higher directionality of the beacon signals can be achieved by using optical signals, such as a laser, as described in reference 18.

Errors will also be caused by nonidentical angular speeds and errors in estimating the initial phase references θ_1, θ_2, and θ_3. However, these are relatively easy to correct by appropriate infrastructure control mechanisms.

Experimental Implementation. To provide a deeper understanding of the issues concerning implementing such a scheme, details of an experimental prototype development of this scheme are described in this section. The prototype was designed with the goal of enabling Crossbow manufactured *mica2* wireless sensor nodes (the Berkeley motes) to self-localize with minimum additional cost in a laboratory setting. Each *mica2* mote consists of the MPR410CB processor and radio platform that is equipped with an Atmel ATmega128L processor, 128-kB program flash memory, 512-kB measurement flash, a 10-bit analog to digital converter (ADC), and 433-MHz radio interface. The motes were equipped with the MTS310 sensor board that has a photosensor along with a number of other analog sensors such as microphone, accelerometer, thermistor, and a magnetometer. The processor runs the Tinyos software operating system developed by UC Berkeley, which supports large-scale self-configuring sensor networking [21].

The beacon generators use optical lasers that are suitable for indoor use and are also cost effective. The primary element in these modules is an Apinex 3-mW 650-nm semiconductor laser equipped with a diffraction grating for line generation and a plastic lens that can be used to adjust the width and fan angle of the generated optical line. In order to make the three beacon signals identifiable, a varying number of laser line generators were used for each beacon—that is, a single laser line generator for beacon-1, a double line for beacon-2, and a triple line for beacon-3.

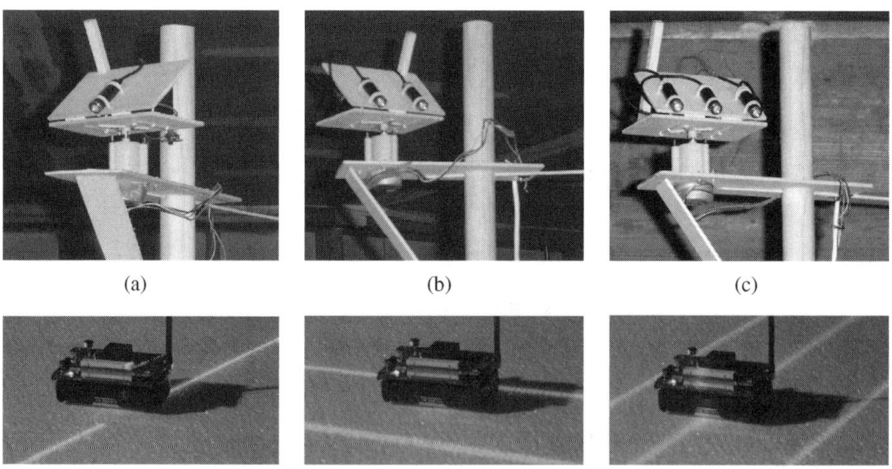

(a) (b) (c)

(d) (e) (f)

Figure 12.8. Beacon generator assemblies for (a) beacon-1, (b) beacon-2, and (c) beacon-3. The corresponding beacons signals are depicted in (d), (e), and (f), respectively.

The problem of maintaining identical angular rotations for all three beacon generator assemblies was solved by using stepper motors, all driven by a common controller. The assembly of laser sources, gearbox for speed control, and stepper motor at each beacon generator were mounted on a pole at a height of 6 feet with the axes of the lasers pointing down at an approximate angle of 45° (see Figure 12.8). Since initial phase errors are critical for location accuracy and it is very difficult to align three independent sources perfectly, a *pilot sensor node* is located at the farthest corner of the area for *estimating* the angles θ_1 and θ_2. Initially all sensor nodes are set to standby until the pilot node determines these phase references. Once that is done, the pilot node broadcasts the estimates, which triggers the nodes in the network to start self localizing by measuring the angles α and β at their locations. Each mote is programmed to identify single, double, and triple beams, by counting optical signals detected within a small window of time. Adequate measures are incorporated to reduce missed and false detections from overlapping or interfering signals. The details are shown in Figure 12.9.

```
/* Wait for pilot message and then start sampling and localization */
if packet received from pilot then
   begin
      read θ1 and θ2 from pilot message
      start obtaining periodic samples s[i]
      /* Begin beam counter window when first signal peak is detected*/
      if ((s[i]>s[i-1]+8) AND (win_flag==0)) then
         begin beam++; win_flag:=1; end
      timer++;
      /* Count beams within beam counter window */
      if (win_flag==1) then
         begin
            win_time++;
            if ((s[i],s[i-1]-8) AND (s[1-1]<s[i-2])) then
               begin beacon++ end
            if end of beam counter window then
               begin
                  detected[beam]:=beacon; t[beam]:=timer;
                  win_flag:=0; beacon:=0; win_time:=0;
         end
      /* At the end of cycle, check for errors and localize */
      if ((beam==4) AND (win_flag==0)) then
         begin
         if (detected[1]+detected[2]+detected[3]==6) then
            begin
               ω:=2π/(t[4]-t[1]);
               if (detected[1]==1) then
                  begin
                     α:=ω(t[2]-t[1])-θ1; β:=ω(t[3]-t[2])-θ2;
                  end
               if (detected[1]==2) then
                  begin
                     α:=ω(t[4]-t[3])-θ1; β:=ω(t[2]-t[1])-θ2;
                  end
               if (detected[1]==3) then
                  begin
                     α:=ω(t[3]-t[2])-θ1; β:=ω(t[4]-t[3])-θ2;
                  end
               (xp,yp)=locate(α,β)   /* apply equation (5) */
            end
            beam:=0; timer:=0
         end
   end
```

Figure 12.9. Algorithm for self-localization using angle estimation used at the sensor nodes.

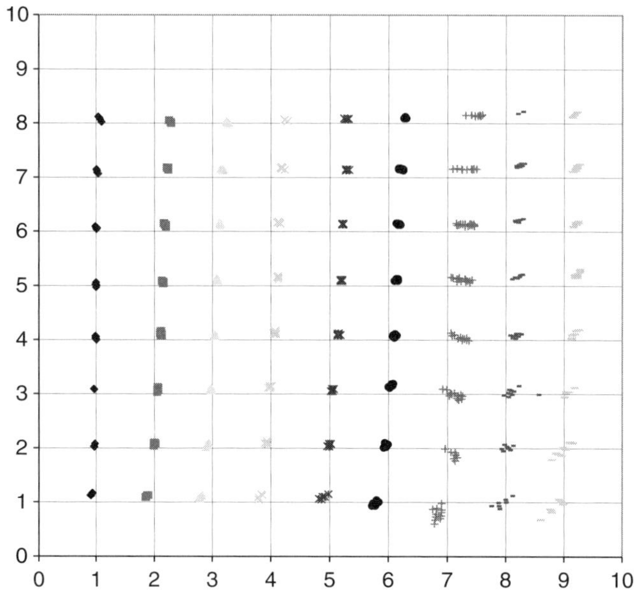

Figure 12.10. Experimentally obtained locations of sensors that are placed in 8 × 8 grid at separations of 1 ft. The angular speed of rotation of beacons was fixed at 9.5 rpm.

Experimental evaluation of the performance of the system shows that when used in a 10-ft × 10-ft area, errors in localization is less than 4 inches and the maximum time required for any sensor to localize is 45 seconds. Localization results of nodes located 1 ft apart in the area are shown in Figure 12.10.

Although this scheme needs special considerations for designing the beacon generators, the attractive feature is that the sensors do not need any special hardware except for being able to detect the beacon signal (i.e., photodetector), which is usually present in most sensor nodes. The primary drawback of this scheme is its requirement for line of sight. Within indoor environments, this can be mostly achieved for most locations by mounting the beacons at the ceiling. For those nodes that do not have line of sight from all three beacons, additional methods may be used, such as atomic or iterative multilateration, to be performed after a sufficient number of nodes in the network have self-localized.

12.6 OTHER LOCALIZATION TECHNIQUES

While the previous sections described the key approaches to localization in sensor networks, the list of new ideas keep growing and it is difficult to capture all the existing work in this rapidly developing topic. Here, samples of some other well-known approaches are presented.

12.6.1 Range-Free Position Estimation

Range-free techniques are those where location estimates are obtained without using concrete distance or angle measurements from specific nodes. The principle here is to determine an unknown location from the *proximity* to beacon nodes or nodes with known locations. This avoids problems arising from errors in range or angle estimation, and it typically requires less costly hardware and simpler calculations. However, they generally have lower precision of localization than those using range-based methods.

A popular example of range-free localization is the *centroid method* proposed in reference 6. Here a node counts the number of beacon signals received from a set of pre-positioned beacon nodes and achieves localization by obtaining the centroid of the received beacon generators. Precision is dependent on the density and locations of the beacon generators.

The *DV-HOP* solution uses locations of anchor nodes (which are special nodes that broadcast control packets to enable localization), the hop counts from anchors, and the average distance per hop for localization. It uses a mechanism that is similar to classical distance vector routing. The anchor nodes broadcast beacon packets that are flooded throughout the network. The beacon packets carry hop counts and the locations of the corresponding anchor. Each receiving node maintains the minimum counter value from each anchor, thereby allowing them to determine the shortest hop distance to each anchor. The location is estimated using average hop distance from anchors.

In reference 22, the authors present a *Point-in-Triangulation* (PIT) test, where nodes use a set of signal strength measurements from neighboring beacon nodes to determine the closest set of three nodes forming a triangle within which it is located. This is repeated with different anchor combinations until all combinations from nodes that are within range are exhausted. Localization is then performed using the center of gravity of the intersection of all the triangles within which the node is located.

A *cluster-based* distributed localization scheme is presented in reference 23. This method avoids using distance or AOA measurements and long-range beacons by utilizing the regularity of clusters in the network. The localization algorithm starts with the development of regular-shaped clusters of nodes, each with a cluster-head node. Such regular clusters can be formed using an algorithm like ACE [24]. Initially, some of the cluster heads are location-aware *anchor nodes* (assuming that these nodes have GPS or are manually deployed). The locations of other location-unaware cluster heads are determined by a process of self-calibration, which utilizes the uniformity of cluster shapes and the average edge length of clusters. The regularity of cluster shapes is also utilized to refine early location estimates of cluster heads at a second stage of the algorithm. When all cluster heads are calibrated, other follower nodes can calibrate using the same range-free principle.

A localization scheme that uses RF connectivity and centralized computations is presented in reference 25. The authors show that given a set of convex proximity constraints and connectivity information under the constraints, fairly accurate location estimates can be obtained using this scheme. This scheme targets large-scale networks of small sensors such as those using the smart dust mote [10].

12.6.2 Probabilistic Position Estimation

Another approach to solving the uncertainties caused by errors in range or angle estimation for localization is using a probabilitistic approach [26–29].

The work in reference 27 models range measurements by a set of density functions. Nodes obtain initial range measurements using RSSI from a set of beacon nodes that are location-aware, either using on-board GPS or from manual deployment. The nodes then compute their position estimates that are represented as three-dimensional density functions. This requires prior knowledge of the density function of range that is obtained from collected sample values. Position estimates from different nodes are then combined by intersection of density functions. When nodes improve their position estimates from new information (i.e., the estimated density function becomes sharper), they broadcast this information to other nodes.

A similar approach is applied in reference 28 to probabilistic localization using AOA measurements. As usual, the error for AOA measurement is modeled probabilistically from offline experiments. The nodes utilize this information to refine position estimates obtained from beacons.

In reference [26], the authors proposed a method applied to mobile sensors that is based on Monte Carlo localization, which uses particle filter combined with probabilistic modeling. The basic idea is to first obtain a posterior distribution of possible locations using a set of weighted samples. Nodes use a mobility model to refine the probability distribution obtained at an initial step. In a second step, the nodes use filtering to eliminate impossible locations based on new observations. This approach has been found to work better with higher node mobility.

An algorithm that applies probabilistic modeling for localization using RSSI based ranging from a mobile node is proposed in reference 29. Here, it is assumed that all sensor nodes are static while a mobile beacon node travels through the network area broadcasting beacon packets. Beacon packets carry updated location of the beacon node. As the beacon packets are received at a sensor node, it obtains new position estimates and combines them probabilistically to refine its position estimate.

12.7 CONCLUSION

Localization is a challenging problem to solve for successful implementation in small, low-cost wireless sensor nodes. Although a significant amount of research has been devoted to this problem, a single practical solution is hard to find. It is likely that solutions would have to be application-specific, because sensors are used in a large number of application scenarios and environments. Also, solutions for any application may require multiple principles for addressing the needs for all nodes. For instance, while some nodes may be able to self-localize, others may need to employ collaborative or centralized computations, depending on the situation at which it is in. Multiple methods would also improve the robustness of the localization system.

This chapter presents the basic principles of localization, the particular challenges with respect to solving the problem in wireless sensor networks, and some existing techniques to solve the problem. This topic is likely to generate more research ideas in future because of its importance in almost all sensor network applications.

12.8 EXERCISES

1. State some of the different techniques that can be used to measure or estimate distances from a wireless beacon generator or base station to a wireless node for localization. Give examples of systems that use each technique. For each technique, explain the technical challenges, if any, of their usage for localization in wireless sensor networks.
2. The locations of three wireless bases stations are as follows: $BS_1 = (0, 0)$, $BS_2 = (20, 5)$, and $BS_3 = (6, 25)$. The wireless system estimates that the distance from a wireless node to BS_1 is 15 units, that from BS_2 is 7 units, and that from BS_3 is 19.4 units. Determine the location (x, y) of the wireless node using triangulation. Write all the equations and show your work. If you use a software tool such as MATLAB, attach printouts of your code. Is it really necessary to get distances from three base stations to estimate the location? Explain.
3. Given that in an angle-based localization scheme as shown in Figure 12.6, the measured angle α can have errors up to 5°, determine an expression for the corresponding error in the location estimate for a sensor node located at an arbitrary location (x, y). Assume that the region of operation is a 10-ft $\times$ 10-ft area and the beacon generators are located at the three corners as shown in Figure 12.7a. Plot the error distribution within the square area.
4. Repeat the above problem when the beacon generators are located on a straight line along the base of the square—that is, at locations $(0, 0)$ (the left bottom corner), $(5, 0)$, and $(10, 0)$.
5. How is the approach for probabilistic localization different from that of estimation of location coordinates under measurement errors? Explain by comparing an existing probabilistic localization scheme based on ranging with that of atomic multilateration.

BIBLIOGRAPHY

1. D. Estrin, R. Govindan, J. Heidemann, and S. Kumar. Next century challenges: Scalable coordination in sensor networks. In *ACM/IEEE International Conference on Mobile Computing and Networking (MOBICOM)*, August 1999, pp. 263–270.
2. I. F. Akyildiz, W. Su, Y. Sankarasubramaniam, and E. Cayirci. Wireless sensor networks: A survey. *Computer Networks*, **38**:393–422, 2002.
3. A. Rao, S. Ratnasamy, C. Papadimitriou, S. Shenker, and I. Stoica. Geographic routing without location information. In *Proceedings of ACM MOBICOM*, September 2003, pp. 96–108.

4. B. Karp and H. T. Kung. GPSR: Greedy perimeter stateless routing for wireless networks. In *ACM/IEEE International Conference on Mobile Computing and Networking (MOBICOM)*, August 2000, pp. 243–254.
5. A. D'Costa and A. M. Sayeed. Data versus decision fusion in wireless sensor networks. In *Proceedings of ICASSP)*, April 2003, pp. 832–835.
6. N. Bulusu, J. Heidemann, and D. Estrin. GPS-less low cost outdoor localization for very small devices. *IEEE Personal Communications Magazine*, October 2000, pp. 28–34.
7. T. S. Rappaport. *Wireless Communications Principles and Practice*. Prentice Hall PTR, Englewood Cliffs, NJ, 2002.
8. C. D. McGillem and T. S. Rappaport. A beacon navigation method for autonomous vehicles. *IEEE Transactions on Vehicular Technology*, **38**(3):132–139, August 1989.
9. http://www.arundel.net/xplane/tml/vor.html.
10. B. Wameke, B. Atwood, and K. S. J. Pister. Smart dust mote forerunners. In *Proceedings of the International Conference on MEMS*, January 2001, pp. 357–360.
11. N. Priyantha, A. Chakraborthy, and H. Balakrishnan. The cricket location support system. In *ACM/IEEE International Conference on Mobile Computing and Networking (MOBICOM)*, August 2000, pp. 32–43.
12. A. Harter, A. Hopper, P. Steggles, and A Ward. The anatomy of a context aware application. In *Proceedings of MOBICOM)*, August 1999, pp. 59–68.
13. P. Bahl and V. Padmanabhan. RADAR: An in-buidling RF-based user location and tracking system. In *Proceedings of INFOCOM*, March 2000, pp. 775–784.
14. A. Savvides, C. C. Han, and M. B. Srivastava. Dynamic fine-grained localization in ad hoc networks of sensors. In *ACM/IEEE International Conference on Mobile Computing and Networking (MOBICOM)*, July 2001, pp. 166–179.
15. D. Niculescu and Badri Nath. Ad hoc positioning system (APS). In *Proceedings of GLOBECOM)*, November 2001, pp. 2926–2931.
16. K. Lorincz and M. Welsh. Motetrack: a robust decentralized approach to RF-based location tracking. In *Proceedings of the International Workshop on Location and Context-Awareness*, May 2005, pp. 63–82.
17. N. B. Priyantha, A. Miu, H. Balakrishna, and S. Teller. The cricket compass for context-aware mobile applications. In *ACM/IEEE International Conference on Mobile Computing and Networking (MOBICOM)*, July 2001, pp. 1–14.
18. A. Nasipuri and R. E. Najjar. Experimental evaluation of an angle based indoor localization system. In *IEEE International Workshop on Wireless Network Measurement (WiNMee)*, April 2006, pp. 1–9.
19. D. Niculescu and B. Nath. VOR base stations for indoor 802.11 positioning. In *ACM/IEEE International Conference on Mobile Computing and Networking (MOBICOM)*, September 2004, pp. 58–69.
20. A. Nasipuri and K. Li. A directionality based location discovery scheme for wireless sensor networks. In *Proceedings of Workshop on Sensor Networks and Applications WSNA'02*, 2002, pp. 105–111.
21. http://webs.cs.berkeley.edu/tos.
22. T. He, C. Huang, B. M. Blum, and J. A. Stankovic. Range-free localization schemes for large scale sensor networks. In *Proceedings of MOBICOM)*, August 2003, pp. 81–95.

23. H. Chan, M. Luk, and A. Perrig. Using cluster information for sensor network localization. In *Proceedings of International Conference on Distributes Computing and Sensor Systems (DCOSS)*, 2005, pp. 109–125.

24. H. Chan and A. Perrig. ACE: An emergent algorithm for highly uniform cluster formation. In *Proceedings of EWSN*, 2004, pp. 154–171.

25. L. Doherty, K. S. J. Pister, and L. El-Ghaoui. Convex position estimation in sensor networks. In *Proceedings of INFOCOM*, April 2001, pp. 1655–1663.

26. L. Hu and D. Evans. Localization for mobile sensor networks. In *ACM/IEEE International Conference on Mobile Computing and Networking (MOBICOM)*, September-October 2004, pp. 45–57.

27. R. Peng and M. L. Sichitiu. Probabilistic localization for wireless sensor networks. *ACM Mobile Computing and Communications Review*, **11**(1):53–64, 2007.

28. R. Peng and M. L. Sichitiu. Angle of arrival localization for wireless sensor networks. In *Proceedings of IEEE Conference on Sensor and Ad Hoc Communications and Networks)*, September 2006, pp. 374–382.

29. M. L. Sichitiu and V. Ramadurai. Localization of wireless sensor networks with a mobile beacon. In *Proceedings of IEEE Conference on Mobile Ad Hoc and Sensor Systems MASS)*, October 2004, pp. 174–183.

CHAPTER 13

QoS-Based Communication Protocols in Wireless Sensor Networks

SERDAR VURAL, YUAN TIAN, and EYLEM EKICI

Department of Electrical and Computer Engineering, Ohio State University, Columbus, OH 43210

13.1 INTRODUCTION

Wireless sensor networks (WSN) have been one of the main research foci in wireless networks area over the last decade [1]. With the evolution of the MEMS technology and the availability of low cost communication and computation hardware, WSNs have been transformed from conceptual paradigms to reality in this short period. Prototype and deployed WSNs currently serve as enablers of several applications such as environmental monitoring, urban safety, traffic monitoring, smart spaces, and surveillance of hostile and inaccessible areas. The majority of the early research efforts have focused on enabling technologies for WSNs, including development of communication protocols, localization methods, and simple application-dependent information processing.

Major research efforts in WSN area have resulted in many deployed testbeds and other implementations. The driving motivation of delivering solutions that can be implemented in short time resulted in systems that can provide only best effort service. Minimization of energy consumption or achieving "high efficiency" (in its versatile definition) has been the objective of many communication protocols designed for WSNs. Their performance strictly depends on the configuration of the network and the load it carries. While the boosting effect of the existing solutions on the research community should be acknowledged, these solutions fall short of addressing requirements of *all* WSN applications and deployment scenarios. More specifically, mission critical and real-time applications suffer from unpredictable performance levels when such communication protocols are used.

Mission critical and real-time applications require performance guarantees from the system on which they are implemented. As an example, a real-time

Algorithms and Protocols for Wireless Sensor Networks, Edited by Azzedine Boukerche

intrusion detection application running on a WSN may require the event detection decision to be made within a given delay bound [2]. Other applications may require that the network deliver data packets to the sink with a given probability. Similarly, the overall energy consumption of all communication events may be subject to energy consumption constraint. Applications running on video sensor networks [3–5], which form a new and emerging class of WSNs, inherently require service guarantees from the communication network due to real-time nature of its multimedia source content. These requirements are classified as *quality of service* (QoS) in other wired and wireless networks, which we also adopt for WSN environments. We classify a communication protocol as QoS-based protocol if it can guarantee one or more performance metrics to upper layers or the application. Under this classification, solutions that simply minimize/maximize a performance metric (delay, energy consumption, packet loss probability, network lifetime, etc.) without performance guarantees are considered as non-QoS-based solutions. We should note that most of the existing proposals for WSNs fall into non-QoS-based category.

In this chapter, we first contrast QoS provisioning in WSNs and other network types and introduce a QoS provisioning framework for WSNs. Then we outline and discuss proposed QoS-based communication protocols for WSNs. We also outline methods that support QoS-based in-network processing along with communication for WSNs. QoS-based capacity estimation methods are also discussed within the proposed framework. We then conclude the chapter with future research directions.

13.2 QoS IN WIRELESS SENSOR NETWORKS

13.2.1 General Principles

QoS has been the target of many communication protocols for numerous network types. In its broadest form, quality of service refers to the contract between the service provider (i.e., the network) and customers (i.e., applications) [6]. In wired networks, one of the main motivations for QoS solutions is the real-time multimedia applications that need bandwidth, delay, and jitter guarantees. ATM networks [7] were proposed to support such requirements from ground up. Although ATM networks are not as widely in use as originally imagined, QoS support mechanisms proposed for ATM networks still inspire new solutions. In IP networks, QoS support of individual flows have been proposed to be handled through IntServ [8] mechanism, which has not gained widespread acceptance due to its scalability problems.[1] In cellular networks, the motivation for QoS support is also inherent to the primary application of such networks: Voice (and recently) video calls are subject to stringent constraints to be commercially viable. In these networks, QoS support are provided through *resource reservation* mechanisms. To accomplish resource reservation, the following steps are followed:

[1]The DiffServ [9] architecture will be proposed later on to support differentiation of groups of flows rather than individual flows to overcome the scalability problem. However, DiffServ mechanism does not provide absolute performance guarantees and therefore cannot be classified as a QoS solution.

- *Available Resource Estimation.* The first step in QoS support is the knowledge of available resources. The estimation of available resources is performed using the *network state* and the communication protocols employed in the network. The network state is comprised of the network connectivity information, maximum capacity of nodes and links, and allocated resources.
- *Calculation of Required Resources.* Given that the performance requirements of applications are known, resources required to sustain the QoS expectations are calculated in the network. Both performance metric conversion and the resource requirement estimation depend on the protocols used in the network. The calculation also involves the selection of the resources in the network for the information flow.
- *Resource Allocation.* Calculated resources are reserved in the network entities. The reservation of such resources is performed via auxiliary protocols such as RSVP [10] or as an integral part of the communication protocol.
- *Resource Deallocation.* When a session terminates, resources are returned to the general pool. The deallocation can be done either explicitly or implicitly through timeout mechanism.

13.2.2 QoS in Ad Hoc Networks

The above-outlined steps work well in networks where resources are separated from each other with well-defined boundaries: In wired networks, link and node resources are clearly separated from each other. As an example, two node-disjoint links can be treated as independent resources even though the load on them may depend on each other. Similarly, in cellular networks, point-to-point links between mobile stations and base stations are separated from each other in time, frequency, code, space, or a combination thereof. Hence, QoS provisioning in both types of networks can easily follow the aforementioned steps.

In wireless ad hoc networks, the resource allocation strategy faces a roadblock at a very fundamental level [11]: Estimation of available resources is a nontrivial task even for simplest multihop ad hoc networks. Wireless resources are shared by multiple nodes that do not have an inherent coordination infrastructure. The contention resolution and resource bidding procedures usually propagate over long distances and affect far-away nodes. The lack of isolation of resources and the dynamic nature of the network make resource estimation and allocation very challenging tasks. This fact, coupled with the weak motivation for QoS-demanding applications for ad hoc networks, have limited the acceptance of QoS-based communication proposals for ad hoc networks.

13.2.3 QoS in Wireless Sensor Networks

A WSN can be regarded as a special type of ad hoc network with very resource-constrained nodes, lower mobility, and larger scale. With these additional constraints, it is easy to dismiss QoS provisioning in WSNs as implausible. However, there is one important difference between ad hoc networks and WSNs. WSNs are defined by applications they are deployed for. A large set of WSN applications, including security and surveillance, requires QoS guarantees from the network. Note that QoS-based applications were not integral parts of ad hoc networks and were proposed as additional applications that could run in parallel with non-QoS applications. Hence, QoS provisioning is a *requirement*, and not an optional feature, for WSNs.

Being multihop networks, WSNs potentially suffer under the same shortcomings and problems as ad hoc networks if similar QoS provisioning mechanisms are adopted. Furthermore, considering very limited resources in sensor nodes and the large scale of WSNs, per flow resource reservation-based approaches are especially ill-suited for WSNs. The two important differences between both network types suggests new directions in QoS provisioning: First, the mobility of WSNs is very limited when compared with ad hoc networks. Resource availability in WSNs fluctuates as a function of the offered load and not as a function of network connectivity over long periods of time. Therefore, communication decisions do not need to be updated very frequently. Second, the large scale of WSNs can easily be used as an advantage to eliminate explicit resource allocation. Distribution of communication responsibility over larger areas provides gains through diversity and allows local decisions to be made, leading to end-to-end QoS guarantees.

QoS provisioning in WSNs is directly geared toward satisfying application requirements. In Figure 13.1, a generalized framework for information flow in a WSN is depicted. The main components of this framework are the sink, source locality, and relay nodes. The information flow starts with assignment of a particular task to sensor nodes. Upon information retrieval through sensors, source and other nodes nearby preprocess the information. The preprocessing may be simply forming data packets with raw data, data aggregation, or completely processing data and forming end results per application requirements. The information is then communicated via the relay nodes to the sink. In return, sink may optionally give feedback to the source locality and/or relay nodes. As will be presented in the next section, many of the QoS-based communication methods form an almost open loop where the source does not return any feedback to information sources or intermediate nodes.

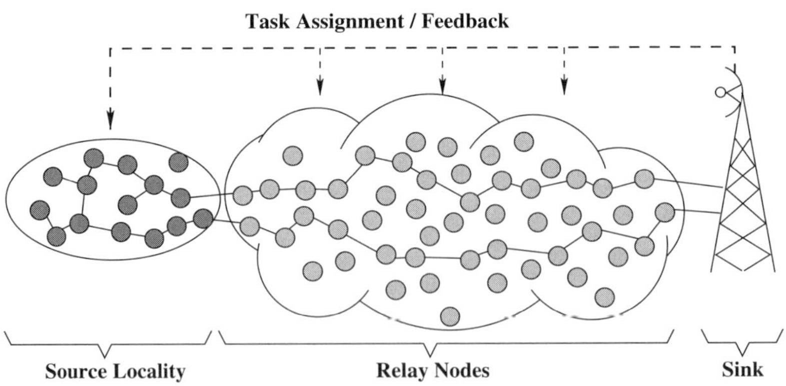

Figure 13.1. Information flow in a WSN.

13.3 QoS-BASED MAC PROTOCOLS FOR WSNs

QoS provisioning in the MAC control layer deals mainly with the scheduling of packets on the wireless channel subject to local constraints. Since local constraints may change based on the needs of individual flows, the decisions are generally very dynamic and must be computed rather fast. The three solutions outlined below consider the requirements of real-time WSNs while scheduling medium access to contending nodes.

13.3.1 QoS-Aware Medium Access Control Protocol [12]

A QoS-Aware Medium Access Control protocol (Q-MAC) is presented in reference [12]. Q-MAC assumes an environment of multihop WSNs where nodes may generate packets with different priorities. The design objective of Q-MAC is to minimize energy consumption and provide QoS guarantees. Q-MAC is composed of intra-node and inter-node QoS scheduling mechanisms. The intra-node QoS scheduling scheme classifies outgoing packets according to their priorities, while the inter-node QoS scheduling solution handles channel access with the objective of minimizing energy consumption via reducing collision and idle listening.

The intra-node scheduling mechanism employs multiple First-In First-Out (FIFO) queues with different priorities, among which an *instant queue* has the highest priority and its enqueued packets are always instantly served. The intra-node scheduling mechanism is outlined in Figure 13.2. Self-generated and relayed packets are classified to different queues with several QoS metrics, such as content importance and number of traveled hops. Data rate allocation between queues and serving packet selection are achieved through the MAX-MIN fairness algorithm [13] and the GPS algorithm [14], respectively.

After a packet is scheduled for transmission, the inter-node scheduling mechanism, Power Conservation MACAW (PC-MACAW), is executed to achieve Loosely Prioritized Random Access (LPRA) between sensor nodes. In PC-MACAW, a successful transmission consists of two periods: the contention period and the packet transmission period. In the contention period, a node sends out RTS after waiting for a certain

Intra-Node_Scheduling:

```
WHILE TRUE
    IF new packet arrives
        Enqueued to different queues after classification
    WHILE instant queue is not empty
        Deliver the first packet in the instant queue
    IF there are queues not empty
        Select q among these queues with the MAX-MIN and GPS algorithms
        Deliver the first packet in q
```

Figure 13.2. The intra-node scheduling Mechanism of Q-MAC.

PC-MACAW:

```
1. /*Enter the contention period*/
2. Backoff for a random number of time slots (ranging in [1,CW])
3. IF channel is not clear during the backoff time
4.    IF the dropping threshold is not reached
5.       GOTO Step 2 with doubled CW
6.    ELSE
7.       Drop the packet
8. ELSE
9.    Send RTS packet
10.   WAIT CTS packet
11.   IF the CTS packet is absent
12.      IF the dropping threshold is not reached
13.         GOTO Step 3 with doubled CW
14.      ELSE
15.         Drop the packet
16.   ELSE
17.      /*Enter the packet transmission period*/
18.      Transmit the packet
19.      WAIT ACK packet
20.      IF the ACK packet is absent
21.         IF the dropping threshold is not reached
22.            GOTO Step 2 with doubled CW
23.         ELSE
24.            Drop the packet
```

Figure 13.3. The inter-node scheduling mechanism (PC-MACAW) of Q-MAC.

duration (contention time) and expects a CTS packet before accessing the channel. The contention time is randomly generated with a contention window size *CW*, where *CW* is determined by each node's transmission urgency including packet criticality, number of transmitted hops, residual energy, and queue's proportional load. After accessing the channel, the node enters the transmission period to send data packets and waits for an ACK packet. In the case of collision, *CW* is doubled and the packet is retransmitted. When the difference between the current time and when the packet is generated exceeds a threshold, the packet is dropped. The PC-MACAW algorithm is outlined in Figure 13.3.

Q-MAC presents a combined effort of intra-node and inter-node QoS scheduling in WSNs. It is shown through simulations that Q-MAC provides the equivalent QoS while consuming less energy in comparison with an existing mechanism, S-MAC. However, complex scheduling mechanisms and relatively loosely defined QoS metrics stand out as shortcomings of this proposal.

13.3.2 Coloring-Based Real-Time Communication Scheduling [15]

The *Co*loring-Based Real-Time *Co*mmunication Scheduling (CoCo) solution is presented in reference [15]. CoCo is designed for multihop WSN environments that use

IEEE 802.11 MAC protocol, where all communication is unicast. It is assumed that node locations are available at all times, and a central scheduler running CoCo is in charge of communication scheduling. CoCo aims to schedule real-time communication avoiding collisions and minimizing the overall packet transmission time.

In CoCo, a set of messages waiting for transmission at various sensors are modeled with a weighted, directed graph $G = (V, E)$, where a vertex denotes a sensor node, a directed edge from vertex v_i to v_j denotes a message to be sent from sensor v_i to v_j, and the weight of an edge denotes the transmission time. The communication problem is equivalent to assigning a color to each edge. Here, each color represents a set of simultaneous communication during disjoint time periods, and the weight of a color equals the maximum weight of the edges assigned with this color. CoCo aims to find edge color assignment such that (i) no adjacent edges share the same color, (ii) no two edges with the same color interfere with each other, and (iii) the overall weight of used colors is minimized.

Since the optimal coloring problem is NP-complete, a coloring heuristic is presented in reference [15]. First, the edges of the vertex with the maximum degree are assigned different colors. Once a color is assigned to an edge, it is removed from the palettes of all adjacent edges, and its weight is updated. Then, the following steps are repeated until all edges are colored: The edge with the smallest palette is chosen. A color from the available palette is assigned to the edge such that no other edge with that color interferes with the chosen edge. Then, the chosen color is removed from the palettes of all uncolored adjacent edges. Three heuristics are presented for selecting a color from an edge's palette: The *Random Color Selection Heuristic* randomly picks a color from the palette that does not cause interference. The *Least Used Color (LUC) Heuristic* chooses the color with the smallest number of colors. The *Minimal Weight Color (MWC) Heuristic* first checks whether there are colors in the palette whose weights are higher than the edge. If so, among these colors, the color with the smallest weight is selected. Otherwise, the color with the maximum weight is assigned from the palette.

CoCo aims to schedule a set of communication events with the minimum communication time in real-time WSNs. According to the simulations, MWC-based CoCo provides performance superior to that of the other two-color selection heuristics, and its performance is close to the optimal solution. The central computation requirement limits the applicability of CoCo in large-scale sensor networks.

13.3.3 Reliability Maintenance Through Activity Management [16]

In reference [16], an activity management mechanism (AMM) is proposed to maintain communication reliability in WSNs. WSNs are considered to work with underlying IEEE 802.15.4 protocol operating in beacon-enabled slotted CSMA/CA mode. Sensors are organized with a star topology, where a node must be admitted by a coordinator to participate in the network. The coordinator is aware of the number of nodes in the network, packet arrival rates, and the desired reliability R. Here, the reliability R is defined as the number of packets delivered to the coordinator per unit time. AMM aims to provide network reliability guarantee through sensor node activity management.

In AMM, the MAC layer exhaustively serves packets in First-Come First-Serve (FCFS) manner. A sensor goes to sleep only after all packets in its buffer are transmitted. During the sleep period, newly arriving packets are enqueued waiting for sensor wake-up to be delivered. In case of queue overflow, the packet at the head of the queue is discarded. The sleep duration is geometrically distributed with the parameter P_{sleep}. A packet is transmitted starting with a random backoff countdown. After the countdown, two Clear Channel Assessments (CCA) are executed by listening to the channel to make sure it is idle. If both CCAs pass, the packet transmission starts and an ACK packet is expected. In the case where an ACK is not received, the transmission is repeated and the countdown exponent is increased. Through theoretical analysis of the throughput of the MAC mechanism above, the network control equation is derived where the network reliability R is a function of several MAC parameters, such as sensor sleep duration, transmission success probability, and basic backoff period. Given desired network reliability R, the coordinator calculates the parameter P_{sleep} and broadcasts it to all nodes to regulate their sleep times accordingly.

AMM provides network reliability thorough activity management and analysis of MAC parameters. It is shown through simulation that through proper activity management, the network reliability is robust against variation of network scale and packet rates. As the author points out, AMM is computationally intensive, and distributed activity management is left to future work.

13.4 QoS-BASED ROUTING PROTOCOLS FOR WSNs

QoS-based routing protocols for WSNs have been proposed in the literature mainly to support two kinds of performance bounds, namely, delay and reliability. The protocols outlined in this section are implemented primarily in the network layer and, in some instances, in the MAC sublayer. These solutions also differ in the management of resources, where some rely on centralized computations while others utilized distributed methods. The common point in all these solutions is that all of them guarantee at least one performance metric to be satisfied in the network.

13.4.1 Sequential Assignment Routing [17]

On-demand multihop routing algorithms such as AODV and TORA eliminate table updates in high-mobility scenarios. However, they introduce high-energy cost during route setup phase. Power-aware routing finds minimum metric paths on two different metrics: minimum energy per packet and minimum cost per packet. The first metric produces substantial energy savings, but performance degradation due to link/node failure is not addressed. The second metric deals with failures by routing traffic away from low-energy nodes at the expense of high path maintenance cost. The Sequential Assignment Routing (SAR) [17] algorithm uses the idea of multiple paths while taking parameters like energy resource, QoS on each path, and the priority of packets into consideration.

In SAR, a table-driven multipath approach is used to improve energy efficiency in a low-mobility sensor network. The failure protection is addressed by having at least k-paths that have no common branches between a node and a sink. This is called a k-disjoint structure. However, the disjoint property creates strong coupling between routing tables, rendering localized recovery schemes ineffective. To reduce this effect, the disjoint requirement is relaxed outside the 1-hop neighborhood of the sink. Furthermore, localized path restoration procedures are used to decrease energy cost in failure recovery. Multiple paths from each node to a sink are created by building multiple trees, each rooted at the 1-hop neighborhood of the sink. Each node uses two parameters to create routing paths:

- Energy resource that is estimated by maximum number of packets that can be routed without energy depletion, assuming that the node has exclusive use of the path.
- Additive QoS metrics where higher metric implies lower QoS.

Path selection is made by nodes that generate packets if no topology change occurs while packets are being routed to their destinations. The energy cost and delay of links are considered as additive QoS metrics. Packet priorities are used in a way that packets with higher priorities use paths with lower latency. In short, for each packet, a weighted QoS metric is computed as the product of a weight coefficient (the priority of the packet) and the additive QoS metric. Hence, QoS is provided to each packet relative to its priority level, where higher QoS is given to higher-priority level packets. The SAR algorithm minimizes the average QoS metric throughout the lifetime of the network.

Periodic metric updates triggered at the sink node are used to account for possible changes in the QoS on individual paths and the changes in energy resources. Simulations show that SAR performs better than a minimum metric algorithm that lowers energy consumption without considering packet priorities. Furthermore, failure recovery is handled by local handshakes between upstream and downstream neighbors in paths. SAR algorithm addresses low-mobility networks, and routes are established at packet sources considering link costs and energy resources as a QoS parameter. Packet priorities are taken into account to relay high-priority packets to popular paths in terms of latency. However, the scheme requires resource-related topology information at packet sources that require frequent parameter updates by a common sink. This incurs high overhead in WSNs with moderate or high mobility and in WSNs carrying high data rates.

13.4.2 Energy-Aware Routing in Mobile and Wireless Ad Hoc Networks [18]

Before focusing on energy-aware routing protocols for WSNs, it is worth focusing on energy-saving routing protocol design for wireless ad hoc networks. These solutions are also directly applicable to WSNs with limited number of nodes and where a number of potentially mobile, high-capability nodes need to communicate possibly over sensor nodes. A good example of modifying existing routing protocols to be energy-aware has been presented in reference [18]. In this paper, two reactive ad hoc routing protocols, namely DSR [19] and TORA [20], are modified to deliver QoS by introducing energy-awareness.

DSR and TORA protocols involve a "route discovery phase" initiated when a mobile source node needs to send data packets to the destination node but does not have an active and valid route to it. Route discovery is performed using the control packets called route request packets (RREQ). The source node broadcasts RREQs and waits for a control packet called the route reply packet (RREP). Using the received RREP, the source node updates its routing table, which is used to keep track of active routes to individual destination nodes in the network. Intermediate nodes, which forward RREQ and RREP packets between source and destination nodes, also use RREP packets to update their routing tables. Although these protocols are effective and robust, the broadcast of RREQ packets leads to unnecessary packet transmissions and inefficient use of limited energy resources.

In reference [18], two modified protocols, EDSR and ETORA, based on the existing DSR and TORA protocols, are introduced, respectively. Both EDSR and ETORA involve an additional RREQ forwarding mechanism that does not exist in DSR and TORA. In an intermediate node, this mechanism considers the current energy level of the node, the energy level of the previous sender node, and the distance to the source node when making routing decisions.

Distance estimation is accomplished using time stamps in RREQ packets. When a node transmits an RREQ packet, it records the time of transmission in the RREQ. Intermediate nodes that receive this RREQ calculate the time difference between transmission and the reception time of the RREQ to estimate their distances to the sender node. Besides transmission times, energy levels of the nodes right before RREQ transmission are also recorded in RREQs.

The RREQ forwarding decision in reference [18] is based on cutoff circles that are placed around each node in a network. Upon the reception of an RREQ packet, an intermediate node calculates the diameter of its "cutoff" circle using its energy level, the energy level of the previous node sending this RREQ packet, and its distance to the previous node calculated by the time stamp in the RREQ packet. The receiving node simply drops the packet, hence does not forward it, if its cutoff circle encircles the previous node. In reference [18], ETORA and EDSR are shown to outperform TORA and DSR, respectively, in terms of overall network throughput, the average number of data packets received at destinations, average data transmission delay, and energy consumption. The pseudocode of the proposed RREQ forwarding algorithm is given in Figure 13.4.

13.4.3 Energy-Aware QoS Routing Protocol for WSNs [21]

The information delivery in video sensor networks requires end-to-end delay guarantees. The QoS-based routing protocol proposed in reference [21] aims to sustain paths that can guarantee such delays for real-time traffic while supporting non-real-time (best-effort) data flows, as well. The network architecture assumed in this proposal

Forward_RREQ:

---

tp: Time of transmission by the previous hop
tr: Time of reception by the intermediate node (IN)
Ep: Energy level of previous node during RREQ transmission
Er: Energy level of this IN when RREQ is received
d_c: Diameter of the cutoff circle of this IN
l: Distance to the previous hop
c: Speed of light

---

1. Calculate time difference $\Delta_t := tr - tp$
2. Estimate distance $l := \Delta_t \times c$
3. Determine diameter of cutoff circle $d_c = 0.4 \cdot Ep + 0.4 \cdot Er + 0.2l$
4. **IF** $d_c \leq l$
5. Drop RREQ packet
6. **ELSE**
7. Forward RREQ packet

Figure 13.4. Energy-efficient RREQ forwarding algorithm.

involves a hierarchical organization: Sensor nodes are grouped into clusters, formed based on criteria such as communication range, number and type of sensors, and geographical location. Clusters are assumed to be controlled by a single command node and have their own gateway nodes, which act as cluster heads. Sensors in a cluster receive commands from and send readings to the cluster gateway node. Gateway nodes of different clusters are able to communicate over long-haul communication links. All sensors are assumed to be stationary. In addition to non-real-time data generated in the network, the network also tries to recognize and track targets in individual clusters using images and video feeds.

In this study, the aim is to find paths within a particular cluster under real-time constraints, without explicitly addressing communication among gateways. Given a cluster of sensor nodes, paths are computed centrally by gateways. To this end, a cost is associated with each link. A link cost is a weighted sum of physical length of the link, residual energy of the sender, expected lifetime of the sender assuming current power consumption rate, and the estimated error rate of the link. Using these link costs, the gateway computes k-shortest paths between individual nodes and the gateway itself. However, this computation is not sufficient to satisfy the delay constraints of flows. Based on precomputed k-paths between sources and the gateway, the gateway also computes the expected delay on every link. Assuming perfect knowledge about the demand of data sources and a fixed ratio r of resources used for real-time flows, a queuing model is formed. This queuing model is used to compute the delay between data sources and the gateway. The path computation algorithm aims to find a common ratio r such that end-to-end delay requirements of all nodes are satisfied for at least one of the k-paths associated with each data source. Once computed, the resource ratio r is broadcast to all nodes along with the selected paths to route information in the network.

This work has an interesting approach to the QoS provisioning in the network where both real-time as well as non-real-time data flows are considered in the network. The proposed solution is expected to perform well when the size of the cluster is small. However, the link cost function helps minimize the energy consumption for appropriate weights for different components of the cost. The authors do not comment on the selection of the weight parameters. Furthermore, the assumption of knowing the data rates of individual sensor nodes as well as the up-to-date state information of relay nodes at a central location is far from being realistic. While this algorithm is a nice attempt to solve a complex problem, it only does so at the expense of limiting simplifications and assumptions of cluster-wide state information.

13.4.4 Energy-Aware Data-Centric Routing Algorithm for WSNs [22]

An energy aware and data-centric routing algorithm (EAD) for WSNs is proposed in reference [22]. The algorithm aims to achieve two performance improving goals: (i) elimination of redundant data by in-network processing and (ii) minimization of overall network energy consumption by using a virtual backbone tree for data forwarding. The EAD algorithm is mainly focused on the construction and maintenance of the forwarding backbone tree rooted at a single data sink with maximal number leaf sensor nodes. The network operation is composed of two major phases, namely the *initialization* and the *data transmit* phases. Initialization and data transmit phases together form a "round." The construction and following updates of the backbone tree are performed during the initialization phase.

The construction of the forwarding tree is based on the following idea: The dominant part of the energy consumption in a sensor node is due to data transmission and reception. To minimize the overall network energy consumption, some nodes should turn their radios off (leaf nodes) while others should remain relaying packets (non-leaf nodes). Hence, to achieve minimal energy consumption, the focus of attention is on the maximization of the leaf nodes. Since this is an NP-complete problem, the EAD algorithm heuristically attempts to achieve this goal. The algorithm periodically updates the distribution of the leaf and non-leaf nodes during the initialization phases by considering a sensor node as a state machine.

The choice of being a non-leaf node depends on two mechanisms, namely neighboring broadcast scheduling and distributed competition among neighbors. These mechanisms ensure that sensors with higher residual power have higher chance to become a non-leaf node, hence conserving the local energy that eventually leads to reduction in overall network energy consumption. Leaf and non-leaf nodes change their states upon the reception of messages from neighbors indicating their parents, energy levels, and distance to the sink node. Nodes sense the channel before transmitting these messages and also have waiting periods to avoid unnecessary local state changes.

In EAD, data relaying and in-network data processing tasks are performed in non-leaf nodes. At each sensor, the local raw data is combined with partially processed data delivered from sensors that are farther away from the sink. (The sensor nodes keep records of their parent nodes and their child nodes, if any, through which they determine their distance to the sink.) Non-leaf nodes summarize and forward the aggregated data to their parents in the tree. To reduce the execution time of EAD processing, a topology-based algorithm is used which preprocesses the network topology to ensure that all sensors are are spanned by the EAD tree even though a subset of sensors participate in EAD execution. The outline of the two algorithms used in EAD are given in Figures 13.5 and 13.6.

Receive_Control_Packet:

n:	Current node
n.nodeId :	Unique ID attribute of *n*
n.EAD_type:	Type attribute of *n* (0: undetermined, 1: leaf, 2: non-leaf)
n.EAD_previous_type:	Previous type attribute of *n*
n.EAD_level:	Current level attribute of *n*
n.EAD_parent:	Distance attribute of *n* to the previous hop
n.EAD_has_child:	Boolean attribute of *n* indicating if it has children
sink:	
P:	Received packet
P.ead_type:	Type of the source of *P*
P.ead_level:	Level of the source of *P*
P.ead_parent :	Parent node of the source of *P*
P.source_add:	Source address of *P*

```
n.EAD_previous_type := n.EAD_type
IF n.EAD_type == 0
    IF P.ead_type == 2
        n.EAD_type := 1
        n.EAD_level := P.ead_level + 1
        n.EAD_parent := P.source_addr
    ELSE IF P.ead_type == 1
        n.EAD_type := 2
        n.EAD_level := P.ead_level + 1
        n.EAD_parent := P.source_addr
    Call finalEADStatusUpdate function
    Send control packet to 1-hop neighbors
ELSE IF n.EAD_type == 1
    IF P.ead_parent == n.nodeId
        n.EAD_type := 2
        Call finalEADStatusUpdate function
        Send control packet to 1-hop neighbors
ELSE IF n.EAD_type == 2
    IF P.ead_type == 2 AND P.ead_parent == n.nodeId
        n.EAD_type := 2
        Call finalEADStatusUpdate function
        Send control packet to 1-hop neighbors
IF P.ead_parent == nodeId
    n.EAD_has_child := TRUE
    n.EAD_type := 2
```

Figure 13.5. Control packet reception procedure used in the EAD algorithm.

finalEADStatusUpdate:

---

n: Current node
P: Received packet

---

1. **IF** $P.ead\_parent == nodeId$
2. $\quad n.EAD\_has\_child := TRUE$
3. **IF** $n.EAD\_has\_child == TRUE$
4. $\quad n.EAD\_type := 2$
5. **ELSE IF** $n.EAD\_previous\_type! = 0$ **AND** $n.EAD\_has\_child == FALSE$
6. $\quad n.EAD\_type := 1$

Figure 13.6. EAD status update procedure used in the EAD algorithm.

The performance of the EAD algorithm is compared with the performance of a simplified AODV without sensor mobility, and LEACH algorithms comparing total number of active nodes, UDP packet throughput, and energy expenditure. The presented results illustrate that EAD outperforms the other two algorithms with respect to these performance metrics. Furthermore, it is shown that there is a tradeoff between system lifetime and system throughput when shorter or longer EAD refresh periods are chosen. Small refresh intervals enable better throughput but require more energy. Moreover, it is concluded that there is no tradeoff between the initialization and data transmit phases of the EAD algorithm.

13.4.5 Reliable Information Forwarding Using Multiple Paths [23]

Data dissemination protocols which are not adaptive to channel error rates and do not support information awareness either spend excessive amount of resources or fail to deliver important information with sufficient reliability. Reliable information forwarding using multiple paths (ReInForM) [23] is a protocol for WSNs to support information awareness, such that the reliability of data transfer depends on the information content despite the presence of significant channel errors. To define the desired reliability levels, ReInForM assigns different priority levels to data packets. Depending on the priority level, multiple copies of the data packets are delivered along multiple paths. Hence, ReInForM relies heavily on the existence of multiple paths between a source and a destination, which is generally available in large-scale WSNs. The simulations investigating the existence and number of edge-disjoint paths show that a network slightly denser than a minimally connected graph is sufficient to have as many edge-disjoint paths as the average node degree. The deviation in the number of hops of these paths is found to be less than two hops, which suggests that the paths have nearly identical lengths. Hence, data delivery on these paths has comparable latency and efficient load balancing among multiple paths is possible.

Under ReInForM, the source node of a packet determines the importance of the information in the packet and decides on a reliability level (r_s). Using the local channel error information (e_s) and the hop distance to the sink (h_s), the source computes the number of paths, $P(h_s, r_s, e_s)$, required to deliver the packet at the chosen reliability level. The neighbors of the source is divided into three subsets, H_s^-,H_s^0 , and H_s^+, designating the neighborhoods at distances of h_s^-, h_s^0, and h_s^+ hops to the sink, where $h_s^- < h_s^0 < h_s^+$ and $h_s = h_s^0$ is the hop distance of the source to the sink. The chosen total number of paths that the source is expected to create, P, is divided into these three sets of neighborhood.

A random node in H_s^- is chosen to be the default node, which always forwards packets. This ensures that we have at least one path toward the sink. Other nodes in H_s^-, using their own local channel error rate e, hop distance to sink h, and reliability r, compute their own P value. If this P value is larger than 1, then the node is chosen to be a forwarding node. If the value is less than 1, then the probability that the node is a forwarding node is simply this local P value. Eventually, a number of forwarding nodes are chosen in set H_s^-. If there are still more paths to be established (meaning that the number of paths over the set H_s^- is less than the total number of paths), then additional paths are created by the nodes in the set H_s^0 in the same way. Paths over the set H_s^+ are created only if there are still more paths needed besides the ones created by H_s^- and H_s^0.

The nodes that decide not to forward a packet simply drop it. Packets carry minimal state information to aid the forwarding decisions. The dynamic local states containing hop distance to sink, reliability, and channel error rate are updated regularly at each forwarding node. After receiving a packet from the source and updating the dynamic states, the node effectively becomes a source. Using the local state information, the node uses the same procedure to compute the path values for its own neighbors and the process continues.

ReInForM is one of the leading examples of multipath routing protocols for WSNs. The gains attained through local multipath forwarding mechanisms lend predictability to applications in terms of reliability. While load balancing is argued to be a natural result of the protocol, the study does not present any the analysis of this issue. Furthermore, not all alternative edge-disjoint paths are of equal length, potentially leading to out-of-order delivery of packets and unpredictable delays.

13.4.6 SPEED Protocol [24]

In large-scale WSNs, the availability of state information of individual nodes cannot always be assumed. Therefore, local decision-based solutions prove to be more flexible as well as feasible for such large-scale networks. SPEED protocol [24] is designed to provide soft end-to-end deadline guarantees for real-time packets in WSN. It uses a geographic forwarding mechanism such that each packet can be routed without global topology information. Thus, it scales well in large sensor networks. More importantly, it ensures a network wide speed of packet delivery for real-time guarantees. The network is assumed to be composed of nodes that have location information. Since the protocol is based on a geographic routing algorithm, nodes are also assumed to gather information about their neighbors' locations.

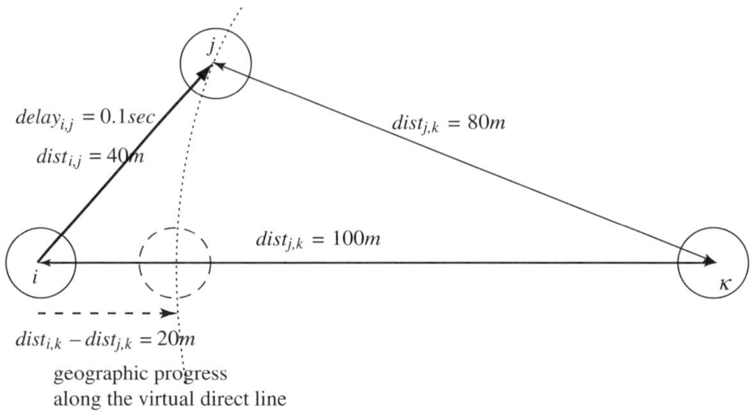

Figure 13.7. Packet progress speed calculation under SPEED protocol.

The concept of soft real-time guarantees in the context of the SPEED protocol refers to the fact that packets travel to their destination at a given propagation speed. For this, each node maintains information about neighboring nodes such as geographic distance and average delay to each neighbor. Using the distance and delay, each node evaluates the packet progress speed of each neighbor node for a packet sent to a specific destination. An example of packet progress speed calculation is shown in Figure 13.7. Let a packet in node i be destined to k, which is 100 m away. Let i forward the packet through a neighbor j, which is 80 m away from k. If this forwarding takes, on the average, 0.1 s, then the packet progresses to its destination k at $\frac{20m}{0.1s} = 200$ m/s. Under SPEED protocol, a packet is forwarded through a neighboring node if and only if the progress speed through that neighbor is higher than the specified lower-bound speed *SetSpeed*.

If each node can find a neighbor that can progress a packet with a speed higher than *SetSpeed*, *SetSpeed* can be guaranteed in the entire network. However, if the load carried in the network is too high, uniform speed guarantees cannot be provided in the network. When a node cannot find any neighbor node whose speed is higher than *SetSpeed*, it probabilistically drops packets to regulate the workload such that at least one neighbor node with a speed higher than *SetSpeed* exists at all times. At the same time, the node sends a back-pressure packet to the previous nodes to prevent them from forwarding any further packets through this congested area. Hence, packet delays can be upper-bounded at the expense of packet losses in the network to sustain network-wide packet progress guarantees. The algorithms used in the SPEED protocol are given in Figures 13.8 and 13.9.

With the SPEED protocol, a uniform packet progress speed is guaranteed in the entire network. Furthermore, the protocol relies only on local information augmented with limited scope feedback, which improves its scalability. However, the SPEED protocol provides only one network-wide speed, which is not suitable for differentiating various flows with different deadlines. Furthermore, it does not provide any guarantees in packet delivery: The fraction of packets lost or dropped in the network

Forward_Packet:

i: Current node
D: Destination node
$d(i, D)$: Destination of node i to D
NS_i: Neighborhood set of node i
FS_i: Set of all nodes in NS_i closer to D than i
$HopDelay(i, j)$: Estimated delay from i to j
$Speed(i, j, D)$: Estimated speed from i to j
p: Packet being processed
BPP: Back-pressure packet
US_i: Upstream Node Set of node i
$S_{setpoint}$: Speed threshold
$FS_{i,set1}$: Nodes in FS_i with sufficient speed
$FS_{i,set2}$: Nodes in FS_i with insufficient speed
e_n: Miss ratio of neighbor n
RR: Relay ratio
K: Proportionality gain

```
FS_i := FS_i,set1 := FS_i,set2 := φ
FOR all k in NS_i
      L_next := d(i, D) − d(k, D)
      IF L_next > 0
        Add k to FS_i
        Speed(i, k, D) := (L − L_next)/HopDelay(i, k)
        IF Speed(i, k, D) > S_setpoint
          Add k to FS_i,set1
        ELSE
          Add k to FS_i,set2
IF FS_i == φ
    Drop p
    FOR all u in US_i
      Send BPP to u
ELSE IF FS_i,set1 ! = φ
    Choose k in FS_i,set1 with max(Speed(i, k, D))
    Forward p to k
ELSE
    R := 1−K × mean (e_n)
    Generate random number RN in [0,1]
    IF RR < RN
      Drop p
      FOR all u in US_i
        Send BPP to u
    ELSE
      Choose k in FS_i with max(Speed(i, k, D))
      Forward p to k
```

Figure 13.8. Stateless nongeographic forwarding algorithm of the SPEED protocol.

Process_BPP:

i:	Current node (recipient of the BPP)
j:	Node with congested downstream neighbors(sender of the BPP)
D:	Destination node
$SenttoDelay_j$:	Delay of downstream node j
$AvgSenttoDelay$:	Average of $SenttoDelay$s of nodes in FS_j

1. **IF** $j \notin FS_i$
2. Drop BPP
3. **ELSE**
4. Adjust $SenttoDelay_j$ with $AvgSenttoDelay$
5. **IF** i is congested for all k in FS_i
6. Send BPP to all upstream nodes of i

Figure 13.9. Back-pressure packet handling algorithm of the SPEED protocol.

cannot be know ahead of time. This, in turn, renders the SPEED protocol not entirely suitable for real-time applications for WSNs.

13.4.7 MMSPEED Protocol [25]

The two important shortcomings of the SPEED protocol [24] are the support of a single propagation level and the lack of support for ensuring end-to-end reliability. The MMSPEED protocol [25] addresses these two shortcomings. This is achieved by creating multiple logical layers in the same physical network. In Figure 13.10a, two logical delay layers that provide two different packet propagation speeds on the same physical network are depicted. To provide delay guarantees, the idea of providing network-wide speed guarantees is adopted [24] for each logical layer. The speed layers are isolated from each other through prioritization in queuing and channel access. Figure 13.10b shows how different reliability levels can be provided in the same network. To guarantee different reliability levels, packets are routed over multiple paths based on the requirements contained in headers and on local statistics on packet loss.

Local forwarding decisions are made considering local statistics and required speed and reliability levels contained in every packet's header. Let a packet x be generated by the source s. The source node s calculates the required speed $S^{\text{req}}(x)$ for x so that x reaches its destination d by its deadline $D(x)$, that is, $S^{\text{req}}(x) = \frac{|s,d|}{D(x)}$, where $\mid s, d \mid$ is the distance between s and d. The source node s includes the required speed $S^{\text{req}}(x)$ in x's header. First, let us consider the delay QoS provisioning in a network that supports L layers of propagation speed $S_1, \ldots, S_L$. At the source, s selects the minimum speed layer l larger than $S^{\text{req}}(x)$, that is, $S_l = \min_{j=1}^{L}\{S_j \mid S_j \geq S^{\text{req}(x)}\}$. Then x is forwarded to one of the neighbor nodes i that has a propagation speed of $S_{s,i}^{d} = \frac{|s,d|-|i,d|}{d_{s,i}}$ with respect to d, where $d_{s,i}$ is the delay estimate from i to s, and $S_{s,i}^{d} \geq S_l$. As the packet is forwarded in the network, it may be propagated slower

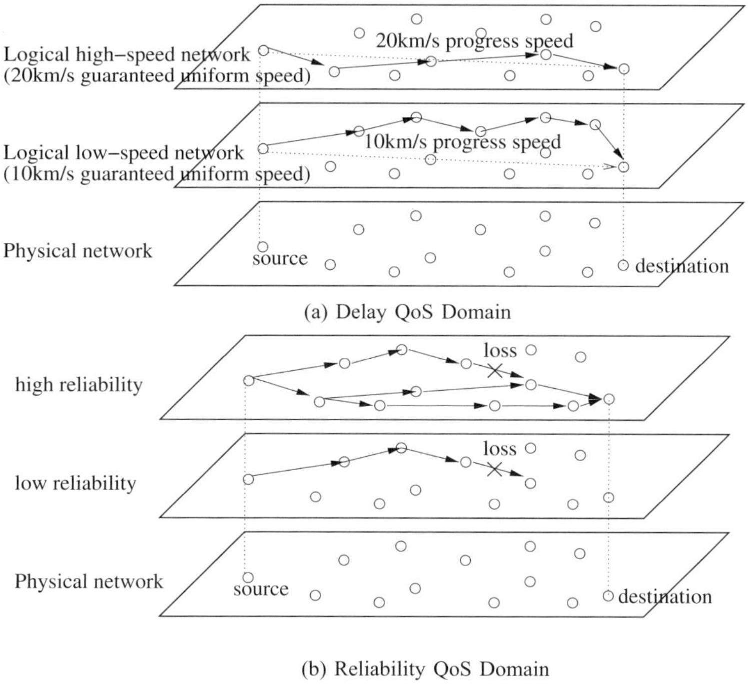

Figure 13.10. Two QoS domains and corresponding layers implemented in the same network.

than anticipated due to random node placement and lack of coordination between distant nodes. If an intermediate node finds that a packet cannot reach its destination at a speed layer, the packet is pushed to a higher speed layer with a new required speed S^{req}. Hence, errors in local decisions are compensated as packets traverse the network.

To provide reliability in reaching the destination, the packet loss rates to neighbors are monitored and the total number of hops to the destination is estimated. The source node s includes the reliability requirement $R^{req}(x)$ of packet x in the header. Consider x being forwarded by node i. All nodes including i keep packet loss statistics for their neighbors j, indicated by $e_{i,j}$. The end-to-end reachability estimate $R^d_{i,j}$ from node i to destination d over neighboring node j is calculated as $R^d_{i,j} = (1 - e_{i,j})^K$, where $K = \lceil \frac{|i,d|}{|i,d|-|j,d|} \rceil$ is the estimate of hop distance from i to d. After determining its neighbors that can sustain the required speed for packet x, node i chooses a subset of neighbors $j_1, \ldots, j_m$ such that $1 - \prod_{n=1}^{m}(1 - R^d_{i,j_n}) \geq R^{req}(x)$. In other words, we try to make sure that the probability of one copy of x to reach d is not smaller than the original required reliability level. At the same time, we make sure that the packet is propagated at the required speed. In each of the copies sent to these neighbors j_n, the required reliability of packet x is updated as $R^{req}(x) = 1 - e_{i,j_n}$. Hence, as the number of paths x is relayed over increases, the individual reliability levels decrease while preserving the total reliability level. Note that the probability of discarding a packet to sustain a speed level increases as the required reliability for the packet decreases.

At this point, we would like to emphasize the interactions between the network layer and the MAC layer. The network layer makes decisions about the selection of forwarding nodes for each packet and maintains a prioritized queue for different speed layers. On the other hand, the MAC layer implements mechanisms for prioritized access to the channel according to speed levels and implements a multicast mechanism. The network layer makes its forwarding decisions based on the information it obtains from the MAC layer. The MAC layer monitors the delay and packet loss probabilities for each neighbor.

The MMSPEED protocol provides delay as well as reliability guarantees to real-time flows. With its dual objective nature, MMSPEED stands out as a multifaceted protocol that cuts across MAC and network layers. These features come at the price of higher complexity in monitoring the network as well as storage requirements. Hence, added storage and processing requirements may potentially increase the cost of individual nodes in the sensor network.

13.4.8 Dynamic Delay-Constrained Minimum Energy Dissemination Protocol [26]

Some real-time applications for WSNs may require multiple sinks to obtain sensory data from a single source. At the same time, the end-to-end data delivery delay between a data source and sinks is required to be upper bounded. The Dynamic Delay-Constrained Minimum Energy Dissemination (DEED) protocol [26] aims to form and maintain multicast trees in the WSN that minimize energy consumption while guaranteeing end-to-end delays. Since the number and locations of the sinks may not be available, it is required to dynamically adapt the data dissemination tree upon the arrival of new sinks and leaving of the existing ones. A single source node in the WSN sends data, and arbitrarily located mobile/stationary sinks request these data. Sensor nodes are assumed to be aware of their own geographic locations and their immediate neighbors. Moreover, each sink has an upper bound for end-to-end message delivery delay (UBED) from the source.

The dissemination tree (d-tree) is rooted at the data source node. New sensors are added to the tree as new sinks request data from the source and register themselves with the tree. Hence, the construction of the d-tree starts from the single source node, and it continues as new sinks arrive. The DEED scheme deals with this construction while minimizing the total energy consumption and meeting UBED requirements. The structure of the d-tree has two main parts, namely, the static part and the mobile part. The static part consists of the source, the sensors (relays) and the multihop edges between the relays. The sensors that connect sinks to the d-tree are called *access relays* (ARs). The mobile part of the d-tree is simply the set of one-to-one connections between sinks and their corresponding access relays (AR).

The end-to-end delay in the DEED protocol (which is upper bounded by the UBED D_m for each destination m) is composed of the end-to-AR delay upper bounded by

P_m and the delay between a sink and its access relay, upper bounded by δ_m, where $\delta_m + P_m = D_m$. The delay through multiple hops is the sum of queuing, transmission, propagation, MAC, and retransmission delays. However, since the queuing, MAC, and retransmission delays are unpredictable, DEED introduces a new parameter, the *average delay per distance* q, which is obtained through tests like ping applications and then given to sensor nodes. Furthermore, since the geometric distance of multihop edges are nearly proportional to hop count in a sensor network, geometric distance is used as a measure of delay along with the parameter q. The end-to-end delay is computed as the sum of the edge delays along an end-to-end path.

The tree-update procedure when a new sink joins is also the mechanism of d-tree construction. The packets for constructing the d-tree are forwarded by greedy forwarding, whereas data packets are broadcast and only the nodes that cache the addresses of the senders receive the packets. The procedure for d-tree construction is as follows: When a sink m wants to join the d-tree, m locates its nearest neighbor a_m and chooses a_m as its access relay. Then, the sink m sends a JOIN query over a_m to the source. Upon the reception of the JOIN query, the source subscribes the sink and its access relay. This procedure is called the *subscription phase*. Then, the source initiates the *gate-relay search* procedure. A gate relay is the relay in the existing tree where a separate branch of the tree is created to reach the access relay. The gate relay is searched recursively over the relays of the d-tree starting from the source node. Let $r[i]$ be the relay that is checked to be a candidate gate relay, where $r[0]$ is the source sensor. Let H be the union of the set of children of the relay node $r[i]$ in the d-tree and the relay node $r[i]$. Furthermore, let s_i be the delay from the source to $r[i]$. The objective in the gate relay search is to minimize $d(h, a_m)$, which is the distance between the access relay and a relay node in H. Moreover, the delay between the source and the access relay over this gate relay should be lower than the end-to-AR delay constraint P_m of the sink as follows:

$$(s_i + q[d(r[i], h) + d(h, a_m)]) < P_m \tag{13.1}$$

Here, h is the element of H with minimum $d(h, a_m)$. If h is found to be the relay node itself (such that $d(r[i], a_m)$ is minimum $d(h, a_m)$) , then $r[i]$ becomes the gate relay. Otherwise, node h with minimum $d(h, a_m)$ is assigned to be the next relay node $r[i+1]$ to run the same algorithm recursively and s_{i+1} is updated accordingly.

The third step of the three construction aims to locally adjust the tree around the gate relay (hence form a branch leading to the access relay) to produce an optimum dissemination tree from source to the destination. A junction node J in the neighborhood of g is searched such that the total of the distances from gate relay g, its closest child c, and the access relay a_m to the junction J is minimized. J should also satisfy the delay constraints of the sink m, as well as all the sinks that node c has as its descendants. If the delay constraints are not satisfied for all neighbors of g, then a_m becomes a direct child of node g. Otherwise, the chosen junction J becomes a relay of the new branch emerging from g toward a_m and J recursively runs the same algorithm until the delay conditions are not met for a junction's all neighbors. At this point, all the relay nodes on the new branch are determined.

If the sink is mobile, the path leading to the sink and the distance between the sink and the access relay change. Hence, the sink notifies its access relay (AR) of its latest nearest-neighbor node in order to continue to communicate with the AR. If the sink-to-AR distance is increased too much and the sink-AR delay constraint is violated, then the sink needs to select another access relay.

DEED is an efficient algorithm to minimize energy consumption and meet delay requirements. However, it does not propose service differentiation among flows toward different sinks. Furthermore, decrease of reliability in data delivery due to possible link errors in wireless channels is not addressed. Additional mechanisms to address path changes due to link errors and congestion are needed. DEED adjusts the delay/distance parameter to handle congestion, but this does not guarantee avoiding highly congested local spots. Furthermore, these additional mechanisms must be local and should not alter the overall tree structure.

13.4.9 QoS for Data Relaying in Hierarchical WSNs [27]

This study presented in reference [27] addresses the selection of one or more routes from sensors to a base station. The routes are chosen such as to satisfy the delay requirements. In this study, large heterogeneous WSNs are considered which are organized in three tiers of hierarchy: a base station (BS), relay nodes (RN), and finally sensor nodes acting as data sources. Relay nodes are placed such that connectivity is maintained. Additional relay nodes are placed to improve energy consumption and reduce interference. However, increasing the number of relay nodes increases the end-to-end delay. Furthermore, relay nodes that are closer to the base station consume more energy compared to others. To address these issues, a hybrid approach is proposed that introduces relay gateways (RG) that receive data from relay nodes and send them directly (in a single hop) to the base station. RGs are pre-deployed in the network and are stationary.

The routing decisions at RN-RG and RN-RN communication level consider the system lifetime as a constraint. System lifetime is defined as the time until at least one RN or RG depletes its energy supply. The lifetime of a node is modeled by the maximum amount of traffic it can handle, which is called the node capacity. The routing from sensors to relay nodes is assumed to be handled by low level protocols. Hence, this protocol only deals with efficient routing of data among relay nodes by delivering it to one of the RGs while meeting the delay requirements. Two different cases of selecting a relay path are proposed: multipath relaying with delay constraints (MPD) and unconstrained multipath relaying (MP). Two different algorithms that are both centralized and optimal are proposed to solve these problems.

In the MP problem, the aim is to minimize the end-to-end communication cost while meeting the capacity constraints of RNs and RGs. However, this problem does not deal with end-to-end delays. The WSN is modeled as a directed graph composed of a set of vertices (composed of RNs and RGs) and a set of arcs which represent the edges between vertices. The cost c of sending a packet over an arc is defined as a function of individual arcs. Furthermore, the flow of data over an arc is defined as a separate function x, which is used to model the routing decisions of every RN.

Hence the total cost of communication over the selected arcs from source RN to RG, $\sum x(a)c(a)$, is minimized, where a is an arc.

The capacity of a node i is represented as $\gamma(i)$, whereas the total amount of data forwarded to a relay from sensors is $\beta(i)$, which is called the demand of i. The set of all arcs entering a node i is $\delta(i-)$, and the set of all arcs leaving i is $\delta(i+)$. The minimization of total cost is subject to $x(\delta(i+)) - x(\delta(i-)) = \beta(i)$, meaning that the demand of a relay node i is equal to the net flow into i, that is, $x(\delta(i-)) - t\beta(i) \leq \gamma(i)$, where it is assumed that transmission requires $t/(1-t)$ more energy than reception.

The MP problem is modeled as a transshipment problem, which is a classical problem in operations research. The transshipment problem can be solved in strongly polynomial time. In the MPD problem, apart from minimizing the total cost and meeting capacity constraints, the total number of intermediary nodes on a path should not be larger than a given value to limit the total delay. Hence, packets that originate at different RNs should be distinguished. The delay constraint can be formulated either using flow functions or using feasible paths. In the former, relay nodes increment a hop-count index that is assigned to individual flows. In the latter, the set of all feasible directed paths $P(r, k)$ originated from an RN r and ending at an RG k is defined. The paths with lengths smaller than a threshold are chosen. However, the demand of every node should be met and the capacity of all nodes should be observed. The MPD with feasible paths formulation can be solved optimally using a linear program. The advantage of feasible path formulation over the flow function formulation is that the number of constraints is linear in the size of the graph. Column generation, which is an implementation of the simplex algorithm for solving linear programs, is used to find a solution to the MPD problem.

This work is based on a graph representation of a WSN, which is used to find QoS-based paths in an hierarchical structure. However, methods to handle network dynamics is not considered. Furthermore, since the algorithm is centralized, it is assumed that every parameter, including the topology, link bandwidths, and link costs, are known at a central location, which is not very practical. A distributed implementation is needed to ensure its practical applicability.

13.5 QoS-BASED COMPUTATION WITH COMMUNICATION SUPPORT

The idea of reducing the communicated data volume through methods like data aggregation has been recognized as a means to reduce energy consumption and prolong WSN lifetime. While majority of in-network processing proposals involve simple operations, more complex algorithms have recently been proposed to process higher data volumes such as in video sensor networks. The main idea behind such proposals is to leverage the collective processing power of individual sensor nodes to run complex processing applications. To fully utilize the collective processing power of sensor nodes, solutions from parallel processing literature have been adopted. The proposed methods also have strong connections with the communication protocols: The exchange of intermediate results occurs over shared wireless channels, which is not considered in wired interconnected networks of processors. The resulting communication schedules determine channel access in sensor clusters, directly affecting the MAC layer in WSNs.

13.5.1 Collaborative Resource Allocation Algorithm [28]

The Collaborative Resource Allocation (CoRAl) algorithm [28] aims to dynamically allocate resources such as bandwidth and CPU time for multiple periodic applications in WSNs. Subject to resource availability, CoRAl aims to adjust application sampling frequencies to meet the temporal constraints and maximize network utility. CoRAl is assumed to be executed in fully-connected single-hop WSNs, where all nodes are synchronized and use Earliest Deadline First (EDF) as the scheduling algorithm. Nodes also implement the implicit EDF algorithm as the underlying wireless network MAC protocol. End-to-end applications are considered in reference [28] that are composed of a chain of tasks already assigned to sensors and sequentially executed in a pipelined manner.

CoRAl achieves its goals by iteratively executing the following steps until the schedule converges: First, the task execution frequencies on each sensor are locally optimized subject to application execution frequency upper bounds, whose initial values are set to be infinite. Then the execution frequency upper bound of each application is reevaluated based on the updated task frequencies and bandwidth allocation.

In CoRAl, the wireless channel is modeled as a dummy node on which only communication can be executed, and the network bandwidth is allocated in the same manner as sensor CPU time allocation. The CoRAl algorithm is presented in Figure 13.11. In each node, an extended version of the SLSS algorithm [29] is implemented to

CoRAl:

T_i: Application i
f_i^{max} : Maximum upper-bound frequency of application T_i
L: Number of Applications
m_k: Sensor node k
f_i^{ld}: Frequency of leader task of application T_i
f_i^{bn}: Frequency of bottleneck task of application T_i

```
1. Initialize maximum upper-bound frequency of each application T_i:
2.     f_i^max = +∞ , i ∈ {1, ..., L}
3. WHILE schedule not converge
4.     FOR each sensor m_k
5.         FOR each task of application T_i assigned on m_k, i ∈ {1, ..., L}
6.             Locally optimize the task subject to f_i^max using the extended SLSS
7.     FOR each application T_i
8.         Reevaluate f_i^max with updated f_i^ld and f_i^bn
```

Figure 13.11. The CoRAl algorithm.

compute locally optimal frequencies subject to node utility constraints. Different from the original SLSS algorithm, the extended SLSS algorithm in reference [28] takes each task's application execution frequency upper-bound into consideration. After each iteration of local optimization, the upper-bound frequency of each application is calculated. Let the *leader task* ld_i and *bottleneck task* bn_i of an application T_i be tasks whose frequency f_i^{ld} and f_i^{bn} are highest and lowest among all tasks of T_i, respectively. The frequency upper bound of T_i is updated as $f_i^{max} = f_i^{bn} + (f_i^{ld} - f_i^{bn})\sigma$, where σ is the factor that controls frequency convergence speed. The optimization procedure terminates when the weighted difference between leader and bottleneck frequencies converges.

CoRAl addresses online resource allocation among multiple applications. According to the simulation results, CoRAl provides performances comparable to the optimal solutions obtained by the nonlinear optimization tool of Matlab at a much higher execution speed. However, in CoRAl, tasks of applications are assumed to be already assigned on sensors, and task mapping remains an open problem. Furthermore, energy consumption is not explicitly considered in reference [28], which is a fundamental problem in WSNs.

13.5.2 EcoMapS Algorithm [30]

A task mapping and scheduling solution, EcoMapS, is presented in reference [30] for energy-constrained applications in single-hop WSNs. It is assumed that networks are composed by homogeneous sensors that can calculate and communicate simultaneously. EcoMapS aims to assign computation tasks and schedule communication events with minimum application execution lengths subject to energy consumption constraints. EcoMapS is composed of two phases: the *Initialization Phase* and the *Quick Recovery Phase*. The Initialization Phase algorithm aims to minimize schedule lengths subject to energy consumption constraints, while the Quick Recovery Phase algorithm handles runtime sensor failures.

In the Initialization Phase, EcoMapS iteratively searches for the schedule with an optimal number of *computing sensors* involved in computation that results in the minimum schedule length under the energy consumption constraint. To exploit the broadcast nature of wireless communication, a hypergraph representation of the Directed Acyclic Graph (Hyper-DAG) is introduced. The Hyper-DAG representation of task dependency explicitly represents communication as well as computation events: The edges between a task and its immediate successors in a DAG is replaced with a *net*, which represents the communication task to send the result of a task to all of its immediate successors in the DAG. The Hyper-DAG extension of the DAG in Figure 13.12a is presented in Figure 13.12b, where r_is are the introduced nets. Similar to CoRAl [28], EcoMapS also models the single-hop wireless channel as a virtual node where only communication tasks can be executed. Based on the virtual node model and Hyper-DAG, a communication scheduling algorithm is developed and embedded into the schedule search algorithm, E-CNPT. E-CNPT is a low-complexity algorithm that first enqueues tasks according to the critical path of a Hyper-DAG, then assigns the enqueued tasks to the node with minimum execution start time. In case

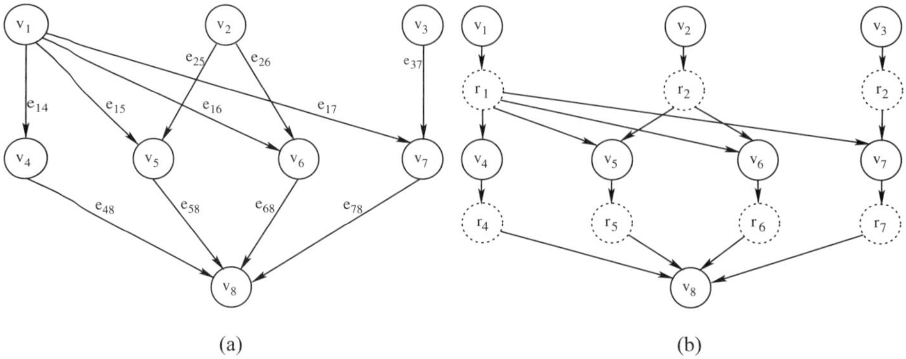

Figure 13.12. (a) DAG and (b) hyper-DAG examples.

communication between sensors is necessary, the proposed communication scheduling algorithm is executed. The Quick Recovery Phase algorithm handles sensor failures by adaptively adjusting the previous schedule. If idle sensors exist, the tasks of the failing sensor are migrated to an idle sensor. Otherwise, they are merged to a sensor that has the most idle time.

EcoMapS is a task mapping and scheduling solution for WSNs that provides application energy consumption guarantee with minimum schedule lengths. According to the simulation results, EcoMapS has superior performance over existing mechanisms in terms of minimizing schedule lengths. Also, the alternative schedules generated after sensor failures are shown to have satisfying performance with small recovery latency. However, EcoMapS has no guarantee of application deadline constraints.

13.5.3 Energy-Balanced Task Allocation Algorithm [31]

An energy-balanced task allocation (EBTA) solution is presented in reference [31]. EbTA assumes single-hop clustered homogeneous WSNs with multiple wireless channels, where sensors are equipped with dynamic voltage scaling (DVS)-enabled processors. EBTA considers real-time applications composed by interdependent tasks. The design objective of EBTA is to map and schedule application tasks to sensors such that balanced energy consumption is minimized subject to deadline constraints. In reference [31], applications are represented with directed acyclic graphs (DAGs) and the scheduling problem is formulated as an integer linear programming (ILP) problem. The exclusive wireless channel access feature is incorporated as additional constraints in the ILP problem.

Because the formulated ILP problem is computationally costly, a three-phase heuristic is proposed in reference [31] to provide a practical solution. In Phase 1, tasks are grouped into clusters to minimize overall application execution time assuming infinite number of sensors. Each task first constitutes a cluster by itself. Then all communication tasks are examined in a nonincreasing order of their data volume. For each communication event $e(i, j)$ between computation task T_i and T_j, the

clusters containing T_i and T_j are merged if it leads to shorter application execution time. When evaluating application execution time, communication events are scheduled to the channel with smallest available time using the First-Come First-Serve (FCFS) policy. In Phase 2, the task clusters from Phase 1 are assigned to sensor nodes with the objective of minimizing the maximum energy expenditure among all sensors. The task clusters from Phase 1 are first sorted in a nondecreasing order of energy consumption and are stored in a queue Π. The clusters in Π are then assigned to the sensor with the minimum normalized energy consumption (task execution energy consumption normalized by sensor residue energy, *norm energy* for short). Each time after a task cluster is assigned to a sensor, the norm-energy of the sensor is updated. This procedure repeats until all task clusters are assigned. Finally, a DVS heuristic is presented for Phase 3 to decrease energy consumption by iteratively adjusting the CPU voltage level of each task. In each iteration, a *critical node* that has the highest norm energy ε is selected. Among the tasks assigned on the critical node, a task is selected such that, by decreasing its CPU supply voltage to the next level, ε is decreased the most. Each time when a task is adjusted, the application schedule is iteratively adjusted accordingly to meet inter-task dependency constraints.

EbTA is one of the first proposals that addresses task allocation in WSNs, where both communication and computation tasks are considered. It is shown through simulations that the three-phase heuristic achieves longer lifetime compared with the baseline without DVS. The performance of the three-phase heuristic is also found to be comparable to that of the ILP-based approach via simulations.

13.5.4 RT-MapS Algorithm [32]

The RT-MapS algorithm [32] is proposed for single-hop clustered WSNs, which are composed of homogeneous DVS sensors with finite number of voltage levels. The design objective of RT-MapS is to provide application deadline guarantees with the minimum energy consumption for WSNs applications. The RT-MapS algorithm contains two phases, namely, *Task Mapping and Scheduling (TMS) Phase* and *DVS Phase*. The flowchart of RT-MapS is shown in Figure 13.13. In the TMS phase, computation and communication events are simultaneously assigned and scheduled with the objective of minimizing energy consumption subject to deadline constraints. To guarantee deadline constraints, sensors are scheduled with highest CPU speed in the TMS phase. Schedules generated in the TMS phase are then further optimized in the DVS phase by reducing CPU speed to decrease energy consumption. Similar to EcoMapS [30], RT-MapS employs Hyper-DAGs to represent applications and utilizes the virtual node model of wireless channels.

The RT-MapS solution is outlined with the pseudocode in Figure 13.14. Here, the communication scheduling algorithm sequentially schedules communication tasks on the virtual channel node to avoid packet collision. Broadcasting is also realized in the communication scheduling algorithm to conserve energy. The communication scheduling algorithm is embedded in the execution of the task mapping and scheduling algorithms, H-CNPT and H-MinMin. H-CNPT is different from

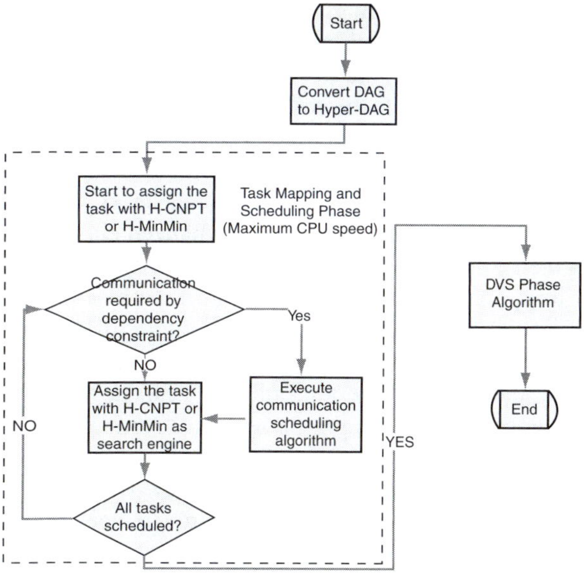

Figure 13.13. Flowchart of RT-MapS.

E-CNPT [30]. Among schedules with different number of computing sensors, the schedule satisfying the deadline constraint with the minimum energy consumption is selected as the optimal solution. The H-MinMin algorithm is an extended version of the Min–Min algorithm [33]. In the core of H-MinMin lies the fitness function that

RT-MapS:

N: Number of sensor nodes in the cluster
v_i: Task i of the application
m_k: Sensor node k
m_k: Sensor node k
f_i^{ld}: Frequency of leader task of application T_i
f_i^{bn}: Frequency of bottleneck task of application T_i

Convert DAG to Hyper-DAG
FOR $n = 0$ to N
    /* Schedule with n computing sensors/*
    Assign tasks using H-MinMin or H-CNPT
    IF communication is needed for an assignment of task v_i on sensor m_k
        Execute the communication scheduling algorithm
Among these candidate schedules, find the optimal schedule H^o:
    H^o has the smallest energy consumption subject to deadline constraints
Adjust the schedule H^o using the DVS algorithm

Figure 13.14. The RT-MapS algorithm.

H-MinMin:

L: Mappable task list
v_i: Task i of the application
m_k: Sensor node k

```
FOR α = 0 step 0.1 to 1
    Assign all entry-tasks
    Initialize the mappable-task list L
    WHILE L is not empty
        FOR task v_i ∈ L
            Find sensor m_k with the minimum fitness of v_i
        Find (v^o, m^o) with the minimum fitness among these combinations
        Assign v^o to m^o
Update L
Among all schedules with different values of α
    Select the optimal schedule
```

Figure 13.15. The H-MinMin algorithm used in the RT-MapS algorithm.

combines schedule length and energy consumption resulting from assigning a task to a sensor. In each iteration of task assignment, each "mappable" task whose immediate predecessors are already assigned is tentatively assigned to different sensors. The suboptimal task-sensor combination with minimum fitness value is kept. Among all suboptimal task–sensor combinations, the pair that gives the minimum fitness value is chosen, and the task is assigned to the corresponding sensor. The pseudocode of H-MinMin is presented in Figure 13.15. The DVS algorithm further reduces energy consumption of the schedules generated in the TMS phase. In the DVS algorithm, the communication tasks assigned on the channel are kept unchanged, and their start time and finish time are taken as the upper and lower bounds to adjust the corresponding sensors' speed during the time interval. During DVS adjustment procedure, CPU speed is reduced in proportion to the CPU utility.

RT-MapS aims to provide application deadline guarantees with minimum energy consumption. Due to the parallelism among sensors and exploiting the broadcast feature of wireless communication, RT-MapS shows superior performance compared with exiting mechanisms including EBTA [31], as presented in the simulation part of reference [32]. However, as other outlined solutions, RT-MapS also fails to extend to multihop WSN cluster. Furthermore, neither execution of concurrent application nor interaction with neighboring clusters is considered, either.

13.6 QoS-BASED CAPACITY ESTIMATION IN WSNs

An important step in end-to-end QoS provisioning in any communication network is the estimation of the network capacity to ensure that admitted flows can be sustained in the network. Such estimations are also important to adjust the data injection rates and other requirements of flows at sources. To provide accurate feedback, sinks must be able to compute the capacity of the network as functions of QoS parameters and the flow characteristics. In the literature, capacity modeling of wireless multihop networks have been investigated with simple metrics. The line of work pioneered by P. R. Kumar aims wireless network capacity in bit-meters per second [34]. This approach has also stimulated similar analysis attempts with increasingly realistic communication models [35–38]. The main shortcoming of these attempts lies in the oversight of an overarching goal of communication networks: Delivery of information to remote devices across a network under realistic conditions. Furthermore, combined metrics do not represent individual metrics such as delay, throughput, and loss rates explicitly.

Motivated by the lack of multimetric capacity modeling work for multihop networks, we have investigated the following problem [39]: *Consider a dense sensor network deployed in a rectangular area, where sources and destinations are located along two opposite edges. Assuming a Rayleigh fading channel, what is the maximum attainable throughput across the network given a delay bound for individual packets and a maximum tolerable bit error rate?* A corollary to this problem is the calculation of the minimum delay attainable if a particular throughput is required. With the results of this analysis, it is possible to estimate the total capacity of the network for a given delay value.

Our analysis is based on a physical layer channel model that accounts for the effects of the noise as well as the interference from other simultaneous transmissions. The bit error rates are calculated as the probability of a signal to be below a given threshold based on the Rayleigh channel model. With this comprehensive interference model, we analyze the effect of packets in the same data flow (intraflow interference) and other data flows (interflow interference).

As a first step, we analyzed a data stream that traverses a linear path. Based on a discrete time model where each packet transmission lasts one time unit, we modeled the behavior of the packets in the same flow. Our analysis confirmed that every packet injected into the network slows down the progress of preceding packets. To solve this problem, waiting times are introduced at the source while injecting packets into the network, allowing earlier packets to move forward before another packet (i.e., an interferer) is introduced. To achieve the lowest end-to-end delay, a packet must exist in the network alone, which reduces the throughput to one packet per end-to-end delay. Similarly, the maximum throughput can be achieved if packets are injected every time unit, which causes the delay to go to infinity as the network density approaches infinity. Intermediate combinations are achievable by adjusting the interpacket waiting times.

The two-dimensional case is modeled as parallel linear flows. In this case, every transmitted packet experiences interflow as well as intraflow interference. To increase the throughput, the distance between flows can be reduced. Small interflow separations increases the interflow interference, causing packets to cover short distances every transmission time, which in turn increases the end-to-end delay. If the flows are separated by larger distances, then the overall throughput decreases along with the end-to-end delay. The parameters that control this behavior are the interpacket delays, relative packet injection times between flows, and the interflow separations. Using nonlinear optimization techniques, it is possible to compute the maximum throughput for a given delay bound and parameters to achieve the desired performance.

This study [39] is merely a first step in this open research area, and it leaves out many important issues such as crossing paths, single destination flows, and non-time-synchronized systems. Furthermore, protocol-dependent effects on capacity are also left out for the sake of simplicity. We are hopeful that more accurate and comprehensive studies will follow this one that will address more realistic settings and that can be directly integrated to admission control and QoS feedback mechanisms.

13.7 CONCLUSIONS AND OPEN RESEARCH PROBLEMS

QoS-based communication in WSNs is a very promising and emerging field to sustain the communication in next wave of truly real-time WSNs. Although there is a significant amount of work that has been performed in the last five years in this area, they are far from addressing all QoS problems in WSNs. While WSNs resemble ad hoc networks and many more solutions for ad hoc networks do exist, it is important to keep in mind that WSNs are fundamentally different than ad hoc networks and have different requirements and constraint. Yet, it is equally important to consider the lessons learned from earlier QoS-based communication protocol proposals to avoid similar pitfalls.

Considering the available pool of solutions, several shortcomings in the development of QoS-based communication protocols stand out: Most of the solutions either rely on hard-to-materialize assumptions or are very complex for implementation on truly resource-constrained sensor nodes. Although the existing and widely used sensor nodes (such as Mica2 sensor nodes [40]) would not have any problems running algorithms outlined in this chapter, their implementation on extremely small devices is questionable. Therefore, simplification of the QoS support mechanisms is very important. Similarly, implementation of select proposals on very simple devices should also be undertaken to assess their value in real operation environments.

A majority of solutions consider only an isolated set of problems (such as only routing, MAC, or processing) or very few QoS parameters (such as delay, reliability, or energy). While the protocol design for WSNs is heavily influenced by applications, protocols that can support multiple metrics are still few and far apart. Support of multiple QoS metrics is an important step to realize many real-time and mission-critical applications. It is clear that such protocols would have to coordinate (if not merge) several networking functions traditionally considered belonging to separate layers. The resulting solutions must still maintain a simple nature for easy and ubiquitous deployment on many platforms.

Another important aspect that has not been studied very extensively is the feedback mechanisms. The solutions presented in this chapter operate in an open loop; no feedback is obtained from end nodes to adjust the behavior of the flows or protocols. Although some examples such as ESRT [41] use feedback mechanisms to sustain specific reporting requirements, they are mostly application-specific solutions and do not apply to generic settings. Furthermore, feedback mechanisms to adjust protocol behavior would improve the application level QoS guarantees as well as improve the total number of applications supported in a real-time WSN.

Finally, the QoS-based capacity estimation is a completely virgin field with very limited work present. Capacity estimation of multihop wireless networks is a challenging topic in itself, and very few proposals can provide actual limits (and not only scaling laws) even with very simplifying assumptions. Clearly, capacity estimation has no low-hanging fruits. Significant and coordinated research efforts are required to derive useful capacity bounds. These bounds should provide not only idealized limits, but also useful estimates that can be used in combination with QoS-based protocols.

13.8 EXERCISES

1. One of the most important reasons for performance degradation in wireless networks in general is the mismatch between solutions deployed together. Considering the protocols and solutions outlined in this chapter, identify

(a) Protocols of different layers that cannot be used in the same node, and

(b) Protocols not necessarily in different layers that cannot be run concurrently in the same network.

Elaborate on reasons for these incompatibilities. Expand your search to more recent work published in the literature.

2. Considering the SPEED protocol, how would the selection of the SetSpeed parameter affect the system performance? Based on these observations, propose a method to select the SetSpeed value for a given overall throughput requirement.

3. Repeat the exercise above for the MMSPEED protocol.

4. How should the RT-MapS algorithm be modified such that it can be used in multihop clusters? Can the presented channel model be sufficient? If not, how should it be updated?

5. The ESRT protocol aims to keep congestion under control by changing the event reporting rates. Identify routing (and MAC if necessary) protocols that are best suited to be used with ESRT. Discuss possible needs for modification for seamless operability. Elaborate on the multihop feedback channel on ESRT performance, and how routing (and MAC) protocols can help with potential problems.

6. Discuss how energy-awareness can be incorporated into multi-path routing protocols in WSNs.

BIBLIOGRAPHY

1. I. F. Akyildiz, W. Su, Y. Sankarasubramaninam, and E. Cayirci. Wireless sensor networks: A survey. *Computer Networks Journal*, **38**(4):393–422, 2002.

2. M. Younis, K. Akkaya, M. Eltoweissy, and A. Wadaa. On handling QoS traffic in wireless sensor networks. In *Proceedings of the 37th IEEE Annual Hawaii International Conference on System Sciences (HICSS'04)*, January 2004, pp. 292–301.
3. M. Gerla and K. Xu. Multimedia streaming in large-scale sensor networks with mobile swarms. *ACM SIGMOD Record*, December 2003.
4. J. Campbell, P. Gibbons, S. Nath, S. Seshan, and R. Sukthankar. IrisNet: An Internet-scale architecture for multimedia sensors. In *Proceedings of ACM Multimedia Conference*, 2005.
5. W. Feng, B. Code, E. Kaiser, W. Feng, and M. L. Baillif. Panoptes: scalable low-power video sensor networking technologies. *ACM Transactions on Multimedia Computing, Communications, and Applications*, January 2005.
6. S. Chen and K. Nahrstedt. An overview of quality of service routing for next-generation high-speed networks: Problems and solutions. *IEEE Network*, **12**(6):64–79, 1998.
7. M. de Prycker. *Asynchronous Transfer Mode: Solutions for Broadband ISDN*. 3rd edition, Prentice-Hall, 1995.
8. R. Braden, D. Clark, and S. Shenker. Integrated services in the Internet architecture: An overview. *RFC 1633*, June 1994.
9. S. Blake, D. Black, M. Carlson, E. Davies, Z. Wang, and W. Weiss. An architecture for differentiated services. *RFC 2475*, December 1998.
10. P. White. RSVP and integrated services in the internet: A tutorial. *IEEE Communications Magazine*, **35**:100–106, 1997.
11. E. Ekici. QoS-based routing in wireless mobile networks. In *Resource Management in Wireless Networking*. Kluwer, 2005.
12. Y. Liu, I. Elhanany, and H. Qi. An energy-efficient QoS-aware media access control protocol for wireless sensor networks. In *Proceedings of International Conference on Mobile Ad Hoc Sensor Systems (MASS'05)*, November 2005, pp. 189–191.
13. D. Bertsekas and R. Gallager. *Data Networks*. Prentice-Hall, 1987.
14. A. Parekh and R. Gallager. A generalized processor sharing approach to flow control in integrated services networks: The single-node case. In *Proceedings of Joint Conference of the IEEE Computer and Communications Societies (INFOCOM'92)*, 1992.
15. H. Li, P. Shenoy, and K. Ramamritham. Scheduling communication in real-time sensor applications. In *Proceedings of the 10th IEEE Real-Time and Embedded Technology and Applications Symposium (RTAS'04)*, May 2004, pp. 10–18.
16. J. Misic, S. Shafi, and V. B. Misic. Maintaining reliability through activity management in 805.15.4 sensor networks. In *Proceedings of 2nd International Conference on Quality of Service in Heterogeneous Wired/Wireless Networks (QShine'05)*, August 2005.
17. K. Sohrabi, J. Gao, V. Ailawadhi, and G. J. Pottie. Protocols for self-organization of a wireless sensor network. *IEEE Personal Communications*, October 2000, pp. 16–27.
18. A. Boukerche and H. Owens. Energy-aware routing protocol for mobile and wireless ad hoc networks. In *Proceedings of IEEE Conference on Local Networks*, 2003, pp. 768–771.
19. D. B. Johnson, D. A. Maltz, Y. Hu, and J. G. Jetcheva. The dynamic source routing protocol for mobile ad hoc networks (DSR). Internet Draft, draft-ietf-manet-dsr-07.txt, 2002.
20. V. D. Park and M. S. Corson. A highly adaptive distributed routing algorithm for mobile wireless networks. In *Proceedings of IEEE INFOCOM*, 1997, pp. 1405–1413.

21. K. Akkaya and M. Younis. An energy-aware QoS routing protocol for wireless sensor networks. In *Proceedings of the 23rd IEEE International Conference on Distributed Computing Systems*, 2003.
22. A. Boukerche, X. Cheng, and J. Li. A performance evaluation of a novel energy-aware data-centering algorithm for wireless sensor networks. *ACM/Springer Wireless Networks*, 2005, pp. 619–636.
23. B. Deb, S. Bhatnagar, and B. Nath. ReInForm: Reliable information forwarding using multiple paths in sensor networks. In *Proceedings of IEEE International Conference on Local Computer Networks*, 2003, pp. 406–415.
24. T. He, J. A. Stankovic, C. Lu, and T. Abdelzaher. SPEED: A stateless protocol for real-time communication in sensor networks. In *Proceedings of IEEE International Conference on Distributed Computing Systems*, 2003, pp. 46–55.
25. A. Felemban, C. G. Lee, and E. Ekici. MMSPEED: Multi-path multi-SPEED protocol for QoS guarantee of reliability and timeliness in wireless sensor networks. *IEEE Transactions on Mobile Computing*, **5**(6):738–754, 2006.
26. H. S. Kim and T. F. Abdelzaher. Dynamic delay-constrained minimum-energy dissemination in wireless sensor networks. *ACM Transactions on Embedded Computing Systems*, **4**(3):679–706, 2005.
27. R. Benkoczi, H. Hassanein, S. Akl, and S. Tai. QoS for data relaying in hierarchical wireless sensor networks. In *Proceedings of the 1st ACM International Workshop on Quality of Service and Security in Wireless and Mobile Networks*, 2005, pp. 47–54.
28. S. Giannecchini, M. Caccamo, and C.-S. Shih. Collaborative resource allocation in wireless sensor networks. In *Proceedings of Euromicro Conference on Real-Time Systems (ECRTS'04)*, June/July 2004, pp. 35–44.
29. D. Seto, J. P. Lehoczky, L. Sha, and K. G. Shin. On task schedulability in real-time control systems. In *Proceedings of the 17th IEEE Real-Time Systems Symposium (RTSS'96)*, December 1996.
30. Y. Tian, E. Ekici, and F. Özgüner. Energy constrained local task mapping and scheduling in wireless sensor networks. In *Proceedings of the 1st IEEE International Workshop on Resource Provisioning and Management in Sensor Networks (RPMSN05)*, November 2005.
31. Y. Yu and V. K. Prasanna. Energy-balanced task allocation for collaborative processing in wireless sensor networks. *ACM/Kluwer Journal of Mobile Networks and Applications*, **10**(1–2):115–131, 2005.
32. Y. Tian, J. Boangoat, F. Özgüner, and E. Ekici. Real-time task mapping and scheduling for collaborative in-network processing in DVS-enabled wireless sensor networks. In *Proceedings of the IEEE International Parallel and Distributed Processing Symposium*, April 2006.
33. S. Shivle, R. Castain, H. J. Siegel, A. A. Maciejewski, T. Banka, K. Chindam, S. Dussinger, P. Pichumani, P. Satyasekaan, W. Saylor, D. Sendek, J. Sousa, J. Sridharan, P. Sugavanam, and J. Velazco. Static mapping of subtasks in a heterogeneous ad hoc grid environment. In *Proceedings of IEEE Parallel and Distributed Processing Symposium*, April 2004.
34. P. Gupta and P. R. Kumar. The capacity of wireless networks. *IEEE Transactions on Information Theory*, **46**(2):388–404, 2000.
35. F. Xue, L.-L. Xie, and P. R. Kumar. The transport capacity of wireless networks over fading channels. *IEEE Transactions on Information Theory*, **51**(3), 2005.

36. G. Barrenechea, B. Beferull-Lozano, and M. Vetterli. Lattice sensor networks: Capacity limits, optimal routing and robustness to failures. In *Proceedings of 3rd International Symposium on Information Processing in Sensor Networks, IPSN 2004*, April 2004.

37. E. S. Sousa and J. A. Silvester. Optimum transmission ranges in a direct-sequence spread-spectrum multihop packet radio network. *IEEE Journal on Selected Areas in Communications*, **8**(5):762–771, 1990.

38. M. Haenggi. Toward a circuit theory for sensor networks with fading channels. In *Proceedings of the 2004 International Symposium on Circuits and Systems ISCAS '04*, Vol 4(IV), May 2004, pp. 908–911.

39. A. Bader and E. Ekici. Throughput and delay optimization in interference-limited multihop networks. In *Proceedings of 7th ACM International Symposium on Mobile Ad Hoc Networking and Computing (MobiHoc 2006)*, May 2006.

40. Crossbow, Inc. MICA2DOT Data Sheet. http://www.xbow.com, 2006.

41. Y. Sankarasubramaniam, O. B. Akan, and I. F. Akyildiz. ESRT: Event-to-sink reliabile transport in wireless sensor networks. In *Proceedings of the 4th ACM International Symposium on Mobile Ad Hoc Networking and Computing (ACM MobiHoc 2003)*, 2003.

CHAPTER 14

Quality of Service in Wireless Sensor Networks

GREGORY J. POTTIE and AMEESH PANDYA

Department of Electrical Engineering, UCLA, Los Angeles, CA 90095

14.1 INTRODUCTION

The purpose of a sensor network is to extract model parameters of the physical world according to the fidelity requirements set by some end user(s), for applications such as basic science, monitoring, or control. The nodes of the sensor network include some combination of sensing, communications, and signal processing/storage capability, and they may additionally possess attributes such as mobility [1]. Early sensor networks were typically arranged in a star topology with (a) very limited processing on the nodes and (b) transport of essentially all data to a central fusion point. Later, nodes with limited processing capabilities (e.g., microcontrollers) and storage, along with simple sensors (e.g., geophones or infrared sensors), were developed [2]. There are now large families of nodes ranging in capabilities from small nodes of this type ("smart dust") [3] to DSP-based nodes with simple operating systems capable of forming ad hoc networks ("motes") [4–6] through to nodes with 32-bit machines running Linux [7], advanced chemical sensors, cameras and 802.11 radios. In deploying nodes of all of these types for a variety of applications, including identification and localization of mobile targets and multimodal sensing for basic science, it has been a challenge to determine in what manner nodes should cooperate to achieve the basic user objectives. This includes questions such as the minimum deployment density, what set of nodes should cooperate for detection/identification of particular phenomena, and how they should cooperate to establish communications among groups performing data fusion and in transporting results to the end user. There is not one best answer that applies to all situations, but rather careful attention needs to be paid to the interplay of objectives, physical models, and network resources.

Algorithms and Protocols for Wireless Sensor Networks, Edited by Azzedine Boukerche

The most important design criterion for any type of network is guaranteeing quality of service, QoS [8]. QoS has become an important issue in various kinds of data networks, because some users are no longer satisfied with resource allocation based on service provisioning. QoS measures include bandwidth, delay, and delivery guarantees. Different classes of traffic (e.g., voice, data, image, video, etc.) have different bandwidth and delay requirements. Many issues of resource allocation for QoS provisioning are discussed in [9–11], although there remains a broad set of optimizations based on user needs to explore.

Some basic sensor network problem types are depicted in Figure 14.1, illustrating a number of QoS issues. The first category of problem is to extract model parameters of some continuous phenomenon within some specified target region (e.g., specification of a temperature field according to some limited spatial frequency range and with some temporal resolution). Some sensors immediately outside the boundary region might be required, while for smoother sections inside the study region some sensors may not be needed. Even with a statistical model of the field, it may not be clear without some preliminary survey what density of nodes is required to gather sufficient data to meet user requirements for fidelity of modeling. An adaptive deployment or node activation strategy might be needed, with the former implying delay and the latter implying potentially excessive use of resources. The second problem category is to determine parameters of some point source (or set of point sources) located within the convex hull of the sensor network (e.g., location and target type). Here, typically a small number of sensors in the immediate neighborhood of the source must cooperate, with the resource optimization problem being to limit the number of nodes involved to those needed to meet the fidelity requirement, subject to delay constraints. Resources will be expended in disseminating queries about such events, collecting, processing, and storing the data, and then transmitting it to the end user, with a number of tradeoffs available between resource efficiency and latency. Similarly to the first problem, without some knowledge of the source statistics, it will be difficult to determine the appropriate node density and cooperative cluster sizes, and so adaptive procedures may be needed (imposing a large initial delay). In the third problem category, sources are located outside the network, with goals similar to the second problem category.

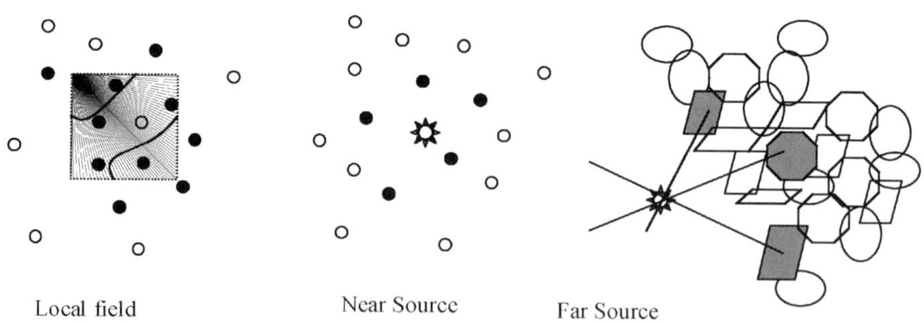

Figure 14.1. Three sensor network problems.

Now, however, that a very large number of nodes will be excited at similar energy levels; and so to avoid unnecessary expenditure of energy, means must be found to limit the number of clusters that are involved in fusing information. Individual clusters might, for example, perform beamforming to establish bearing angles and enhance the SNR, with the longer-range transport being limited to likelihoods of target identity and bearing angle, thus permitting fusion and localization.

In all of these cases, there is also potentially longer-range multihop transport of data to the end user, along with distribution of queries from users to the network that set policies for searching for and storing (fused and compressed) data. Latency in both processes should be within the constraints set by the end user. The transport of both data and queries should be reliable, and in many situations there will also be concerns for the security of the transport. Additionally, there is little point in collecting sensor data unless the sensors are calibrated and metadata relevant to the data interpretation is also included. A variety of propagation models can be considered in these problems, including distance loss laws and discontinuities caused by obstructions, while source models can range from the very simple (low-order differential equations for fields, Gaussian sources) to the more complicated, with varying levels of uncertainty in the models.

The remainder of the chapter is as follows. In Section 14.2 we discuss some background topics in network information theory relevant to the efficient collection, compression, and reliable communication of sensor data. We then discuss how a QoS perspective enables scalability in classical flat sensor networks, and we explore a number of practical approaches for high-fidelity data extraction in large-scale networks. In Section 14.3 we discuss some of the implications of introducing mobile elements and other forms of heterogeneity. In Section 14.4 we describe some of the data integrity concerns in sensor networks, including both calibration and security issues, for both flat and heterogeneous networks. In Section 14.5 we provide our conclusions.

14.2 FIDELITY AND RESOURCE TRADEOFFS IN FLAT NETWORKS

14.2.1 Network Information Theory

The sensor network problem is at a fundamental level governed by network information theory: It is a network rate distortion problem, subject to a number of practical constraints. Sharp information theoretic results could therefore provide guidance on design tradeoffs. Unfortunately, information theory has had relatively little impact in multihop networking compared to its success in point-to-point communications. There are several reasons, among them: It is difficult to compute the capacity of large networks (or even some networks with as few as three nodes), there is no source-channel coding separation theorem for general networks, and the number of possible interactions presents combinatorial difficulties. Additionally, the optimizations posed by Shannon are not as complete a match to the practical issues in networking because they are in point-to-point communications. Latency is less of a concern in single links than when it gets aggregated in multiple links across a network. With multiple users sharing a network, it is possible to assign different QoS values to flows, vastly expanding the number of possible acceptable solutions. Consequently, the usual mode of networking research is to (a) create a hierarchy of abstractions to manage the complexity and (b) then propose (and analyze) protocols within each layer to deal with different QoS, physical resource, and traffic model scenarios. Corresponding to these abstractions are particular physical components and algorithms [12]. This structure enables many of the components to be reused even when the conditions or objectives change, minimizing the effort involved in redesigning the system. Therefore, after reviewing progress in network information theory, we will consider how a QoS perspective for the theory of sensor networks can enable fresh progress.

Information Theory for 1-Hop Neighborhoods. Neglecting cryptography, there are two basic questions asked in classical information theory:

- *Channel Capacity:* For communications between a transmitter and receiver, what is the maximum rate at which information can be conveyed given the power, bandwidth and error rate constraints, and the noise in the channel?
- *Source Coding:* What is the rate of the minimum description of some random process (discrete or continuous) such that we can perfectly reproduce the process from our description (noiseless source coding) or can reproduce it with some bounded distortion (rate distortion coding)?

Enormous progress has been made since Shannon first posed these questions in the context of point-to-point communications systems [13, 14], and subsequently for many-to-one or one-to-many systems. Channel coding systems have been devised for the Gaussian channel and a number of its variants (e.g., the Rayleigh fading channel) that get within a fraction of a decibel of the SNR corresponding to channel capacity. The capacities of multiple access, broadcast, and multi-input multiple-output (MIMO) channels have been characterized [15, 16]. In multiple cell communication systems, users communicate directly with one or more base stations over a common set of channel resources (bandwidth/time) while the base stations use an independent (and low-cost) set of resources to communicate among themselves. One can assume varying degrees of cooperation among the base stations in coordinating their own communications and in assisting in separating the communications of the users, with, for example, cellular capacity more than doubling in going from no cooperation to having all base stations share information [17, 18]. There is also a broad set of practical techniques for managing interference, assuming differing levels of ability to estimate the (dynamic) channel [e.g., 19–21].

In the domain of source coding, practical lossless schemes have been devised that get very close to the entropy limits for a single source. Additionally, Slepian–Wolf coding enables separate lossless coding of multiple sources to the entropy limit assuming modest statistical knowledge has been distributed [22]. In rate distortion (lossy) coding, considerable progress has been made for Gaussian sources. In

Wyner–Ziv coding, a second source is presumed to supply side information about the primary source (e.g., correlated sensor data) that can then be used to lower the overall rate. Here the rate region is known for a single source with side information [23] and for cases with multiple sources where the sources are conditionally independent, given the side information [24]. In the Gaussian channel estimation officer (CEO) problem a single fusion point is assumed, with the objective of minimizing the overall rates from the (correlated) sources in the presence of Gaussian noise. While the rate region is not known for all cases, bounds have been developed [25, 26]. For the m-helper problem, one main node makes observations and communicates with a fusion center on some rate-constrained channel, along with m helpers that provide side information on the source through a nonconstrained channel. The question is then, How much can the helpers reduce the rate from the main node for a given distortion? A general solution is presented in reference 27, which was then applied to problems that more closely match the usual question of minimizing the total rate of all communications, such as the separate coding and Gaussian CEO problems. This has resulted in tight upper and lower bounds for general correlation relationships among the sensor readings [28].

Unicast Communications in Ad Hoc Networks. For unicast communication in ad hoc networks, pairs of source and destination nodes exchange independent messages across a network. Nodes, all of which share the same communication resources, may cooperate in transmitting, receiving, and relaying messages. The communications can cause interference to other communications within the network, degrading its quality of service. The network is thus a combination of relay, interference, broadcast, and multiple access channels, the first two of which have unknown capacities except in the simplest of situations. However, upper bounds on the scalability of the network with respect to the capacity of a single link are known. For a network with uniform source-destination traffic flows and decay of signal strength with distance according to some rule such as d^{-k}, for $k > 2$ the behavior arises from having an average path length (in near-neighbor hops) of $\Theta(\sqrt{n})$, and thus the need to carry traffic of roughly that number of users. The overall capacity available per source-destination pair declines as the square root of the network size, for fixed transmission bandwidth [29]. While antenna arrays, cooperative communications, and so on, all lead to improved communications, they do not sufficiently change the asymptotics [30–36], so that the network is not scalable under these assumptions on the traffic and propagation model. Under the assumption of multihop transmission, it may be shown that delay also scales as $\Theta(\sqrt{n})$ [37].

In the random graph model, it is assumed that the quality of links between nodes does not follow a simple geometric rule, a reasonable model for local links dominated by multipath. The capacity depends strongly on the particular model chosen for link quality, and can in the most favorable circumstances of high probabilities of long links being present result in scalability, even assuming uniform source-destination probabilities [38]. The bounds can be tightened if the source-destination distribution is highly irregular [39, 40]. For example, for the model considered in reference 29, if the traffic pattern is such that the average distance between source and destination

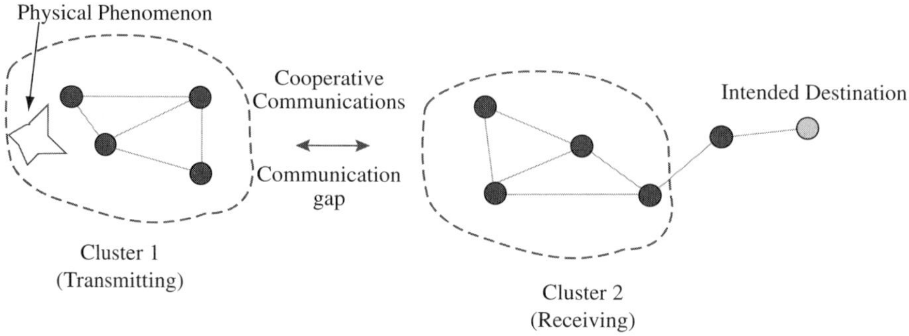

Figure 14.2. Cooperative communication in sensor network.

nodes remains small as the network grows, the throughput per node was derived to be $\Theta(1)$ in reference 41. The mobility model also has a large impact on capacity; an extended discussion is provided in Section 14.3.

Cooperative communication, while in itself insufficient for scalability, can lead to significant improvements in the capacity of individual links and thus potentially the reliability of finding a path through the network at sufficient rate to satisfy QoS requirements. For example, in Figure 14.2, nodes are clustered in the network such that there is a communications gap. The transmitting cluster senses the phenomenon and the measured data needs to be transmitted to the destination. One likely scenario for cooperation is to have two independent transmitters and two independent receivers. For cooperation to be of much benefit, the distance between the two clusters should be much larger than the distances within a cluster. The information theoretic aspects of such a channel models are presented in references 42–46 with explicit derivation of the data rates for each transmitting node and consideration of the effect of transmitter and/or receiver cooperation on the rate region. The system consisting of three nodes was considered in references 42, 45–47. The achievable rates for the channel model with two cooperative transmitters and a receiver is derived in references 45 and 47.

The four-node scenario with two nodes acting purely as relays is considered in reference 48. In references 49 and 50, channel with two cooperating transmitters and noncooperating receivers is considered. Their derivation was based on outage and diversity. The behavior for a fading channel is considered in reference 50 and that of a nonfading channel but with a complicated transmitter cooperation scheme involving dirty paper coding is presented in reference 49. In more recent work, two cooperating receivers along with the two cooperating transmitters are considered in reference 31. However, the model did not consider the transmission of information from a transmitter to both the receivers. In contrast to this, the system with two cooperating transmitters and two cooperating receivers is considered in references 43 and 44, where the data stream from each transmitter is intended for both the receivers. Apart from overcoming the gap in the sensor networks, the sensors cooperate with each other to achieve more reliable and higher rate communications. Power rather than bandwidth is the main constraint. Also, multiple sensors occupy the same channel; hence, standard multiplexing techniques such as TDMA, FDMA, CDMA, and OFDM may not be readily employed. For many cooperative schemes, it is assumed that a high level of synchronism can be maintained; if this cannot be done (see Section 14.4), then considerably lower gains result from noncoherent combining.

Network Channel Coding. In the previous set of scenarios, it was assumed that independent traffic streams would interfere, limiting capacity. In network channel coding, the capability of combining information streams as information is redundantly spread over the network is exploited [51]. There are two ways in which network coding can improve network capacity [52]: (1) When there is one source to be multicast, store and forward may fail to optimize bandwidth, and (2) when there are two or more independent sources to be transmitted in a network, even for a unicast situation, store and forward relaying may fail to optimize bandwidth. Considerable progress has been made on practical schemes for the first situation [53], but less for the second. In some examples, network coding makes the difference between network scalability or not [54]. However, in other situations [55] it can be shown that classical routing achieves the same performance as network coding.

Network Source Coding. We have previously addressed network source coding for the classic instances in which data are to be collected in a single communication hop, and now we focus on when it is aggregated across a multihop network. In data fusion problems, information from multiple sensors, possibly of different types, is combined to produce a decision variable [56, 57]. In classic data fusion problems, reduction of the transmission rate between sensors or fusion points is not considered, but clearly this is an important consideration in sensor networks [1, 58]. Remarkably, while obviously suboptimal from the point of view of rate reduction, it may be shown that the decision of a sensor to participate or not in contributing to fusion is sufficient to enable network scalability whenever imperfect fidelity of source reconstruction is permitted [59]. That is, while total network communications capacity increases with the number of nodes, under a fidelity constraint the amount of information to extract from the network eventually saturates. We will explore the consequences of this observation more fully in Section 14.2.

Better energy efficiency can be achieved by jointly considering coding and routing [60]. Here there are two basic choices to make, given that the optimization of network resources (e.g., communications energy) is NP hard. One can perform a relatively simple form of incremental compression (that is, with all data present at the node where fusion occurs), but then face a difficult routing optimization problem. Alternatively, one can perform (difficult) joint source coding, but can then route data using simple methods such as shortest path tree (SPT). For lossless compression of discrete sources, it has been shown that Slepian–Wolf coding followed by SPT is optimal. Either approach can work in small neighborhoods, but becomes impractical over larger regions. While it has been shown that at high resolution Wyner–Ziv coding resembles Slepian–Wolf coding [61], there are still many open issues in rate-distortion coding in spite of progress on practical coding methods (e.g., 62–66). Reference 67 presents a two-step DPCM encoding process, where first local measurements are encoded and then encoded outputs from other such encoders are combined. While producing higher mean squared error than joint encoding, the problem is more stable. In the combined problem of distributed coding with routing, the above scheme might be used in doing incremental compression. For such problems, it may be shown that routing using trees is suboptimal; and while the combined routing and fusion problem is NP hard (since it embeds a Steiner tree problem), if source correlations are highly localized, then simple heuristics can be quite effective [68, 69], considering both tree and non-tree constructions. However, much remains to be done to quantify the precise conditions under which such simplifications are effective and how much performance is lost.

Also of continuing interest is the problem of efficiently searching a sample volume, whether for reliably detecting intrusions in a surveillance application or characterizing some physical phenomenon. This may be done, for example, to learn a model as a prelude to source coding (e.g., correlations to enable Slepian–Wolf or Wyner–Ziv coding). The search may be conducted in many fashions: through activation of dense sets of static sensor nodes, by multiscale multisensor search, or through the use of mobile elements [70, 71]. Associated with each search procedure will be a source coding/information routing problem. Another important consideration is the construction of the physical model. Radically different conclusions on the efficacy of sampling methods can result from differing assumptions on the underlying model. For example, a smooth surface is far easier to deal with than one with discontinuities, and assuming that a phenomenon is caused by some finite number of sources results in different asymptotics than assuming it is due to high-order differential equations. As cluttered and inhomogeneous environments are the norm rather than the exception, devising information theoretic results for such domains is an interesting challenge.

14.2.2 Fidelity and Scalability

While ad hoc networks are generally not scalable, with particular conditions on the source-destination (S-D) distribution, scalability of sensor networks is possible [72]. The S-D distribution can be made sufficiently localized to enable scalability through either (a) decision-making of the type to be outlined below or (b) inclusion of additional communication layers, which make the S-D distribution appear local in the lower layers. These higher layers may be either traditional telecommunications infrastructure or mobile nodes, which can play the role of infrastructure with the cost of additional delay.

We now outline how local decision-making is sufficient for scalability. The sensor network problem is to extract information concerning some physical phenomenon (point source or field) to within some fidelity, given nodes with some constraint on resources (e.g., bandwidth). As the density of nodes increases, the possibilities for spatial re-use of frequencies improve and so the information volume that can be carried increases. However, the information that must be conveyed will saturate according to the fidelity threshold, if only nodes have a mechanism for determining which ones will be involved in some form of local fusion and which ones will report nothing. For example, in an identification problem, as the density of nodes increases, eventually there will be a node so close to the source that it alone will be able to make the decision. All that is required then is the capability for the higher SNR nodes to suppress activity by nodes that detect the phenomenon at lower fidelity but at SNRs at levels where they might be inclined to engage in cooperative detection. A simple relay strategy will then suffice to get the information out of the network. Notice that the keys to the scalability are: separation of function between source coding and long-range communication (allowing different densities for the logical functions of communications and detection), local decision-making, and a fidelity criterion. Without the last two, the amount of information to convey would outstrip communications capability for certain source models.

Now clearly neither the communications relay strategy nor the local decision rules are optimal in information theoretic senses; they are merely sufficient to ensure scalability. However, this result is highly suggestive of what an optimal strategy might look like under a fidelity (quality of service) constraint. Once a sufficient number of nodes are identified that (with appropriate network source coding) achieve mutual information between observations and source phenomenon above some threshold, then no more nodes need be involved. Long-range transport of data need not further consider interactions of source and channel coding. If the network is to be scalable, local communications must dominate, with longer-range communications becoming much less important (e.g., not requiring heroic cooperation strategies). Cooperation in relaying the data may be more of a secondary matter of achieving reliable communications over local gaps in the network. Given the local interactions dominate, it is in this domain that combined source coding and routing/channel coding are of most interest.

Put another way, while there is no general source-channel separation theorem, when source models, deployment scenarios, and QoS (fidelity) objectives are specified, the above sensor network scalability results imply that there can be separation on particular *scales*. For example, if the objective is to identify and track a number of objects that enter a study region, and the signals decay strongly with distance, then only the nodes that are in close proximity to each object will have sufficient mutual information to be included in the decision-making. While some performance loss may result in separating source and channel coding to the point of data fusion among this group, beyond this the problem reduces to a pure network coding (or routing) problem, with no further consideration of source coding. Furthermore, in the limit of very strong decay of signals with distance, the relative SNRs of the best and next best sensor will in likelihood be quite large so that even optimal fusion buys little performance gain, allowing separation of source and channel coding even at the local level. This suggests a two-scale design approach: For local regions, a solution specific to the underlying physical model is devised, which may involve some combined source coding/routing/channel coding scheme, while for long-range transport only communications considerations apply.

The corresponding optimization problem can be conceived of as a two-step process. The first step is the selection of a cost function that will control the set of nodes that will be locally cooperating, given a cooperation protocol. One may then select the cooperation protocol from a set of candidates. For a significant set of practical scenarios the number of nodes involved and their time of involvement are the most important resource questions. For example, in short-range radio communications, power is consumed simply from the fact of a radio being on, with transmission power being a secondary issue. The power and bandwidth requirements linearly depend upon their duty cycle. Similarly, the time spent processing is related to the amount of data being collected, again a function of the duty cycle of involvement. The amount of processing (or communications) required is often an exponential function of the number of nodes involved, again suggesting that minimization of the number of nodes involved (while meeting the fidelity constraint) is a primary consideration.

This type of cost function also leads to a reasonable set of optimization problems, as follows. The scaling laws for the number of nodes to be involved in fusing information about some source event to the desired fidelity can be relatively easily obtained in many circumstances. In particular, this is true for both maximal ratio (MR) combining and beamforming problems. For example, if the fidelity requirement is that the fused sensor data must achieve an SNR greater than some threshold in detecting some point source, and if AWGN is assumed and there is a second power source propagation loss, then assuming that coherent combining is possible every time the radius of the cooperation region is doubled, the expected SNR improves 6 dB (i.e., it scales with area, assuming a uniform distribution of nodes) [1, Chapter 5]. The relationship between coverage and density may similarly be easily obtained for this set of assumptions. For other models and cooperation protocols, the details of the tradeoff will vary, but the above serves as a good prototypical scenario for assessing the suitability of particular cost functions and also conveniently serves as a lower bound on achievable performance. Assuming that there is some maximum cost in terms of the number of nodes involved in the fusion activity, one would obtain a family of curves of the type illustrated in Figure 14.3 for different fidelity requirements. Below density ρ_l we hit the ceiling $n_{\max}$ on the maximum number n of nodes permitted, resulting in an inability to meet the fidelity requirements, while above density ρ_h only the minimum number of nodes $n_{\min}$ to carry out the task are needed (e.g., one in detection, four for 3-D localization, etc.).

The cost function must include the resource cost for the network as a whole in having a higher density (that is, the total number of nodes in the network, N), which reflects the equipment and maintenance costs. This will drive the solution

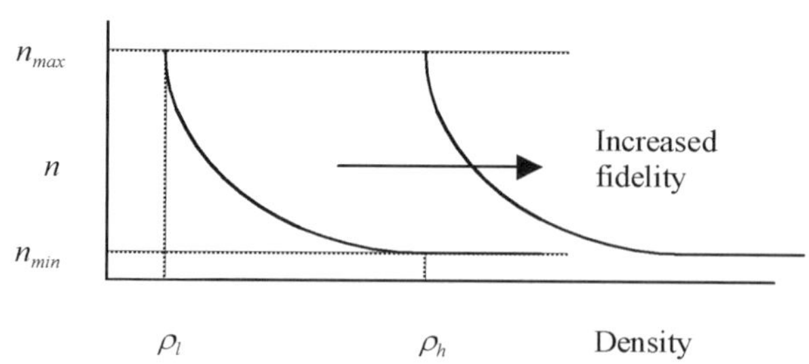

Figure 14.3. Number of cooperating nodes versus node density.

toward lower densities, while the operational costs for each detection event will drive it toward higher densities. Thus, for a given cooperation protocol source model, fidelity requirement there will be an optimal deployment density for at least some cost functions in this family. A broad set of such problems remains to be more fully explored.

14.3 FIDELITY IN HETEROGENEOUS NETWORKS

The basic set of assumptions regarding network configuration in the previous section might be termed "sensor network classic:" a dense set of nodes of the same type (sensor, radios, processors) usually with highly limited resources (especially energy) but which are inexpensive and can be deployed in large numbers to form a multihop network. Apart from the challenge of data extraction at minimal resource cost, other difficult problems for networks of this type that have attracted the attention of the research community will be discussed in Section 14.4. For the remainder of this chapter we will also consider heterogeneous networks, including the possibility of mobile elements and hierarchical networks that have nodes with greater capabilities (e.g., sensing or radio range, more storage, position-location ability, reliable energy supply, etc.). In this section we explore the implications for the fidelity problem, first with mobility and then with hierarchical networks.

14.3.1 Mobility

In MANETs, mobility is one of the major contributors to overhead from the protocol stack, thereby reducing communication efficiency [73–77]. More recently, however, mobility has been found to increase the capacity of wireless ad hoc networks [78–81]. Here, the networks have a mixture of static and mobile nodes. The type of mobility that mobile agents can possess falls into one of the following three categories [43]:

- *Random Mobility.* The nodes are assumed to move in an arbitrarily random pattern typically modeled as uniform Brownian motion for analytical convenience [78–81]. This model has also been used in data mule work [82, 83].
- *Predictable Mobility.* This model assumes that the pattern of mobility of the mobile nodes is known, and this knowledge can be exploited to route data [84–86]. The mobile agents are not moving for the purpose of data transfer and hence their paths may not coincide with the routing requirements.
- *Controlled Mobility.* Here the mobility pattern of the mobiles is completely under the control of the network. Prototypes of such networks have been produced [87–90]. Controlled mobility could also be implemented by providing infrastructure to move the nodes. This infrastructure can then serve additional purposes such as logistical support to the network [158].

There are fundamental differences in the throughput and delay properties when the mobility is controlled as opposed to when the mobility is random or predictable, as we now detail. In reference 81, it is assumed that all nodes are randomly mobile within a unit disc area. The data traffic pattern is assumed to be random as in reference 40. Data travels over only two wireless hops, from the source node to a mobile node that acts as a relay and then from the relay to the destination. With this model the throughput was found to be $\Theta(1)$. A similar result was obtained for mobility constrained to one dimension in [79]. The delay for the above scenarios was found in reference 80, and algorithms were discussed for trading delay and capacity. Delay $D(n)$ is related to throughput $T(n)$ by $D(n) = \Theta(nT(n))$ for the wireless network scenario of [29]. For the model in reference 81 when the nodes are randomly mobile with average velocity $v(n)$, the delay scales as $\Theta(\sqrt{n/v(n)})$. As noted in reference 80, three important features that influence the throughput and delay in ad hoc networks are the number of hops, the transmission range, and the node mobility and velocity. In reference 80, schemes are proposed that exploit these three features to different degrees to obtain different points on the throughput-delay curve in an optimal way, enabling achievement of the results in references 29 and 81 as limiting cases of a larger achievable region. Another scenario for a network with mobile nodes was considered in reference 78. The network consisted of n static nodes, which acted as sources and destinations for data. However, the network also had m randomly mobile nodes which were used as relays. For this model, using the routing scheme proposed in reference 78, the throughout is $\Theta(m/n \log^3 n)$ with an average delay of $2d/v$ where v is the velocity of the mobile nodes.

Now consider the network scenario as in reference 81 where all nodes are mobile except that their motion is controlled rather than random. Here a naïve communication strategy is for each source to move to its destination and communicate at almost zero range. Hence interference among simultaneous transmissions is zero, and each sender–receiver pair can utilize the full available bandwidth W. The per-node throughput is W with constant delay. The delay depends on the traveling time of mobile nodes to reach their destinations, which is constant as network area is constant. This compares to the worst-case delay in reference 81, which is infinite. Clearly, controlled mobility has the potential to yield fundamentally different throughput and delay limits compared to those achieved with random mobility in references 78 and 81. Further details, including mobile routing protocols, are presented in reference 43. We note that, of course, hybrid routing strategies are possible, with some combination of multihopping in the static network within close neighborhoods of mobile nodes, in which case the mobile nodes serve the role of a higher level of communication infrastructure.

In a network where most nodes are static, a reasonable way to view the mobile nodes is as a form of infrastructure for the support of the static elements. One consequence of providing long-range data transport is that the mobile nodes can help save energy in the static nodes, since these must no longer relay data from other nodes over long multihop wireless routes. The extra energy overhead of mobility may not be a major concern because the mobile nodes can periodically recharge themselves.

The reliability of transmission across a network also depends strongly on the mobility assumptions. One of the common problems arising in ad hoc networks is that of having disjoint networks. Mobile agents can be used to connect such sparse and disjoint networks. Also, by using mobile agents with longer-range radios, the number of wireless hops traveled by a data packet is reduced. This, in turn, reduces the possibility of packet error, thereby helping enhance goodput performance, and it reduces delays due to retransmission. The reliability analysis in reference 37 considered a number of possible metrics for network dependability, and it posed a number of graph theoretic optimizations to model the behavior of heterogeneous mobile ad hoc networks. Given that the optimizations are NP-hard, simulations were used to compare protocols for a number of different network configurations.

Mobile components can provide additional benefits. It was shown in reference 91 that the time synchronization error increases with an increasing number of hops between two nodes. Using the mobiles for time synchronization reduces the hop distance between nodes, and hence much finer time synchronization is possible than in a multihop case. Controlled mobility also helps improve the performance of localization systems [92]. Mobile components can support other system activities such as delivering required resources [93, 94]. Further details on some of these uses of mobility are provided in Section 14.4.

Other Forms of Hierarchy. Much of the literature on sensor networking is devoted to optimizations that consider variously computation, storage, energy, and communications bandwidth to be the scarce resources. Yet often logistical tasks are also limiting, including the planning for deployment, maintenance in the field, and time spent securing the network. This is part of the reason why, in practice, most telecommunication networks are hierarchical. Specialization of components allows limiting the demands placed on different levels of the hierarchy, simplifying design of each task and permitting different components/algorithms to be employed. A particular advantage is that it is easier to compose, debug, and re-use software in such settings, in that the design choice has been made to provide the higher levels of the network with far greater resources. When considering the large volumes of code required to get large networks to do anything useful, along with the frequency of updates required as application demands change, the challenge of having resource-constrained nodes perform such tasks is obvious. With hierarchical networks, one can have the advantage of re-using code and the potentially abundant bandwidth for the upper levels (e.g., not having to reinvent the capabilities of the TCP/IP stack or SQL databases), together with the capabilities of covering local regions with dense collections of nodes with deployments matched to the physical phenomenon of interest. To date, the largest deployments of sensor networks have also been hierarchical.

A basic principle of design in such situations is to take maximum advantage of the additional resources provided by the higher network layers. This can include some combination of low-latency long-range high-bandwidth data transport, reliable time and position information, high-capacity storage, longer-range and/or well-calibrated sensors, and so on. For example, in the Tenet protocol suite [95] the asymmetry of resources is exploited to enable simplified retasking and debugging of the network, thereby off-loading (for example) computation of new routes to the master nodes in the network. Clearly also with such a topology the data in the lower level need go only a few hops to reach a base station, thereby mitigating the reliability and scalability problems.

14.4 DATA INTEGRITY IN SENSOR NETWORKS

In this section we consider the question of how to trust the data that the network produces, given the problem of component drift and miscalibration and also malicious attack upon the network. These are referred to collectively as data integrity and are clearly prime QoS concerns. Data integrity must be maintained at multiple levels [96]. Errors can be introduced at the level of the physical events, the sensors, and middleware services (e.g., location, synchronization, data aggregation) in addition to the usual network consideration of communications losses. Confusion can arise in the events being monitored either due to malicious activity or the occurrence of some confounding set of events that are beyond the model assumed. Sensors can be faulty or can be captured. Services may yield less accuracy than promised (or needed) due to simplistic modeling or failure/capture of nodes. The solutions to these problems may be categorized according to how trust is embedded in the network. One may place trust in some physical model (for example, resulting from a sequence of experiments with carefully calibrated instruments), some set of trusted nodes (e.g., a base station or cluster heads), or the relative reliability of redundant deployments of nodes compared to sparse ones. Usually some combination of these trusted elements is required to provide a robust solution, although different elements are emphasized. For example, while intuitively one would expect that deployments of large numbers of nodes should enable greater collective reliability than the reliability of the individual nodes, this rests on particular assumptions regarding both the physical process under study and the corruptive processes. In particular, it is generally assumed that the process is oversampled, so that correlated readings are available to provide consistency checks, and that the corruptive processes are independent (in location and/or time). The quality of the data integrity assurances depends on how reliable these models are, notwithstanding the analysis of protocols designed assuming particular models. In the following we consider a number of specific problems in ensuring data integrity in sensor networks: calibration, location, synchronization, secure key distribution, reputation systems, and reliable query systems.

14.4.1 Calibration of Sensors

Calibration of sensors can in certain circumstances be quite easy. If the resolution requirements are loose and the sensor is stable for the environmental conditions under which it is deployed, a one-time calibration (comparison to a trusted standard) is sufficient. Temperature sensors deployed indoors fall into this category. However, usually the situation is more difficult, in that some of the following uncertainties can apply:

- The coupling to the medium may be unknown, causing the laboratory calibration to be largely irrelevant; and the medium may be dynamic, making a one-time field calibration ineffectual.
- There may be confounding physical effects (e.g., sunlight heating up the package).
- The sensor may be subject to biofouling (a particular problem for biochemical sensors), or otherwise become degraded by the environment.
- The electronic components can be subject to drift (e.g., droop in battery supply voltage).
- There can be hard faults (e.g., stuck-at values).

These uncertainties are often poorly modeled. While manufacturers will often design sensors to deal with some of the largest confounding effects, at other times the calibration curves provided are fictional with respect to field deployments [97, 98]. There are two basic ways to deal with these problems: redundancy of deployment (including multiplicity of sensing modes) and an explicit maintenance/calibration schedule. We deal with them in turn.

First consider the limit of a very large numbers of sensors, such that the phenomenon of interest is greatly oversampled with respect to the spatial and temporal fidelity requirements. Suppose the sensors are all initially perfectly calibrated and in working order. If sensor failures are uniformly distributed over some spatiotemporal window, then consensus-based mechanisms can be used to identify outliers [99] and, over time, exclude those nodes that are farthest from the group consensus of the remaining set. Supposing for now that these mechanisms are perfect, absent sensor drift the performance of the remaining ensemble will meet fidelity requirements until such time as their density is above the target, and the nodes will be able to reliably report their performance with respect to this target to enable decisions such as adding nodes to the network. Hence, the essential requirements are (a) ability to recognize the quality of individual readings with respect to the ensemble and (b) ability to recognize the quality of the decision of the ensemble. Now suppose that all the measurements are subject to independent drifts. With redundant deployment of sensors, either the law of large numbers or the Cramer–Rao bound can be invoked to argue that the consensus measurement will be more reliable than that of individual sensors. Unfortunately, over time, this consensus will deteriorate (e.g., as a result of the random walk of the combined noise and drift); thus, to have the same network lifetime as when there are random faults alone, an even greater level of redundancy is needed. Note that here as well what is fundamentally required to give reliability guarantees is some means for measuring the quality of the decision of the ensemble. Examples of such measures include (a) the variance of the measurements used to compute the ensemble and (b) its history over a sequence of measurements.

In the limit that the number of sensors goes to infinity, relatively simple mechanisms can be used to provide reliable decisions. Suppose that the objective is to determine the location of some source using coherent acoustic beamforming, and also suppose that the nodes have been synchronized using radios. The nodes can learn their own positions if sources at known locations and times emit sounds; this may be done simultaneously or sequentially. Then for sources at unknown locations, a random selection of a small number of nodes close to the source can compute the apparent position. The results for many such groups of nodes can be averaged, with groups that produce results far from the consensus excluded. Two basic approaches are to employ thresholds based upon (a) absolute distance of measurement values or (b) distance normalized by the average spread of the remaining members of the (local) ensemble. The choice of decision metric depends on the objectives, the prior knowledge of the source phenomenon and sensor behavior (e.g., models), and the dynamics of the physical situation (e.g., deterministic or rapidly and randomly changing).

With finite numbers of nodes, matters become more complicated in that it is more difficult to construct reliable reference ensembles; errors and drift in a small number of sensors can significantly alter results. With only two sensors, for example, it is impossible to tell which is more reliable, absent some side information such as a reliable model of the source phenomenon. When there are more sensors, clearly there will be greater certainty in the consensus, but statistical tests designed to gauge the reliability of the consensus or to make exclusion decisions will themselves be subject to considerable uncertainty. With modest redundancy, one must therefore fall back upon some source of trust beyond the primary sensors, such as the physical model, reference events, complementary sensors, or auditing by trusted nodes. Many possible combinations of these can be explored at varying levels of redundancy of deployment to produce effective protocols.

For example, trusted mobile elements may be employed to periodically calibrate nodes (e.g., a person or robot carrying trusted equipment). A variety of strategies are possible:

- Generation of standardized physical events at known locations and times (e.g., beacon signals)
- Measurement of physical phenomena in close proximity to node(s) being calibrated
- Replacement of fielded nodes with freshly calibrated ones
- Hybrid combinations of the above

The mobile patrols can be according to some fixed schedule, responsive to reported node failures or other indications from the network that performance has become unacceptable, or adaptive, using actual drift/faults to adjust the schedule of visits. That is, the patrol/audit period depends on the perceived drift and fault models. Additionally, the density at which standardized events are generated or the proximity within which calibration is required depends on model of the source phenomenon. If the events to be detected are rare, the second strategy might not be sufficient, and so the more aggressive tactic of generating events on a schedule may be needed.

In all these cases the conclusions made based on the data between calibration events are tentative, and thus use of the data might need to wait until the next calibration event. If the calibration takes place frequently enough for data to be usually within the desired fidelity bounds, then fewer such post-facto adjustments are needed. Having operated for some time in the field, one may also build up a better model of sensor drift and use this to predict offsets so that the objective of calibration is then to correct the model prediction error. With a good model, this may be considerably less frequently than otherwise. Similarly, one may employ redundant deployment to reduce the frequency of calibration events.

14.4.2 Reliable Sensor Location

Sensor self-localization or localization of sources may be attempted using a wide variety of algorithms. Most simply, sensor nodes can be told their locations, obtained using standard survey techniques, which themselves rely upon knowing some combination of orientation and the positions of landmarks or beacons such as in GPS [157]. With dense deployments of nodes, physical phenomena may be located by the expedient of identifying which set of nodes has signal strength above some threshold and then selecting either (a) the node with the strongest signal as the location or (b) some weighted combination of the nodes above the threshold. These techniques have low spatial fidelity, being on the order of the node separation. While it may be argued that if the propagation environment is known, then a weighted combination of signal strengths should be highly accurate, in practice most environments in which we would employ sensor networks do not have such simple propagation laws, or else we would simply have deployed a very small number of trusted long-range sensors. A better estimate of source position may be obtained by coherent beamforming [100, 101], in which either (a) several clusters of nodes form bearing angles toward the source and then a least squares estimate is made of the intersection point or (b) a single cluster directly estimates position, also usually with some type of least squares estimate. The latter technique is accurate only when the source lies within the convex hull of the nodes performing the sensing, while the former can be successful at considerable range. Since a wavefront will typically only be coherent within some local region, this implies that multiple clusters are usually needed for distant sources. Note that the price paid for the superior capabilities of coherent beamforming is that the nodes must be tightly synchronized, with absolute error among the clocks in the cluster being a small fraction of the period of oscillation of the waves in question.

Large collections of nodes can locate themselves with respect to a relative coordinate system without outside assistance, but require some landmarks or known positions to establish absolute location with respect to absolute coordinates. A noncoherent approach to (approximate) localization of nodes begins with certain nodes with known positions recording the number of communication hops to other nodes in the network. If these are at the periphery of the network, then nodes can estimate position with respect to these fixed nodes, assuming that the line of propagation to each is straight and that the propagation losses are constant. A least squares estimate can be used to deal with inconsistencies of the estimates, and refinements can be made based upon actual propagation losses between links [102, 103]. However, in reality a relatively high density of landmarks will typically be required due to the typically large variations in near-ground propagation.

If nodes are synchronized, then time distance of arrival techniques can be used to systematically determine the actual distances of nodes from each other, and then geometric relations can be used to snap together the links in a consistent fashion, again using least squares techniques to resolve the imperfections of the distance estimates. Such multilateration techniques can construct a relative coordinate system or, given some landmarks, an absolute coordinate system [104]. Note, however, that the greater the number of links that are built outward from known coordinates, the greater the chance that errors will accumulate in the position estimates. The errors can be controlled by sprinkling nodes with known (surveyed) positions within the network volume.

The above techniques all assume that nodes are working and are reliably reporting information. Implicitly, errors are small and random and the propagation conditions are reasonably uniform, at least within neighborhoods. Algorithms become complicated in practice because these conditions are seldom met: Some nodes are faulty, obstructions produce multipath and complicate range estimates, and the system may even be subject to malicious attack. The first level of defense, as in calibration, is to deploy a redundant set of nodes and then perform statistical tests that seek to identify and discard outliers. The next level, as already indicated, is to explicitly include more nodes whose position is trusted. In the SeRLoc and HiRLoc protocols [68, 69], multiple secure nodes produce beacons that are used by lower-cost sensors to passively estimate their positions, eliminating interactions among insecure nodes and thus many attacks. In the SPINE protocol [105], multiple beacons are also employed, and certain facts regarding propagation are used to verify distances; in particular, it is shown that attackers may be able to lie about position to extend distances, but cannot produce viable attacks that shorten distances.

One might accomplish a similar aim with mobile nodes with accurate location capability passing through the network in one sweep, in a position calibration exercise. Mobile elements can also provide timing beacons successively to different parts of the network to enable accurate distance estimates, while sets of reliable elements can better estimate the propagation model parameters for use in refining position estimates. Some of the benefits possible with controlled mobility are described in reference 92. Turning things around, static elements, once having determined their positions, can greatly assist mobile elements in estimating their own positions as they traverse a network.

14.4.3 Reliable Synchronization

As noted in the prior section, synchronization is usually a requirement for reliable position location, and of course the time that events take place is often of interest on its own. The level of synchronization required very much depends on the applications. In practical terms, one may distinguish among synchronism mediated by the operating system of the sensor node as compared to synchronism for physical layer communications. Very different types of algorithms are required; and in practice, multiple layers of synchronism are often present in sensor networks.

Explicit synchronism algorithms are required because clocks that are initially aligned will not stay that way. Even if they are physically identical, small temperature differences will be manifested as frequency offsets, and thermal noise will cause jitter. Radios typically present the most stringent synchronism requirements in sensor networks since the carrier frequency and phase must usually be aligned, using some combination of frequency and phase-locked loops [106], wherein the receiver locks to the reference of the transmitter. Motion of either the communicating parties or strong reflecting media will also affect timing loops; in general, high channel dynamics are also problematic in that it is difficult to achieve high-precision estimates before the situation changes. The data bearing signals themselves are used to establish symbol synchronism, while preambles of packets or frames can be used to establish frame synchronism. It is this level of alignment that usually forms the basis of more coarse levels of network synchronism, as may be required for time stamping data, duty-cycling nodes to conserve energy, or performing coherent combining of low frequency signals such as acoustic or seismic waves.

Synchronism within the communications range of a single master clock is relatively straightforward, with the clocks of all the subordinate nodes "slaved" to it, offset by some delay. The propagation delay is important only for applications such as coherent combining of radio signals; for other applications it will be a negligible fraction of the total timing offset produced by lower-cost procedures. Chief among the delays are (a) the time taken to interrupt a processor to record a timing event and (b) delays experienced when there is queuing of packets over congested links. These delays will then accumulate in a multihop network as timing references propagate outward from the master clock. In the NTP protocol used in wired networks [107], queuing delays are mitigated by sending large numbers of packets and taking the early arrivals and returns as being more reliably indicative of the actual delay than other statistical quantities. The RBS protocol [108] shares this feature and additionally is designed to reduce software delays that are typical when a processor controls a radio. In small networks it produces sufficient accuracy to enable acoustic beamforming [109], but in larger networks it suffers from the problem of accumulation of random error as more links are traversed. This error can be mitigated to some extent through explicit estimation of clock skews; but even so, the error variance will generally grow linearly with the number of hops since the problem is well-modeled as a random walk. Interestingly, this growth in synchronization error with network size is not fundamental but rather an algorithmic artifact. It has been shown that there exist distributed algorithms for which the synchronization error variance is $O(1)$ as network size grows [110]. Quite a number of protocols have been proposed to establish synchronism in networks with resource-constrained nodes including TPSN [91], LTS [111], Mini/Tiny Sync [112], and RATS [113].

If there is GPS at every node, synchronism to a high level is quite easy, but there are many physical situations in which GPS is either unavailable (indoors, mountainous terrain, heavy tree cover, etc.) or simply too expensive in cost or energy consumption. Nonetheless, if timing beacons are available, they can greatly simplify network synchronism, and to achieve phase alignment of radios there is no practical alternative. Just as a set of nodes with known locations can anchor network location algorithms, a network of nodes slaved to a common timing base can similarly stabilize network synchronism. Thus, even if not all nodes have GPS or lock with some other timing beacon, the remaining nodes can benefit through limitation of the number of hops over which clock synchronism is propagated.

Again as in the calibration and position location problems, provision must be made for misbehaving nodes, with explicit estimation of outliers. This is only possible with redundant sources of timing information (e.g., communication of time by multiple neighbors), or the presence of trusted elements such as timing beacons [114]. The RANSAC [115] and LMS [116] algorithms have been proposed for detection of outliers, whether due to node failures or malicious attack. One may additionally exploit certain physical facts concerning propagation time, which can be exploited to detect inconsistencies in timing for individual links [96]. Secure routing protocols may then be used as the basis for secure multihop synchronism.

14.4.4 Key Distribution in Sensor Networks

Another component of data integrity is some security in the communications system. Many solutions that apply to MANETs also apply to sensor networks, with the caveat that some of the nodes in the network may have low computational and storage capabilities. Since the nodes are potentially deployed for long periods of time, they may be captured and subverted. This can result in insertion of bad data, interception of good data, and failure to forward information. While it is unrealistic to expect that full recovery from some events is possible, what is desired is that the damage should be detectable and limited in scope in its effects on the network. Since the managers of the network may be in a position to add new nodes to the network over time (e.g., to recover from regions of failures/subversion), it is desirable therefore for the security protocol to be able to accommodate variable numbers of nodes, added at different times.

A basic requirement for any security protocol is some mechanism for distribution of cryptographic keys. An ideal protocol would provide strong encryption, scalability, low vulnerability to the capture of small numbers of nodes, adaptability to adding new nodes, low computational cost, and low communications overhead. Obviously, these are difficult to all satisfy, and so protocols have been proposed using a variety of approaches. TinySec, among its various security features, provides symmetric keys and is designed specifically for resource constrained motes [117]. Keys may be deterministically preassigned, as in the SPINS [118] and LEAP [119] protocols. This has the virtue of perfect security in that capture of one node does not impact the security of other links, but relies upon sharing of keys with one secure base station and requires considerable work prior to deployment. Such strategies are thus suitable for modest-sized stable deployments, but are not well-suited to large-scale systems or rapidly changing topologies. Random key distribution schemes [120–122] yield more flexibility but are vulnerable to the capture of a relatively small number of nodes in the network. Post-deployment derivation of keys as in the LEAP, Secure Pebblenet [123], and PIKE [124] protocols uses a shared secret in combination with pseudorandom number generators implemented in software which take as parameters some physical attribute of the node (e.g., location or ID number) to create the keys. Since nodes erase the shared secret after creation of their key, the network is not vulnerable to node capture, but nodes are also unable to generate new keys in the future.

Recently, trusted platform modules (TPM) have appeared, which provide strong security primitives in a tamper-resistant and compact hardware module. While perhaps at the moment too expensive for the lowest-cost nodes, the presence of such trusted devices in a subset of nodes in the network can be used to dramatically improve performance. In reference 96, such modules are exploited to produce a scalable and adaptable protocol, with strong security. This is a particular instance of algorithms that rely upon centralized (or multiple) trusted authorities [125–128]. In general, where there are sources of trust available, be they gateways or TPMs, it is very much to the advantage of the network designer to make use of them, so that the many difficulties inherent in fully distributed key generation are mitigated. In this, the situation is parallel to that of calibration; the more sources of trust available, the easier it is to construct reliable procedures.

14.4.5 Reputation Systems

So far we have discussed means for making lower-level services related to data integrity more robust. One may also attack the problem at higher levels such as aggregation of decisions from nodes. Clearly, a system in which data integrity is assured at multiple levels will in general be more robust than one that relies upon just one technique. Here as well, protocols may be categorized according to the sources of trust invoked, namely, trusted authorities versus some combination of redundant deployment and trust in models.

Aggregation of decisions or measurements can be very negatively impacted by nodes that have failed or been subverted. In reputation systems the idea is to rate the reliability of current information based upon some combination of past reliability and plausibility of the current information given the group consensus. Since trusted authorities may be assigned a higher reputation, such systems can be designed for both flat and hierarchical networks. Similarly, systems can also be designed to deal with fusion of information from multiple sensing modes. Information from nodes/sensors with poor reputations can then either receive a low weight or be ignored.

Reputation in sensor network decisions is different in some respects from other domains such as e-commerce and pure communication networks [129–135]. To begin with, for scalability of the protocol there will not necessarily be a single repository of reputation information, although of course there can be some hierarchy in reputation storage. Second, unlike MANETs, the relative stability of at least the static portions of the network enables time to build up information on the performance of near neighbors, with whom most interactions will likely take place (e.g., for the reasons articulated in Section 14.2). Third, in ad hoc networks, reputation is often used as a means to incentivize rational nodes to pass on data, but faulty nodes will not behave rationally. Finally, for reputation to be checked with respect to sensor data, some redundant deployment will be required, making implicit use of at least the correlation of data from nodes in close proximity. In this the situation is very similar to our discussion of the benefits of redundancy with respect to calibration, localization, and synchronization, in that trust is placed in some combination of the physical model and the opinions of other nodes in order to compute some parameter from the data. Reputation systems are, however, somewhat more general because the decisions to be made can come from multiple sensor streams, and reputation scores can to some extent be abstracted away from the particular physical models.

Part of the commonality with the calibration problem is the need for an outlier detection protocol. With strong prior information concerning the model of the physical system, model-based outlier detection can be employed in sensor networks [87, 136, 137]. The reputation rating of data points is made proportional to its conformance to this model. The model could be deterministic or statistical, or even constructed using adaptive techniques in a phase where all nodes are trusted (e.g., following calibration). With only partial knowledge of the physical situation, one may apply simple criteria such as rejection of values outside particular ranges, or sequences which change too quickly, or sequences that change too little (indicating stuck-at faults).

Apart from the difficulty of dealing with dynamic environments, model-based techniques suffer from the difficulty of modeling the fault behavior of nodes. An alternative is to use consensus-based techniques, which use much weaker assumptions concerning models. In particular, it is assumed that neighboring nodes have correlated readings and that nonfaulty nodes are the large majority of those still permitted to participate in decisions. To plan a sensor node density sufficient for the first of these conditions to be met of course requires some knowledge of the physical model. In operation, readings are compared to the consensus of neighboring nodes in order to compute a reputation score, rather than being compared to the model predictions. There are two basic consensus criteria: (a) outlier designation based on distance from the consensus or (b) density-based outlier detection. In distance-based mechanisms, the absolute difference in sensor readings from the (reputation-weighted) average of the remaining nodes is used as the basis for outlier rejection [138, 139]. To set the outlier detection thresholds, some knowledge of the statistical model of the system (phenomenon and sensor behavior) is thus required. Density-based outlier rejection [140, 141] operates similarly, except that now the data can be clustered into groups with differing spreads, for which different distance thresholds then apply.

Any outlier detection scheme [142] can be used to produce the reputation score. Having accomplished this, there are then several possibilities for how these scores will be accumulated and used. One can assign scores from any transaction as being binary (satisfactory/unsatisfactory) or give more levels. A particularly convenient framework is to accumulate reputation metrics using a beta distribution [96, 143] since only two parameters are needed to characterize the entire probability distribution. Reputations can age over time, so that more weight is given to recent behaviors. Formulation of reputation as being related to the probability of yielding a useful input is a flexible framework that allows for fully trusted nodes to be included (e.g., a trusted central or mobile authority), and for fusion of information collected by very different types of sensor. It is also amenable to using results from decision theory [143–144] to deal with various malicious attacks [96].

14.4.6 Efficient Query Systems

Sensor networks are fundamentally a means to actively answer queries about the physical world. They differ from traditional databases in some important respects. For example, streaming data from sensors presents a problem of scaling. Even without energy and communication bandwidth constraints, a network of thousands of sensors will quickly overrun whatever fixed storage media are supplied (and especially if the bandwidth is large). For scalability it is required that queries that direct the collection, storage, and transport of data should specify not only geographic and temporal scope, but also spatial, temporal, and accuracy fidelity requirements. An expression of priority or interest is also needed because network resources are finite. On the other hand, problems of concurrency, integrity, security, and efficient access to perform logical inferences remain, and so it is desirable to re-use as much of the scaffolding from traditional databases as possible, particularly when multiple applications are intended to use the same system.

A useful abstraction for such problems is diffusion of interests. A user indicates interests in data with particular attributes, with multiple such interests interacting within the network to determine the flow of information. At the bottom (physical) layer they might include sensor type, spatial resolution (or location), and temporal resolution (or time). Higher-level queries such as a request for a map of the temperature gradients in some particular region must then be translated into a set of requests for information from particular nodes, a strategy for communicating the information, and selection of the means for performing the signal processing, all consonant with resource constraints. Interests interact in increasing the likelihood of particular information being stored, processed, and communicated, but also in creating possibilities for congestion and other resource contention. These interests thus compete with each other and against the inhibition built into the network for usage of its resources. Examples of protocols that fairly directly implement this abstraction to establish routes in a resource-efficient fashion for both queries and responses are push-directed diffusion [145] (oriented toward situations where there are many data sources and sinks) and two-phase pull diffusion [146] (few sinks and many data sources).

In general, which routing mechanism to employ depends on (a) how data are to be stored and used and (b) constraints such as latency for query dissemination and network reply. Very different traffic volumes result, depending upon the choices made. In external storage systems, data are automatically sent to a sink whenever some standing conditions have been met (generically, an event has occurred); data transport clearly dominates the traffic in this case. In local storage systems, data are only sent to a sink in response to a particular request. In this case, there can be many queries. In rendezvous systems, data regarding events and queries meet at intermediate points, which may shift in order to balance the load, as for example in data-centric storage (DCS) [147]. This allows trading storage requirements against query response time, as well as trading query against data traffic, at the cost of requiring geographic information for routing.

Clearly, establishment of routes is but part of the problem. There are possibilities for data fusion and aggregation along the way [148, 149], data resolution may be reduced if resources are deficient [e.g., 150], and consideration must be given to the timeliness of response to queries and the reliability of the data storage in the network (e.g., replication to deal with node failure) [147]. An example of a system that enables network operators to balance some of this concerns is TinyDB, a SQL-like utility developed for operation on networks of (resource-constrained) sensor nodes running TinyOS [151]. It has a number of features that allow data-driven applications to be developed and deployed with far less effort than attempting to use fully custom code, including meta-data management, high-level query support, network topology management (including routing tables), support for multiple queries with different fidelity criteria, and adaptation to nodes entering/leaving the network.

Hierarchy can be used to considerable benefit. For example, in the aggregration system proposed in reference 148, a base station with additional resources is required for computing the routing trees and aggregation strategy. Nodes with longer-range communication links and additional storage are natural locations for rendevous points, allowing for faster response to queries and greater fidelity in the data records. More generally, with powerful higher-level nodes, standard communication and database might be re-used, with novel algorithms required only for control of the information flows to and from the lower-level nodes. On the other hand, rapid network topology changes will, as for routing in MANETs, result in considerably more overhead if response times are to be kept reasonable.

14.5 CONCLUSIONS

Sensor networks come in a wide variety of forms, from flat networks of highly resource-constrained nodes to hierarchical networks with mobile elements and powerful communications, processing, and storage capability. The networks themselves may be designed for one special purpose, or as a tool to answer a broad variety of questions about the physical world. Which resource constraints apply and the scope of the network objectives lead to vastly different optimization problems; and thus in considering QoS requirements for sensor networks, the assumptions concerning the technology, physical model, and usage must be made explicit.

In this chapter, we have attempted to show that while classical network information theory problems are at best a partial fit to problems of resource optimization in sensor networks, a QoS perspective at once makes many of the problems both more relevant and tractable. Mobility in network elements at the same time expands the richness of the problem set and gives rise to simple solutions to many otherwise difficult problems in the domains of data communication and integrity. This is a yet another manifestation of the truism that for robust/simple design there must be at least some quantity available in abundance. If everything is constrained, then models must be highly accurate and the system tends to be both overoptimized and brittle. With a little headroom (e.g., additional trusted beacons, sources of security, communications paths, computational cycles) the system can be more easily adapted to deal with uncertainties in models and objectives. Much of the sensor network community continues to be engaged in the project of exploring how having an abundance of nodes (redundancy with respect to the physical phenomena) can lead to robust and flexible design, while there is an emerging trend to look at how hierarchical overlays of more capable nodes (including mobile ones) can simplify design and expand capabilities. While considerable progress has been made in both domains, a voluminous design space remains to be explored.

One such example is presented below. Consider mobile ad hoc networks, in which nodes must balance a variety of tasks including sensing and communications. Nodes might thus change location or trajectory for sensing purposes, with constraints on disruption to the QoS of the network. For example, a sensor node may be required to have its position change for improved source identification without disrupting the preexisting communications. Similarly, in an overlay network, a node might have to move to replace a failed backbone node. This gives rise to following two problems:

1. Maximization of non-communication application QoS (such as node motion to facilitate sensing) with communication QoS constraints (packet delay, etc.).
2. Maximizing throughput for the newly formed links with communication QoS and link capacities constraints at the new position of the node.

One solution to these problems is the formulation of a geometric program [43, 59]. Because the mobile radio channel is fast-changing and the number of user nodes is large, a fast and robust decision-making algorithm is needed that accommodates a large number of variables for dynamic resource allocation to be feasible [9, 11]. A substantial literature is available on resource allocation problems that use techniques such as game theory, cross-layer optimization, and so on [152–156]. Many other problems involving a mix of controls, sensing, and networking can be formulated.

14.6 EXERCISES

1. **Mixed Deployments of Fixed and Mobile Nodes**: Discuss the design tradeoffs among deploying fixed and mobile nodes with costs n_f and n_m respectively, where the objective is to reproduce a field (such as temperature) to within some specified spatiotemporal fidelity. How is the choice affected by prior knowledge of the model parameters of the field? With high initial uncertainty in the model, why might a sequence of deployments be desirable?

2. **Sensor Network Asymptotics**: Various results on the scalability of the network with respect to communications capacity, reconstruction under fidelity constraints, and synchronization error have been presented. What are the benefits of asymptotic analysis in the design of practical networks, and what are the limitations?
3. **Small-group Cooperation**: With small groups of nodes, it is often possible to formulate QoS optimization problems that while having poor scalability are nonetheless tractable due to the small numbers. Given that centralized computation is possible in such cases, why are heuristics and distributed algorithms still often preferred solutions? (*Hint*: Consider the various sources of uncertainties in an optimization problem such as localization of a source in an obstructed environment.)
4. **Data Integrity**: Suppose a set of static sensor nodes is used to track a slowly varying phenomenon with known spatial correlation function. The sensors drift according to a random walk with known rate. A number of higher-cost mobile nodes are available that are fully reliable, but limited to some maximum velocity. Discuss the design tradeoffs for creating a hybrid system of static and mobile nodes to study the phenomenon, where spatiotemporal fidelity requirements must be met.
5. **Sources of Trust**: In order to trust an experiment, one must have confidence in a combination of the apparatus, experimental procedure, and the model of the physical phenomenon (and failure modes of the apparatus). The experiment will typically be designed to answer some small set of questions well. Suppose that one has available a trusted experimental procedure for determining seismic response at one particular spatial scale, but the procedure is expensive, resulting in low numbers of nodes being feasible. Suppose now that lower-cost accelerometers are available that provide lower resolution. How might these be used in combination with the trusted procedure, possibly in an iterated fashion, to extend a physical model, either in spatial density or to cover larger scales?
6. **Design Trends**: The great project of the scientific revolution is the extension of mathematical models to explain physical phenomena. Sensor networks of different varieties provide new tools for this enterprise. Choose an application domain. How do current technological trends support the future use of sensor networks in this domain, and what are the principal design barriers to overcome in order for there to be widespread adoption of the technology? Do the trends favor homogeneous or hierarchical networks? What nontechnological barriers must also be overcome, and how do these affect the choices of technology?

BIBLIOGRAPHY

1. G. J. Pottie and W. J. Kaiser. *Principles of Embedded Networked Systems Design*, Cambridge University Press, New York, 2005.

2. K. Bult et al. Wireless integrated microsensors. In *Solid-State Sensor and Actuator Workshop*, Hilton Head Island, SC, June, 1996, pp. 205–210.
3. B. Warneke, M. Last, B. Liebowitz, and K. Pister. Smart dust: Communicating with a cubic-millimeter computer. ***Computer***, **34**(1):44–51, 2001.
4. G. J. Pottie and L. P. Clare. Wireless integrated network sensors: Towards low-cost and robust self-organizing security networks. In *Proceedings of SPIE*, Boston, November 3–5, 1998, Vol. 3577, pp. 86–95.
5. J. Agre, L. P. Clare, and G. J. Pottie. Development platform for self-organizing sensor networks. In *Aerosense'99*, Orlando, FL, April 1999.
6. Mica wireless measurement system. *Datasheet*, Crossbow Technology Inc. http://www.xbow.com/Products/Product pdf files/Wireless pdf/MICA.pdf.
7. "Stargate: Xscale processor platform," http://www.xbow.com/Products/Product pdf files/Wireless pdf/6020-0049-01 C Stargate.pdf. Data Sheet.
8. E. Mingozzi. Scheduling algorithms for guaranteeing QoS in wireless Networks. In *Proceedings of MTM Workshop*, February 2000.
9. M. Chiang, D. O'Neill, D. Julian, and S. Boyd. Resource allocation for QoS provisioning in wireless ad hoc networks. *IEEE Globecom*, **5**:2911–2915, 2001.
10. D. Julian, M. Chiang, D. O'Neill, and S. Boyd. QoS and fairness constrained convex optimization of resource allocation for wireless cellular and ad hoc networks. In *IEEE INFOCOM*, June 2002, pp. 477–486.
11. D. Julian, D. O'Neill, and M. Chiang. Robust and QoS constrained optimization of power control in wireless cellular networks. In *IEEE Vehicular Technology Conference*, October 2001, pp. 1932–1936.
12. M. Schwartz. *Telecommunication Networks: Protocols, Modeling, and Analysis*. Addison Wesley, Reading, MA, 1987.
13. C. E. Shannon. A mathematical theory of communication. *Bell System Technical Journal*, **27** (July and October): 379–423 and 623–656, 1948.
14. C. E. Shannon. Communication in the presence of noise. *Proceedings IRE*, **37**:10–21, 1949.
15. T. M. Cover and J. A. Thomas. *Elements of Information Theory*. John Wiley & Sons, New York, 1991.
16. I. E. Teletar. Capacity of multi-antenna Gaussian channels. *European Transactions on Telecommunications*, **10**:585–596, 1999.
17. S. V. Hanly and P. A. Whiting. Information-theoretic capacity of multireceiver networks. *Telecommunication Systems*, **1**:1–42, 1993.
18. A. Vanelli-Coralli, R. Padovani, J. Hou, and J. Smee. Capacity of cell clusters with coordinated processing. In *Information Theory and Its Applications Workshop (ITA '06)*, San Diego, February 6–10, 2006.
19. S. Verdu. Multiuser Detection. In *Advances in Statistical Signal Processing*, **2**:369–409, 1993.
20. A. Duel-Hallen. Decorrelating decision-feedback multi-user detector for synchronous code-division multiple access channel. *IEEE Transactions and Communications*, **41**:285–290, 1993.
21. U. Madhow and M. L. Honig. MMSE interference suppression for direct-sequence spread-spectrum CDMA, *IEEE Transactions on Communications*, **42**:3178–3188, 1994.

22. D. Slepian and J. K. Wolf. Noiseless coding of correlated information sources. *IEEE Transactions on Information Theory*, **19**:471–480, 1973.
23. A. D. Wyner and J. Ziv. The rate-distortion function for source coding with side information at the decoder. *IEEE Transaction on Information Theory*, **22**:1–11, 1976.
24. M. Gastpar. The Wyner–Ziv problem with multiple sources. *IEEE Transactions on Information Theory*, **50**(11):2762–2768, 2004.
25. T. Berger, Z. Zhang, and H. Vishwanathan. The CEO problem. *IEEE Transactions on Information Theory*, **42**:887–902, 1996.
26. H. Vishwanathan and T. Berger. The quadratic Gaussian CEO problem. *IEEE Transactions on Information Theory*, **43**:1549–1559, 1997.
27. A. Pandya, A. Kansal, G. J. Pottie, and M. Srivastava. Lossy source coding of multiple Gaussian sources: m-helper problem. In *Proceedings of Information Theory Workshop 2004*, San Antonio, TX, October 24–29, 2004.
28. A. Pandya, A. Kansal, G. J. Pottie, and M. B. Srivastava. Fidelity and resource sensitive data gathering. In *42nd Allerton Conference*, Allerton, IL, September 2004.
29. P. Gupta and P. R. Kumar. The capacity of wireless networks. *IEEE Transactions on Information Theory*, **46**:388–404, 2000.
30. P. Gupta and P. R. Kumar. Towards an information theory of large networks: An achievable rate region. *IEEE Transactions on Information Theory*, **49**, 1877–1894, 2003.
31. M. Gastpar and M. Vetterli. On the capacity of wireless networks: The relay case. In *Proceedings of the 21st INFOCOM*, New York, June 2002, pp. 1577–1586.
32. N. Jindal, U. Mitra, and A. Goldsmith. In Capacity of ad-hoc networks with node cooperation. In *Proceedings of IEEE International Symposium Information Theory*, Adelaide, June 2004, p. 271.
33. M. Uppal, Z. Liu, V. Stankovic, A. Host-Madsen, and Z. Xiong. Capacity bounds and code designs for cooperative diversity. In *Information Theory and Its Applications Workshop (ITA '06)*, San Diego, February 6–10, 2006.
34. C. T. K. Ng and A. J. Goldsmith. Capacity and cooperation in wireless networks. In *Information Theory and Its Applications Workshop (ITA '06)*, San Diego, February 6–10, 2006.
35. A. Host-Madsen. On the achievable rate for receiver cooperation in ad-hoc networks. In *Proceedings of IEEE International Symposium on Information Theory*, Adelaide, 2004, p. 272.
36. A. Host-Madsen and J. Zhang. Capacity bounds and power allocation for wireless relay channel. *IEEE Transactions on Information Theory*, **51**(6):2020–2040, 2005.
37. M. Ahmed. Decentralized Information Processing in Wireless Peer to Peer Networks. Ph.D. dissertation, Department of Electrical Engineering, UCLA, 2002.
38. R. Gowaikar, B. Hochwald, and B. Hassibi. Throughput scaling in random wireless networks. In *Information Theory and Its Applications Workshop (ITA '06)*, San Diego, February 6–10, 2006.
39. A. Sriniva and E. Modiano. Finding minimum energy disjoint paths in wireless ad hoc networks. In *ACM Wireless Networks*, November **11**:201–417, 2005.
40. M. Neely and E. Modiano. Capacity and delay tradeoffs for ad hoc mobile networks. In *IEEE Transactions on Information Theory*, **51**(6):1917–1937, 2005.

41. J. Li, C. Blake, D. De Couto, H. Lee, and R. Morris. Capacity of ad hoc wireless networks. In *7th Annual International Conference on Mobile Computing and Networking*, ACM Press, New York, 2001, pp. 61–69.
42. Y. S. Tu, Cooperative communications among wireless sensor networks. Ph.D. Dissertation, Department of Electrical Engineering, UCLA, 2003.
43. A. Pandya. Fundamental limits of scalable sensor networks. Ph.D. Dissertation, Department of Electrical Engineering, UCLA, 2004.
44. A. Pandya and G. Pottie. Bounds on achievable rates for cooperative channel coding. In *38th Annual Asilomar Conference on Signals, Systems, and Computers*, Pacific Grove, CA, November 2004.
45. A. Sendonaris, E. Erkip, and B. Aazhang. User cooperation diversity—Part I: system description. *IEEE Transactions on Communications*, **51**(11):1927–1938, 2003.
46. A. Sendonaris, E. Erkip, and B. Aazhang. User cooperation diversity—Part II: Implementation aspects and performance analysis. *IEEE Transactions on Communications*, **51**(11):1939–1948, 2003.
47. K. Yazdi, H. Gamal, and P. Schniter. On the design of cooperative transmission schemes. In *Allerton Communications, Computing, and Control*, October 2003.
48. U. Mitra and A. Sabharwal. Complexity constrained sensor networks: Achievable rates for two relay networks and generalizations. In *Third International Symposium on Information Processing in Sensor Networks*, ACM Press, New York, 2004, pp. 301–310.
49. A. Host-Madsen. A new achievable rate for cooperative diversity based on generalized writing on dirty paper. In *IEEE International Symposium on Information Theory*, June 2003, p. 317.
50. J. Laneman, D. Tse, and G. Wornell. Cooperative diversity in wireless networks: Efficient protocols and outage behavior. *IEEE Transactions on Information Theory*, **50**(12):3062–3080, 2004.
51. R. Ahlsede, N. Cai, S.-Y. R. Li, and R. W. Yeung. Network information flow. *IEEE Transactions on Information Theory*, **46**:1204–1216, 2000.
52. R. W. Yeung and N. Cai. Network coding, algebraic coding, and network error correction. In *Information Theory and its Applications Workshop (ITA '06)*, San Diego, February 6–10, 2006.
53. R. Koetter and M. Medard. An algebraic approach to network coding. *IEEE/ACM Transactions on Networking*, **11**:782–795, 2003.
54. A. G. Dimakis, D. Petrovic, and K. Ramchandran. From dumb wireless networks to smart wireless networks using network coding. In *Information Theory and Its Applications Workshop (ITA '06)*, San Diego, February 6–10, 2006.
55. S. Savari and G. Kramer. Capacity bounds for relay networks. In *Information Theory and Its Applications Workshop (ITA '06)*, San Diego, February 6–10, 2006.
56. J. W. Fisher, T. Darrell, W. T. Freeman, and P. Viola. Learning joint statistical models for audio-visual fusion and segregation. In T. K. Leen, T. G. Dietterich, and V. Tresp, editors, *Advances in Neural Information Processing Systems*, Vol. 13, MIT Press: Cambridge, MA, 2001, pp. 772–778.
57. M. Ahmed and G. Pottie. Fusion in the context of information theory. In S. R. Iyengar and R. R. Brooks, editors, *Distributed Sensor Networks*, CRC Press, Boca Raton, FL, 2005.

58. G. J. Pottie and W. J. Kaiser. Wireless integrated network sensors. *Communications of ACM*, **43**:51–58, 2000.
59. A. Pandya and G. J. Pottie. On scalability and source/channel coding decoupling in large scale sensor networks. *CENS Technical Report 17*, 2003.
60. R. Cristescu, B. Beferull-Lozano, and M. Vetterli. On network correlated data gathering. In *IEEE Infocom*, 2004, pp. 2571–2582.
61. R. Zamir and T. Berger. Multiterminal source coding with high resolution. *IEEE Transactions on Information Theory*, **45**(1):106–117, 1999.
62. J. Garcia-Frias and W. Zhong. LDPC codes for compression of multi-terminal source with hidden Markov correlation. *IEEE Communication Letters*, **7**(3):115–117, 2003.
63. S. S. Pradhan, J. Kusuma, and K. Ramchandran. Distributed compression in a dense microsensor network. *IEEE Signal Processing Magazine*, **19**:51–60, 2002.
64. S. S. Pradhan and K. Ramchandran. Distributed source coding using syndromes (DISCUS): Design and construction. *IEEE Transactions on Information Theory*, **49**:626–643, 2003.
65. Z. Xiong, A. D. Liveris, and S. Cheng. Distributed source coding for sensor networks. *IEEE Signal Processing Magazine*, **21**:80–94, 2004.
66. R. Zamir, S. Shamai, and U. Erez. Nested linear/lattice codes for structured multiterminal binning. *IEEE Transactions on Information Theory*, **48**(6):1250–1276, 2002.
67. H. Luo, Y. Tong, and G. Pottie. A two-stage DPCM scheme for wireless sensor networks. In *Proceedings on IEEE Intlernational Conference on Acoustics, Speech and Signal Processing (ICASSP '05)*, Philadelphia, Vol. 3, 2005, pp. 661–664.
68. L. Lazos and R. Poovendran. SeRLoc: robust localization for wireless sensor networks. *ACM Transactions on Sensor Networks (TOSN)*, **1**:73–100, 2005.
69. L. Lazos and R. Poovendran. HiRLoc: High resolution localization for wireless sensor networks. *IEEE Journal on Selected Areas in Communications (JSAC)*, **24**(2):233–246, 2006.
70. M. Rahimi, M. Hansen, W. J. Kaiser, G. S. Sukhatme and D. Estrin. Adaptive sampling for environmental field estimation using robotic sensors. In *IEEE/RSJ International Conference on Intelligent Robots and Systems*, August 2005, pp. 3692–3698.
71. X. Kong, R. Pon, W. Kaiser, and G. Pottie. Environmental sampling with multiscale sensing. In *ICASSP 2006*, Toulouse, Vol. 4, May 2006, pp. 873–876.
72. A. Pandya and G. Pottie. QoS in ad hoc networks. *IEEE Vehicular Technology Conference*, October 2003.
73. S. Baatz, M. Frank, R. Gopffarth, D. Kassatkine, P. Martini, M. Scetelig, and A. Vilavaara. Handoff support for mobility with IP over Bluetooth. In *25th Annual Conference on Local Computer Networks (LCN00)*, Tampa, FL, November 2000.
74. A. Campbell, J. Gomez, C. Wan, and S. Kim. Cellular IP. http://comet.ctr.columbia.edu/cellularip/pub/draft-ietf-mobileipcellularip-00.txt. Internet draft, January 2000.
75. J. Kempf and P. Yegani. Openran: A new architecture for mobile wireless internet radio access networks. *IEEE Communications Magazine*, **40**:118–123, 2002.
76. C. Lin and M. Gerla. Adaptive clustering for mobile wireless networks. *IEEE Journal of Selected Areas in Communications*, **15**(7):1265–1275, 1997.
77. C. Perkins. editor, *In Ad Hoc Networking*, Addison Wesley, Boston, 2001, pp. 1–28.

78. N. Bansal and Z. Liu. Capacity, delay and mobility in wireless ad-hoc networks. In *IEEE INFOCOM 2003*, San Francisco, CA, April 1–3, 2003, pp. 1553–1563.

79. S. Diggavi, M. Grossglauser, and D. Tse. Even one-dimensional mobility increases ad hoc wireless capacity. In *Proceedings of ISIT*, Lausanne, Switzerland, June 2002, p. 352.

80. A. Gamal, J. Mammen, B. Prabhakar, and D. Shah. Throughput-delay tradeoff in wireless networks. In *IEEE INFOCOM 2004*, Hong Kong, Vol. 1, March 7–11, 2004, pp. 464–475.

81. M. Grossglauer and D. Tse. Mobility increases the capacity of wireless of ad hoc wireless networks. *IEEE/ACM Transactions on Networking*, **10**, pp. 477–486, 2002.

82. S. Jain, R. Shah, G. Borriello, W. Brunette, and S. Roy. Exploiting mobility for energy efficient data collection in sensor networks. In *Modeling and Optimization in Mobile, Ad Hoc and Wireless Networks* (*WiOpt*), 2004.

83. R. Shah, S. Roy, S. Jain, and W. Brunette. Datamules: Modelling a three tiered architecture for sparse sensor networks. In *First IEEE International Workshop on Sensor Network Protocols and Applications (SNPA)*, May 2003.

84. A. Chakrabarty, A. Sabharwal, and B. Aazhang. Using predictable observer mobility for power efficient design of a sensor network. In *Second International Workshop on Information Processing in Sensor Networks (IPSN)*, Palo Alto, CA, April 2003, pp. 129–145.

85. S. Jain, K. Fall, and R. Patra. Routing in a delay tolerant network. In *SIGCOMM*, 2004, pp. 145–157.

86. A. Pentland, R. Fletcher, and A. Hasson. Daknet: Rethinking connectivity in developing nations. *IEEE Transactions on Computer*, **37**(1):78–83, 2004.

87. F. Koushanfar, M. Potkonjak, and A. Sangiovanni-Vincentelli. On-line fault detection of sensor measurements. In *IEEE International Conference on Sensors*, October 2003, pp. 974–980.

88. A. Kansal, A. Somasundara, D. Jea, M. Srivastava, and D. Estrin. Intelligent fluid infrastructure for embedded networks. In *Second International Conference on Mobile Systems, Applications, and Services (MobiSys)*, June 2004.

89. W. Zhao and M. Ammar. Message ferrying: Proactive routing in highly partitioned wireless ad hoc networks. In *Ninth IEEE Workshop on Future Trends of Distributed Computing Systems (FTDCS)*, May 2003.

90. W. Zhao, M. Ammar, and E. Zegura. A message ferrying approach for data delivery in sparse mobile ad hoc networks. In *Fifth ACM International Symposium on Mobile Ad Hoc Networking and Computing (MobiHoc)*, 2004, pp. 187–198.

91. J. Elson, L. Girod, and D. Estrin. Fine-grained network time synchronization using reference broadcasts. *SIGOPS Operating System Review*, **36**(1):147–163, 2002.

92. J. Scott and M. Hazas. User-friendly surveying techniques for location-aware systems. In *Proceedings ubicomp*, published in LNCS Vol. 2864, Oct. 2003, pp.45–54.

93. A. LaMarca, W. Brunette, D. Koizumi, M. Lease, S. Sigurdsson, K. Sikorski, D. Fox, and G. Borriello. Plantcare: An investigation in practical ubiquitous systems. In *Ubicomp*, September 2002.

94. M. Rahimi, H. Shah, G. Sukhatme, J. Heidemann, and D. Estrin. Studying the feasibility of energy harvesting in a mobile sensor network. In *IEEE International Conference on Robotics and Automation*, 2003, pp. 19–24.

95. R. Govindan, E. Kohler, D. Estrin, F. Bian, K. Chintalapudi, O. Gnawali, S. Randwala, R. Gummadi, and T. Sathopoulos. Tenet: An architecture for tiered embedded networks. Center for Embedded Networked Sensing Technical Report 56, November 2005.
96. Saurabh Ganeriwal. Trustworthy sensor networks. Ph.D. dissertation, Department of Electrical Engineering, UCLA, 2006.
97. P. Buonadonna, D. Gay, J. Hellerstein, W. Hong, and S. Madden. TASK: Sensor network in a box. Intel Research Berkeley Technical Report IRB-TR-04-021, January 2005.
98. Y. H. Hu and B. Benson. Sensor network data quality assurance. Technical Report, University of Wisconsin, Madison, 2005.
99. V. Barnett and T. Lewis. *Outliers in Statistical Data*, Wiley, 1994.
100. J. C. Chen, K. Yao, and R. E. Hudson. Acoustic source localization and beamforming: Theory and practice. *Eurasian Journal on Applied Signal Processing*, **4**:359–370, 2003.
101. J. C. Chen, K. Yao, T. L. Tung, C. W. Reed, and D. Chen. Source localization and tracking of a wideband source using a random distributed beamforming sensors array. *International Journal of High Performance Computing Applications*, **16**:259–272, 2002.
102. N. Bulusu. Self-Configuring Localization Systems. Ph.D. dissertation. Computer Science Department, UCLA, 2001.
103. D. Niculescu and B. Nath. Position and orientation in ad hoc networks. *Elsevier Journal of Ad Hoc Networks*, **2**(2):133–151, 2004.
104. A. Savvides. Design, implementation and analysis of ad-hoc localization methods. Ph.D. Dissertation, Electrical Engineering Department, UCLA, 2003.
105. S. Capkun and J. P. Hubaux. Secure positioning in wireless networks. *IEEE Journal on Selected Areas in Communications*, **24**(2):221–232, 2006.
106. J. G. Proakis. *Digital Communications*, McGraw-Hill, New York, 2001.
107. http://www.ntp.org.
108. J. Elson, Time synchronization in wireless sensor networks, Ph.D. dissertation, Computer Science Deptartment, UCLA, 2003.
109. J. C. Chen, L. Yip, J. Elson, H. Wang, D. Maniezzo, R. E. Hudson, K. Yao, and D. Estrin. Coherent acoustic array processing and localization on wireless sensor network. In *Proceedings of the IEEE*, Vol. 91, no. 8, August 2003, pp. 1154–1162.
110. A. Giridhar and P. R. Kumar. Distributed clock synchronizatin over wirelss networks: Algorithms and analysis. In *Proceedings of the 45th IEEE Conference on Decision and Control*, San Diego, December 13–15, 2006, pp. 4915–4920.
111. J. van Greunen and J. Rabaey. Lightweight time synchronization for sensor networks. In *2nd ACM International Conference on Wireless Sensor Networks and Applications (WSNA)*, 2003, pp. 11–19.
112. M. Sichitiu and C. Veeraraittiphan. Simple, accurate time synchronization for wireless sensor networks. In *IEEE Wireless Communication and Networking Conference (WCNC)*, 2003, pp. 1266–1273.
113. S. Ganeriwal, D. Ganesan, H. Sim, V. Tsiatsis, and M. B. Srivastava. Estimating clock uncertainty for efficient duty-cycling in sensor networks. In *ACM Conference on Networked Sensor Systems (Sensys)*, 2003, pp. 138–149.

114. K. Sun, P. Ning, and C. Wang. Secure and resilient clock synchronization in wireless sensor networks. *IEEE Journal on Selected Areas in Communications*, **24**(2):395–408, 2006.

115. J. P. Rousseeuw and M. A. Leroy, Robust Regression and Outlier Detection, John Wiley & Sons, New York, 1987.

116. M. Manzo, T. Roosta, and S. Sastry. Time synchronization attacks in sensor networks. In *ACM Workshop on Security of Ad Hoc Sensor Networks (SASN)*, 2005, pp. 107–116.

117. C. Karlof, N. Sastry, and D. Wagner. Tinysec: A link layer security architecture for wireless sensor networks. In *ACM Sensys '04* 2004, pp. 162–175.

118. A. Perrig, R. Szewczyk, V. Wen, D. Culler, and J. D. Tygar. SPINS: Security protocols for sensor networks. In *7th ACM Conference on Mobile Computing and Networking (Mobicom)*, 2001, pp. 189–199.

119. S. Zhu, S. Setia, and S. Jajodia. LEAP: Efficient security mechanisms for large-scale distributed sensor networks. In *10th ACM Conference on Computer and Communication Security*, ACM Press, New York, 2003, pp. 62–72.

120. L. Eschenauer and V. D. Gligor. A key-management schemes for distributed sensor networks. In *ACM Conference on Computer and Communications Security*, ACM Press, New York, 2002, pp. 41–47.

121. H. Chan, A. Perrig, and D. Song. Random key predistribution schemes for sensor networks. *IEEE Symposium on Research in Security and Privacy*, IEEE Computer Society, May 2003, p. 197.

122. D. Liu and P. Ning. Efficient distribution of key chain commitments for broadcast authentication in distributed sensor networks. In *10th Annual Network and Distributed System Security Symposium*, 2003.

123. S. Basagni, K. Herrin, D. Bruschi, and E. Rosti. Secure pebblenets. In *ACM International Symposium on Mobile Ad Hoc Networking and Computing (MOBIHOC)*, ACM Press, New York, 2001, pp. 156–163.

124. R. Anderson, H. Chan, and A. Perrig. Key infection: Smart trust for smart dust. *IEEE Conference on Network Protocols*, Berlin, Germany, October 2004.

125. W. Du, J. Deng, Y. S. Han, S. Chen, and P. K. Varshney. A key management scheme for wireless sensor networks using deployment knowledge. *IEEE Conference on Computer Communications (Infocom)* 2004, pp. 586–597.

126. W. Du, J. Deng, Y. S. Han, and P. K. Varshney. A pairwise key predistribution schemes for wirelss sensor networks. In *ACM Conference on Computer and Communications Security*, October 2003, pp. 42–51.

127. J. Hwang and Y. Kim. Revisiting random key pre-distribution for sensor networks. In *ACM Workshop on Security of Ad Hoc and Sensor Networks (SASN)* 2004, pp. 43–52.

128. D. Liu and P. Ning. Establishing pairwise keys in distributed sensor networks. In *ACM Conference on Computer Communications Security*, October 2003, pp. 52–61.

129. M. Blaze, J. Feigenbaum, J. Ioannidis, and A. Keromytics. The keynote trust management system. In *IETF RFC 2704*, 1999.

130. M. Blaze, J. Feigenbaum, and J. Lacy. Decentralized trust management. In *IEEE Conference on Security and Privacy*, 1996, pp. 164–173.

131. S. Buchegger and J. Y. LeBoudec. Performance analysis of the CONFIDANT protocol. In *ACM International Symposium on Mobile Ad Hoc Networking and Computing (MobiHoc)* 2002, pp. 226–236.
132. P. Michiardi and R. Molva. Core: A COllaborative REputation mechanism to enforce node cooperation in mobile ad hoc networks. In *Communication and Multimedia Security*, September 2002.
133. P. Resnick, K. Kuwabara, R. Zeckhauser, and E. Friedman. Reputation systems: facilitating trust in e-commerce systems. *Communications of the ACM* **43**(12):45–48, 2000.
134. P. Resnick and R. Zeckhauser. Trust among strangers in Internet transactions: Empirical analysis of eBay's reputation system. Working paper for the NBER Workshop on Empirical Studies of Electronic Commerce, 2000.
135. L. Xiong and L. Liu. A reputation-based trust model for peer-to-peer ecommerce communities. In *IEEE Conference on e-commerce*, 2003.
136. F. Koushanfar and M. Potkonjak. Markov-cain based models for missing and faulty data in mica2 sensor networks. In *IEEE International Conference on Sensors*, October 2005, pp. 576–579.
137. F. Koushanfar, M. Potkonjak and A. Sangiovanni-Vincentelli. Fault tolerance in wireless ad hoc sensor networks. In *IEEE International Conference on Sensors*, Oct. 2003, pp. 974–980.
138. E. M. Knorr and R. T. Ng. Algorithms for mining distance-based outliers in large datasets. In *24th International Conference on Very Large Databases*, 1998, pp. 392–403.
139. E. M. Knorr and R. T. Ng. Finding intensional knowledge of distance-based outliers. In *25th International Conference on Very Large Databases*, 1999, pp. 211–222.
140. M. M. Breunig, H. P. Kriegel, R. T. Ng, and J. Sander. LOF: Identifying density-based local outliers. In *ACM SIGMOD Conference* 2000, pp. 93–104.
141. S. Papdimitriou, H. Kitawaa, P. B. Gibbons, and C. Faloutsos. LOCI: Fast outlier detection using the local correlation integral. In *Proceedings of 19th International Conference on Data Engineering*, March 2003, pp. 315–326.
142. V. Barnett and T. Lewis. *Outliers in Statistical Data*, John Wiley & Sons, New York, 1994.
143. A. Josang and R. Ismail. The beta reputation system. In *15th Bled Electronic Commerce Conference*, June 2002.
144. G. Shafer. *A Mathematical Theory of Evidence*, Princeton University Press, Princeton, NJ, 1976.
145. J. Heidemann, F. Silva, and D. Estrin. Matching data dissemination algorithms to application requirements. In *Proceedings of ACM SenSys '03*, Los Angeles, November 5–7, 2003.
146. C. Intanagonwiwat, R. Govindan, and D. Estrin. Directed diffusion: A scalable and robust communication paradigm for sensor networks. In *Proceedings of Mobicom 2000*, Boston, MA, August, 2000.
147. S. Shenker, S. Ratnasamy, B. Karp, R. Govindan, and D. Estrin. Data-centric storage in sensor networks. *SIGCOMM Computer Communication Review*, **33**(1):137–142, 2003.
148. S. Madden, M. J. Franklin, J. M. Hellerstein, and W. Hong. TAG: A tiny aggregation service for ad-hoc sensor networks. *SIGOPS Operating Systems Review*, **36**(SI):131–146, 2002.

149. S. Madden, R. Szewczyk, M. J. Franklin, and D. Culler. Supporting aggregate queries over ad-hoc wireless sensor networks. In *4th IEEE Workshop on Mobile Computing Systems and Applications*, 2002, pp. 49–58.

150. D. Ganesan, B. Greenstein, D. Perelyubskiy, D. Estrin, and J. Heidemann. An evaluation of multi-resolution storage in resource-constrained sensor networks. *Proceedings ACM Sensys'03*, Los Angeles, November 5–7, 2003.

151. S. Madden, J. Hellerstein, and W. Hong. *TinyDB: In-Network Query Processing in TinyOS*. UC Berkeley EECS Department, 2003. http://telegraph.cs.berkeley.edu/tinydb.

152. Y. Xue, B. Li, and K. Nahrstedt. Optimal resource allocation in wireless ad hoc networks: A price-based approach. *IEEE Transactions on Mobile Computing*, **5**(4):347–364, 2006.

153. P. Marbach and R. Berry. Downlink resource allocation and pricing for wireless networks. In *INFOCOM*, 2002, 1470–1479.

154. S. Cui, R. Madan, A. Goldsmith, and S. Lall. Cross-layer energy minimization in Sensor Networks, *Proceedings: Allerton Conference on Communications, Control, and Computing*, Monticello, IL, September 2004, pp. 1891–1900.

155. S. Shakkottai, T. S. Rappaport, and P. C. Karlsson. Cross-layer design for wireless networks. *IEEE Communications Magazine*, **41**:74–80, 2003.

156. V. Kawadia and P. R. Kumar. A cautionary perspective on cross-layer design. *IEEE Transactions on Wireless Communications*, **12**(1):3–11, 2005.

157. B. Hofmann-Wellenhof, H. Lictenegger, and J. Collins *GPS: Theory and Practice*, Springer-Verlag, Berlin, 1997.

158. W. Kaiser, G. Pottie, M. Srivastava, G. Sukhatme, J. Villasenor, and D. Estrin. Networked infomechanical systems (NIMS) for ambient intelligience. In W. Weber, J. M. Rabaey, and E. Aarts, editors, In *Ambient Intelligence*, Springer, New York, 2005.

159. H. Luo and G. J. Pottie. Routing explicit side information ofr data compression in wireless sensor networks. In *DCOSS 2005*, Marina del Rey, CA, June 30–July 1, 2005.

160. H. Luo and G. J. Pottie. Balanced aggregation trees for routing correlated data in wireless sensor networks. In *IEEE ISIT 2005*, Adelaide Australia, September 4–9 2005, 14–18.

CHAPTER 15

Energy-Efficient Algorithms in Wireless Sensor Networks

AZZEDINE BOUKERCHE

School of Information Technology and Engineering, University of Ottawa, Ottawa, Ontario K1N 6N5, Canada

SOTIRIS NIKOLETSEAS

Department of Computer Engineering and Informatics, University of Patras, Patras, Greece; and Computer Technology Institute (CTI), Patras 26500, Greece

15.1 INTRODUCTION

Recent dramatic developments in microelectromechanical (MEMS) systems, wireless communications, and digital electronics have already led to the development of small-sized, low-power, low-cost sensor devices. Such extremely small devices integrate sensing, data processing, and wireless communication capabilities. Current devices have a size at the cubic centimeter scale, a CPU running at 4 MHz, some memory, and a wireless communication capability at a 4-kbps rate. Also, they are equipped with a small but effective operating system and are able to switch between "sleeping" and "awake" modes to save energy. Pioneering groups (like the "Smart Dust" Project at Berkeley, the "Wireless Integrated Network Sensors" Project at UCLA, and the "Ultra low Wireless Sensor" Project at MIT) pursue further important goals, like a total volume of a few cubic millimeters and extremely low energy consumption, by using alternative technologies, based on radio frequency (RF) or optical (laser) transmission.

Examining each such device individually might appear to have small utility; however, the effective *distributed coordination* of large numbers of such devices may lead to the efficient accomplishment of large sensing tasks. Large numbers of sensor nodes can be deployed in areas of interest (such as inaccessible terrains or disaster places) and use *self-organization and collaborative methods* to form a sensor network.

Algorithms and Protocols for Wireless Sensor Networks, Edited by Azzedine Boukerche

Their wide range of applications is based on the possible use of various sensor types (i.e., thermal, visual, seismic, acoustic, radar, magnetic, etc.) in order to monitor a wide variety of conditions (e.g., temperature, object presence and movement, humidity, pressure, noise levels, etc.). Thus, sensor networks can be used for continuous sensing, event detection, location sensing, and as microsensing. Hence, sensor networks have important applications, including (a) military (like forces and equipment monitoring; battlefield surveillance; targeting; nuclear, biological, and chemical attack detection), (b) environmental applications (such as fire detection, flood detection, precision agriculture), (c) health applications (like telemonitoring of human physiological data), and (d) home applications (e.g., smart environments and home automation). For an excellent survey of wireless sensor networks see, references 1–3.

15.1.1 Critical Challenges

The efficient and robust realization of such large, highly dynamic, complex, non-conventional networking environments is *a challenging algorithmic and technological task*. Features including the huge number of sensor devices involved, the severe power, computational, and memory limitations, their dense deployment, and frequent failures pose *new design, analysis, and implementation aspects* that are essentially different with respect to not only distributed computing and systems approaches but also ad-hoc networking techniques.

We emphasize the following characteristic differences between sensor networks and ad hoc networks:

- The number of sensor particles in a sensor network is extremely large compared to that in a typical ad hoc network.
- Sensor networks are typically prone to faults.
- Because of faults as well as energy limitations, sensor nodes may (permanently or temporarily) join or leave the network. This leads to highly dynamic network topology changes.
- The density of deployed devices in sensor networks is much higher than in ad hoc networks.
- The limitations in energy, computational power, and memory are much more severe in sensor networks.

Because of the above rather unique characteristics of sensor networks, efficient and robust distributed protocols and algorithms should exhibit the following critical properties:

Scalability. Distributed protocols for sensor networks should be highly scalable, in the sense that they should operate efficiently in extremely large networks composed of huge numbers of nodes. This feature calls for an urgent need to prove by analytical means and also validate (by large-scale simulations) certain efficiency and robustness (and their tradeoffs) guarantees for asymptotic network sizes.

Efficiency. Because of the severe energy limitations of sensor networks and also because of their time-critical application scenarios, protocols for sensor networks should be efficient, with respect to both energy and time.

Fault-Tolerance. Sensor particles are prone to several types of faults and unavailabilities and may become inoperative (permanently or temporarily). Various reasons for such faults include (a) physical damage during either the deployment or the operation phase and (b) permanent (or temporary) cease of operation in the case of power exhaustion (or energy saving schemes, respectively). The sensor network should be able to continue its proper operation for as long as possible despite the fact that certain nodes in it may fail.

15.1.2 The Energy Efficiency Challenge

Since one of the most severe limitations of sensor devices is their limited energy supply, one of the most crucial goals in designing efficient protocols for wireless sensor networks is minimizing the energy consumption in the network. This goal has various aspects, including (a) minimizing the total energy spent in the network, (b) minimizing the number (or the range) of data transmissions, (c) combining energy efficiency and fault-tolerance, by allowing redundant data transmissions which, however, should be optimized to not spend too much energy, (d) maximizing the number of "alive" particles over time, thus prolonging the system's lifetime, and (e) balancing the energy dissipation among the sensors in the network, in order to avoid the early depletion of certain sensors and thus the breakdown of the network.

We note that it is very difficult to achieve all the above goals at the same time. There even exist tradeoffs between some of the goals above. Furthermore, the importance and priority of each of these goals may depend on the particular application. Thus, it is important to have a variety of protocols, each of which may possibly focus on some of the energy efficiency goals above (while still performing well with respect to the rest of the goals).

In this chapter, we present four energy-efficient protocols:

- *The Local Target Protocol (LTP)*, which performs a local optimization trying to minimize the number of data transmissions.
- *The Probabilistic Forwarding Protocol (PFR)*, which creates redundant data transmissions that are probabilistically optimized, into perform a tradeoff between energy efficiency and fault-tolerance.
- *The Energy-Balanced Protocol (EBP)*, which focuses on guaranteeing the same per sensor energy dissipation, in order to prolong the lifetime of the network.
- *The Variable Transmission Range Protocol (VTRP)*, which dynamically adapts the transmission range to bypass sensors that tend to be overused, in order to avoid possible obstacles.

Because of the complex nature of a sensor network (that integrates various aspects of communication and computing), protocols, algorithmic solutions, and design schemes for all layers of the networking infrastructure are needed. Far from being exhaustive, we mention the need for frequency management solutions at the physical layer, for Medium Access Control (MAC) protocols to cope with multihop transmissions at the data link layer. The interested reader may use the excellent survey by Akyildiz et al. [4] for a detailed discussion of design aspects of all layers of the networking infrastructure.

We focus in this chapter on algorithms and protocols for the network layer [references 23–32]. We believe that a complementary use of rigorous analysis and large-scale simulations is needed to fully investigate the performance of data propagation protocols in wireless sensor networks. In particular, asymptotic analysis may lead to provable efficiency and robustness guarantees toward the desired scalability of protocols for sensor networks that have extremely large size. On the other hand, simulation allows us to investigate the effect of a great number of detailed technical specifications of real devices, a task that is difficult (if possible at all) for analytic techniques which, by their nature, use abstraction and model simplicity.

15.2 LTP: A HOP-BY-HOP DATA PROPAGATION PROTOCOL

15.2.1 The Model

The LTP Protocol was introduced by Chatzigiannakis, Nikoletseas, and Spirakis [5]. The authors adopt a two-dimensional (plane) framework: A *smart dust cloud* (a set of particles) is spread in an area (for a graphical presentation, see Figure 15.1).

Let d (usually measured in numbers of particles/m^2) be the *density* of particles in the area. Let $\mathcal{R}$ be the maximum (radio/laser) transmission range of each grain particle.

A *receiving wall* $\mathcal{W}$ is defined to be an infinite line in the smart-dust plane. Any particle transmission within range $\mathcal{R}$ from the wall $\mathcal{W}$ is received by $\mathcal{W}$. $\mathcal{W}$ is assumed to have very strong computing power, able to collect and analyze received data, and has a constant power supply and so it has no energy constraints. The wall represents in fact the authorities (the fixed control center) to whom the realization of a crucial event should be reported. The wall notion generalizes that of the sink and may correspond to multiple (and/or moving) sinks. Note that a wall of appropriately big (finite) length suffices.

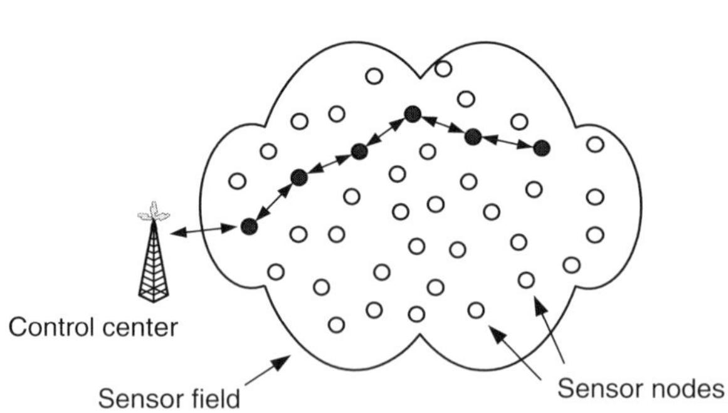

Figure 15.1. A smart dust cloud.

The notion of multiple sinks that may be static or moving has also been studied in reference 6, where Triantafilloy et al. introduce "NanoPeer Words," which are merging notions from Peer-to-Peer Computing and Smart Dust.

Furthermore, there is a setup phase of the smart dust network, during which the smart cloud is dropped in the terrain of interest; when using special control messages (which are very short, inexpensive, and transmitted only once), each smart dust particle is provided with the direction of $\mathcal{W}$. By assuming that each smart dust particle has individually *a sense of direction* and by using these control messages, each particle is aware of the general location of $\mathcal{W}$.

15.2.2 The Protocol

Let $d(p_i, p_j)$ be the distance (along the corresponding vertical lines toward $\mathcal{W}$) of particles p_i, and p_j, and let $d(p_i, \mathcal{W})$ the (vertical) distance of p_i from $\mathcal{W}$. Let *info*($\mathcal{E}$) be the information about the realization of the crucial event $\mathcal{E}$ to be propagated. Let p be the particle sensing the event and starting the execution of the protocol. In this protocol, each particle p' that has received *info*($\mathcal{E}$) does the following:

- *Search Phase.* It uses a periodic low-energy directional broadcast in order to discover a particle nearer to $\mathcal{W}$ than itself (i.e., a particle p'' where $d(p'', \mathcal{W}) < d(p', \mathcal{W})$).
- *Direct Transmission Phase.* Then, p' sends *info*($\mathcal{E}$) to p''.
- *Backtrack Phase.* If consecutive repetitions of the *search phase* fail to discover a particle nearer to $\mathcal{W}$, then p' sends *info*($\mathcal{E}$) to the particle from which it originally received the information.

Note that one can estimate an a priori upper bound on the number of repetitions of the search phase needed, by calculating the probability of success of each search phase as a function of various parameters (such as density, search angle, and transmission range). This bound can be used to decide when to backtrack.

For a graphical representation see Figures 15.2 and 15.3.

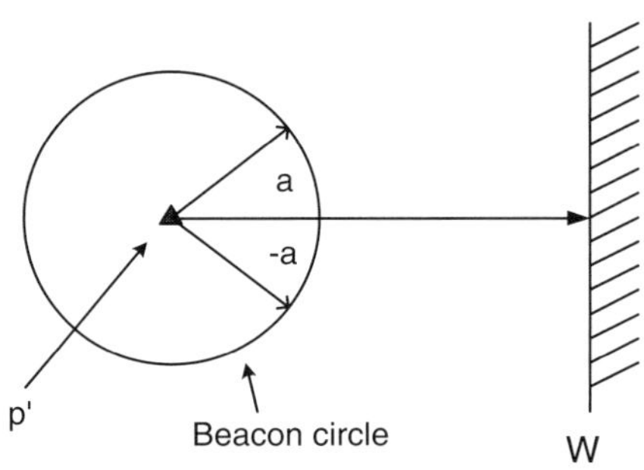

Figure 15.2. Example of the search phase.

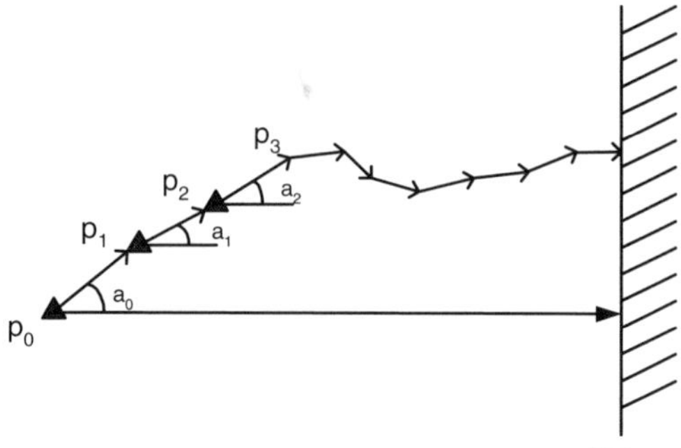

Figure 15.3. Example of a transmission.

15.2.3 Theoretical Analysis

Reference 5 first provides some basic definitions.

Definition 15.2.1. Let $h_{\text{opt}}(p, \mathcal{W})$ be the (optimal) number of "hops" (direct, vertical to $\mathcal{W}$ transmissions) needed to reach the wall, in the *ideal* case in which particles always exist in pairwise distances $\mathcal{R}$ on the vertical line from p to $\mathcal{W}$. Let Π be *a smart dust propagation protocol*, using *a transmission path* of length $L(\Pi, p, \mathcal{W})$ to send information about event $\mathcal{E}$ to wall $\mathcal{W}$. Let $h(\Pi, p, \mathcal{W})$ be the actual number of hops (transmissions) taken to reach $\mathcal{W}$. The *"hops" efficiency* of protocol Π is the ratio

$$C_h = \frac{h(\Pi, p, \mathcal{W})}{h_{opt}(p, \mathcal{W})}$$

Clearly, the number of hops (transmissions) needed characterizes the energy consumption and the time needed to propagate the information $\mathcal{E}$ to the wall. Remark that $h_{\text{opt}} = \lceil \frac{d(p, \mathcal{W})}{\mathcal{R}} \rceil$, where $d(p, \mathcal{W})$ is the (vertical) distance of p from the wall $\mathcal{W}$.

In the case where the protocol Π is randomized, or in the case where the distribution of the particles in the cloud is a random distribution, the number of hops h and the efficiency ratio C_h are random variables and one wishes to study their expected values.

The reason behind these definitions is that when p (or any intermediate particle in the information propagation to $\mathcal{W}$) "looks around" for a particle as near to $\mathcal{W}$ as possible to pass its information about $\mathcal{E}$, it may not get any particle in the perfect direction of the line vertical to $\mathcal{W}$. This difficulty comes mainly from three causes: (a) Due to the initial spreading of particles of the cloud in the area and because particles do not move, there might not be any particle in that direction. (b) Particles of sufficient remaining battery power may not be currently available in the right direction. (c) Particles may temporarily "sleep" (i.e., not listen to transmissions) in order to save battery power.

Note that any given distribution of particles in the smart dust cloud may not allow the ideal optimal number of hops to be achieved at all. In fact, the least possible number of hops depends on the input (the positions of the grain particles). Reference 5, however, compares the efficiency of protocols to the ideal case. A comparison with the best achievable number of hops in each input case will of course give better efficiency ratios for protocols.

To enable a first step toward a rigorous analysis of smart dust protocols, reference 5 makes the following simplifying assumption: *The search phase always finds a* p'' (of sufficiently high battery) in the semicircle of center the particle p' currently possessing the information about the event and radius R, in the direction toward $\mathcal{W}$. Note that this assumption on always finding a particle can be relaxed in the following ways: (a) By repetitions of the search phase until a particle is found. This makes sense if at least one particle exists but was sleeping during the failed searches. (b) By considering, instead of just the semicircle, a cyclic sector defined by circles of radiuses $\mathcal{R} - \Delta\mathcal{R}$, $\mathcal{R}$ and also taking into account the density of the smart cloud. (c) If the protocol during a search phase ultimately fails to find a particle toward the wall, it may *backtrack*.

Reference 5 also assumes that the position of p'' is uniform in the arc of angle $2a$ around the direct line from p' vertical to $\mathcal{W}$. Each data transmission (one hop) takes constant time t (so the "hops" and time efficiency of our protocols coincide in this case). It is also assumed that each target selection is stochastically *independent* of the others, in the sense that it is always drawn uniformly randomly in the arc $(-\alpha, \alpha)$.

The above assumptions may not be very realistic in practice; however, they can be relaxed and in any case allow us to perform a first effort toward providing some concrete analytical results.

Lemma 1 [5]. The expected "hops efficiency" of the local target protocol in the a-uniform case is

$$E(C_h) \simeq \frac{\alpha}{\sin \alpha}$$

for large h_{opt}. Also

$$1 \leq E(C_h) \leq \frac{\pi}{2} \simeq 1.57$$

for $0 \leq \alpha \leq \frac{\pi}{2}$.

Proof. Due to the protocol, a sequence of points is generated, $p_0 = p$, p_1, $p_2, \ldots, p_{h-1}$, p_h, where p_{h-1} is a particle within $\mathcal{W}$'s range and p_h is part of the wall. Let α_i be the (positive or negative) angle of p_i with respect to p_{i-1}'s vertical line to $\mathcal{W}$. It is

$$\sum_{i=1}^{h-1} d(p_{i-1}, p_i) \leq d(p, \mathcal{W}) \leq \sum_{i=1}^{h} d(p_{i-1}, p_i)$$

Since the (vertical) progress toward $\mathcal{W}$ is then $\Delta_i = d(p_{i-1}, p_i) = \mathcal{R} \cos \alpha_i$, we get

$$\sum_{i=1}^{h-1} \cos \alpha_i \leq h_{\text{opt}} \leq \sum_{i=1}^{h} \cos \alpha_i$$

From Wald's equation for the expectation of a sum of a random number of independent random variables (see reference 7), then

$$E(h-1) \cdot E(\cos \alpha_i) \leq E(h_{\text{opt}}) = h_{\text{opt}} \leq E(h) \cdot E(\cos \alpha_i)$$

Now, $\forall i$, $E(\cos \alpha_i) = \int_{-\alpha}^{\alpha} \cos x \, \frac{1}{2\alpha} dx = \frac{\sin \alpha}{\alpha}$. Thus

$$\frac{\alpha}{\sin \alpha} \leq \frac{E(h)}{h_{\text{opt}}} = E(C_h) \leq \frac{\alpha}{\sin \alpha} + \frac{1}{h_{\text{opt}}}$$

Assuming large values for h_{opt} (i.e., events happening far away from the wall, which is the most interesting case in practice since the detection and propagation difficulty increases with distance), we have (since for $0 \leq \alpha \leq \frac{\pi}{2}$ it is $1 \leq \frac{\alpha}{\sin \alpha} \leq \frac{\pi}{2}$) and the result follows. ■

15.2.4 Local Optimization: The Min-two Uniform Targets Protocol (M2TP)

Reference 5 further assumes that the search phase always returns *two points* p'', and p''', each uniform in $(-\alpha, \alpha)$, and that a modified protocol M2TP selects the best of the two points, with respect to the local (vertical) progress. This is in fact an optimized version of the Local Target Protocol.

In a similar way as in the proof of the previous lemma, the authors prove the following result:

Lemma 2 [5]. The expected "hops efficiency" of the "min-two uniform targets" protocol in the a-uniform case is

$$E(C_h) \simeq \frac{\alpha^2}{2(1 - \cos \alpha)}$$

for $0 \leq \alpha \leq \frac{\pi}{2}$ and for large h.

Now remark that

$$\lim_{\alpha \to 0} E(C_h) = \lim_{\alpha \to 0} \frac{2\alpha}{2 \sin a} = 1$$

and

$$\lim_{\alpha \to \frac{\pi}{2}} E(C_h) = \frac{(\pi/2)^2}{2(1-0)} = \frac{\pi^2}{8} \simeq 1.24$$

Thus, reference 5 proves the following:

Lemma 3 [5]. The expected "hops" efficiency of the min-two uniform targets protocol is

$$1 \leq E(C_h) \leq \frac{\pi^2}{8} \simeq 1.24$$

for large h and for $0 \leq \alpha \leq \frac{\pi}{2}$.

Remark that, with respect to the expected hops efficiency of the local target protocol, the min-two uniform targets protocol achieves, because of the one additional search, a relative gain that is $(\pi/2 - \pi^2/8)/(\pi/2) \simeq 21.5\%$. Reference 5 also experimentally investigates the further gain of additional (i.e., $m > 2$) searches.

15.3 PFR—A PROBABILISTIC FORWARDING PROTOCOL

The LTP protocol, as shown in the previous section, manages to be very efficient by always selecting exactly one next-hop particle, with respect to some optimization criterion. Thus, it tries to minimize the number of data transmissions. LTP is indeed very successful in the case of dense and robust networks, since in such networks a next hop particle is very likely to be discovered. In sparse or faulty networks, however, the LTP protocol may behave poorly, because of many backtracks due to frequent failure to find a next hop particle. To combine energy efficiency and fault-tolerance, the Probabilistic Forwarding Protocol (PFR) has been introduced. The tradeoffs in the performance of the two protocols implied above are shown and discussed in great detail in reference 8.

15.3.1 The Model

The PFR protocol was introduced by Chatzigiannakis et al. [9]. They assume the case where particles are *randomly deployed* in a given area of interest. Such a placement may occur, for example, when throwing sensors from an airplane over an area.

As a special case, they consider the network being a lattice (or grid) deployment of sensors. This grid placement of grain particles is motivated by certain applications, where it is possible to have a pre-deployed sensor network, where sensors are put (possibly by a human or a robot) in a way that they form a *two-dimensional lattice*. Note indeed that such sensor networks, deployed in a structured way, might be useful in precise agriculture, where humans or robots may want to deploy the sensors in a lattice structure to monitor, in a rather homogeneous and uniform way, certain conditions in the spatial area of interest. Certainly, exact terrain monitoring in military applications may also need some sort of a grid-like-shaped sensor network. Note also that Akyildiz et al. in a recent state-of-the-art survey [1] do not exclude the pre-deployment possibility. Also, reference 10 explicitly refers to the lattice case. Moreover, as the authors of reference 10 state in an extended version of their work ([11]), they consider, for reasons of "analytic tractability," a square grid topology.

Let N be the number of deployed grain particles. There is a single point in the network area which we call the sink S, and it represents a control center to which data should be propagated.

We assume that each grain particle has the following abilities:

1. It can estimate the direction of a received transmission (e.g., via the technology of direction-sensing antennae).
2. It can estimate the distance from a nearby particle that did the transmission (e.g., via estimation of the attenuation of the received signal).
3. It knows the direction toward the sink S. This can be implemented during a setup phase, where the (very powerful in energy) sink broadcasts the information about itself to all particles.
4. All particles have a common coordinates system.

   Notice that GPS information is not needed for this protocol. Also, there is no need to know the global structure of the network.

15.3.2 The Protocol

The PFR protocol is inspired by the probabilistic multipath design choice for the Directed Diffusion paradigm mentioned in reference 10. Its basic idea of the protocol (introduced in reference 9) lies in probabilistically favoring transmissions toward the sink within a *thin zone* of particles around the line connecting the particle sensing the event $\mathcal{E}$ and the sink (see Figure 15.4). Note that transmission along this line is energy optimal. However, it is not always possible to achieve this optimality, basically because certain sensors on this direct line might be inactive, either permanently (because their energy has been exhausted) or temporarily (because these sensors might enter a sleeping mode to save energy). Further reasons include (a) physical damage of sensors, (b) deliberate removal of some of them (possibly by an adversary in military applications), (c) changes in the position of the sensors due to a variety of reasons (weather conditions, human interaction, etc). and (d) physical obstacles blocking communication.

The protocol evolves in two phases:

Phase 1: The "Front" Creation Phase. Initially the protocol builds (by using a limited, in terms of rounds, flooding) a sufficiently large "front" of particles, in order to guarantee the survivability of the data propagation process. During

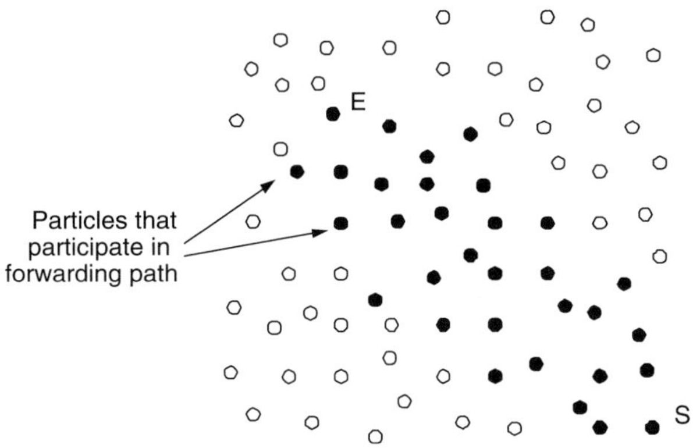

Figure 15.4. Thin zone of particles.

this phase, each particle having received the data to be propagated, deterministically forwards them toward the sink. In particular, and for a sufficiently large number of steps $s = 180\sqrt{2}$, each particle broadcasts the information to all its neighbors, toward the sink. Remark that to implement this phase, and in particular to count the number of steps, we use a counter in each message. This counter needs at most $\lceil \log 180\sqrt{2} \rceil$ bits.

Phase 2: The Probabilistic Forwarding Phase. During this phase, each particle P possessing the information under propagation calculates an angle ϕ by calling the subprotocol "ϕ-calculation" (see description below) and broadcasts $info(\mathcal{E})$ to all its neighbors with probability $\mathbf{P}_{\text{fwd}}$ (or it does not propagate any data with probability $1 - \mathbf{P}_{\text{fwd}}$) defined as follows:

$$\mathbf{P}_{\text{fwd}} = \begin{cases} 1 & \text{if } \phi \geq \phi_{\text{threshold}} \\ \frac{\phi}{\pi} & \text{otherwise} \end{cases}$$

where ϕ is the angle defined by the line EP and the line PS and $\phi_{\text{threshold}} = 134°$ (the selection reasons of this $\phi_{\text{threshold}}$ will become evident in Section 15.3.4).

In both phases, if a particle has already broadcast $info(\mathcal{E})$ and receives it again, it ignores it. Also the PFR protocol is presented for a single event tracing. Thus no multiple paths arise and packet sizes do not increase with time.

Note that when $\phi = \pi$, then P lies on the line ES and vice versa (and always transmits).

If the density of particles is appropriately large, then for a line ES there is (with high probability) a sequence of points "closely surrounding ES" whose angles ϕ are larger than $\phi_{\text{threshold}}$, and thus successive points are within transmission range. All such points broadcast and thus essentially they follow the line ES (see Figure 15.4).

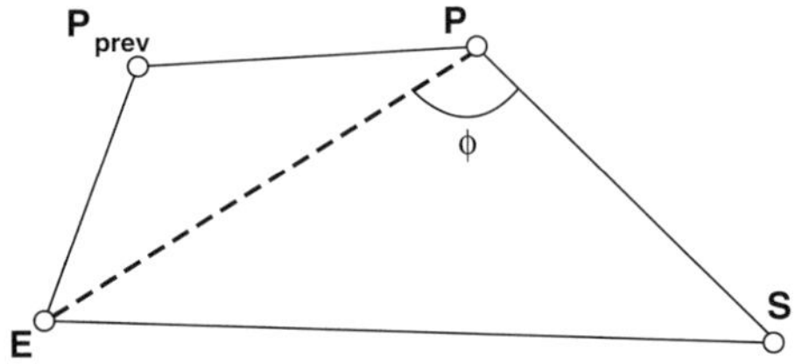

Figure 15.5. Angle ϕ calculation example.

The ϕ-Calculation Subprotocol (see Figure 15.5). Let P_{prev} be the particle that transmitted $info(E)$ to P.

1. When P_{prev} broadcasts $info(E)$, it also attaches the info $|EP_{\text{prev}}|$ and the direction $\overrightarrow{P_{\text{prev}}E}$.
2. P estimates the direction and length of line segment $P_{\text{prev}}P$, as described in the model.
3. P now computes angle $(\widehat{EP_{prev}P})$, and computes $|EP|$ and the direction of $\overrightarrow{PE}$ (this will be used in further transmission from P).
4. P also computes angle $(\widehat{P_{prev}PE})$ and by subtracting it from $(\widehat{P_{prev}PS})$ it finds ϕ.

Notice the following:

i. The direction and distance from activated sensors to E is inductively propagated (i.e., P becomes P_{prev} in the next phase).
ii. The protocol needs only messages of length bounded by $\log A$, where A is some measure of the size of the network area, since (because of (i) above) there is no cumulative effect on message lengths.

Essentially, the protocol captures the intuitive, deterministic idea "if my distance from ES is small, then send, else do not send." Chatzigiannakis et al. [9] have chosen to enhance this idea by random decisions (above a threshold) to allow some local flooding to happen with small probability and thus to cope with local sensor failures.

15.3.3 Properties of PFR

Any protocol Π solving the data propagation problem must satisfy the following three properties:

- *Correctness*. Π must guarantee that data arrive to the position S, given that the whole network exists and is operational.

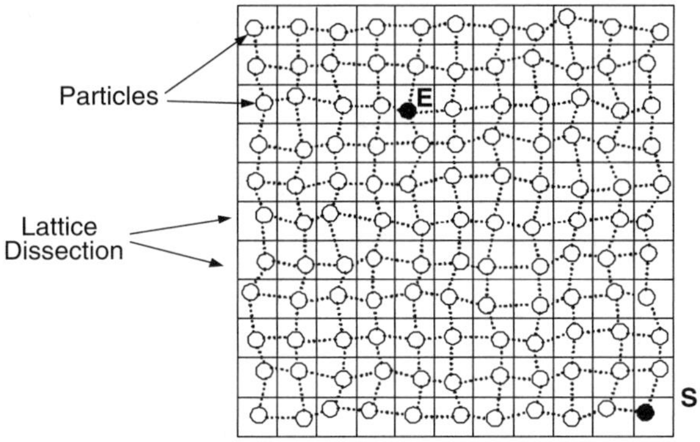

Figure 15.6. A lattice dissection G.

- *Robustness*. Π must guarantee that data arrive at enough points in a small interval around S, in cases where part of the network has become inoperative.
- *Efficiency*. If Π activates k particles during its operation, then Π should have a small ratio of the number of activated over the total number of particles $r = \frac{k}{N}$. Thus r is an energy efficiency measure of Π.

Reference 9 shows that this is indeed the case for PFR.

Consider a partition of the network area into small squares of a fictitious grid G (see Figure 15.6). Let the length of the side of each square be l. Let the number of squares be q. The area covered is bounded by ql^2. Assuming that we randomly throw in the area at least $\alpha q \log q = N$ particles (where $\alpha > 0$ a suitable constant), then the probability that a particular square is avoided tends to 0. So with very high probability (tending to 1), all squares get particles.

Reference 9 conditions all the analysis on this event, call it F, of at least one particle in each square.

15.3.4 The Correctness of PFR

Without loss of generality, we assume that each square of the fictitious lattice G has side length 1.

In reference 9 the authors prove the correctness of the PFR protocol, by using a geometric analysis. We sketch their proof below.

Consider any square Σ intersecting the ES line. By the occupancy argument above, there is with high probability a particle in this square. Clearly, the worst case is when the particle is located in one of the corners of Σ (since the two corners located most far away from the ES line have the smallest ϕ-angle among all positions in Σ).

By some geometric calculations, reference 9 finally proves that the angle ϕ of this particle is $\phi > 134°$. But the initial square (i.e., that containing E) always broadcasts, and any intermediate intersecting square will be notified (by induction) and thus will broadcast because of the argument above. Thus the sink will be reached if the whole network is operational.

Lemma 4 [9]. PFR *succeeds with probability 1* in sending the information from E to S given the event F.

15.3.5 The Energy Efficiency of PFR

Reference 9 considers the fictitious lattice G of the network area and let the event F hold. There is (at least) one particle inside each square. Now join all nearby particles of each particle to it, thus forming a new graph G' which is "lattice-shaped," but its elementary "boxes" may not be orthogonal and may have varied length. When G's squares become smaller and smaller, then G' will look like G. Thus, for reasons of analytic tractability, Chatzigiannakis et al. [9] assume that particles form a lattice (see Figure 15.7). They also assume length $l = 1$ in each square, for normalization purposes. Notice, however, that when $l \to 0$, then "$G' \to G$" and thus all results in this section hold for any random deployment "in the limit."

The analysis of the energy efficiency considers particles that are active but are as far as possible from ES. Thus the approximation is suitable for remote particles.

Reference 9 estimates an upper bound on the number of particles in an $n \times n$ (i.e., $N = n \times n$) lattice. If k is this number, then $r = \frac{k}{n^2}$ $(0 < r \leq 1)$ is the "energy efficiency ratio" of PFR.

More specifically, Chatzigiennakis et al. [9] prove the (very satisfactory) result below. They consider the area around the ES line, whose particles participate in the propagation process. The number of active particles is thus, roughly speaking, captured by the size of this area, which in turn is equal to $|ES|$ times the maximum distance from $|ES|$ (where maximum is over all active particles).

This maximum distance is clearly a random variable. To calculate the expectation and variance of this variable, the authors in reference 9 basically "upper bound" the stochastic process of the distance from ES by a random walk on the line, and subsequently they "upper bound" this random walk by a well-known stochastic process (i.e., the "discouraged arrivals" birth and death Markovian process; see, e.g., reference 12). Thus they prove the following:

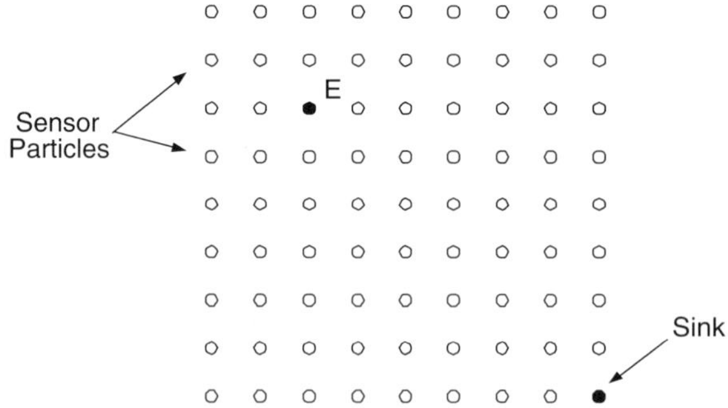

Figure 15.7. A lattice sensor network.

Theorem 15.3.1 [9]. The energy efficiency of the PFR protocol is $\Theta((\frac{n_0}{n})^2)$, where $n_0 = |ES|$ and $n = \sqrt{N}$, where N is the number of particles in the network. For $n_0 = |ES| = o(n)$, this is $o(1)$.

15.3.6 The Robustness of PFR

To prove the following robustness result, the authors in reference 9 consider particles "very near" to the ES line. Clearly, such particles have large ϕ-angles (i.e., $\phi > 134°$). Thus, even in the case where some of these particles are not operating, the probability that none of those operating transmits (during the probabilistic phase 2) is very small. Thus, reference 9 proves the following.

Lemma 5 [9]. PFR manages to propagate the crucial data across lines parallel to ES, and of constant distance, with *fixed* nonzero probability (not depending on n, $|ES|$).

15.4 THE ENERGY BALANCE PROBLEM

In order to save energy and keep the network functional for as long as possible, various approaches, including hop-by-hop transmission techniques [5, 10, 11], as well as clustering techniques [13] and alternating power-saving modes [14], have been proposed.

All such techniques do not *explicitly* take care of the possible overuse of certain sensors in the network. As an example, note that in hop-by-hop transmissions toward the sink, the sensors lying closer to the sink tend to be utilized exhaustively (since all data pass through them). Thus, these sensors may die out very early, thus resulting in network collapse, although there may be still significant amounts of energy in the other sensors of the network. Similarly, in clustering techniques the cluster heads that are located far away with respect to the sink tend to spend a lot of energy.

In this chapter, we present two protocols trying to balance energy dissipation among the sensors in the network: (a) the EPB (Energy Balance) protocol, introduced in reference 15, which probabilistically chooses between either propagating data one hop toward the sink or sending directly to the sink. The first choice is more energy-efficient, while the latter bypasses the critical (close to the sink) sectors. The appropriate probability for each choice in order to achieve energy balance is calculated in reference 15. (b) VTRP (Variable Transmission Range Protocol), proposed in reference 16, which implicitly contributes to energy balance by appropriately adapting (increasing) the transmission range, thus bypassing critical sensors and avoiding possible obstacles.

15.5 EBP: THE ENERGY BALANCE PROTOCOL

15.5.1 The Model and the Problem

We assume that crucial events, which should be reported to a control center, occur in the network area. Furthermore, we assume that these events are happening at random uniform positions. Let N be their total number in a certain period (i.e., during the execution of the protocol).

The sensors are spread in the network area randomly uniformly, so their number in a certain area is proportional to the area's size. There is a single point in the network, which we call the "sink" S, that represents the fixed control center, to which data about event realization should be reported. The sink is very powerful, in terms of both energy and computing power. Sensors can be aware of the direction (and position) of the sink, as well as of their distance to the sink. Such information can be easily obtained during a setup phase, by having the (powerful) sink broadcast control messages to the entire network area. We assume that the transmission range of sensors can vary with time (in fact, for each sensor our protocol may use only two different ranges: R and $i \cdot R$, where i is a measure of the sensor's distance to the sink). The sensors do not move.

We virtually "cover" the network area by a cycle sector of angle ϕ (see Figure 15.8). The cycle sector is divided into n ring sectors or "*slices*." The first slice has radius R (i.e., the sensors' transmission range). Slice i $(2 \leq i \leq n)$ is defined by two cycles sectors, one of radius $i \cdot R$ and the other of radius $(i - 1) \cdot R$. Taking a sufficiently large angle ϕ and/or by taking multiple sectors, we can cover the whole area.

As far as energy dissipation is concerned, we assume that the energy spent at a sensor when transmitting data is proportional to the square of the transmitting distance. Our protocol's performance analysis can, however, be extended to any energy cost model. Note that the energy dissipation for receiving is not always negligible with respect to the energy when sending, such as in the case when transmissions are too short and/or radio electronics energy is high (see reference [13]). In the analysis, we only count (for simplicity) energy spent during transmissions. Since, however, in our protocol (see next section) there is one receipt for each transmission, it is clear that even when energy during receipt is more or less the same as energy during transmissions, the analysis can be extended easily to the full case (counting both transmissions and receipts).

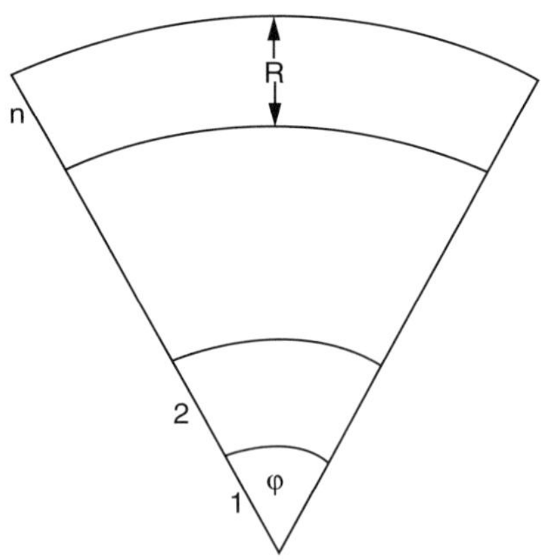

Figure 15.8. Sensor network with n ring sectors, angle ϕ, and ring "width" R.

Definition 15.5.1. The area between two consecutive cycle sectors is called *a ring sector* (or "slice"). Let T_i $(1 \leq i \leq n)$ be the ith ring sector of the network.

T_1 stands for the ring sector with center the sink and radius equal to R.

Definition 15.5.2. Let S_i be the area size of the ring sector T_i of the network $(1 \leq i \leq n)$.

We wish to solve the "*energy balanced data propagation problem*",—that is, propagate data to the sink in such a way that the "average" energy dissipation in each sensor is at each time the same. The average energy dissipation per sensor is taken to be the fraction of the total energy spent by sensors in a ring sector over the number of sensors in that sector. Because of our assumption that the number of sensors in an area is proportional to the area size, the average energy dissipation per sensor is calculated by dividing the total energy spent in a sector by the sector size.

Here we do not study medium access aspects, assuming the existence of underlying MAC protocol.

15.5.2 The Protocol

We assume that each event is sensed by only one sensor. This assumption is not restrictive since we may consider multiple sensing and propagation of an event by various sensors as sensing and propagation of many different events. A sensor sensing an event generates a data message that should be eventually delivered to the sink. On each ring sector, T_i, a number of events occur and a corresponding number of messages (one for each event) is generated.

Randomization is used to achieve some "load balancing" by evenly spreading the "load" (energy dissipation). In particular, on ring sector T_i each event is propagated to T_{i-1} (i.e., the "*next*" sector towards the sink) with probability p_i, while with probability $1 - p_i$ it is propagated directly to the sink S. Each message in T_i is handled stochastically independently of the other events' messages.

The choice of probability p_i for T_i is made so that the average energy consumption per area unit (and thus per sensor) is the same for the whole network. There is a tradeoff from choosing p_i: If p_i increases, then transmissions tend to happen locally, thus energy consumption is low; however, sensors closer to the sink tend to be overused since all data passes through them. On the other hand, if p_i decreases, there are distant transmissions (thus a lot of energy is consumed); however, particles closer to the sink are bypassed. Calculating the appropriate probability p_i for each T_i and solving the problem of energy balance is very important since it combines efficient data propagation with increased network's lifetime.

By using an underlying subprotocol [5, 10], we can guarantee that only one "next hop" sensor receives the transmitted message. Note also that data messages are of fixed size; that is, no further information is added to a message on its route toward the sink.

Our protocol is (a) *distributed*, since each sensor chooses propagation probability independently of other sensors, and (b) uses only *local* information, in the sense that

p_i depends only on i, that is, a parameter related to the distance from the sink. Note that distance from the sink information for each sensor can be easily obtained—that is, during a setup phase where the sink broadcasts control messages to the network. Several techniques (including signal attenuation evaluation) can be used to estimate each sensor's distance for the sink. (c) The protocol is very *simple*, since it just uses a random choice based only on parameter i.

15.5.3 Basic Definitions—Preliminaries

We aim at calculating probability p_i for each i in order to ensure the energy balance property. Using simple geometry, one can easily prove the following lemmas.

Lemma 6. The area size, S_1, of the ring sector T_1 is $S_1 = \frac{\phi}{2} \cdot R^2$.

Lemma 7. The relation between the area size of the ring sector T_i and that of T_1 is $S_i = (2i - 1) \cdot S_1$.

Definition 15.5.3. Let λ_i the probability that an event will occur on the ring sector T_i.

There are n ring sectors in the network.

Lemma 8. Assuming a random uniform generation of events in the network area, the probability λ_i of an event occurring on the ring sector T_i $(1 \leq i \leq n)$ is

$$\lambda_i = \frac{(2i - 1)}{n^2}$$

Let us now consider sector T_i.

Definition 15.5.4. An area T_i *"handles"* an event generated in ring sector j if either the message was generated in the area T_i (i.e. $j = i$) or the message was propagated to T_i from the ring sector T_{i+1}.

Definition 15.5.5. Let h_i be the number of the messages that are "handled" by the area T_i.

We now define energy ϵ_{ij} spent for message j when sector i handles it.

Definition 15.5.6. Let ϵ_{ij} a random variable which measures the energy that dissipates the sector T_i so as to handle the message j. For ϵ_{ij} we have that

$$\epsilon_{ij} = \begin{cases} cR^2 & \text{with probability } p_i \\ c(iR)^2 & \text{with probability } 1 - p_i \end{cases}$$

where cR^2 is the energy dissipation for sending a message j from T_i to its adjacent ring sector T_{i-1} and c is a constant.

Thus, the expected energy dissipation in sector i for handling a message is

$$E[\epsilon_{i,j}] = cR^2 \cdot [i^2 - p_i(i^2 - 1)] \tag{15.1}$$

Note: The expected energy above is the same for all messages; we use j just for counting purposes.

Definition 15.5.7. Let $\mathcal{E}_i$ the *total energy* spent by sensors in T_i. Clearly,

$$\mathcal{E}_i = \sum_{j=1}^{h_i} \epsilon_{ij} \tag{15.2}$$

Energy balance is defined as follows:

Definition 15.5.8. The network is energy balanced if the average per sensor energy dissipation is the same for all sectors, i.e. when

$$\frac{E[\mathcal{E}_i]}{S_i} = \frac{E[\mathcal{E}_j]}{S_j} \qquad i, j = 1, \ldots, n \tag{15.3}$$

15.5.4 The General Solution

We next provide a lemma useful in the estimation of the total energy dissipation in a sector.

Lemma 9. The expected total energy dissipation in sector i is

$$E[\mathcal{E}_i] = E[h_i] \cdot E[\epsilon_{ik}]$$

Proof.

$$\begin{aligned}
E[\mathcal{E}_i] &= E\left[\sum_{k=1}^{h_i} \epsilon_{ik}\right] \\
&= \sum_{n=0}^{N} E\left[\sum_{k=0}^{h_i} (\epsilon_{ik} \cap h_i = n)\right] \\
&= \sum_{n=0}^{N} E\left[\sum_{k=1}^{h_i} \epsilon_{ik} \middle| h_i = n\right] \cdot \mathbf{P}\{h_i = n\}
\end{aligned}$$

Furthermore,

$$E\left[\sum_{k=1}^{h_i} \epsilon_{ik} \middle| h_i = n\right] = E\left[\sum_{k=1}^{n} \epsilon_{ik}\right] = \sum_{k=1}^{n} E[\epsilon_{ik}]$$
$$= n \cdot E[\epsilon_{ik}]$$

Thus, we have from the above that

$$E[\mathcal{E}_i] = \sum_{n=0}^{N} E\left[\sum_{k=1}^{h_i} \epsilon_{ik} \middle| h_i = n\right] \cdot \mathbf{P}\{h_i = n\}$$
$$= E[\epsilon_{ik}] \cdot E[h_i]$$

■

Definition 15.5.9. Let g_i be the number of the messages that are *generated* in the area T_i.

Note that messages are generated in an area only when events occur in this area.

Definition 15.5.10. Let f_i be the number of the messages that are *forwarded to* the area T_i.

We note that messages are forwarded to a ring sector (say i) only because of an event generation at a sector $j > i$ and successive one-hop propagations from sector j to sector i.
We notice the following important relation:

$$h_i = g_i + f_i \tag{15.4}$$

which means that the number of messages that area T_i handles equals the number of the messages that are generated in T_i, plus the number of messages that are forwarded to it. Because of event generation according to a probability distribution and also because of the probabilistic nature of message propagation in, all three quantities above are random variables. By linearity of expectation, we get the following lemma.

Lemma 10. $E[h_i] = E[g_i] + E[f_i]$.

We establish a relationship between $E[f_i]$ and $E[h_{i+1}]$.

Lemma 11. $E[f_i] = p_{i+1} \cdot E[h_{i+1}]$.

Proof. Let $\delta_{i,j}$ an indicator random variable that is equal to 1 if area T_i forwards the message j to the area T_{i-1} and 0 otherwise. Thus

$$\delta_{i,j} = \begin{cases} 1 & \text{with probability } p_i \\ 0 & \text{with probability } 1 - p_i \end{cases}$$

Clearly, $\delta_{i,j}$ depends only on i, but we add j for counting purposes. Obviously, $E[\delta_{i,j}] = p_i$. It is

$$f_i = \sum_{j=0}^{h_{i+1}} \delta_{i+1,j}$$

Similarly to the proof of Lemma 9, we get

$$E[f_i] = \sum_{n=0}^{N} E\left[\sum_{j=0}^{h_{i+1}} \delta_{i+1,j} \middle| h_{i+1} = n\right] \cdot \mathbf{P}\{h_{i+1} = n\}$$

and the proof is completed. ■

Recall that, according to Definition 15.5.8, to achieve the same on the average energy dissipation *per area unit* (and thus per sensor) in the network area, the following equality should hold:

$$E\left[\frac{\sum_{k=1}^{h_i} \epsilon_{ik}}{S_i}\right] = E\left[\frac{\sum_{k=1}^{h_j} \epsilon_{jk}}{S_j}\right] \quad \forall i, j \in \{1, \ldots, n\} \tag{15.5}$$

that is, the average energy consumption per sensor should be equal in any two ring sectors. By induction, it suffices to guarantee this for any two adjacent sectors. In what follows, we guarantee the above balance property, requiring a certain recurrence relation to hold. This recurrence basically relates three successive terms of the $E[f_i]$ sequence (the $E[g_i]$ terms depend only on i and on input parameters).

Theorem 15.5.1. To achieve energy balance in the network, the following recurrency equation should hold:

$$\begin{aligned} a_{i+1}E[f_{i+1}] - (d_i + a_i)E[f_i] + d_{i-1}E[f_{i-1}] \\ = a_i E[g_i] - a_{i+1}E[g_{i+1}] \end{aligned}$$

where

$$a_i = \frac{i^2}{2i-1}, \qquad d_i = \frac{(i+1)^2-1}{2i+1}$$

Proof. For the case $j = i + 1$ of Eq. (15.5) and using Lemmas 7 and 9, we have

$$\frac{E[h_i]E[\epsilon_{i,j}]}{S_i} = \frac{E[h_{i+1}]E[\epsilon_{i+1,j}]}{S_{i+1}} \Leftrightarrow$$

$$\frac{E[h_i][i^2 - p_i(i^2-1)]}{(2i-1)} = \frac{E[h_{i+1}]\{(i+1)^2 - p_{i+1}[(i+1)^2-1]\}}{(2i+1)} \Leftrightarrow$$

$$\frac{i^2}{2i-1}E[h_i] - p_i E[h_i]\frac{i^2-1}{2i-1} = \frac{(i+1)^2}{2i+1}E[h_{i+1}] - p_{i+1}E[h_{i+1}]\frac{(i+1)^2-1}{2i+1}$$

Let a_i, and d_i be as defined in the theorem statement above. By Lemma 11 we know that $p_i E[h_i] = E[f_{i-1}]$, and by Lemma 10 it is $E[h_i] = E[g_i] + E[f_i]$; thus the last equation becomes

$$a_{i+1}E[f_{i+1}] - (d_i + a_i)E[f_i] + d_{i-1}E[f_{i-1}]$$
$$= a_i E[g_i] - a_{i+1}E[g_{i+1}]$$

■

To solve the above recurrency we must compute $E[g_i]$.

Lemma 12. If N is the total number of events that are generated in the network, the mean value of g_i is given by the following relationship:

$$E[g_i] = N \cdot \lambda_i$$

Proof. Because the position of each event is independent of other events and because for each sector i, probability λ_i is the same, clearly g_i is binomial with parameters N and λ_i. ■

In order to have a simpler recurrence involving only two (successive in fact) terms of the $E[f_i]$ sequence, we will transform the recurrency relation of Theorem 15.5.1. into the following (easier to solve) relation:

Lemma 13. The recurrency relation

$$t_i - t_{i-1} = a_i \cdot E[f_i] - a_{i+1} \cdot E[f_{i+1}] \qquad \text{for} \quad i = 1, \ldots, n-1 \quad \text{and}$$
$$t_0 = a_1 \cdot E[f_1]$$

has as a solution the function

$$t_i = \sum_{j=1}^{i} \left(a_j E[g_j] - a_{j+1} E[g_{j+1}]\right) + a_1 \cdot E[f_1]$$

Proof. The proof is done by induction on i. For $i = 0$, it is obviously true. Let it be true for $i - 1$. For i we have

$$t_i = t_{i-1} + a_i \cdot E[g_i] - a_{i+1} \cdot E[g_{i+1}]$$

By the induction hypothesis, we get the solution

$$t_i = \sum_{j=1}^{i} \left(a_j E[g_j] - a_{j+1} E[g_{j+1}]\right) + a_1 \cdot E[f_1]$$

■

Now the recurrency relation of Theorem 15.5.1. is simplified:

$$a_{i+1} \cdot E[f_{i+1}] - d_i \cdot E[f_i] = t_i, \qquad i = 1, \ldots, n-1$$

Thus, we get a recurrence for sequence $E[f_i]$ involving only two successive terms of the sequence.

Theorem 15.5.2. The recurrency relation

$$a_{i+1} E[f_{i+1}] - d_i E[f_i] = t_i, \qquad i = 1, \ldots, n-1$$

where t_i is defined in Lemma 13, has the following solution:

$$E[f_{n-i}] = -\sum_{k=1}^{i} \frac{\prod_{j=k}^{i-1} a_{n-j}}{\prod_{j=k}^{i} d_{n-j}} \cdot t_{n-k}$$

The proof is rather complex, so we omit it here. The interested reader may find it in reference 15.

The full expression for $E[f_i]$ can be expressed by substituting i with $n - i$, thus

$$E[f_i] = -\sum_{k=1}^{n-i} \frac{\prod_{j=k}^{n-i+1} a_{n-j}}{\prod_{j=k}^{n-i} d_{n-j}} \cdot \left(\sum_{j=1}^{n-k} (a_j E[g_j] - a_{j+1} E[g_{j+1}]) + a_1 \cdot E[f_1] \right)$$

where $\prod_i^{i-1} a_i = 1$.

We note that all the parameters of the recurrency solution above are expressed as a function of $E[f_1]$ and i. So as to compute them, we firstly compute the value of $E[f_1]$. Then we can compute all the other parameters by replacing the already computed $E[f_1]$.

Now, the calculation of the probabilities p_i is quite easy.

Theorem 15.5.3. The energy balance property is achieved if any ring sector (say T_i) propagates each message it handles with probability p_i to the next ring sector, T_{i-1}, and with probability $1 - p_i$ it propagates the message directly to the sink. The value of each p_i is given by the following relation:

$$p_i = \frac{E[f_{i-1}]}{E[g_i] + E[f_i]}$$

where the values of $E[f_i]$ and $E[g_i]$ are obtained from Theorem 15.5.2. and Lemma 12, respectively.

Proof. From Eq. (15.11) we know that $E[f_{i-1}] = p_i E[h_i]$ and also by Lemma 10 we know that $E[h_i] = E[g_i] + E[f_i]$. ■

Remark. Note that, interestingly, p_i's are *independent* of the number N of the events that occur in the network, since p_i's depend only on i and the number of ring sectors n (which is broadcast to sectors by the sink). Thus the protocol assumes only local information.

We note that the analysis above allows the exact derivation of probabilities p_i's as a function of i and n which (although complicated and not obviously leading to a closed form) can be easily calculated by the sensors in the network by carrying out very simple calculations.

The authors of reference 15 also prove the correctness of the protocol.

Theorem 15.5.4. Given that the energy is each time on the average the same in all network sensors, each message will finally get to the sink.

15.5.5 A Closed Form

Under specific assumptions (that we discuss and motivate), we can make the calculation of probabilities p_i simpler. Combining Lemmas 10 and 12, we have that

$$E[h_i] = \lambda_i N + E[f_i] = \frac{2i-1}{n^2} N + E[f_i] \tag{15.6}$$

By the corresponding relation for $E[h_{i-1}]$, it must be

$$\frac{\frac{2i-1}{n^2}N + E[f_i]}{(2i-1)\frac{\phi}{2}R^2}[p_i + (1-p_i)i^2]R^2 = \frac{\frac{2i-3}{n^2}N + E[f_{i-1}]}{(2i-3)\frac{\phi}{2}R^2}[p_{i-1} + (1-p_{i-1})(i-1)^2]R^2$$

But $2i - 1 \simeq 2i - 3$ and $\frac{\phi}{2}$, R^2 cancel. Dividing by N, we get

$$\left[1 + \frac{E[f_i]}{N}\right]\left[p_i + (1-p_i)i^2\right] \simeq \left[1 + \frac{E[f_{i-1}]}{N}\right]\left[p_{i-1} + (1-p_{i-1})(i-1)^2\right]$$

If $E[f_i] \simeq E[f_{i-1}]$, then the previous relation becomes

$$p_i + (1 - p_i)i^2 = p_{i-1} + (1 - p_{i-1})(i - 1)^2 \tag{15.7}$$

In reference 15 we show how to solve the above recurrence.

Theorem 15.5.5. If $E[f_i] \simeq E[f_{i-1}]$, $3 \leq i \leq n$, then the one-hop forwarding probability, guaranteeing energy balance, is

$$p_i = 1 - \frac{3x}{(i+1)(i-1)}$$

where $p_2 = x \in (0, 1)$ a free parameter and $p_1 = 0$.

We note that the assumption $E[f_i] \simeq E[f_{i-1}]$ is quite reasonable and well-motivated. We provide the following intuitive explanation of why this happens. Note indeed that the area sizes of adjacent sectors (and thus the number of events generated in such sections) are more or less the same, especially when i increases. Furthermore, the probability p_i of forwarding to the adjacent (towards the sink) sector increases very fast with i and becomes 1 in most sectors of the network (in the middle territory).

15.6 VTRP: THE VARIABLE TRANSMISSION RANGE PROTOCOL

15.6.1 Introduction

The VTRP protocol has been proposed by T. Antoniou et al. [16]. They study the problem of *multiple event detection and propagation*—that is, the local sensing of a series of crucial events and the energy-efficient and fault-tolerant propagation of data reporting the realization of these events to a (fixed or mobile) control center. The control center could in fact be some human authorities responsible for taking action upon the realization of the crucial event. We use the term *sink* for this control center. We note that this problem *generalizes* the single event propagation problem (with respect to references 5, 8, and 15) and poses new challenges for designing efficient and fault-tolerant data propagation protocols. The new protocol we present here can also be used for the more general problem of data propagation in sensor networks (see reference 10).

The basic innovation in our protocol is to vary the range of data transmissions. This feature aims at better performance, compared to typical fixed transmission range data propagation, in some rather frequently occurring situations such as the following:

(a) The case of low densities of sensor particles. In such networks, fixed-range protocols may trap in backtracking actions when no particles toward the sink are found. Our protocol, by increasing the transmission range, may find such particles and avoid extensive backtracking.

(b) Because of the possibility to increase transmission range, VTRP performs better in cases of obstacles or faulty/sleeping sensors. Also, it bypasses certain critical sensors (like those close to the sink) that tend to be overused, thereby prolonging the network lifetime.

To demonstrate the above properties of VTRP, we compare it to a typical fixed-range protocol: the Local Target Protocol (LTP).

The ability of LTP to propagate information regarding the realization of a crucial event to the control center depends on the particle density of the network. The experiments conducted in reference 5 indicate that for low particle densities, LTP fails to propagate the messages to the control center (while for high particle densities the failure rate drops very fast to zero; that is, the messages are almost always reported correctly). The new protocol that we propose in this chapter successfully overcomes this problem by increasing the transmission range of the particles that fail to locate an active neighboring particle toward the sink. In fact, the experiments conducted in this chapter (see Section 15.6.7) demonstrate the superiority of VTRP over LTP even for sensor networks with very low particle densities.

Further note that this is the first time that the LTP protocol is evaluated under the setting of multiple events. Our findings indicate that LTP has a fundamental design flaw in this case, because the success of the propagation process heavily depends on the lifetime of the particles that are located around the control center. As soon as these particles exhaust their power supplies, the whole network becomes inoperable. The new protocol that we present here successfully overcomes this problem by adjusting the transmission range of the particles as soon as the particles closer to the control center "die". Our experiments indicate that VTRP increases the ability of the network to report multiple events up to 100%, compared to LTP.

We propose four different mechanisms for varying the transmission range of the particles that aim at different types of smart dust networks regarding particles densities and energy-saving criteria. Our experimental results show that VTRP can be easily modified to further improve its performance. Actually, $VTRP_p$ (where range is increased aggressively) and $VTRP_r$ (which randomizes between the various range change functions toward a better average case performance) successfully propagate about 50% more events than the "basic" VTRP and almost 200% more events than the original LTP protocol.

15.6.2 The Model

The model considered here is an extension of the one presented in Section 15.2.1. In our model here (Figure 15.9), we assume that the transmission range (R) can vary (i.e., by setting the transmission power at appropriate levels) while the transmission angle (let it be α) is fixed and cannot change throughout the operation of the network (since this would require a modification or movement of the antenna used). Note that the protocols we study in this work can operate even under the broadcast communication mode (i.e., $\alpha = 2\pi$). The laser possibility is added for reducing energy dissipation in long-distance transmissions.

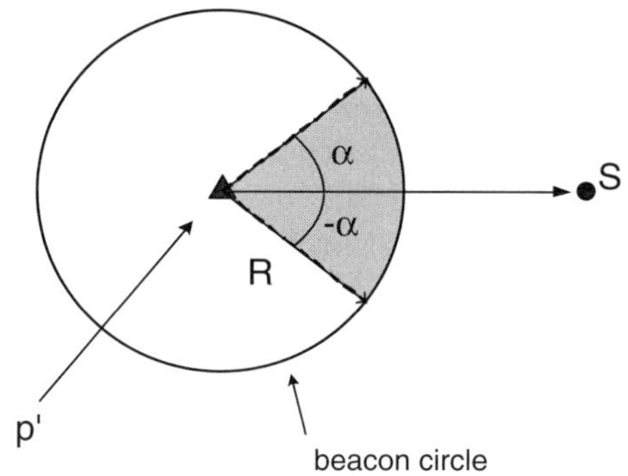

Figure 15.9. Directed transmission of angle α.

Each particle can be in one of three different modes at any given time, regarding the energy consumption. These modes are (a) transmission of a message, (b) reception of a message, and (c) sensing of events.

For the case of transmitting and receiving a message, we assume the following simple model where the radio dissipates E_{elec} to run the transmitter and receiver circuitry and ϵ_{amp} for the transmit amplifier to achieve acceptable SNR (signal-to-noise ratio). We also assume an r^2 energy consumption due to channel transmission at distance r. Thus to transmit a `k-bit` message at distance r in our model, the radio expends

$$E_T(k, r) = E_{T-\text{elec}}(k) + E_{T-\text{amp}}(k, r)$$
$$E_T(k, r) = E_{\text{elec}} \cdot k + \epsilon_{\text{amp}} \cdot k \cdot r^2$$

and to receive this message, the radio expends

$$E_R(k) = E_{R-\text{elec}}(k)$$
$$E_R(k, r) = E_{\text{elec}} \cdot k$$

where $E_{T-\text{elec}}$, $E_{R-\text{elec}}$ stand for the energy consumed by the transmitter's and receiver's electronics, respectively.

Concluding, there are three different kinds of energy dissipation:

- E_T: Energy dissipation for transmission.
- E_R: Energy dissipation for receiving.
- E_{idle}: Energy dissipation for idle state.

For the idle state, we assume that the energy consumed for the circuity is constant for each time unit and equals E_{elec} (the time unit is 1 second).

We note that in our simulations we *explicitly measure the above energy costs*. We feel that our model, although simple, depicts accurately enough the technological specifications of real smart dust systems. Similar models are being used by other researchers in order to study sensor networks (see reference 17). In contrast to references 10 and 18, our model is weaker in the sense that *no geolocation abilities* are assumed (e.g., a GPS device) for the smart dust particles, leading to more generic and thus stronger results. In reference 19 a thorough comparative study and description of smart dust systems is given, from the technological point of view.

15.6.3 The Problem

Assume the realization of a series of K crucial events $\mathcal{E}_i$, with each event being sensed by a single particle p_i $(i = 1, 2, \ldots, K)$. Then the *multiple event propagation problem* $\mathcal{P}$ is the following:

> How can each particle p_i $(i = 1, 2, \ldots, K)$, via cooperation with the rest of the grain particles, in an efficient (mainly with respect to energy and time) and fault-tolerant way, propagate information *info*($\mathcal{E}_i$) reporting realization of event $\mathcal{E}_i$ to the sink $\mathcal{S}$?

We note that this problem is a *generalization* of the single event propagation problem, which is more difficult to cope with because of the severe energy restrictions of the particles.

Certainly, because of the dense deployment of sensor particles close to each other, communication between two particles is much more energy-efficient than direct transmission to the sink. Furthermore, short-range hop-by-hop transmissions can effectively overcome some of the signal propagation effects in long-distance transmissions and may help to smoothly adjust propagation around obstacles. Finally, the low-energy transmission in multihop communication may enhance security, protecting from undesired discovery of the data propagation operation.

On the other hand, long-range transmissions require the participation of few particles and therefore reduce the overhead on particle resources and provide better network response times. Furthermore, long-range communication permits the deployment of clustering and other efficient techniques, developed for ad hoc wireless networks. In particular, a clustering scheme enables cluster heads to reduce the amount of transmitted data by aggregating information.

The above suggest that many diverse approaches exist to the solution of the *multiple event propagation problem* $\mathcal{P}$. Further to choosing between long or short transmissions, certain additional tradeoffs are introduced by choosing between fixed or varying transmission range. In particular, we wish to focus on the following important properties:

(a) *Obstacle Avoidance:* This may be achieved by increasing transmission range when an obstacle is encountered.

(b) *Fault Tolerance:* Increasing range may reach active sensors when the current range does not succeed, because of either faulty or "sleeping" sensors close to sensor which is currently transmitting or in the case of very low network densities.

(c) *Network Longevity:* An interesting aspect of the problem under investigation is the lifetime of particles, since it affects the ability of the network to propagate data to the sink, because available routes are reduced as more particles consume their energy resources and "die." Varying transmission range may bypass the sensors lying close to the sink, which tend to be overused in the case of fixed range transmissions, since all data pass through them in this case. The same holds also in the case of a geographical concentration of event generation.

15.6.4 The Variable Transmission Range Protocol (VTRP)

In this protocol, each particle p' that has received *info*($\mathcal{E}$) from p (via, possibly, other particles) does the following:

Phase 1: The Search Phase. It uses a periodic low-energy broadcast of a beacon in order to discover a particle nearer to $\mathcal{S}$ than itself. Among the particles returned, p' selects a unique particle p'' that is "best" with respect to progress toward the sink. More specifically, the particle p''_E that among all particles found achieves the bigger progress on the $p'S$ line should be selected (see Figure 15.2).

Phase 2: The Direct Transmission Phase. Then, p' sends *info*($\mathcal{E}$) to p'' and sends a *success* message to p (i.e., to the particle from which it originally received the information).

Phase 3: The Transmission Range Variation Phase. If the *search phase* fails to discover a particle nearer to $\mathcal{S}$, p' enters the *transmission range variation phase*. More specifically, each particle maintains a local counter τ, with initial value $\tau = 0$. Every time the *search phase* fails, this counter is increased by 1. Thus τ is an indication of the number of failures to locate an active particle. Based on τ, the particle modifies its transmission range $\mathcal{R}$ according to a change function $\mathcal{F}(\tau)$. We here consider four different functions for varying the transmission range:

(a) *Constant Progress.* This choice is more suitable in the case where the network is comprised of a large number of particles and thus, a small increment of the transmission range will probably suffice to locate an active particle. Based on this assumption, the change function is defined as follows:

$\mathcal{F}(\tau) = \mathcal{R}_{\text{new}} = \mathcal{R}_{\text{init}} + c \cdot \tau,$ where c is a constant set to a small value (i.e., $c = 10$)

This is considered as the "basis" VTRP and is denoted as VTRP_c.

(b) *Multiplicative Progress.* In this case, the transmission range of the particle is increased mode drastically. We call this variation of our protocol VTRP_m.

$\mathcal{F}(\tau) = \mathcal{R}_{\text{new}} = \mathcal{R}_{\text{init}} + \mathcal{R}_{\text{init}} \cdot m \cdot \tau,$ where m is a constant set to a small value (i.e., $m = 3$)

This drastic change has bigger probability of finding an active particle; however, it leads to higher energy consumption.

(c) *Power Progress.* In this case, the transmission range of the particle is increased even faster using the following scheme:

$$\mathcal{F}(\tau) = \mathcal{R}_{\text{new}} = \mathcal{R}_{\text{init}} + \mathcal{R}_{\text{init}}^{\sqrt{(\tau+1)}}$$

We call this protocol VTRP_p.

(d) *Random Progress.* When the density of the network is not known in advance, we use randomization to avoid bad behavior due to the worst-case input distributions for each choice above (i.e., small modifications to the transmission range in VTRP_c in case of low densities and big modifications resulting from VTRP_p in high particle densities). We call this variation VTRP_r which is defined as follows:

$$\mathcal{F}(0) = \mathcal{R}_{\text{init}}$$
$$\mathcal{F}(\tau) = \mathcal{F}(\tau - 1) + \mathcal{R}_{\text{init}} \cdot r, \qquad \text{where } r \in (0, 8], \text{ a random value}$$

At any given time there could be more than one event being propagated toward the sink. In order to avoid repeated transmissions and infinite loops, each particle is provided with a limited "cache memory." In this cache, the particle registers the event IDs for each distinct event it has "heard of." Each event `ID`'s uniqueness is guaranteed, by choosing it to be a concatenation of the source particle `ID` and the timestamp of the sensed event. Upon the receival of a message, a particle checks whether the pertinent event is enlisted in its cache. If the event that it is not in the particle's cache, it is registered and then the particle proceeds to the proper actions defined by the VTRP protocol. However, if the event was already seen, the message is dropped and no further action is taken.

Presumably, a relatively small amount of memory (e.g., up to 2 MB) would be adequate for such purpose. Note that in the future the particle cache could enforce a policy of limited lifetime for each of its contents, thus reducing the space requirements to a minimum. Data aggregation also poses a challenge for further study and efficiency assessment.

15.6.5 Implementation Details

To implement the protocols presented in the previous sections, we have used simDust [16], which operates in `Linux` using `C++` and the `LEDA` [20] algorithmic and data structures library. An interesting feature for our simulator is its ability to experiment with very large networks of thousands of nodes. In fact, two major reasons for creating this simulator (a) the complexity of extending existing networks simulators (b) and their (in cases of large instances) time-consuming execution. simDust enables the protocol designer to implement the protocol using just `C++` and avoids complicate procedures that involve the use of more than one programming language. Additionally, simDust generates all the necessary statistics at the end of each simulation based on a wide variety of metrics that are implemented (such as delivery percentage, energy consumption, delivery delay, longevity, etc.)

The key points in simDust's implementation are the following:

Operation in Rounds. A basic concept used in the simulator is that its operation is divided into discrete rounds. One round represents a time interval in which a particle can transmit or receive a message and process it according to the protocol that is being simulated.

MAC Layer Assumptions. simDust leaves transmission collisions to be handled by lower MAC layer protocols and does not take them into account. It is our intention to consider them in future versions of this simulator.

Energy Assumptions. We have included a detailed energy dissipation scheme for both protocols implemented. In particular, we have assumed that a particle consumes a standard amount of energy E_{elec} per round while being awake. Furthermore, in each transmission, energy consumption is proportional to the square of the transmission distance. For each receive, a node is credited with an amount of energy that practically reflects the power needed to run the transceiver circuit, namely, E_{elec}. Finally, a particle can switch to the sleep state to save energy. No energy consumption virtually takes place while the particle remains in the sleep mode, since it keeps its transceiver and its sensors shut down.

Size of Messages: Regarding the communication cost in terms of the bits transmitted per message, we assume that information messages require 1 Kbyte, plus a 40-bits header, containing a 32-bit identifier for the sender particle and an 8-bit code that determines the message type.

15.6.6 Efficiency Measures

On each execution of the experiment, let K be the total number of crucial events $(\mathcal{E}_1, \mathcal{E}_2, \ldots, \mathcal{E}_K)$ and let k be the number of events that were *successfully* reported to the sink $\mathcal{S}$. Then, we define the *success rate* as follows:

Definition 15.6.1. The success rate, $\mathbf{P}_s$, is the fraction of the number of events *successfully* propagated to the sink over the *total number of events*, that is, $\mathbf{P}_s = k/K$.

Another crucial efficiency measure of our comparative evaluation of the two protocols is the average available energy of each particle in the network over time:

Definition 15.6.2. Let E_i be the available energy for the particle i. Then $E_{\text{tot}} = \sum_i^n E_i$ is the total energy available in the smart dust network, where n is the number of the total particles dropped. Note that E_i and E_{tot} vary with time.

Clearly, the less energy a protocol consumes, the better, but we have to notice that the comparison, in order to be fair, should be done in cases where the other parameters of efficiency should be similar (i.e., satisfy certain quality of service guarantees).

Finally, we consider as a measure of efficiency of the two protocols the number of alive particles, capturing the network survivability in each case. As in the case of the energy, the more particles that are alive, the better. A source of crucial information is also the particular manner in which particles die over time, such as (a) the geographical distribution of the nodes that die out earlier and (b) the evolution of energy consumption in critical sensors such as those lying close to control center.

Definition 15.6.3. Let h_A (for "alive") be the number of "alive" sensor particles participating in the sensor network.

15.6.7 Experimental Results

We start our experimentation by evaluating the effect of the particle density on the performance of the new protocol VTRP when compared to the already existing one, LTP. We generate a variety of sensor fields in a 2000-m by 2000-m square; in these fields, we drop $n \in [1000, 8000]$ particles uniformly distributed on the smart-dust plane. In each execution, *we generate a single event by randomly selecting a particle in the network*. The results of this experiments are shown in Figure 15.10.

It is evident that the effect of particle density has significant impact on the performance of LTP. We observe that for low densities (i.e., $n \leq 2000$) the protocol almost always fails to report the event to $\mathcal{S}$, while when $n \geq 5000$ the success rate increases approaching very fast one. This can be justified by taking into account the average degree of each particle for various network sizes n. Note that similar obser-

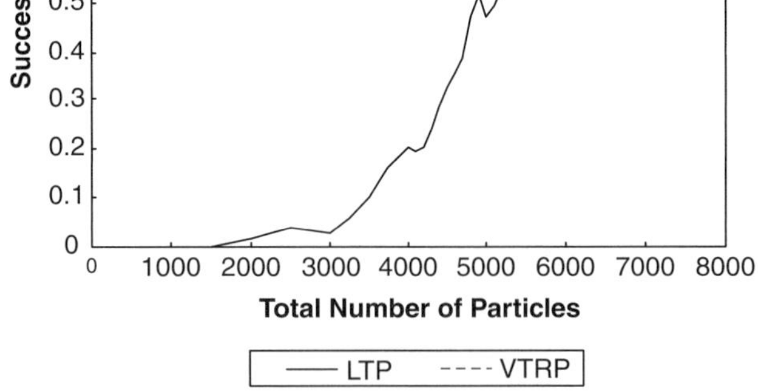

Figure 15.10. Success rate ($\mathbf{P}_s$) for LTP and VTRP for various particle densities ($n \in [1000, 8000]$).

vations for LTP have been made in reference 5. On the other hand, the mechanism of VTRP that increases the transmission range of the particles successfully overcomes these problems. Even for the cases of very low particle densities, VTRP manages to propagate the information reporting the realization of the event to the sink, with high probability.

We continue our experimentation by investigating the performance of the protocols in the case of multiple events. For this set of experiments we drop $n = 5000$ particles uniformly distributed in a 2000-m by 2000-m square field. Then in each simulation round, we generate one event at a random location in the sensor field that is sensed by only one particle (given that this particle has enough power to sense it); that is, we use a high event generation rate. This is repeated until a total of 9000 events are generated. Note that this is the first time that the LTP protocol is evaluated under the setting of multiple events.

Figure 15.11 depicts the success rate of the two protocols as the multiple events are generated. Clearly, VTRP achieves better results than LTP and in fact manages to propagate almost two times more events. The superiority of VTRP is explained by the fact that in LTP the particles that are closer to $\mathcal{S}$ will always participate in the propagation of the messages. The continuous transmissions of messages will eventually exhaust the power of this small group of (highly critical) particles, rendering the rest of the network useless (although there are still energy supplies available) since no further events can be reported to $\mathcal{S}$. VTRP overcomes this problem by activating the *Transmission Range Variation Phase*. As soon as the particles close to $\mathcal{S}$ "die," the neighboring nodes will sense it (during the *Search Phase*) and adjust their transmission range appropriately, thus bypassing them and reaching the sink directly.

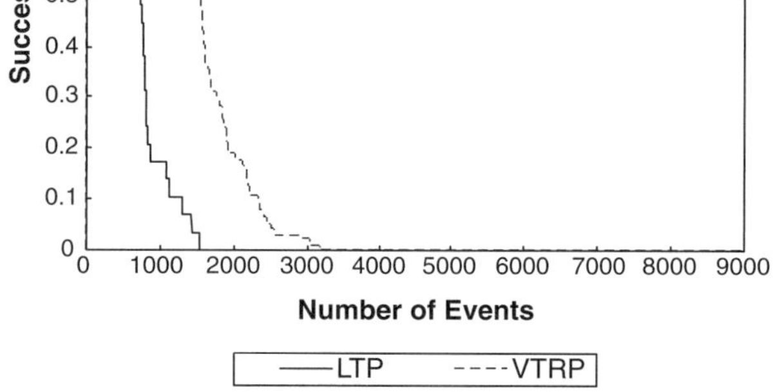

Figure 15.11. Success rate ($\mathbf{P}_s$) for LTP and VTRP for multiple events ($n = 5000$).

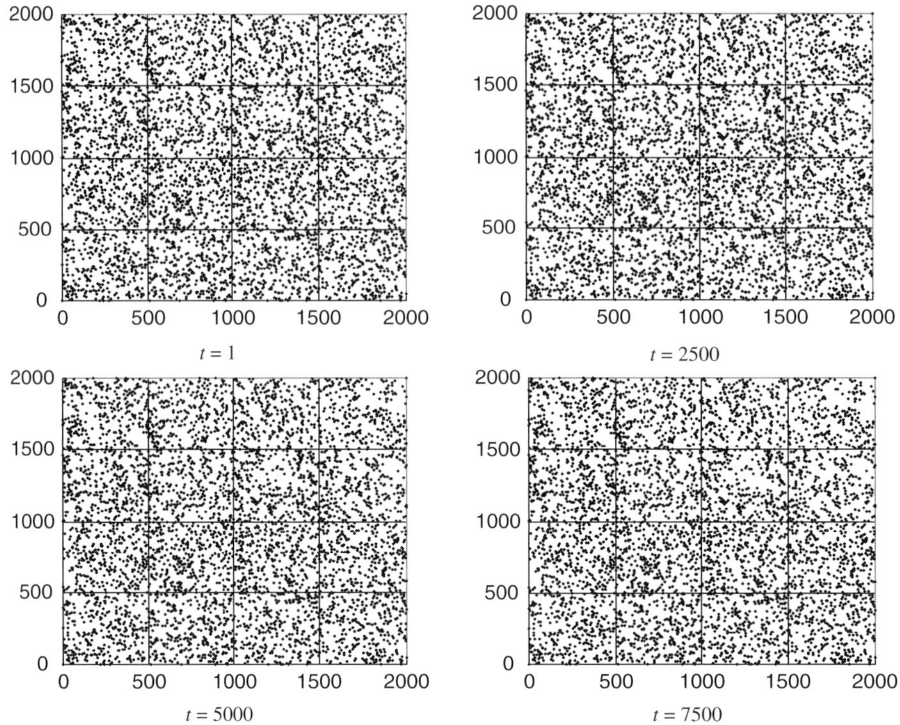

Figure 15.12. Snapshots of the network showing alive particles when executing LTP at different time instances ($n = 5000$).

This is clearly seen in Figure 15.12, where snapshots of the network are taken for different time instances. As soon as some particles around $\mathcal{S}$ "die," LTP fails to deliver the remaining events.

Essentially, VTRP manages the energy of the network in a more efficient way. By examining Figure 15.13, we observe that VTRP *ends up using slightly more energy than* LTP *in order to propagate more events to the control center*. In fact, VTRP will force the particles to spend more energy so that their transmissions manage to reach $\mathcal{S}$ even if this will exhaust their power supplies. Again, this is clearly seen in Figure 15.14, where snapshots of the network are taken for different time instances.

To get a more complete view on how each protocol manages the energy resources of the particles, Figures 15.15 and 15.16 show the number of alive particles based on their distance from the sink. In these figures we have grouped the particles in 32 sets based on the division of the diagonal line connecting (0, 0) with (2000, 2000) in 32 sectors. We observe that for different time instances, the total number of alive particles that are close to the sink (for sections 1–10) drops as the time increases while the particles further away almost always remain active until the end of the experiment. Observe how VTRP forces the particles close to $\mathcal{S}$ to sacrifice their battery supplies in order to propagate more messages.

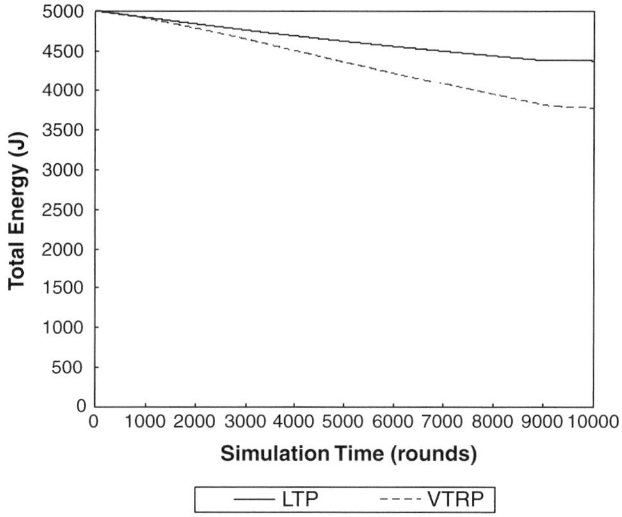

Figure 15.13. Total energy (E_{tot}) for LTP and VTRP for multiple events ($n = 5000$).

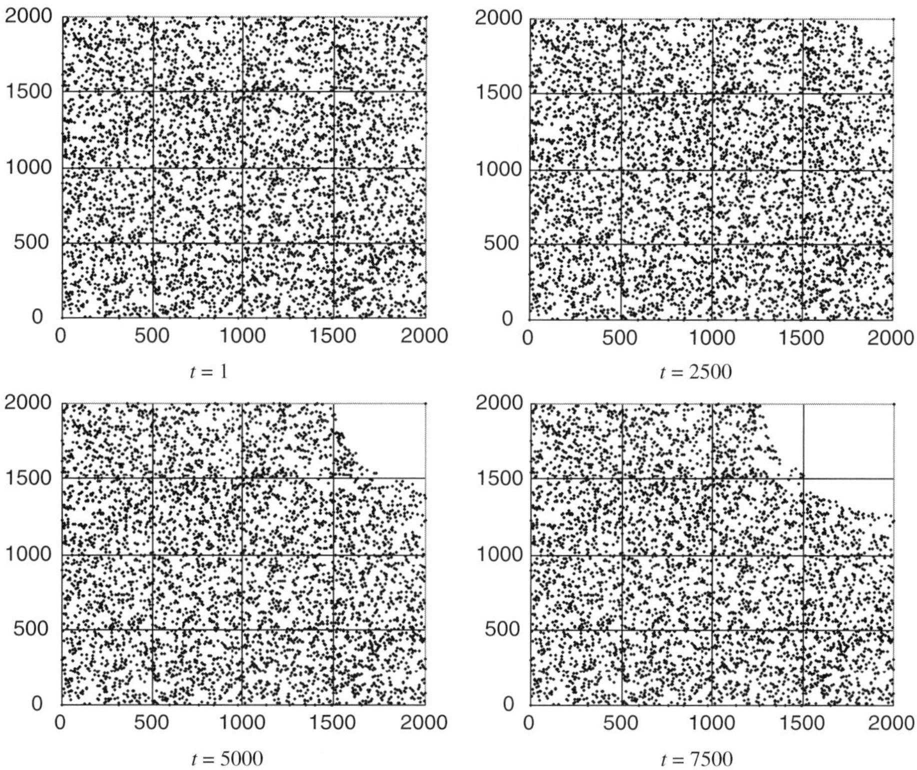

Figure 15.14. Snapshots of the network showing alive particles when executing VTRP at different time instances ($n = 5000$).

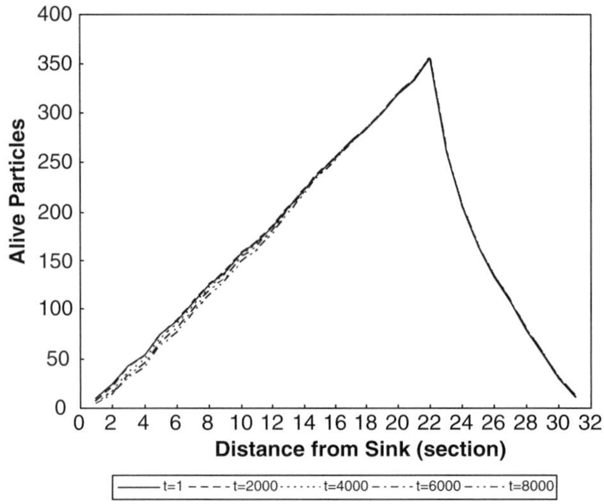

Figure 15.15. Alive particles (h_A) for LTP at different time instances ($n = 5000$).

In the last set of experiments we evaluate the performance of the four different functions for varying the transmission range of the particles when *phase 3* is activated. We use a similar setting as in the previous experiments; that is, the field size is 2000 m by 2000 m, and we deploy $n = 5000$ sensors and generate 9000 events. The result of this set of experiments are shown in Figures 15.17–15.22.

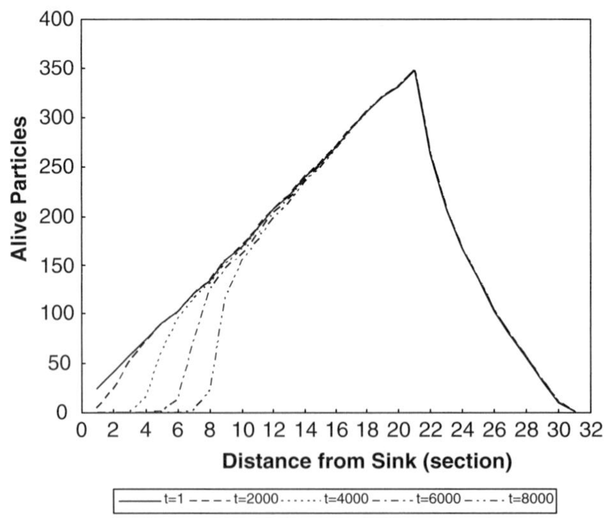

Figure 15.16. Alive particles (h_A) for VTRP at different time instances ($n = 5000$).

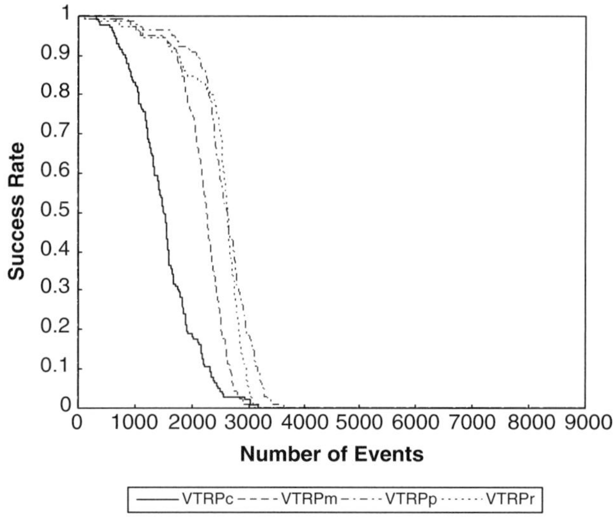

Figure 15.17. Success rate ($\mathbf{P}_s$) for the VTRP variations for multiple events ($n = 5000$).

The results indicate that the *constant progress* seems to be the least efficient function regarding the *success rate* metric (Figure 15.17), while for the other three functions the achieved success rate seems to be at similar levels. In fact, this is also the case for the *total energy* consumption (Figure 15.18). The *constant progress* function seems to be the most conservative; however; as in the case of LTP, it actually implies that VTRP$_C$ just fails to reach the sink.

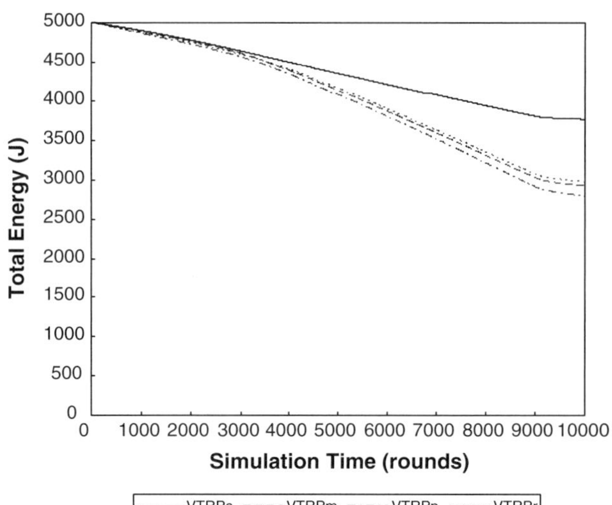

Figure 15.18. Total energy (E_{tot}) for the VTRP variations for multiple events ($n = 5000$).

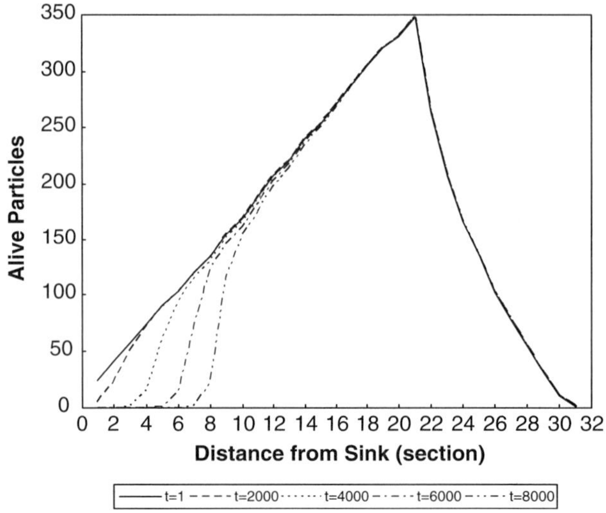

Figure 15.19. Alive particles (h_A) for VTRP$_c$ at different time instances ($n = 5000$).

A possible explanation to this behavior of VTRP$_C$ is the way the protocol modifies the transmission range by making small, constant steps. At the early stages of the network's operation, when only a small number of particles have "died," these small steps suffice to reach the sink. However, as the distance of the closest still-active

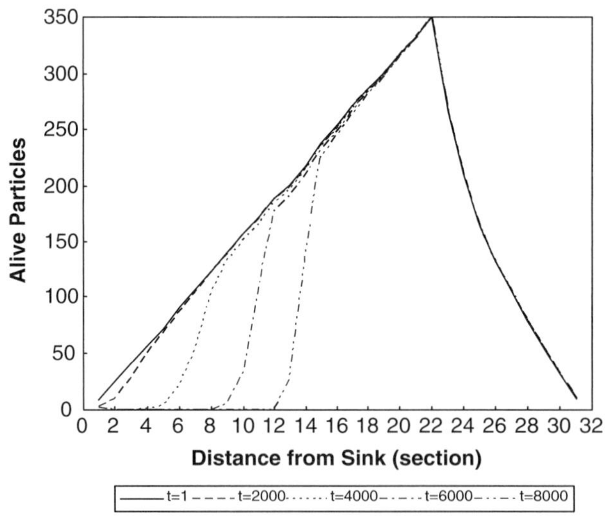

Figure 15.20. Alive particles (h_A) for VTRP$_m$ at different time instances ($n = 5000$).

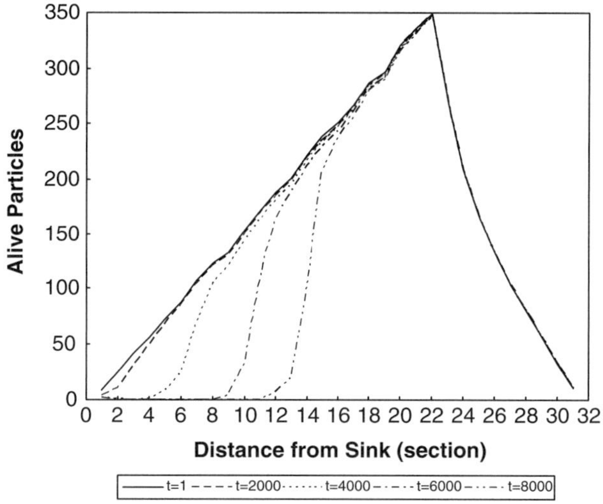

Figure 15.21. Alive particles (h_A) for VTRP$_p$ at different time instances ($n = 5000$).

particle to $\mathcal{S}$ increases (see Figure 15.14), the strategy of making small steps becomes inefficient. The series of small increments in the transmission range and failed searches waste the power sources of the particles and eventually cause the "death" of the particle before the information reaches the sink.

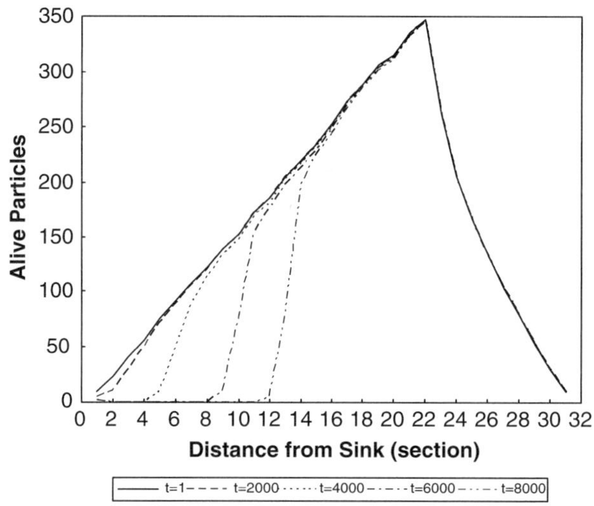

Figure 15.22. Alive particles (h_A) for VTRP$_r$ at different time instances ($n = 5000$).

15.7 SOME RECENT RELEVANT WORK

Nikoletseas et al. [21] propose *extended versions* of two data propagation protocols: the Sleep–Awake Probabilistic Forwarding Protocol (SW-PFR) and the Hierarchical Threshold-Sensitive Energy-Efficient Network Protocol (HTEEN). These nontrivial extensions aim at improving the performance of the original protocols by (a) introducing *sleep–awake periods* in the PFR protocol to save energy and (b) introducing a *hierarchy of clustering* in the TEEN protocol to better cope with large networks areas. The authors have *implemented* the two protocols and performed an *extensive experimental comparison* (using simulation) of various important measures of their performance with a focus on energy consumption. They investigate in detail the *relative advantages and disadvantages* of each protocol and discuss and explain their behavior. As a result of the above, they propose and discuss a possible *hybrid combination of the two protocols* toward optimizing certain goals.

Recently, Boukerche et al. [22] proposed a novel and efficient energy-aware distributed heuristic, which they refer to as EAD, to build a special rooted broadcast tree with many leaves that is used to facilitate data-centric routing in wireless microsensor networks. The EAD algorithm makes no assumption on local network topology and is based on residual power. It makes use of a *neighboring broadcast scheduling* and *distributed competition among neighboring nodess.*

EAD basically computes a tree with many leaves. With the transceivers of all leaf nodes being turned off, the network lifetime can be greatly extended. Boukerche et al. [22] implement an EAD scheme and present an extensive simulation experiments to study the its performance. The experimental results indicate clearly that the EAD scheme outperforms previous schemes, such as LEACH among other protocols.

ACKNOWLEDGMENTS

The work of Azzedine Boukerche has been supported by the Canada Research Chair Program, Canada Foundation Innovation Grant and Ontario Distinguished Researcher Award #201722.

The work of Sotiris Nikoletseas has been supported by the IST/FET programme of the European Union under contract numbers IST-1999-14186 (ALCOM-FT), IST-2001-33116 (FLAGS) and IST-2001-33135 (CRESCCO).

BIBLIOGRAPHY

1. I. F. Akyildiz, W. Su, Y. Sankarasubramaniam, and E. Cayirci. Wireless sensor networks: A survey. *Journal of Computer Networks*, **38**:393–422, 2002.
2. D. Estrin, R. Govindan, J. Heidemann, and S. Kumar. Next century challenges: Scalable coordination in sensor networks. In *Proceedings of the 5th ACM/IEEE International Conference on Mobile Computing—MOBICOM*, 1999.

3. J. M. Kahn, R. H. Katz, and K. S. J. Pister. Next century challenges: Mobile networking for smart dust. In *Proceedings of the 5th ACM/IEEE International Conference on Mobile Computing*, September 1999, pp. 271–278.
4. I. F. Akyildiz, W. Su, Y. Sankarasubramaniam, and E. Cayirci. A survey on sensor networks. *IEEE Communications Magazine*, pp. 102–114, August 2002.
5. I. Chatzigiannakis, S. Nikoletseas, and P. Spirakis. Smart dust protocols for local detection and propagation. Distinguished paper. In *Proceedings of the 2nd ACM Workshop on Principles of Mobile Computing*—POMC '2002, pp. 44–51.
6. P. Triantafilloy, N. Ntarmos, S. Nikoletseas, and P. Spirakis. NanoPeer networks and P2P worlds. In *Proceedings of the 3rd IEEE International Conference on Peer-to-Peer Computing*, 2003. IEEE CD-ROM.
7. S. M. Ross. *Stochastic Processes*, 2nd edition, John Wiley & Sons, New York, 1995.
8. I. Chatzigiannakis, T. Dimitriou, M. Mavronicolas, S. Nikoletseas, and P. Spirakis. A comparative study of protocols for efficient data propagation in smart dust networks. In *Proceedings of the International Conference on Parallel and Distributed Computing — EUPOPAR 2003*, pp. 436–440.
9. I. Chatzigiannakis, T. Dimitriou, S. Nikoletseas, and P. Spirakis. A probabilistic algorithm for efficient and robust data propagation in smart dust networks. In *Proceedings of the 5th European Wireless Conference on Mobile and Wireless Systems Beyond 3G (EW 2004)*, 2004.
10. C. Intanagonwiwat, R. Govindan, and D. Estrin. Directed diffusion: A scalable and robust communication paradigm for sensor networks. In *Proceedings of the 6th ACM/IEEE International Conference on Mobile Computing—MOBICOM*, 2000.
11. C. Intanagonwiwat, R. Govindan, D. Estrin, J. Heidemann, and F. Silva. Directed diffusion for wireless sensor networking. Extended version of reference [10].
12. L. Kleinrock. *Queueing Systems, Theory*, Vol. I, John Wiley & Sons, New York, 1975, p. 100.
13. W. R. Heinzelman, A. Chandrakasan, and H. Balakrishnan. Energy-efficient communication protocol for wireless microsensor networks. In *Proceedings of the 33rd Hawaii International Conference on System Sciences—HICSS*, 2000.
14. C. Schurgers, V. Tsiatsis, S. Ganeriwal, and M. Srivastava. Topology management for sensor networks: Exploiting latency and density. In *Proceedings of MOBICOM*, 2002.
15. H. Euthimiou, S. Nikoletseas, and J. Rolim. Energy balanced data propagation in wireless sensor networks. In *Proceedings of the 4th International Workshop on Algorithms for Wireless, Mobile, Ad-Hoc and Sensor Networks (WMAN '04)*, IPDPS, 2004.
16. T. Antoniou, A. Boukerche, I. Chatzigiannakis, G. Mylonas, and S. Nikoletseas. A new energy efficient and fault-tolerant protocol for data propagation in smart dust networks using varying transmission range. In *Proceedings of the 37th Annual ACM/IEEE Simulation Symposium (ANSS '04)*, 2004.
17. A. Manjeshwar and D. P. Agrawal. TEEN: A routing protocol for enhanced efficiency in wireless sensor networks. In *Proceedings of the 2nd International Workshop on Parallel and Distributed Computing Issues in Wireless Networks and Mobile Computing*, satellite workshop of *16th Annual International Parallel & Distributed Processing Symposium—IPDPS*, 2002.
18. B. Karp. Geographic routing for wireless networks. Ph.D. dissertation, Harvard University, Cambridge, MA, 2000.

19. S.E.A. Hollar. COTS dust. M.Sc. Thesis in Engineering, Department of Mechanical Engineering, University of California, Berkeley, 2000.

20. K. Mehlhorn and S. N'aher. *LEDA: A Platform for Combinatorial and Geometric Computing*, Cambridge University Press, New York, 1999.

21. S. Nikoletseas, I. Chatzigiannakis, H. Euthimiou, A. Kinalis, A. Antoniou, and G. Mylonas. Energy efficient protocols for sensing multiple events in smart dust networks. In *Proceedings of the 37th Annual ACM/IEEE Simulation Symposium* (*ANSS '04*), 2004.

22. A. Boukerche, X. Cheng, and J. Linus. Energy-aware data-centric routing in microsensor networks. In *Proceedings of ACM Modeling, Analysis and Simulation of Wireless and Mobile Systems*, September 2003, pp. 42–49.

23. I. Chatzigiannakis and S. Nikoletseas. A sleep–awake protocol for information propagation in smart dust networks. In *Proceedings of the 3rd Workshop on Mobile and Ad-Hoc Networks (WMAN)—IPDPS Workshops*, IEEE Press, New York, 2003, p. 225.

24. B. Chen, K. Jamieson, H. Balakrishnan, and R. Morris: SPAN: An energy efficient coordination algorithm for topology maintenance in ad-hoc wireless networks. In *Proceedings of the ACM/IEEE International Conference on Mobile Computing—MOBICOM*, 2001.

25. W. R. Heinzelman, J. Kulik, and H. Balakrishnan. Adaptive protocols for information dissemination in wireless sensor networks. In *Proceedings of the 5th ACM/IEEE International Conference on Mobile Computing—MOBICOM*, 1999.

26. C. Intanagonwiwat, D. Estrin, R. Govindan, and J. Heidemann. Impact of network density on data aggregation in wireless sensor networks. Technical Report 01-750, Computer Science Department, University of Southern California, November 2001.

27. μ-Adaptive Multi-domain Power Aware Sensors: http://www-mtl.mit.edu/research/icsystems/uamps, April, 2001.

28. C. E. Perkins: *Ad Hoc Networking*. Addison-Wesley, Boston, 2001.

29. TinyOS: A Component-Based OS for the Network Sensor Regime. http://webs.cs.berkeley.edu/tos/, October, 2002.

30. W. Ye, J. Heidemann, and D. Estrin. An energy-efficient MAC protocol for wireless sensor networks. In *Proceedings of the 12th IEEE International Conference on Computer Networks*—INFOCOM, 2002.

31. Wireless Integrated Sensor Networks: http:/www.janet.ucla.edu/WINS/, April, 2001.

32. Y. Xu, J. Heidemann, and D. Estrin. Geography-informed energy conservation for adhoc routing. In *Proceedings of the ACM/IEEE International Conference on Mobile Computing—MOBICOM*, 2001.

CHAPTER 16

Security Issues and Countermeasures in Wireless Sensor Networks

TANVEER ZIA and ALBERT Y. ZOMAYA

School of Information Technologies, The University of Sydney, Sydney, NSW 2006, Australia

16.1 INTRODUCTION

Sensor networks pose unique security challenges because of their inherent limitations in communication and computing. The deployment nature of sensor networks makes them more vulnerable to various attacks. Sensor networks are deployed in applications where they have physical interactions with the environment, people, and other objects, making them more vulnerable to security threats. We envision that sensor networks would be deployed in mission critical applications such as battlefield, security of key land marks, building and bridges, measuring traffic flow, habitat monitoring, and farming. Inherent limitations of sensor networks can be categorized as node and network limitations. The privacy and security issues in sensor networks raise rich research questions. Dense deployment of sensor networks in an unattended environment makes sensor nodes vulnerable to potential attacks. Attackers can capture the sensor nodes and compromise the network to accept malicious nodes as legitimate nodes. Once within the network, attackers can wage a variety of attacks. We expect Moore's law to be applied to drive down the cost of sensor nodes instead of improving its resources and performance.

Hardware and software improvements will address these issues to some extent, but complete secure sensor networks require deployment of countermeasures such as secure key management, secure routing, and lightweight encryption techniques. This chapter provides an overview of security issues known so far in sensor networks, along with the countermeasures against these issues followed by key management schemes and secure routing protocols. The key management and routing protocols

Algorithms and Protocols for Wireless Sensor Networks, Edited by Azzedine Boukerche

discussed in this chapter are the resilient solutions toward the attacks identified so far; however, research toward a complete secure sensor network is still in its infancy stages.

16.2 LIMITATIONS IN SENSOR NETWORKS

The following sections list the inherent limitations in sensor networks which make the design of security procedures more complicated.

16.2.1 Node Limitations

A typical sensor node processor is of 4–8 MHz, having 4 Kbyte of RAM, 128 Kbyte flash and ideally 916 MHz of radio frequency. Heterogeneous nature of sensor nodes is an additional limitation that prevents one security solution. Due to the deployment nature, sensor nodes would be deployed in environments where they would be highly prone to physical vandalism.

16.2.2 Network Limitations

Besides node limitations, sensor networks bring all the limitations of a mobile ad hoc network where they lack physical infrastructure, and they rely on insecure wireless media.

16.2.3 Physical Limitations

Sensor networks' deployment nature in public and hostile environments in many applications makes them highly vulnerable to capture and vandalism. Security sensor nodes with tamper-proof material increases the node cost.

16.3 SENSOR NETWORKS AND MANETS

Wireless sensor networks characteristics are common with Mobile Ad-hoc Networks (MANETs). Both have limitations in terms of memory, power, and computational capabilities, and they rely on wireless communication via radio frequency (RF). Both include data collection and data aggregation to make it available for processing. In both networks, some nodes are statically deployed while most have a high level of mobility after deployment. However, sensor networks differ from MANETs in many areas: Sensor networks are very densely deployed (thousands as compared to hundreds in MANETs), they are dynamic [1, 2] (i.e., they allow addition or deletion of nodes after deployment to extend the network or to eliminate failed nodes without physical contact), and they are highly susceptible to capture and manipulation by an adversary.

16.4 SECURITY IN SENSOR NETWORKS

Security goals in sensor networks depend on the need to know what we are going to protect. We determine four security goals in sensor networks: confidentiality, integrity, authentication, and availability (CIAA).

- *Confidentiality.* This is the ability to conceal a message from a passive attacker, where the message communicated on sensor networks remains confidential.
- *Integrity.* This is the ability to confirm that the message has not been tampered with, altered, or changed while it was on the network.
- *Authentication.* If the messages are from the node it claims to be from, we need to determine the reliability of the message's origin.
- *Availability.* If a node has the ability to use the resources, the network is available for the messages to move on.

16.4.1 Security Classes

Pfleeger [3] has identified four classes of security in computing systems. We integrate these four threat classes in sensor networks. In computing systems the major assets are hardware, software, and data. While in sensor networks, our goal is to protect the network itself, the nodes, and communication among the sensor nodes. There are four classes of threats which exploit the vulnerability of our security goals. Figure 16.1 shows these four threat classes:

1. In an *interruption*, a communication link in sensor networks becomes lost or unavailable. Examples of this sort of threat are node capture, message corruption, addition of malicious code, and so on.
2. An *interception* means a sensor net has been compromised by an adversary where the attacker gains unauthorized access to a sensor node or to data in it. An example of this type of attack is node capture.
3. *Modification* means that an unauthorized party not only accesses the data but tampers with it—for example, modifying the data packets being transmitted, causing a denial of service attack; flooding the network with bogus data; and so on.
4. In *fabrication*, an adversary adds false data, making the whole network unreliable.

16.4.2 Security Threats in Sensor Networks

Having built a foundation of security threats in computing, the following sections list the possible security threats in sensor networks identified by Undercoffer et al. [4].

Passive Information Gathering. An adversary with powerful resources collecting information from sensor networks if information is not encrypted.

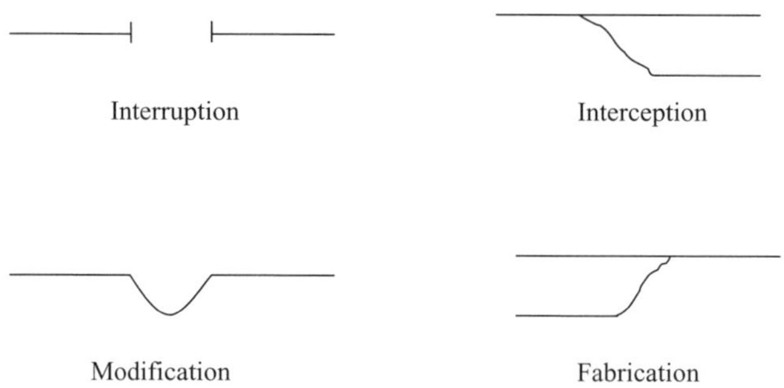

Figure 16.1. Pfleeger's four classes of systems security threats.

Node Subversion. Capture of a node may reveal its information including disclosure of cryptographic keys, hence compromising the whole sensor net.

False Node. Addition of a malicious node by an adversary to inject the malicious data, false node would be computationally robust to lure other nodes to send data to it.

Node Malfunction. A malfunctioning node will generate inaccurate data that would jeopardize the integrity of a sensor net, especially when that node is a data aggregating node—for example, a cluster leader.

Node Outage. What happens when a cluster leader stops functioning? Sensor net protocols should be robust enough to mitigate the effects of node outages by providing an alternate route.

Message Corruption. When contents of a message are modified by an attacker, it compromises the message integrity.

Traffic Analysis. Even if the message transfer is encrypted in sensor networks, its still leaves the high probability of analysis of communication patterns and sensor activities revealing enough information to enable an adversary to cause more malicious harm to sensor networks.

16.4.3 Security Attacks in Sensor Networks

Chris Karlof et al. [5] have presented detailed attacks in sensor networks which are described in the following section. Table 16.1 lists these attacks.

TABLE 16.1. Security Attacks in Wireless Sensor Networks

Spoofed, altered, or replayed routing information	Create routing loop, attract or repel network traffic, extend or shorten source routes, generate false error messages, etc.
Selective forwarding	Either in-path or beneath path by deliberate jamming, allows to control which information is forwarded. A malicious node acts like a black hole and refuses to forward every packet it receives.
Sinkhole attacks	Attracting traffic to a specific node—for example, to prepare selective forwarding.
Sybil attacks	A single node presents multiple identities, allows us to reduce the effectiveness of fault-tolerant schemes such as distributed storage and multipath.
Wormhole attacks	Tunneling of messages over alternative low-latency links to confuse the routing protocol, thereby creating sinkholes, etc.
Hello floods	An attacker sends or replays a routing protocols hello packets with more energy.

Routing Loops. In sensor networks, routing loops attack the information exchanged between nodes. False error messages are generated when an attacker alters and replays the routing information. Routing loops attract or repel the network traffic and increase node-to-node latency.

Selective Forwarding. Selective forwarding is a way to influence the network traffic by believing that all the participating nodes in network are reliable to forward the message. In selective forwarding attack, malicious nodes simply drop certain messages instead of forwarding every message. Once a malicious node cherry picks on the messages, it reduces the latency and deceives the neighboring nodes that they are on a shorter route. Effectiveness of this attack depends on two factors. First is the location of the malicious node. the closer it is to the base station, the more traffic it will attract. Second is the percentage of messages it drops. When a selective forwarder drops more messages and forwards less, it retains its energy level, thus remaining powerful to trick the neighboring nodes.

Sinkhole Attacks. In sinkhole attacks, an adversary attracts the traffic to a compromised node. The simplest way of creating a sinkhole is to place a malicious node where it can attract most of the traffic, possibly closer to the base station or malicious node itself, which deceptively acts as a base station. One reason for sinkhole attacks is to make selective forwarding possible to attract the traffic toward a compromised node. The nature of sensor networks where all the traffic flows toward one base station makes this type of attack more susceptible.

Sybil Attacks. This is, a type of attack where a node creates multiple illegitimate identities in sensor networks either by fabricating or stealing the identities of legitimate nodes. Sybil attacks can be used against routing algorithms and topology maintenance; it reduces the effectiveness of fault-tolerant schemes such as distributed storage and dispersity. Another malicious factor is geographic routing where a Sybil node can appear at more than one place simultaneously.

Wormholes. In wormhole attacks, an adversary positioned closer to the base station can completely disrupt the traffic by tunneling messages over a low-latency link. Here an adversary convinces the nodes which are multihop away that they are closer to the base station. This creates a sinkhole because an adversary on the other side of the sinkhole provides a better route to the base station.

Hello Flood Attacks. This involves broadcasting a message with stronger transmission power and pretending that the HELLO message is coming from the base station. Message receiving nodes assume that the HELLO message sending node is the closest one and they try to send all their messages through this node. In this type of attack, all nodes will be responding to HELLO floods and wasting the energies. The real base station will also be broadcasting the similar messages but will have only a few nodes responding to it.

DoS Attacks. Denial of service attacks occur at a physical level, causing radio jamming, interfering with the network protocol, battery exhaustion, and so on.

16.4.4 Layering-Based Security Approach

Application Layer. Data are collected and managed at application layer therefore it is important to ensure the reliability of data. Wagner [6] has presented a resilient aggregation scheme that is applicable to a cluster-based network where a cluster leader acts as an aggregator in sensor networks. However, this technique is applicable if the aggregating node is in the range with all the source nodes and there is no intervening aggregator between the aggregator and source nodes. In the hierarchical clustering approach, a communication channel between the aggregator and base station has potentially limited bandwidth because the cluster leader as an aggregator itself is a sensor node [6, 7]. To prove the validity of the aggregation, cluster leaders use the cryptographic techniques to ensure the data reliability. We will discuss the cryptography in key management section.

Network Layer. The network layer is responsible for routing of messages from node to node, node to cluster leader, cluster leaders to cluster leaders, cluster leaders to the base station, and vice versa.

Routing protocols in sensor networks are of two types: (1) ID-based protocols, in which packets are routed to the destination based on the IDs specified in the packets, and (2) data-centric protocols [8] in which packets contain attributes that specify the type of data being provided. Law and Havinga [7] have described Karlof and Wagner [5] routing attacks in sensor networks as below:

1. Packets are dropped completely, or selectively.
2. The network is flooded with global broadcasts.
3. Some sensor nodes in the network are misguided into believing that nodes are either multiple hops away or do not exist at all in the neighbors.
4. A significant proportion of the traffic is tunneled from one place in the network to another distant place of the network depriving other parts of the network that under normal circumstances would have received the traffic themselves.
5. Sometimes traffic is lured to a particular node or a small group of nodes, depriving other parts of the network that normally would have received the traffic themselves.

Security of routing protocols depends on the location of nodes and the encryption techniques.

Data Link Layer. The data link layer does the error detection and correction, as well as encoding of data. The link layer is vulnerable to jamming and DoS attacks. TinySec [9] has introduced link layer encryption, which depends on a key management scheme. However, an attacker having better energy efficiency can still wage an attack. Protocols like LMAC [10] have better anti-jamming properties, which are viable countermeasures at this layer.

Physical Layer. The physical layer emphasizes the transmission media between sending and receiving nodes; the data rate, signal strength, and frequency types are also addressed in this layer. Ideally, the FHSS frequency hopping spread spectrum is used in sensor networks.

Table 16.2 summarizes the attacks and countermeasures in a layering model in sensor networks.

TABLE 16.2. Layering Approach in Sensor Network Attacks and Countermeasures

	Attack Types	Countermeasures
Application layer	Subversion and malicious nodes	Malicious node detection and isolation
Network layer	Wormholes, sinkholes, Sybil, routing loops	Key management, secure routing
Data link layer	Link layer jamming	Link layer encryption
Physical layer	DoS attacks, Radio jamming, node capture	Adaptive antennas, spread spectrum

16.5 CRYPTOGRAPHY IN SENSOR NETWORKS

To achieve security in any communication model, it is important to encrypt messages among the sensor nodes on an agreed key management scheme. In traditional networks we use the complex key management schemes such as Public and Diffie–Hellman [11] keying schemes. However, providing a secure key management scheme in sensor networks is difficult due to ad hoc, dynamic topology and resource limitations.

The general key distribution problem refers to the task of distributing secret keys between communicating parties to provide security properties such as secrecy and authentication.

In sensor networks, key distribution is usually combined with initial communication establishment to bootstrap a secure communication infrastructure from a collection of deployed sensor nodes.

To bootstrap the security in sensor networks, nodes must be able to establish a secure node-to-node communication such as the following:

- Additional legitimate nodes deployed at a later time should be able to form secure connections with already-deployed nodes.
- Unauthorized nodes should not be able to gain entry into the network, either through packet injection or masquerading as legitimate node.
- The scheme must work without prior knowledge of which nodes will come into the communication range of each other after deployment.
- The computational and storage requirement of the key management scheme must be low, and the scheme should be robust to DoS attacks from out-of-network source.

Figure16.2 shows a keying process to establish secure communication between node A and node B.

16.5.1 Asymmetric Cryptography

Asymmetric cryptography, also known as public key, is the most commonly used keying method in computing. However, the following three factors make public keying not feasible for sensor networks:

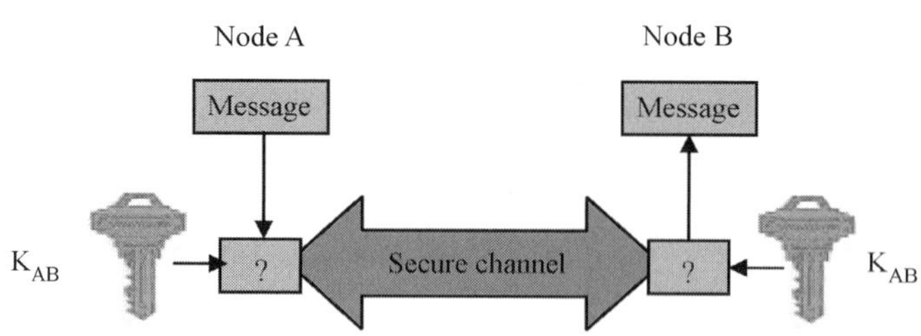

Figure 16.2. Secure channel between nodes A and B.

First, asymmetric cryptography requires involves extensive hardware and software and mathematical functions that are beyond the processing power of tiny sensor nodes. In order to deploy asymmetric cryptography on sensor nodes, it is necessary to have dedicated cryptographic hardware that would either increase the nodes cost or embed mathematical functions in software, which would be much slower than using symmetric keying. *Second*, asymmetric cryptography involves intensive computation that may take up to minutes of processing for sensor node to complete one signature generation, making nodes vulnerable to battery exhaustion and denial of service attacks where they are continuously requested to generate signatures. *Third*, there is no resistance against node captures.

16.5.2 Symmetric Cryptography

In symmetric cryptography, every node in a sensor network shares a unique symmetric key with every other node in the network. According to Chan and Perrig [12], using symmetric cryptography will have n nodes in a sensor network. Every node stores $n - 1$ keys, one of each of the other nodes in the network. After deployment, nodes must perform key discovery to verify identity of the node that they are communicating with. Symmetric keying in sensor networks has two benefits. *First*, any node captured reveals no information about the communication, providing stronger resilience against node capture attack. *Second*, due to the broadcast nature of sensor networks, compromised keys can be revoked. For example, a captured node's entire $n - 1$ keys can be broadcasted, and a node hearing this broadcast will stop using the keys.

16.6 KEY MANAGEMENT SCHEMES

This section discusses various key management schemes introduced in sensor networks, along with the secure triple key management scheme we have introduced.

16.6.1 Master-Key-Based Key Predistribution Scheme

A Master-key-based predistribution scheme is based on a single master key that is predeployed in all the sensor nodes. Each sensor node uses part of the memory to store the master keys. A pair of sensor nodes exchanges random nonce values. The pair uses a master key to establish the session keys. Compromise of master key makes the whole sensor net insecure. Dutertre et al. [13] have presented a lightweight key management system where they have proposed more than one master key on an assumption that sensor networks are deployed in various phases. Each sensor node stores a group authentication key *bk1* and a key generation key *bk2*. If two sensors *S1* and *S2* are from the same phase, they authenticate each other by using the authentication keys *bk1*. They exchange random nonce values *RN1* and *RN2* and establish the session key. If the nodes are from two different phases, a sensor node from an old phase stores a random nonce and a secret key for each new phase and authenticates itself with that secret key. Once authenticated, both nodes generate a pairwise key to establish the secure communication.

16.6.2 Basic Random Key Pre-distribution Scheme

A probabilistic key pre-distribution scheme is presented by Eschenauer and Gligor [1] where each sensor node receives a random subset of keys from a large key pool before deployment. To agree on a key for communication, two nodes find one common key within their subsets and use that key as their shared key.

The random key pre-distribution scheme denotes m as the number of distinct cryptographic keys that can be stored in the key ring of a sensor node. In this scheme, sensor nodes pick a random pool of keys Q out of the total possible key space in an initialization phase. Nodes store m number of keys randomly selected from the pool Q. This set of m keys is called a node's *key ring*. The number of keys in the key pool Q is chosen such that two random subsets of m in Q will share at least one key with some probability p. In this scheme all nodes use the same key pool Q. That means security of the network would be weak if keys from pool Q are compromised. Basic random key pre-distribution contains five offline steps:

i. Generation of a large pool of P keys and their key identifiers
ii. Random drawing of k keys out of P without replacement to establish the key ring of a sensor
iii. Loading of the key ring into the memory of each sensor
iv. Saving of the key identifiers of a key ring and associated sensor identifier on a trusted controller node
v. For each node, loading the ith controller node with the key shared with that node

The key pre-distribution phase ensures that only a small number of keys need to be placed on each sensor node's *key ring*, ensuring that any two nodes share at least one key with a chosen probability.

In the *shared-key discovery* phase, every node discovers its neighbors in wireless communication range with which it shares the key. Each node broadcasts in plain text the list of identifiers of the keys on their *key ring*. Shared-key discovery takes place during sensor network initialization, and it establishes the topology of the sensor array by the routing layer of the sensor network. A link exists between two sensor nodes only if they share a key; and if a link exists between two nodes, all communication on that link is secured by link encryption.

During the *path-key establishment* phase a path key is assigned to a set of selected pairs of sensor nodes in wireless communication range that do not share a key but are connected by two or more links at the end of the shared-key discovery phase.

Revocation is used to eliminate the *key ring* of nodes which have been compromised. To execute revocation, a controller node (which is comparatively more powerful and mobile in terms of range) broadcasts a single revocation message containing a signed list of K key identifiers for the *key ring* to be revoked. To sign the list of key identifiers, the controller generates a signature key K_e and unicasts it to each node by encrypting it with a key K^{ci}. After obtaining the signature key, each node verifies the signature of the signed list of key identifiers, locates those identifiers in its key ring, and removes the corresponding keys. Once the keys are removed from key rings, some links may disappear and the affected nodes need to reconfigure those links by restarting the *shared-key discovery* and possibly *path-key establishment*.

Re-keying is described as self-revocation of a key by a node if its life is expired. After removal of the expired key, affected nodes restart the shared-key discovery and path-key establishment process.

Resiliency to Sensor Node Capture. There are two levels of threats posed by the node capture. The first level of threat is manipulation of sensor's data when an adversary injects the fake data in sensor network. Detecting this nature of attack requires offline data correlation analysis and data anomaly detection by collection and processing nodes. The second level of threat occurs when physically a sensor node is in control of the adversary. This level of active threat includes data manipulation of captured node and other nodes on the network. Countermeasures to node capture threats are tamper-detection technologies [14] to shield the sensors in such a way that vandalism of nodes causes the erasure of a sensor's key ring, eventually disabling the sensor's operation.

16.6.3 Extended Random Key Pre-distribution Scheme

Chan et al. [15] extended the idea of Eschenauer and Gligor [1] and developed three key pre-distribution schemes; *q-composite, multipath reinforcement*, and *random-pairwise* keys schemes to over come the weakness of basic random key pre-distribution scheme.

In the q-composite scheme, q common keys ($q > 1$) are needed, instead of just one. In this scheme the key pool size S is reduced and multiple keys are used to establish communications instead of just one.

To compute the key pool size, let $p(i)$ be the probability that any two nodes have exactly i keys in common. Let P_{connect} be the probability of any two nodes sharing sufficient keys from a secure connection. $P_{\text{connect}} = 1-$(probability that the two nodes share insufficient keys to form a connection), hence $P_{\text{connect}} = 1 - (p(0) + p(1) + \ldots + p(q-1))$.

For a given key ring size m, minimum key overlap q, and minimum connection probability p, the largest $|S|$ is chosen such that $P_{\text{connect}} \geq p$.

To initialize the key setup, a set of S of random keys out of the total key space is picked. For each node, m random keys are selected from S (where m is the number of keys each node can carry in its key ring) and store them into the node's key ring. In the key setup phase, each node discovers all common keys it possesses with each of its neighbors.

q-composite key scheme strengthens the network resilience against node capture when the number of nodes captured is small. However, if the large number of nodes have been captured, the q-composite keys scheme tends to reveal a larger fraction of network to the adversary.

In the *multipath* key reinforcement scheme, Chan et al. [15] present a method to strengthen the security of an established link key by establishing the link key through multiple paths. Assuming that key setup has been completed, there are now many secure links formed through the common keys in the various nodes key rings. Suppose node A has a secure link to node B after key setup. This link is secured using a single key k from the key pool S. k may be residing in the key ring memory of some other nodes elsewhere in the network. If any of those nodes are captured, the security of the link between nodes A and B is jeopardized. To address this, the communication key is updated to a random value after key setup. However, key update cannot be coordinated using the direct link between nodes A and B since if the adversary has been recording all key-setup traffic, it could decrypt the key update message after it obtained k and still obtain the new communication key. In this approach, key update is coordinated over multiple independent paths. Assume that enough routing information can be exchanged such that node A knows all disjoint paths to node B created during initial key setup that are h hops or less. The more paths between two nodes A and B, the more security multipath key reinforcement provides from the link between A and B. However, for any given path, the probability that the adversary can eavesdrop on the path increases with the length of the path since if any one link on the path is insecure, then the entire path is made insecure. Furthermore, it is increasingly expensive in terms of communication overhead to find multiple disjoint paths that are very long. To address this issue, a 2-hop multipath key reinforcement scheme is presented where paths of only 2 links are considered. The effectiveness of 2-hop multipath key reinforcement is evaluated by simulating the random deployment of 10,000 sensor nodes on a square planar field. Successfully implementing *multipath* key reinforcement on the key management scheme of Eschenauer and Gligor [1] enables it to outperform the *q-composite* scheme for $q \geq 2$ even when the q-composite scheme is supplemented by key reinforcement. However, compounding both schemes compounds their weaknesses: The smaller key pool size of the *q-composite* keys scheme undermines the effectiveness of multipath key reinforcement by making it easier to build up a critically larger collection of keys.

The third mechanism introduced in this scheme is a *random pairwise scheme* where node-to-node authentication is established. The random pairwise scheme introduces the following properties:

- Resilience against node capture
- Node-to-node identity authentication
- Distributed node revocation
- Resistence to node replication and generation
- Comparable scalability

In random key pool distribution schemes like *q-composite* and *multipath*, keys can be issued multiple times out of the key pool, and the node-to-node authentication is not possible. *Pairwise key distribution* assigns a unique key to each pair of nodes.

Recall that the size of each node's key rings is m key, and the probability of any two nodes being able to communicate securely is p. The random pairwise keys scheme proceeds as follows:

i. In the pre-deployment initialization phase, a total of ($n = m/p$) unique node identities are generated. The actual size of the network may be smaller than n. Unused node identities will be used if additional nodes are added to the network in the future. Each node identity is matched up with m other randomly selected distinct node IDs, and a pairwise key is generated for each pair of nodes. The key is stored in both nodes key rings, along with the ID of the other node that also knows the key.

ii. In the post-deployment key-setup phase, each node first broadcasts its node ID to its intermediate neighbors. By searching for each other's IDs in their key rings, the neighboring nodes can tell if they share a common pairwise key for communication.

16.6.4 Multiple Space Key Pre-distribution Scheme

Du et al. [16, 17] propose a key pre-distribution scheme that is based on Blom [18] key pre-distribution method introduced in 1985. Blom's key pre-distribution method was not for sensor networks; however, it allows any pair of nodes in a network to be able to find a pairwise secret key. As long as no more than X nodes are compromised, the network is perfectly secure (this is called the *X-secure* property).

In this scheme, the concept is taken from graph theory where we draw an edge between two nodes if and only if they can find a secret key between themselves. The hypothesis here is that by requiring the graph to be only connected, each sensor node needs to carry less key information.

During the key pre-distribution phase, key information is assigned to each node, such that after deployment, neighboring sensor nodes can find a secret key between them. In Du et al. [16] the key pre-distribution phase generates G and D matrices followed by selection of a *key space*; then in the key agreement phase, after deployment, each node discovers whether it shares any key space with its neighbors. To achieve this, each node broadcasts a message containing the node's ID, the indices of key spaces it carries, and the seed of the column of G it carries. In the previous two schemes, an adversary only needs to compromise less than 100 nodes in order to compromise 10% of the rest of the secure links, whereas in this scheme the adversary needs to compromise 500 nodes, thus lowering the payoff to the adversary of smaller-scale network breaches.

16.6.5 Key Management Scheme Using Deployment Knowledge

Du et al. [19] make an extension on the key management scheme developed by Eschenauer and Gligor [1]. Their scheme describes the deployment knowledge where sensors are prearranged in a sequence before deployment. Once dropped from a plane or helicopter, sensors have better probability to recognize the prearranged sequence of neighbor sensors. This prior deployment knowledge is useful for key pre-distribution. When neighbor sensors are known, key pre-distribution becomes trivial; for each node n we just need to generate a pairwise key between n and each of its neighboring nodes, and we save these keys in n's memory. This guarantees that each node can establish a secure channel with each of its neighbors after deployment.

Deployment knowledge in this scheme is modeled using *probability density functions (pdf)*. When the *pdf* is uniform, no information can be gained regarding where a node is more likely to reside. Du et al. [19] focus on nonuniform *pdf* functions. Since the distribution is different from uniform distribution, we can assume that a sensor is more likely to be deployed in certain areas.

16.6.6 Trusted Base Station as Key Distribution Center

A base station is used as a trusted third party to distribute the keys to provide link keys to sensor nodes such as Kerberos [20]. In this keying mechanism a base station becomes the single point of vulnerability if compromised. Due to extensive communication with the base station, nodes closer to the base station will lose their battery early. This approach assumes some level of reliable communication mode between the node and the base station before any key scheme is deployed.

16.6.7 A Secure Triple Key Management Scheme

Our secure triple-key management scheme [21] consists of three keys: two pre-deployed keys in all nodes and one in-network generated cluster key for a cluster to address the hierarchical nature of sensor network.

Otherwise, the packet is dropped by the cluster leader. The node builds the message using the fields below:

K_n	MAC	ID	TS	S	message	Level 2

Cluster Leader to Next-Hop Cluster Leader Key Calculation. A cluster leader aggregates the messages received from its nodes and forwards it to a next-level cluster leader; or if the cluster leader is one hop closer to the base station, it directly sends the message to the base station. The receiving cluster leader checks its routing table and constructs the following packet to be sent to the next-level cluster leader or base station. The cluster leader adds its own ID_{CLn}, its network, and its cluster key in incoming packet and rebuilds the packet as follows:

$$\{\mathrm{ID_{CLn}}, K_n, [\mathrm{ID_{sn}}, \mathrm{Kn}, \mathrm{TS}, \mathrm{MAC}, \mathrm{S(Aggr\ message)}]\}$$

K_n	MAC	ID	TS	S	Aggr message	Level 1

Here ID is the ID of the receiving cluster leader which wraps the message and sends it to the next-hop cluster leader or to the base station if directly connected. The next-hop cluster leader receives the packet and checks the ID. If the ID embedded in the packet is the same as it holds, it updates the ID for the next hop and broadcasts it; otherwise the packet is discarded. *Aggr message* refers to the message aggregated by the cluster leader.

Cluster Leader to Base Station Key Calculation. A base station receives the packet from its directly connected cluster leader; it checks the ID of the sending cluster leader, and it verifies the authentication and integrity of the packet through MAC. The cluster leader directly connected with base station adds its own ID along with the packet received from the sending cluster leader. The packet contains the following fields:

$$\{ID_{CL2}[ID_{CL4}, K_n, [ID_{s10}, Kn, TS, MAC, S\,(Aggr\ message)]]\}$$

Figure 16.3 illustrate the hierarchal structure of our secure triple-key management scheme where we have shown sensor nodes *S1..S11*, cluster leaders *CL1..CL4,* and base station communication using the triple-key management scheme.

K_n *(Network Key).* This is generated by the base station, pre-deployed in each sensor node, and shared by the entire sensor network. Nodes use this key to encrypt the data and pass onto next hop.

K_s *(Sensor Key).* This is generated by the base station, pre-deployed in each sensor node, and shared by the entire sensor network. A base station uses this key to decrypt and process the data, and a cluster leader uses this key to decrypt the data and send it to base station.

K_c *(Cluster Key).* This is generated by the cluster leader, and it is shared by the nodes in that particular cluster. Nodes from a cluster use this key to decrypt the data and forward it to the cluster leader.

The secure triple-key management scheme is a much resilient solution against many of sensor networks attacks. The next section describes how triple keys are used to send messages from base station to node, nodes to cluster leader, cluster to cluster leader and base station.

Base Station to Node Key Calculation. The base station uses K_n to encrypt and broadcast data. When a sensor node receives the message, it decrypts it by using its K_s. In Figure 16.3, the base station uses $K_{n1..nn}$ to broadcast the message. This

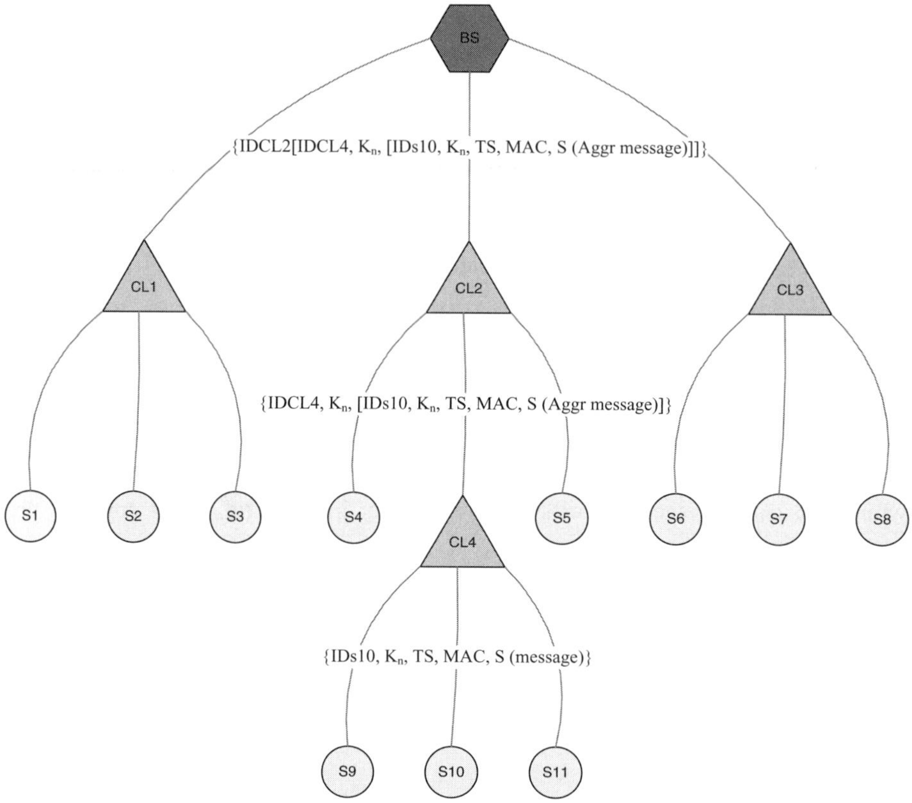

Figure 16.3. A secure triple-key management.

process is as follows: The base station encrypts its own ID, a current timestamp TS, and its K_n as a private key.

The base station generates a random seed S and assumes itself at level 0. The packet contains the following fields:

K_n	MAC	ID	TS	S	message	Level 0

The sensor node decrypts the message received from the base station using K_s. MAC is a message authentication code for a message m.

Nodes to Cluster Leader Key Calculation. When a node sends a message to cluster leader, it constructs the message as follows

$$\{\text{ID}_{sn},\ K_s,\ \text{TS},\ \text{MAC},\ \text{S}\,(\text{message})\}$$

The cluster leader checks the ID from the packet, checks if the ID in the packet matches the ID it holds, and verifies the authentication and integrity of the packet through MAC.

The cluster key K_c is used for message decryption with in a cluster where a cluster leader needs to decrypt the message sent by a node and prepare that message to forward the next hop.

This secure triple-key management scheme provides added resilience toward susceptible attacks on sensor networks by keeping in mind the resource-starved nature of sensor nodes.

16.7 SECURE ROUTING

Secure routing in sensor networks is challenging due to the unique characteristics sensor networks have compared to wired and wireless ad hoc networks. Traditional IP-based routing is not a viable solution due to a relatively large number of sensor nodes because the overhead of IP maintenance is very high. Following are the issues that need to be kept in mind while designing a secure routing protocol in sensor networks.

1. Sensor nodes are self-organizing due to the ad hoc deployment, and nodes are left unattended after the deployment.
2. In sensor networks, most of the time flow of data would be from nodes to cluster leader and base station.
3. Careful route management due to nodes limitations.
4. Frequent changes in network topology due to the dynamic nature of sensor networks.
5. Sensor networks are application-specific and data-centric.
6. Secure location of sensor nodes because Global Posting Systems (GPS) are not suitable for sensor networks.

There are very few routing protocols proposed to address the secure routing issues in sensor networks. In the following sections we present some of these solutions discussed by Saraogi [22].

16.7.1 SPINS: Security Protocols for Sensor Networks

SPINS is a suite of security building blocks proposed by Perig et al. [23]. It is optimized for resource constrained environments and wireless communication. SPINS has two secure building blocks: SNEP and μTESLA (the micro version of TESLA). SNEP provides data confidentiality, two-party data authentication, and data freshness. μTESLA provides authenticated broadcast for severely resource-constrained environments. All cryptographic primitives (i.e., encryption, message authentication code (MAC), hash, random number generator) are constructed out of a single block cipher for code reuse. This, along with the symmetric cryptographic primitives used, reduces the overhead on the resource constrained sensor network. In a broadcast medium such as a sensor network, data authentication through a symmetric mechanism cannot be applied because all the receivers know the key. μTESLA constructs authenticated broadcast from symmetric primitives, but introduces asymmetry with delayed key disclosure and one-way function key chains.

SNEP. SNEP uses encryption to achieve confidentiality and uses *message authentication code* (MAC) to achieve two-party authentication and data integrity. Apart from confidentiality, another important security property is semantic security, which ensures that an eavesdropper has no information about the plaintext, even if it sees multiple encryptions of the same plaintext [24]. The basic technique to achieve this is randomization: Before encrypting the message with a chaining encryption function (i.e., DESCBC), the sender precedes the message with a random bit string (also called the *initialization vector*). This prevents the attacker from inferring the plaintext of encrypted messages if it knows plaintext–ciphertext pairs encrypted with the same key. To avoid adding the additional transmission overhead of these extra bits, SNEP uses a shared counter between the sender and the receiver for the block cipher in counter mode (CTR). The communicating parties share the counter and increment it after each block. SNEP offers the following unique features:

Semantic Security. Since the counter value is incremented after each message, the same message is encrypted differently each time. The counter value is long enough that it never repeats within the lifetime of the node.

Data Authentication. If the MAC verifies correctly, a receiver can be assured that the message originated from the claimed sender.

Replay Protection. The counter value in the MAC prevents replaying old messages. Note that if the counter were not present in the MAC, an adversary could easily replay messages.

Data Freshness. If the message verified correctly, a receiver knows that the message must have been sent after the previous message it received correctly (that had a lower counter value). This enforces a message ordering and yields weak freshness.

Low Communication Overhead. The counter state is kept at each end point and does not need to be sent in each message.

μTESLA. Most of the proposals for authenticated broadcast are impractical for sensor networks, because they rely on asymmetric digital signatures for the authentication. The TESLA protocol provides efficient authenticated broadcast [25, 26];

however, it is not designed for limited computing environments. μTESLA solves the following three inadequacies of TESLA in sensor networks:

1. TESLA authenticates the initial packet with a digital signature, which is too expensive for sensor nodes. μTESLA uses only symmetric mechanisms.
2. Disclosing a key in each packet requires too much energy for sending and receiving. μTESLA discloses the key once per epoch.
3. Sensor node is unable to store the one-way key chain making TESLA infeasible for sensor networks.

μTESLA restricts the number of authenticated senders. μTESLA uses symmetric authentication but introduces asymmetry through a delayed disclosure of the symmetric keys, which results in an efficient broadcast authentication scheme. For the base station to broadcast authenticated information to the nodes, μTESLA requires that the base station and nodes are loosely time synchronized, and each node knows an upper bound on the maximum synchronization error. To send an authenticated packet, the base station simply computes a MAC on the packet with a key that is secret at that point in time. When a node gets a packet, it can verify that the corresponding MAC key was not yet disclosed by the base station (based on its loosely synchronized clock, its maximum synchronization error, and the time schedule at which keys are disclosed). Since a receiving node is assured that the MAC key is known only by the base station, it assures that no adversary could have altered the packet in transit. The node stores the packet in a buffer. At the time of key disclosure the base station broadcasts the verification key to all receivers. When a node receives the disclosed key, it can easily verify the correctness of the key. If the key is correct, the node can now use it to authenticate the packet stored in its buffer. Each MAC key is a key of a key chain, generated by a public one-way function F. To generate the one-way key chain, the sender chooses the last key K_n of the chain randomly, and repeatedly applies F to compute all other keys: $K_i = F(K_i + 1)$. Each node can easily perform time synchronization and retrieve an authenticated key of the key chain for the commitment in a secure and authenticated manner, using the SNEP building block.

SPINS leaves some questions such as security of compromised nodes, DoS issues, network traffic analysis issues. SPINS assumes the static network topology ignoring the ad hoc and mobile nature of sensor nodes.

16.7.2 INSENS: Intrusion-Tolerant Routing in Wireless Sensor Networks

INSENS [26, 27] tolerates intrusion by bypassing the malicious nodes instead of detecting the intrusion. Even if a malicious node exists in the network, INSENS mitigates the impact of that intrusion. INSENS design is based on the following three principles:

1. To prevent DoS-style attacks, the type of communication is constrained; that is, individual nodes are not allowed to broadcast. Only the base station does

broadcasting. Authentication of base station is used via one-way hashes so that individual nodes cannot spoof the base station and thereby flood the network. This feature exploits redundancy to tolerate intrusions without any need for detecting the nodes where intrusions have occurred. INSENS operates correctly in the presence of undetected intruders.

2. To prevent advertisement of false routing data, control routing information must be authenticated. Due to localized intrusions, the base station might not receive the entire topology discovery and control information and will farm an incorrect network. INSENS performs all complex computation at the base station and minimizes the overheads from sensor nodes in building routing tables. INSENS also addresses other issues of sensor nodes limitations such as memory, computation, and bandwidth.
3. INSENS addresses the resource limitations by using symmetric key cryptography for confidentiality and authentication between base station and nodes. To address the issues of compromise nodes, redundant multipath routing is built in INSENS. Even if a node is compromised, a secondary path will exist to forward the packet to the right destination. This feature limits the damage done by adversaries by limiting flooding and using right authentication mechanisms such as symmetric key cryptography.

INSENS does not rely on conventional anomaly-based intrusion detection techniques; instead it uses an intrusion-tolerant mechanism that reduces the harm caused by presence of a small number of undetected intruders in the network by incorporating redundancy in routing. INSENS is comprised of two phases: (1) a route discovery phase and (2) a data forwarding phase

The route discovery phase builds the topology of the network and creates forwarding tables on various nodes. Route discovery is further divided into three rounds:

(a) *A Request Message.* This is broadcasted by the base station to all the sensor nodes.
(b) *A Feedback Message.* Each sensor node sends its neighborhood topology information back to the base station.
(c) *Routing Update Message.* The base station authenticates the neighborhood information, constructs the topology of the network, computes the forwarding table for each sensor node, and sends the tables to each sensor node.

The data forwarding phase enables data forwarding from each sensor node to the base station.

16.7.3 TinySec: A Link Layer Security Architecture

TinySec [9] is a lightweight, generic security architecture that can be integrated into sensor network applications. It is incorporated into the official TinyOS release.

Ganesan et al. [8] reason why link layer security is ideal for sensor networks. Sensor networks use in-network processing such as aggregation and duplicate elimination [28, 29] to reduce traffic and save energy. Since in-network processing requires the intermediate nodes to access, modify, and suppress the contents of messages, end-to-end security mechanisms between each sensor node and the base station cannot be used to guarantee the authenticity, integrity, and confidentiality of messages.

End-to-end security mechanisms are also vulnerable to certain denial of service attacks. If message integrity is only checked at the final destination, the network may route packets injected by an adversary many hops before they are detected. This kind of attack will waste energy and bandwidth. Link layer security architecture can detect unauthorized packets when they are first injected into the network. TinySec provides the basic security properties of message authentication and integrity (using MAC), message confidentiality (through encryption), semantic security (through an initialization vector), and replay protection. TinySec supports two different security options: authenticated encryption (TinySec-AE) and authentication only (TinySec-Auth). With authenticated encryption, TinySec encrypts the data payload and authenticates the packet with a MAC. The MAC is computed over the encrypted data and the packet header. In authentication-only mode, TinySec authenticates the entire packet with a MAC, but the data payload is not encrypted.

TinySec uses an 8-byte *initialization vector* (IV) and *cipher block chaining* (CBC) [30]. The structure of the IV is $dst||AM||l||src||ctr$, where *dst* is the destination address of the receiver, *AM* is the active message (AM) handler type, *l* is the length of the data payload, *src* is the source address of the sender, and *ctr* is a 16-bit counter. The counter starts at 0 and the sender increases it by 1 after each message sent. A stream cipher uses a key K and IV as a seed and stretches it into a large pseudorandom keystream GK(IV). The keystream is then xored against the message: C=(IV, GK(IV) *xor* P). The fastest stream ciphers are faster than the fastest block ciphers, which might make them look tempting in a resource-constrained environment. However, stream ciphers have a failure mode: If the same IV is ever used to encrypt two different packets, then it is often possible to recover both plaintexts. Guaranteeing that IVs are never re-used requires IVs to be fairly long, say at least 8 bytes. Since an 8-byte overhead in a 30-byte packet is unacceptable in the resource-constrained sensor network, TinySec uses block cipher.

Using a block cipher for encryption has an additional advantage. Since the most efficient message authentication code (MAC) algorithms use a block cipher, the nodes will need to implement a block cipher in any event. Using this block cipher for encryption conserves code space. The advantage of using CBC is that it degrades gracefully in the presence of repeated IVs. If we encrypt two plaintexts *P1* and *P2* with the same IV under CBC mode, then the cipher texts will leak the length (in blocks) of the longest shared prefix of *P1* and *P2* and will leak nothing more. For instance, if the first block of *P1* is different from the first block of *P2*, as will typically be the case, then the cryptanalyst learns nothing apart from this fact. CBC mode is provably secure when IVs do not repeat. However, CBC mode was designed to be used with a random IV, and it has a separate leakage issue when used with a counter as the IV (note that the TinySec IV has a 16-bit counter). To fix this issue, TinySec pre-encrypts the IV.

The developers of TinySec give reasons behind their choice of cipher in reference 9. Initially they found AES and Triple-DES to be slow for sensor networks. They found RC5 and Skipjack to be most appropriate for software implementation on embedded microcontrollers. Although RC5 was slightly faster, it is patented. Also, for good performance, RC5 requires the key schedule to be pre-computed, which uses 104 extra bytes of RAM per key. Because of these drawbacks, the default block cipher in TinySec is Skipjack.

TinySec always authenticates messages, but encryption is optional. TinySec uses a cipher block chaining construction, CBC-MAC, for computing and verifying MACs. CBC-MAC is efficient and fast, and the fact that it also relies on a block cipher minimizes the number of cryptographic primitives that we must implement in the limited memory available. However, the standard CBC-MAC construction is not secure for variably sized messages. Adversaries can forge a MAC for certain messages.

TinySec creates confusion because of its three characteristics, such as no TinySec, TinySec-Auth, and TinySec-AE. Also, TinySec assumes a message length of 8 bytes or more, and it does not address the smaller messages. TinySec does not provide a secure routing mechanism while our framework addresses these issues.

16.8 SUMMARY

Sensor networks have become promising future to many applications. In the absence of adequate security, deployment of sensor networks is vulnerable to a variety of attacks. A sensor node's limitations and nature of wireless communication pose unique security challenges. Researchers have introduced secure communication using secure key management and secure routing techniques to address the unique security needs in sensor networks. Current research in sensor network security is mostly built on a trusted environment [31]; however, several research challenges remain before we can trust on sensor networks. In this chapter we have listed the prominent research efforts to meet the security goals of confidentiality, integrity, authentication, and availability in sensor networks. We have discussed cryptographic needs of tiny sensors and emphasized that only cryptography is not enough to secure sensor networks. Security issues such as presence of a malicious node and compromise of a legitimate node raise concern to look beyond cryptography. On the basis of our observation, we motivate the need for a security framework to address the secure key management, secure routing, malicious node detection, and secure location in sensor networks.

16.9 EXERCISES

1. Why is security different in wireless sensor networks? What makes sensor networks more vulnerable compare to wireless networks and mobile ad hoc networks?
2. List and discuss at least five security threats in wireless sensor networks.
3. Explain the difference between sinkhole and wormhole attacks.

4. Discuss the attacks and countermeasures in application, network, data link, and physical layers in sensor networks.

5. Summarize basic and random key pre-distribution schemes. How is multiple space key pre-distribution scheme different from basic and random key pre-distribution schemes?

6. What is SPINS? Discuss the features of SNEP and μTESLA.

7. What is TinySec? What are the two security options offered by TinySec?

BIBLIOGRAPHY

1. L. Eschenauer and V. Gligor. A key-management scheme for distributed sensor networks. In *Proceedings of the 9th ACM conference on Computer and Communication Security 2002*, Washington, DC.
2. A. Boukerche. *Handbook of Algorithms for Wireless Networking and Mobile Computing*, CRC/Chapman Hall, Boca Raton, FL, 2006.
3. C. P. Fleeger. *Security in Computing*, 3rd edition, Prentice-Hall, Upper Saddle River, NJ, 2003.
4. J. Undercoffer, S. Avancha, A. Joshi, and J. Pinkston. Security for sensor networks. In CADIP *Research Symposium*, 2002.
5. C. Karlof and D. Wagner. Secure routing in wireless sensor networks: Attacks and countermeasures. *Elsevier's Ad Hoc Networks Journal, Special Issue on Sensor Network Applications and Protocols*, **1**(2–3):293–315, 2003.
6. D. Wagner, Resilient aggregation in sensor networks. In *Proceedings of the 2nd ACM Workshop on Security of Ad Hoc and Sensor Networks*, ACM Press, New York, 2004, pp. 78–87.
7. Y. W. Law and P. J. M. Havinga. How to secure sensor network. In *Proceedings of the 2005 International Conference on Sensor Networks and Information Processing*, 5–8, December 2005, pp. 89–95.
8. D. Ganesan, A. Cerpa, Y. Yu, and D. Estrin. Networking issues in wireless sensor networks, *Journal of Parallel and Distributed Computing (JPDC), Special issue on Frontiers in Distributed Sensor Networks*, **64**, 2004.
9. C. Karlof, N. Shastry , and D. Wagner. TinySec: A link layer security architecture for wireless sensor networks, *SenSys '04*, November 3–5, 2004, Baltimore. MD.
10. L. V. Hoesel and P. Havinga. A lightweight medium access protocol (LMAC) for wireless sensor networks: Reducing preamble transmissions and transceiver state switches. In *Proceedings of INSS*, June 2004.
11. W. Diffie and M. E. Hellman. New directions in cryptography. *IEEE Transactions on Information Theory*, **22**:644–654, 1976.
12. H. Chan and A. Perrig. Security and privacy in sensor networks. *IEEE Journal of Computing*, **36**(10):103–105, 2003.
13. B. Dutertre, S. Cheung, and J. Levy. Lightweight key management in wireless sensor networks by leveraging initial trust, *SDL Technical Report SRI-SDL-0402*, April 6 2004.
14. F. Stajano. *Security for Ubiquitous Computing*, John Wiley & Sons, New York, 2002.

15. H. Chan, A. Perrig, and D. Song. Random key predistribution schemes for sensor networks. In *Proceedings of the IEEE Symposium on Security and Privacy*, Oakland, CA.
16. W. Du, J. Deng, Y. S. Han, and P. K. Varshney. A pairwise key pre-distribution scheme for wireless sensor networks. In *ACM CCS*, 2003.
17. W. Du, J. Deng, Y. S. Han, S. Chen, and P. K. Varshney. A key management scheme for wireless sensor networks using deployment knowledge. *IEEE InfoCom*, 2004.
18. R. Blom. An optimal class of symmetric key generation systems. Advances in cryptology. In *Proceedings of EUROCRYPT 84*, 1985.
19. W. Du, J. Deng, Y. S. Han, and P. K. Varshney. A key predistribution scheme for sensor networks using deployment knowledge. *IEEE Transactions on Dependable and Secure Computing*, **3**(1): pp. 62–77, 2006.
20. B. C. Neuman and T. Tso. Kerberos: An authentication service for computer networks. *IEEE Communications*, **32**(9):33–38, 1994.
21. T. Zia and A. Zomaya. A secure triple-key management scheme for wireless sensor networks. In *IEEE Infocom Workshop 2006*, Barcelona, Spain.
22. M. Saraogi. Security in wireless sensor networks, Department of Computer Science, University of Tennessee, Knoxville, TN, 2005.
23. A. Perrig, R. Szewczyk, V. Wen, D. Cullar, and J. Tygar. SPINS: Security protocols for sensor networks. In *Proceedings of the 7th Annual ACM/IEEE International Conference on Mobile Computing and Networking (MobiCom)*, pp. 189–199.
24. F. Stajano. *Security for Ubiquitous Computing*, John Wiley & Sons, New York, 2002.
25. A. Perrig , R. Canetti, J. D. Tygar, and D. Song. Efficient authentication and signing of multicast streams over lossy channels. In *IEEE Symposium on Security and Privacy*, 2000.
26. J. Deng, R. Han, and S. Mishra. INSENS: Intrusion-tolerant routing in wireless sensor networks. University of Colorado, Department of Computer Science, *Technical Report CU-CS-939-02*, 2003.
27. J. Deng, R. Han, and S. Mishra. A performance evaluation of intrusion-tolerant routing in wireless sensor networks. In *IPSN 2003, LNCS 2634*, Springer-Verlag, Berlin, pp. 349–364. 2003.
28. S. R. Madden, M. J. Franklin, J. M. Hellerstein, and W. Hong. TAG: A tiny aggregation service for ad-hoc sensor networks. In *The Fifth Symposium on Operating Systems Design and Implementation (OSDI 2002)*, 2002.
29. S. R. Madden, R. Szewczyk, M. J. Franklin, and D. Culler. Supporting aggregate queries over ad-hoc wireless sensor networks. In *Workshop on Mobile Computing and Systems Applications*, 2002.
30. M. Bellare, A. Desai, E. Jokipii, and P. Rogaway. A concrete security treatment of symmetric encryption: Analysis of the DES modes of operation. In *Proceedings of 38th Annual Symposium on Foundations of Computer Science (FOCS '97)*, 1997.
31. E. Shi and A. Perrig. Designing secure sensor networks, *Journal of IEEE Wireless Communications*, **11**(6):38–43, 2004.

CHAPTER 17

A Taxonomy of Secure Time Synchronization Algorithms for Wireless Sensor Networks

AZZEDINE BOUKERCHE

School of Information Technology and Engineering, University of Ottawa, Ottawa, Ontario K1N 6N5, Canada

DAMLA TURGUT

School of Electrical Engineering and Computer Science, University of Central Florida, Orlando, FL 32816-2362

17.1 INTRODUCTION

Wireless sensor networks naturally sense the desired event or phenomena in the real-world environment, and they communicate the sensed data to the global processing unit via the intermediate sensor nodes or gateway nodes for processing to draw relevant conclusions. Most of the time, the data received from multiple sensors are aggregated before reaching the final processing unit. In order to carry out the tasks discussed above, the physical time of the sensor nodes has to be synchronized with each other. The distributed wireless sensor networks heavily depend on the time synchronization for various reasons such as determining location and proximity of the deployed sensor nodes, intra-network coordination among different sensor nodes, temporal message ordering, security, time division multiplexing in wireless communication, improving energy-efficiency of sensor nodes by scheduling the sleep times of the sensor nodes, and so on [1].

In the computer synchronization history, Lamport's work [2] is considered a pioneering approach that emphasizes the need of virtual clocks in computer systems in which causality is more important than the absolute time. Even though the total ordering of the events was the focus of Lamport's method, this work has influenced the way the sensor networks have emerged today.

Algorithms and Protocols for Wireless Sensor Networks, Edited by Azzedine Boukerche

Most computer devices contain an internal clock, usually designed to be synchronized with the exact real-world time at the specific location of the computer, although many functionalities depend on the clock even on desktop computers—for instance, the scheduled Friday afternoon virus checks, or the popular "make" program, which determines whether a file needs to be recompiled comparing the timestamp of the source and object files. However, in practice, a desktop computer can function correctly even if its internal clock is minutes or even years away from the correct time.

Let us first define the ways in which the clocks of two nodes A and B might be out of sync. Let us note the clock of a node X with a function $C_X(t)$, which returns the reading of the clock at real time t. The first type of difference is the *offset*: $\delta_{AB} = C_A(t) - C_B(t)$. That is, the two clocks are identical, except that the clock of node A is early (if $\delta_{AB} > 0$). Now, we might fix this by setting the clock of node A back; however, that would create a problem, because the same time slice would appear twice for node A. This creates major problems for a number of protocols. It is better to set the clock of node B forward; however, most protocols simply require the nodes to keep track of their offsets without actually changing the internal clock.

The second type of synchronization difference is the *clock skew*; that is, one of the clocks is running faster than the other. This can be expressed as a difference in the derivatives of the clock function with respect to the time: $\eta_A B = \frac{\partial C_A(t)}{\partial t} - \frac{\partial C_B(t)}{\partial t}$. While this appears to be a more difficult problem, if a node is aware of its clock skew, it can very easily account for it.

Neither clock offset nor clock skew requires periodic synchronizations. If we know the offset and the skew of a node's clock, we can calculate the time difference at any moment in time. However, the frequency of the clocks can change randomly because of environmental conditions such as temperature difference or the aging of the hardware, a condition called *drift error*. The drift error appears as a nonzero second derivative in one or both clocks $\lambda_A B = \frac{\partial^2 C_A(t)}{\partial t^2} - \frac{\partial^2 C_B(t)}{\partial t^2}$. The clocks of sensor nodes usually accumulate several seconds of drift error per day; because the drift is not predictable, it needs to be solved using clock synchronization.

However, for sensor networks, the correct synchronization of the clocks is frequently a necessary component of the ability of the sensor network to function correctly. Unsynchronized clocks can yield invalid observations, can create uncovered areas and timeslots, and in the worst case can disable the communication architecture of the network.

Let us consider several examples. The individual nodes of the sensor network are sending their timestamped observations to the sink.

In Figure 17.1, an intruder is sensed consecutively by sensors S_1 and S_2, and their reports are sent to the sink. Based on the reports (intruder, S_1, t_1) and (intruder, S_2, t_2), knowing the locations of the sensors S_1 and S_2, and noticing that $t_1 < t_2$ and $t_2 - t_1 < 1$ second, the sink can correctly infer that the observations refer to the same intruder who is moving from left to right (in certain cases, this inference can be performed through in-network processing). However, this inference is valid only under the assumption that the clocks of the two sensors are synchronized at the level of tenths of seconds. Let us explain this further.

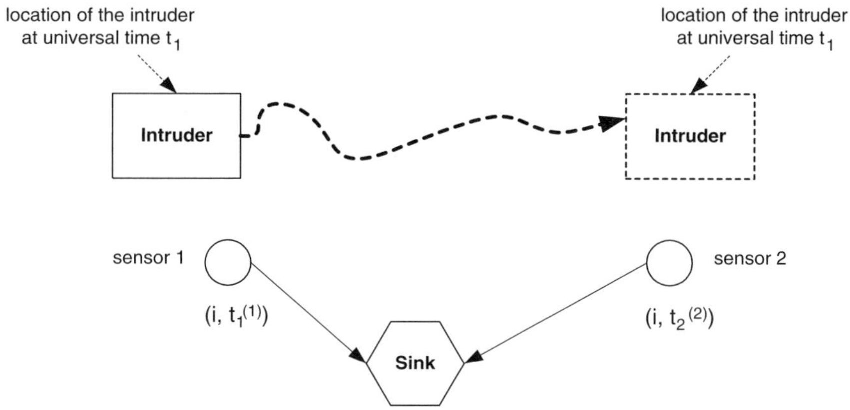

Figure 17.1. Intruder movement detected by sensors 1 and 2 at times $t_1^{(1)}$ according to clock 1 and $t_2^{(2)}$ according to clock 2.

If clocks 1 and 2 are synchronized—that is, they have the same offset $\delta^{(1)} = \delta^{(2)}$ compared to a universal time t—then $t_2^{(2)} - t_1^{(1)} = t_2 + \delta^{(2)} - t_1 - \delta^{(1)} = t_2 - t_1 > 0$. So, $t_2 - t_1$ indicates the correct order of the arrival to the sensors. However, if there is a large offset between these two, then we have $\delta^{(1)} - \delta^{(2)} \ll 0$; that is, the clock of $\delta^{(1)}$ is early. We might have a situation where $t_2^{(2)} - t_1^{(1)} = (t_2 - t_1) + \delta^{(2)} - \delta^{(1)} < 0$; that is, the sink will infer incorrectly that the intruder is moving from right to left. This is case where if the clock of the sensor S_1 is two seconds late, the inference would be that the intruder moves from right to the left. As a drift of several seconds per day is a normal occurrence for the internal oscillators of the devices, we cannot rely on the initial setting of the clocks at deployment time. The clocks need to be synchronized periodically in the field.

Notice that the faster the intruder moves, the smaller $(t_2 - t_1)$ and the more accurate synchronization is needed in order to make the correct inferences.

Our second example concerns the wake-up time of the sensors. Sensors have limited power resources. To extend the lifetime of a deployed network, the sensor nodes are frequently selectively put to sleep. The idea of the method is that the set of currently active nodes at any given moment in time covers the area to be surveyed and form a connected network. If an attacker can modify the internal clock of certain sensor nodes, such that these nodes, for instance, do not wake up in time, certain areas might not be surveilled by the sensors for a certain amount of time, allowing an intruder to operate unreported.

Finally, Time Division Multiple Access (TDMA)-based channel sharing protocols rely on the participating nodes to transmit at well-defined timeslots. Relatively small time drifts in the clock of the individual nodes can make the transmission intrude on the adjacent time slot, causing a collision. Repeated collisions can significantly disrupt the network. A detailed survey on clock synchronization protocols can be found in references 3 and 4.

The rest of the chapter is organized as follows. The challenges and design issues of time synchronization protocols in sensor networks are summarized in Section 17.2. In Section 17.3, we survey some of the time synchronization protocol including possible attacks and proposed countermeasures against these attacks. The types of attackers and attacks are presented in Section 17.4. Section 17.5 gives detailed discussion about the approaches for secure time synchronization. We conclude in Section 17.6.

17.2 TIME SYNCHRONIZATION IN SENSOR NETWORKS

17.2.1 Challenges

Different applications will have different synchronization requirements, and almost all the synchronization methods rely on message exchange between nodes. The nondeterminism in the network operations can cause delays in the message delivery and, in turn, contributes to the error in synchronization. Let us see how the source of a message's latency is decomposed in Figure 17.2. According to references 5 and 6, the time synchronization schemes have four basic packet delay components: send time, access time, propogation time, and receive time.

- *Send Time*. This is the time it takes to construct a message at the sender including the overhead of the operating system and the time to transfer the message to the network interface.
- *Access Time*. This is the delay encountered at the medium access control (MAC) layer prior to accessing the transmission channel due to contention, collisions, and so on. This delay is dependent on the MAC protocol in place. For instance, carrier sensing multiple access (CSMA) [7] requires every node to sense the carrier before transmitting, and it does not start a transmission if the medium is busy. CSMA/CA consists of both carrier sensing and a collision avoidance; the IEEE 802.11 standard [8] is the best-known instance of CSMA/CA.
- *Propogation Time*. This is the time required in propagation of the message from sender to the receiver. The propagation time varies, depending on the location of the sender and the receiver. For instance, if they are one-hop neighboring nodes in an ad hoc network, the propagation time equals to the physical propagation time

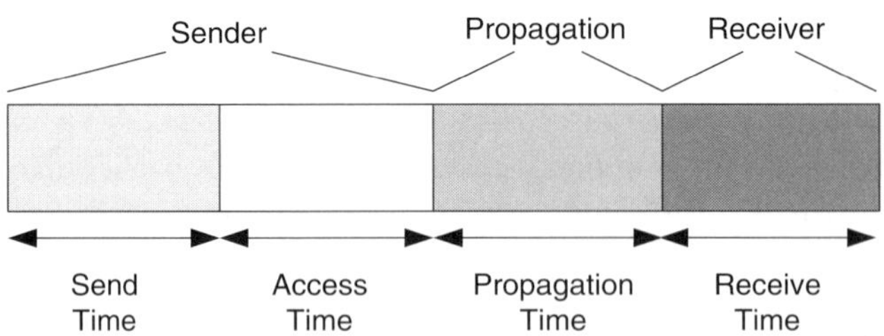

Figure 17.2. Packet delay components.

of the message traveling in the media. On the other hand, the propagation time can be much larger if we add switching and queueing delays into the formula.

- *Receive Time*. This is the time for the receiver node to process the message and acknowledge the host the arrival of the message. Depending on the level in which the arrival time was timestamped, the receive time may or may not include the overhead of the transfering of the message from the network interface to the host.

The challenge is not only the existence of packet delay but also the difficulty of prediction of the time needed on each delay component even though Figure 17.2 shows each component of equal length.

17.2.2 Design Issues

Here, we present a set of metrics for evaluating time synchronization protocols in sensor networks. Naturally, there are tradeoffs between these metrics; therefore, it is difficult to find a protocol that can satisfy them all.

- *Availability and Scope*. All the nodes in the network can be synchronized based on a global time, or a set of nodes locally in close proximity to each other can be synchronized based on a local time. Due to mainly the energy and bandwidth constraints in large-scale sensor networks, global synchronization is not only difficult to implement but can also be costly. Only a few applications would require global synchronization. Completeness of coverage within the specific region of nodes determines the availability requirement.
- *Cost and Size*. Since most sensor nodes considered are small and low-cost devices, the secure synchronization algorithms must be designed to meet the cost and size requirement. For example, it is not a viable solution to attach a global positioning system (GPS) device to a sensor node.
- *Efficiency*. Due to the low weight and small size of the sensor nodes used in many sensor network applications, they possess limited resources such as energy, memory, computation power, and so on. The security mechanisms developed within the time synchronization protocols must adapt to the limited computation power and memory of these sensor nodes since most security protocols are complex and require extensive resources to run efficiently. Additionally, communication capabilities are also limited due to these resource constraints. GPS or Universal Time (UTC) are generally used to synchronize the network to an accurate time in traditional protocols such as Network Time Protoocol (NTP) [9]. Using GPS requires high energy consumption; therefore, it is not considered a viable solution. Reducing the energy consumption can be achieved by transmitting over sequences of hops of short distances rather than a single one-hop long path. The secure time synchronization algorithms should also take into consideration the time needed to synchronize the nodes.

- *Infrastructure*. Sensor networks may have to be deployed in a random fashion to remote or dangerous regions. In this case, the sensor nodes are expected to self-organize themselves into a network since they cannot rely on an existing infrastructure such as NTP [9] in which the precise time is available to any node in the network from only a few hops away. NTP achieves this by providing the reference time at various points in the network. The task of the secure time synchronization becomes complex due to the many factors such as scalability issues, mobility, and of course the lack of infrastructure to rely on.
- *Lifetime*. The duration of the synchronization can be one of the two forms: (i) almost instantenous or (ii) persistent, in which case the synchronization lasts as long as the lifetime of the network itself. The nearly instantenous synchronization can be appropriate for applications in which an immediate action needs to be triggered based on the nodes' detection times of a specific event.
- *Network Dynamics*. Regardless of the type of deployment (random versus pre-engineered), sensor nodes face a high degree of network dynamics such as frequent changes in the network topology due to mobility, network partitioning, or energy depletion of the nodes. Designing secure time synchronization protocol becomes an even more complex task when one needs to consider ways to ensure the running of the network operations smoothly under these dynamic changes.
- *Precision*. The level of precision or accuracy required generally depends on both the application and the objective of the synchronization. There are three basic types of synchronization methods. The first one relies on the ordering of messages and events; it is considered the simplest. The next method allows the nodes to keep track of the drift and offset with respect to their neighboring nodes. This is the most common type of synchronization encountered in the time synchronization protocol applications. The most strict and complex one is the global synchronization in which all the nodes are synchronized according to a global time across the network. This method is the least used one since it is the most difficult to implement and also precise time synchronization is not always essential.

17.3 SECURE TIME SYNCHRONIZATION

These examples show us that time synchronization is vital for the correct operation of a sensor network. Because relatively small time drifts can cause significant disruption, we cannot rely on the precision of the hardware; we need to use external synchronization protocols. Furthermore, it was found that because relatively small changes in the clocks can disturb the operation of the sensor network or even cause it to make erroneous inferences about the observed event, the time synchronization protocols are a convenient target for malicious attackers. Most time synchronization protocols were not designed with security in mind. Recently, however, several research groups performed an analysis of various vulnerabilities and proposed countermeasures against them.

In the following, we survey some of the time synchronization protocols, discuss their benefits, and outline possible attacks and proposed countermeasures.

17.3.1 Reference Broadcast Synchronization

A pioneering approach called *post facto synchronization* was proposed by Elson and Estrin [10]. This is a low-power method of synchronization of clocks when the timestamps must be accurate for desired events. The nodes' clocks are are normally unsynchronized. When a phenomenon is sensed, each node records the time of sensing according to its own local clock. Shortly after, a beacon node sends a synchronization message to the nodes within its transmission range. The nodes receiving this message can adjust their timestamps of the sensed phenomena to the time of the receipt of this synchronization message. The post facto synchronization forms the basis for the reference broadcast synchronization method.

A sender-to-receiver synchronization method is used in most time synchronization protocols. In this method, the sender transmits the timestamp information and the receiver synchronizes.

The Reference Broadcast Synchronization (RBS) [11] protocol differs from the sender-to-receiver synchronization since it uses receiver to receiver synchronization method. Basically, the RBS is based on a synchronization signal broadcasted by an external unit. The receivers record their local time when they receive this reference message, and then they exchange this information among themselves (see Figure 17.3). The recording of a message is not 100% exact, because of hazards such as the propagation time of the message or the processing time of the packet at the lower protocol layers. To improve the precision, a number of reference messages can be broadcasted, and the nodes exchange the arrival times for each message and then find the best approximation using a least squares fit.

The keypoint of the RBS is that it uses a broadcast technique within a wireless medium to minimize latency and nondeterminism issues related to latency in the time synchronization protocol. On the other hand, the most notable drawback of the RBS is its requirement of a network with a physical broadcast channel [11].

Let us consider a scenario shown in Figure 17.4 a where a group of nodes lie in the range of multiple broadcast instead of a range of a single broadcast. Both sender

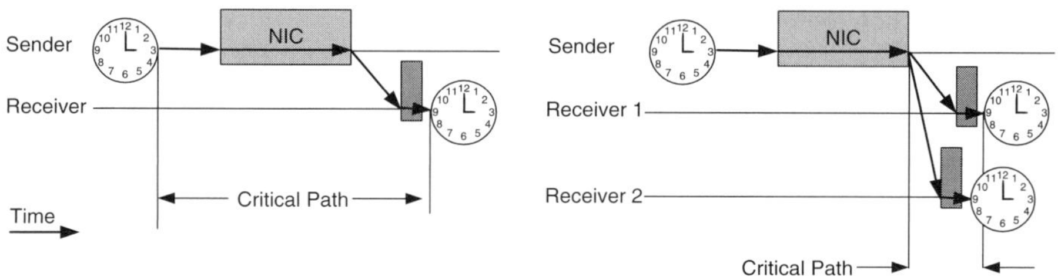

Figure 17.3. A critical path analysis for traditional time synchronization protocols (left) and RBS (right). (Adapted from reference 11).

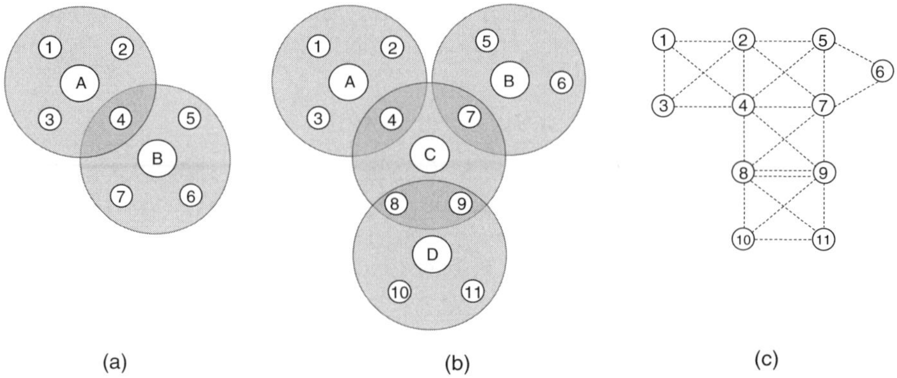

Figure 17.4. Simple (a) versus complex (b) multihop network topology with the corresponding logical topology (c) of the complex topology. (Adapted from reference 11).

nodes A and B send a synchronization message. According to the transmission ranges, nodes 1–4 will be able to synchronize with node A and nodes 4–7 can synchronize with node B; however, nodes A and B cannot hear each other's synchronization message. Among the nodes, node 4 is the only one that can hear the synchronization messages from both nodes A and B. This means that node 4 can correlate the clocks in node A's neighborhood to the clocks in node B's neighborhood or vice versa. In Figure 17.4 b, we have 3-hop network topology and the corresponding logical topology in Figure 17.4 c. The dotted lines—that is, the graph edges—are drawn between two nodes whose clocks are known to each other—for instance, node pairs (1–4), (7–9),(8–9), and so on. Note that there are two links between nodes 8 and 9, meaning that they have two receptions in common since they are located in the overlapping region of C and D.

Possible Attacks on RBS. In RBS, upon receiving a broadcast signal, two nodes exchange their local clock time. An attack can happen if one of the receiver nodes is compromised with an incorrect time. The compromised node then can send the incorrect time information to its neighbor, causing the uncompromised node calculating an incorrect offset.

The multihop version of RBS can face attacks as well. If a compromised node is located in any of the overlapping regions, it can inject an incorrect value into the clock conversion process, affecting multiple regions at once. This miscalculation in the clock conversion can be propagated across the network.

17.3.2 Time Synchronization Protocol Sensor Networks (TPSN)

TPSN [12] has two phases: *level discovery* and *synchronization*. In the level discovery phase, a spanning tree is created for the sensor network where each node is assigned a level. The root of the tree is usually a base station and assigned level 0. It is assumed

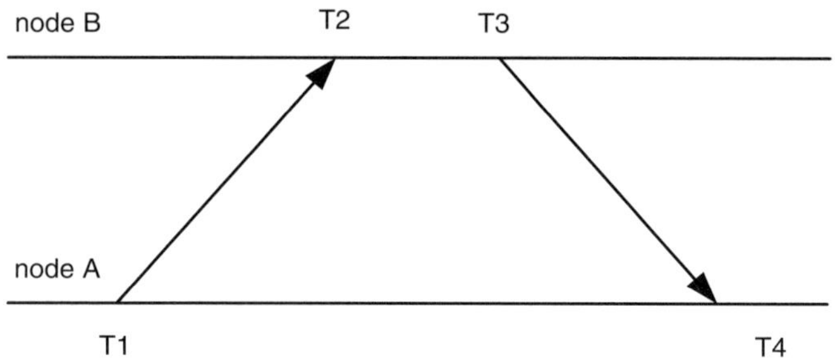

Figure 17.5. Two-way message exchange between pair of nodes. T2 and T3 are measured in node B's clock while T1 and T4 are measured in node A's clock. (Adapted from reference 12.)

that a node at level n can communicate with a node or a set of nodes at level $n - 1$. In the synchronization phase, the child nodes are synchronized to the parent. The synchronization is initiated by the child node, which sends a synchronization packet at time t_1. This is received by the parent at t_2, and an acknowledgment packet is sent in response at time t_3. The values of t_2 and t_3 will be included in the acknowledgment packet. This packet is received by the child node at time t_4 (see Figure 17.5). Knowing these four time values, the child node can calculate its clock offset relative to the parent node as well as propagation delay as follows:

$$\Delta t = \frac{(t_2 - t_1) - (t_4 - t_3)}{2} \tag{17.1}$$

$$d = \frac{(t_2 - t_1) + (t_4 - t_3)}{2} \tag{17.2}$$

The decomposition of the packet delay is depicted in Figure 17.6 similar to references 5 and 6 as shown in Figure 17.2. The transmission and reception time are more detailed here. The transmission time is the time it takes to transmit a packet on a bit-by-bit basis at the physical layer via the wireless link [12]. The reception

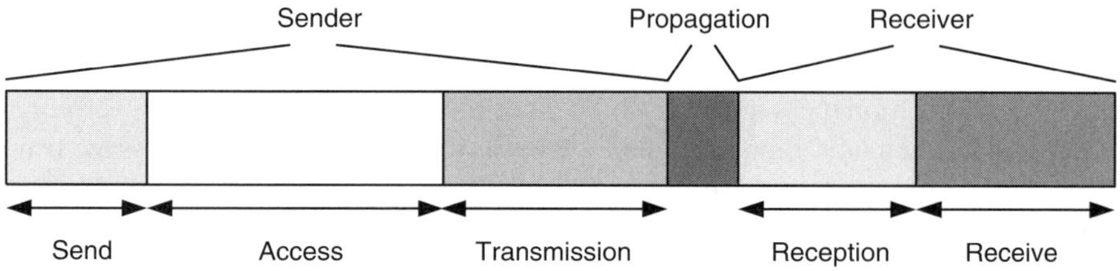

Figure 17.6. Decomposition of packet delay over a wireless link in TPSN. (Adapted from reference 12).

time is the time required to receive and forward all the bits to the link layer. Both transmission time and reception time are generally considered deterministic; however, variations can occur, depending on the underlying hardware structure.

The nodes are assumed to have unique ids, and each node has knowledge of its neighboring nodes. TPSN takes advantage of only the symmetric links for pairwise synchronization between nodes even though the network may also have asymmetric links. This can be considered one of the drawbacks of the protocols. Other drawbacks may include its limited suitability for applications serving highly mobile networks and its lack of support for multihop communication. On the other hand, TPSN is scalable and its computation overhead is less than some other protocols such as NTP [9]. The root selection and the tree construction mechanisms need to be reinvoked when topology changes occur due to node failures or other factors.

Possible Attacks on TPSN. A compromised node cannot cause any problem by requesting a time synchronization message because the message will reach to the parent node only. On the other hand, the compromised node can send erroneous time information to its children.

Naturally, the child node is relying on the parent for its clock synchronization; by providing incorrect values for t_2 and t_3, the parent can set an arbitrary offset on its child node. What is more, this incorrect offset will then be propagated down the tree. Therefore, the number of nodes whose synchronization can be affected by the compromised node depends on the location of the compromised node on the tree. One way for a malicious attacker to compromise a larger number of nodes is to reposition itself in a higher location on the tree or to answer queries instead of the proper parent. This is surprisingly easy to do in the original algorithm.

Yet another type of attack can surface when the compromised node misinforms its level in the tree, basically announcing a lower level than its current level. The compromised node can also attempt to trick other nodes at its level in requesting synchronization updates from itself. Furthermore, the compromised node can disconnect a number of nodes from being included in the tree by simply not participating in the level discovery phase.

17.3.3 Flooding Time Synchronization Protocol (FTSP)

In FTSP [13], the nodes are participating in a process in which a *root node* is elected. The root is the origin of the time synchronization messages. If a node does not hear a time synchronization message for a while, it declares itself the new root. The protocol requires that if at a later time the node receives a time synchronization message from a node with a lower id than itself, it gives up its root status. When a node receives a time synchronization message from the root, it adjusts its clock and broadcasts its own time to its neighbors. In the message broadcast, the preamble bytes are transmitted first, followed by *sync* bytes, message descriptor, the actual message, and finally *crc* bytes (see Figure 17.7).

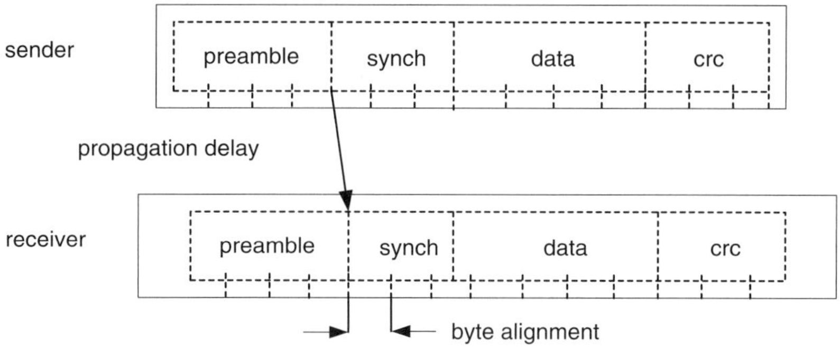

Figure 17.7. Data packets transmitted over the radio channel. (Adapted from reference 13.)

Possible Attacks on FTSP. The FTSP protocol is more robust for node failures than the TPSN protocol, because there is no need to maintain a tree structure that is notoriously vulnerable to single-point failures: The failure of a single node can disconnect the whole subtree. The weak point of the FTSP protocol is the election process. Any node can declare itself a root, and the protocol relies on the node to step back if a lower id root appears. A compromised node can easily masquerade as a root; and by declaring a very low id, it can actually dislodge the existing legitimate root. Then, by sending a synchronization message with a fake timestamp, it can make the nodes synchronize to an incorrect time.

17.3.4 Countermeasures for the Attacks

In this section, we describe the possible countermeasures for time synchronization attacks in both single-hop and multihop networks [14]. In single-hop networks, transmission range of each node reaches to every other node in the network. Let us consider a specific case of a single-hop network where there is a base station and its transmission range covers all the nodes in the network. The challenge for this type of networks is to preclude the malicious node(s) from compromising the base station and immediately start injecting the network with invalid timing information. In this case, the message from the base station must be authenticated in order to carry out the correct sequence of time synchronization methods. This can be achieved by utilizing a broadcast authentication scheme such as μTESLA [15]. Another approach can be the use of different private keys between the sender and receiver nodes.

In the multihop networks, many nodes may need to communicate with each other via intermediate nodes due to the limited transmission ranges. It makes sense for nodes to receive the global time of their immediate neighbors instead of neighbors several hops away. In that case, we need to take into consideration the incurred network delays between these nodes and their far-away neighbors. An approximation approach can be used find an upper bound on the error produced by the malicious node. Introduction of redundancy to the network is another viable approach. For instance, both FTSP and TPSN compute the offset and skew of their clock based on the timing information obtained from only one neighboring node. This redundancy approach can be easily applied to FTSP where a set of nodes can be used for the synchronization computations. The nodes can take the median of these received multiple updates. The private keys can be set up between the nodes and their neighbors at the beginning such that a malicious node can be precluded from injecting erroneous updates into the network. Furthermore, if a node becomes aware of one of its neighbors' update value being substantially different than the updates from its other neighbors, the node can refrain itself from including the updates from the suspicious neighbor into its computations of clock skew and offset. This approach of containment works, however, under the assumptions that the nodes have a sufficient number of sources for the updates. For instance, in TPSN, children of the nodes nearby the malicious or compromised node must find another parent, which may not always be feasible. Essentially, maintaining multiple trees comes with a price whereas in FTSP, no such cost exists. Utilizing the LS linear regression by each node in the network to compute the skew of its clock can further enhance security in time synchronization protocols. Algorithms such as RANSAC [16] can be used for this purpose.

17.4 ATTACKERS AND ATTACKS

The possible attacks against time synchronization protocols depend on the nature and capabilities of the attackers. We will first identify three different types of attackers, outline the types of attacks they are capable of, and then discuss the various types of defenses proposed against these types of attacks.

We will describe the system with the characters regularly used in the description of cryptographic protocols. We assume that Alice and Bob (and, potentially, additional nodes Carol, Dave, and so on) are engaging in a time synchronization process.

The **malicious outsider** Malory is a wireless device inserted in the range of the nodes of the sensor network, which has the ability to send and receive packets. We assume that the attacker can eavesdrop on any ongoing transmission; we can also assume that the attacker can eavesdrop on any ongoing transmission and that the attacker can transmit messages which are physically indistinguishable from the other nodes messages. However, this type of attacker does not have access to keys or other confidential information, other than what it can infer from eavesdropping on transmissions.

Attacker with Jamming and Replay Ability. Jimmy is an attacker which has the ability to jam the message, record it, and possibly replay it at later time. This type of attack is called a *pulse delay attack*. Although, in principle, the existence of jamming can be detected, it requires significant resources; by default, most nodes are not prepared for it.

A **compromised node** is a node that was taken over by the attacker. We will call this Zach (for zombie). One example of this is the physical capture of the node by an attacker, although a node can, in principle, be compromised with purely software methods. Compromised nodes have access to all the keys and other information of the original node, and they represent the most difficult type of attacker to defend against.

The challenge of secure time synchronization is to defend against all three types of attackers. An attacker is considered successful if it succeeds in making the nodes calculate an incorrect offset. By default, all three time synchronization protocols we described are vulnerable to all three types of attacker.

17.5 APPROACHES FOR SECURE TIME SYNCHRONIZATION

Malicious outsiders can affect all types of protocol. The primary defense against a malicious outsider is cryptographic techniques of the authentication of messages. If the sender and the receiver is sharing a key K_B, they can use it to sign the messages. To prevent an attacker from capturing a valid message and inserting a copy of it later in a different synchronization round, the sender sends a random *nonce* in the initial message, which then needs to be signed by the synchronization partner. While the attacker can still replay the same message in the same synchronization round, by simply considering only the first arrived message, the receiver can ignore the malicious outsider.

Ganeriwal et. al. [17, 18] proposed a series of secure time synchronization protocols based on this idea. The protocols are adapted to pairwise single-hop and multihop synchronization and for group synchronization. The protocols can also detect the existence of a pulse delay attack by calculating the end-to-end delay d of the message. If the delay is larger than a predetermined threshold $d*$, the protocols assume that an attack is in progress and abort the synchronization. We should note that key exchange is a major problem for these types of algorithm, due to the ways in which sensor nodes are deployed, which does not always permit the exchange of the keys in a secure environment.

Notice that cryptographic methods are not feasible against a compromised node, which has all the keys and knowledge to correctly answer all the challenges and appropriately sign its messages. If the time synchronization protocol happens only between Alice and Zach, it is impossible for Alice to detect or mitigate the attack. However, for protocols with a larger number of participants, we can use the redundancy in the synchronization messages to identify the malicious participants or messages. We note that delay attacks can be performed by either Zach or Jimmy, but not Mallory.

Song et al. [19] propose a method for making time synchronization protocols resilient to delay attacks based on techniques of *outlier detection*. The essential assumption behind this method is that the synchronization signals received from compromised nodes will be "much different from one another." Thus, messages coming from compromised nodes can be identified using statistical techniques as outliers and then eliminated from the package, and the synchronization can be performed with the remaining nodes.

The authors propose two alternative methods. One of them uses the generalized extreme studentized deviate (GESD), a generalization of the well-known Grubb's test from statistics. GESD can identify multiple outliers in a sample drawn from a normal distribution. GESD requires as one of its outputs the estimated number of malicious nodes.

A somewhat simpler approach is based on a delay threshold. At the setup time of the system, the nodes determine the maximum amount of time offsets they will tolerate, based on information about the typical drift rate of the nodes. A received offset that is higher than this value will be considered to come from a malicious node and discarded.

As an observation, naturally, Zach, the compromised node, would have exact knowledge about the thresholds used (but not Jimmy). Therefore, Zach has the possibility to remain undetected, by setting the delay such that it will put it just below the threshold (or, in the GESD case, such that it will not be identified as an outlier), but still have a distorting effect on the time synchronization process. Thus, the statistical techniques can only reduce, but not necessarily eliminate, the effect of delay attacks by nodes with insider knowledge.

Sun et al. [20] propose a statistical method for secure and resilient clock synchronization in the presence of compromised nodes. The techniques are applied for both *level-based clock synchronization*, where a hierarchical structure of nodes is developed which determines which node is synchronized with whom, and diffusion-based clock synchronization, which does not use such a structure and simply relies on the reachability information of the network. Naturally, the level-based approach allows for a more disciplined control of the synchronization flow, and thus a higher accuracy, whereas the diffusion method has the advantage that it can be applied to dynamic sensor networks with mobile nodes.

Furthermore, the authors consider both (a) the case with a single source of synchronization information and (b) the case with multiple sources. For instance, in the single-source case, the goal is assumed to be to find the clock offset δ_{iS} from the node to the source. The technique assumes that at every level, a normal node collects $2t + 1$ candidate source clock differences from its $2t + 1$ neighbors and chooses the *median* of them. Thus, the node can tolerate up to t compromised nodes, while retaining correct synchronization. Similar considerations apply to the diffusion-based approach. In the case of multiple sources, the node can receive synchronization information from $2s + 1$ sources, synchronized to the same external standard (such as a GPS signal) and tolerate up to s compromised sources by selecting the median.

Note that this approach uses the whole redundancy of the system to defend against an external attack: Out of $2t + 1$ recorded offsets, the method will pick a single one, the offsets median. Approaches that assume a benign environment usually select the *mean* of these measurements, therefore improving accuracy; however, the mean is vulnerable to even a single malicious node.

In addition, this approach requires a unique pairwise-key-based authentication of the nodes. Otherwise, the malicious node could impersonate multiple nodes (the so-called Sybil attack).

A significantly improved version of this technique was presented in reference 21. In the approach called TinySeRSync, time synchronization is performed in two phases. While we will call them Phase I and Phase II for convenience, these two processes are taking place asynchronously in the sensor network. In the first phase, single-hop pairwise synchronization is performed. The main feature of the pairwise synchronization process is that it relies on a hardware-enhanced authenticated MAC

TABLE 17.1. A Summary of the Various Techniques Proposed for Secure Time Synchronization

Approach	Protects Against	Uses Crytographic Techniques?	Uses Statistical Techniques?
Ganeriwal et al. [17]	Jimmy, Mallory	Yes	Yes
Song et al. [19]	Zach, Jimmy, Mallory	No	Yes
Sun et al. [20]	Zach	No	Yes
TinySeRSync, Sun et al. [21]	Zach, Mallory	Yes	Yes
Manzo et al. [14]	Zach	Yes	Yes

layer timestamping. The hardware is programmed to add a timestamp authenticated with a message integrity code (MIC) to every MAC packet transmitted. This is especially challenging for the newer radios, such as the ones on the newer-generation MICAz motes, where the time required to authenticate the timestamp can interfere with the transmission rate of the radio. The authors propose a prediction-based approach where the authenticated timestamp includes a prediction of the time required to calculate the MIC.

Through these techniques (see Table 17.1 for summary), the nodes achieve a sufficiently good local-level synchronization, which is exploited in the second phase. The second phase implements a global synchronization using the μTESLA broadcast authentication protocol. This protocol relies on the loose time synchronization between the nodes, and it uses a unidirectional keychain. Messages received need to be stored by the receiver and will be authenticated only after several timeslots. This prevents an attacker from forging messages, but it opens the doors for a denial-of-service attack. The messages received need to be buffered for future authentication; and because the memory of sensor nodes is limited, Mallory can create fake messages, which will not pass the authentication test, but will fill the buffer, preventing the node from receiving legitimate messages. To prevent denial-of-service attacks, the authors propose a modified version of μTESLA. To reduce the timeslots when the adversary nodes can flood the node with messages based on captured keys (which the receiver node needs to store for future authentication), TinySeRSync uses an implementation with very short delays r (made possible by the good local synchronization achieved in phase I). However, such short delays would require the generation of a large number of keys; the implementation utilized short intervals r used for message broadcasting, alternated with long intervals R used for broadcasting the disclosed keys.

The global synchronization in TinySeRSync still relies on the selection of the median from the $2t + 1$ candidate offsets, therefore tolerating the presence of at most t compromised nodes.

Notice that the approach presented in references 20 and 21 uses the *median* rather than the *mean* as a choice of the estimated time offset, thus obtaining a high protection against malicious nodes (provided that the technique is coupled with cryptographic defenses). However, it sacrifices the ability to improve precision through multiple independent observations.

We can attack the general problem of finding the best estimation of the time offset δ_{best} from a set of candidate offsets $\{\delta_1, ..., \delta_n\}$ by applying the principles of *robust estimation*. We note that the individual offset measurements δ_i might have natural noise, but some of them might be a result of a malicious attack. For any estimation method, the *breakdown point* is the smallest number of contaminated values that can move the estimate arbitrarily far from the correct value. Unfortunately, the most frequently used estimators, the average and the least squares estimators, have a very low breakdown point; a single malicious value can modify the estimate arbitrarily far. Manzo et al. [14] propose the use of the least mean squares (LMS) estimator for a more robust modeling. The generalized extreme studentized deviate (GESD) used by reference 19 is another example of the application of the techniques of robust estimation.

17.6 CONCLUSIONS

Among the many challenges in designing and employing wireless sensor networks is the clock synchronization between the sensor nodes. Agreeing on a common time is needed and even required by many of the sensor applications to carry out the sensing, communication, and processing of the sensed data. The time synchronization protocols in traditional wired networks cannot simply be re-used in the wireless sensor networks domain due to the inherent characteristics and limited resources of these networks. Therefore, several time synchronization protocols have been proposed recently; however, most of them do not consider security aspect during the design stages. There are only handful of protocols where the security has been taken into the consideration.

In this chapter, we reviewed the three most common secure time synchronization protocols: (i) Reference Broadcast Synchronization (RBS), (ii) Time Synchronization Protocol Sensor Networks (TPSN), and (iii) Flooding Time Synchronization Protocol (FTSP). We then evaluated these algorithms based on factors such as their countermeasures against various attacks and the types of techniques used (cryptographic versus statistical).

BIBLIOGRAPHY

1. K. Römer. Time synchronization and localization in sensor networks. Ph.D. thesis, Swiss Federal Institute of Technology Zurich (ETH Zurich), Zurich, Switzerland, 2005.
2. L. Lamport. Time, clocks, and ordering of events in a distributed system. *Communications of ACM*, **21**(4):558–565, July 1978.
3. B. Sundararaman, U. Buy, and A. D. Kshemkalyani. Clock synchronization for wireless sensor networks: A survey. *Ad Hoc Networks*, **3**(3):281–323, 2005.
4. F. Sivrikaya and B. Yener. Time synchronization in sensor networks: A survey. *IEEE Network Magazine's special issue on Ad Hoc Networking: Data Communications & Topology Control*, **18**(4):45–50, 2004.

5. H. Kopetz and W. Ochsenreiter. Clock synchronization in distributed real-time systems. *IEEE Transactions on Computers*, **C-36**(8):933–939, 1987.
6. H. Kopetz and W. Schwabl. Global time in distributed real-time systems. *Technical report 15/89*, Technische Universität Wien, 1989.
7. L. Kleinrock and F. A. Tobagi. Packet switching in radio channels: Part I—carrier sense multiple-access modes and their throughput-delay characeteristics. *IEEE Transactions on Communications*, **COM-23**(12):1400–1416, 1975.
8. IEEE std 802.11b-1999. part 11: Wireless LAN medium access control (MAC) and physical layer (PHY) specifications, August 1999.
9. D. L. Mills. Internet time synchronization: The network time protocol. *IEEE Transactions on Communications*, **39**(10):1482–1483, 1991.
10. J. Elson and D. Estrin. Time synchronization for wireless sensor networks, In *Parallel and Distributed Processing Symposium (IPDPS)*, April 2001, pp. 1–6.
11. J. Elson, L. Girod, and D. Estrin. Fine-grained network time synchronization using reference broadcasts. In *Proceedings of the Fifth Symposium on Operating Systems Design and Implementation (OSDI 2002)*, December 2002, pp. 147–163.
12. S. Ganeriwal, R. Kumar, and M. B. Srivastava. Timing-sync protocol for sensor networks. In *ACM Proceedings of the 1st International Conference on Embedded Networked Sensor Systems (SenSyS 2003)*, November 2003, pp. 138–149.
13. M. Maróti, B. Kusy, G. Simon, and Á. Lédeczi. The flooding time synchronization protocol. In *ACM Proceedings of the 2nd International Conference on Embedded Networked Sensor Systems (SenSyS 2004)*, November 2004, pp. 39–49.
14. M. Manzo, T. Roosta, and S. Sastry. Time synchronization attacks in sensor networks. In *The Third ACM Workshop on Security of Ad Hoc and Sensor Networks (SASN 2005)*, November 2005, pp. 107–116.
15. A. Perrig, R. Szewczyk, V. Wen, D. Culler, and J. D. Tygar. Spins: Security protocols for sensor networks. In *International Conference on Mobile Computing and Networking*, July 2001, pp. 189–199.
16. M. A. Fischler and R. C. Bolles. Random sample consensus: A paradigm for model fitting with applications to image analysis and automated cartography. *Communications of ACM*, **24**(6):381–395, 1981.
17. S. Ganeriwal, S. Capkun, C.-C. Han, and M. B. Srivastava. Secure time synchronization service for sensor networks. In *Wireless Security Workshop (WISE)*, September 2005, pp. 97–106.
18. S. Ganeriwal, S. Capkun, and M. B. Srivastava. Secure time synchronization in sensor networks. In *ACM Transactions on Information and Systems Security*, March 2006.
19. H. Song, S. Zhu, and G. Cao. Attack-resilient time synchronization for wireless sensor networks. *Ad Hoc Networks Journal*, **5**(1):112–125, 2007.
20. K. Sun, P. Ning, and C. Wang. Secure and resilient clock synchronization in wireless sensor networks. *IEEE Journal on Selected Areas in Communications*, **24**(2):395–408, 2006.
21. K. Sun, P. Ning, C. Wang, A. Liu, and Y. Zhou, TinySeRSync: Secure and resilient time synchronization in wireless sensor networks. In *Proceedings of the 13th ACM Conference on Computer and Communications Security (CCS '06)*, November 2006, pp. 36–42.

CHAPTER 18

Secure Localization Systems: Protocols and Techniques in Wireless Sensor Networks

AZZEDINE BOUKERCHE

School of Information Technology and Engineering, University of Ottawa, Ottawa, Ontario K1N 6N5, Canada

HORACIO A. B. F. OLIVEIRA

School of Information Technology and Engineering, University of Ottawa, Ottawa, Ontario K1N 6N5, Canada; Federal University of Minas Gerais, Brazil; and Federal University of Amazonas, Brazil

EDUARDO F. NAKAMURA

Federal University of Minas Gerais, Brazil; and FUCAPI—Analysis, Research, and Technology Innovation Center, Brazil

ANTONIO A. F. LOUREIRO

Department of Computer Sciences, Federal University of Minas Gerais, Belo Horizonte, Brazil

18.1 INTRODUCTION

In Chapter 1, wireless sensor networks (WSNs) [1–8] are defined has a network composed of a large number of sensor nodes that cooperate among themselves to monitor an area of interest. This type of network has become popular due to its wide applicability, including many different areas such as the environmental, medical, industrial, and military fields. For military applications, WSNs show a number of desirable characteristics such as (a) being autonomous systems able to be deployed in remote—possibly hostile—environments and (b) their ability to perform tasks such as battlefield surveillance or enemy tracking as well as security monitoring of military facilities. However, one of the main challenges in these critical applications, as shown in Chapter 16, is the *security* issue. We need to avoid attacks on networks located in hostile environments

Algorithms and Protocols for Wireless Sensor Networks, Edited by Azzedine Boukerche

as well as attacks to sensor networks that monitor the security of military facilities in order to guarantee the reliability of the provided data and requested tasks.

As mentioned in Chapter 11, a WSN is able to monitor a number of physical properties such as temperature, humidity, pressure, ambient light, and movement. However, all these gathered data—and, thus, the sensor nodes themselves—need to be localized in space in order to identify the events' location. This positioning task is the main goal of the *localization systems* [6, 8–18]. Since positioning data are used not only to locate events but also as base information for routing [19–22], density control, tracking, and a number of other protocols, the localization systems are considered as key parts of WSNs. By performing key essential functions, localization systems can be used as a security target that could compromise the whole functioning of a WSN, which could lead to incorrect military plans and decision-making, among other problems. Thus, these systems should be secured in order to protect the whole network and avoid the consequences of having hostile nodes and untrusted data traffic affecting the performance and function of the WSN. In this chapter, we show how current localization systems are vulnerable to these security attacks and how proposed techniques can be used to prevent or make these attacks difficult to perform in WSNs.

As we will see, for each proposed technique, a set of network resources (e.g., nodes' energy, media access, processor, and memory consumption) are required in order to implement and maintain the provided security. While in most networks this is not a problem, in WSNs this issue becomes a little bit more complicated due to their resource limitations. In this case, as we show in this chapter, we need to decide on the required level of security, which is application-dependent, and how many resources can be spent in providing these levels of security. Depending on this cost–benefit analysis, we can decide which solution or what security techniques will be used to secure the WSN.

The remainder of this chapter is organized as follows. In the next section, we briefly present an overview and definition of a secure localization system. Section 18.3 identifies the vulnerabilities to which localization systems are exposed, while in Section 18.4 we show some techniques used to eliminate these vulnerabilities. Finally, Section 18.5 presents our conclusions and future directions for secure localization systems.

18.2 PROBLEM STATEMENT

Before defining *secure localization systems*, we will first take a look at some general concepts and definitions used in normal localization systems. From the viewpoint of localization systems, we have basically two types of nodes: regular nodes and beacons. *Regular nodes*, also known as unknown, free, or dumb nodes, refer to the nodes in the network that have no knowledge of their position and have no special hardware to acquire this information. The *beacon nodes*, also known as landmarks, anchors, or locators, are the nodes that do not need a localization system in order to estimate their physical positions; in fact, they form the base of these systems. Their position is obtained by manual placement or by external means such as GPS.

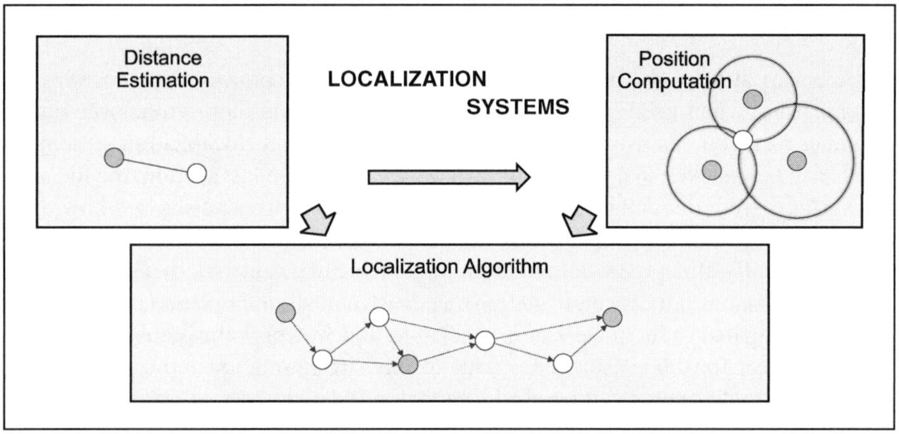

Figure 18.1. The division of localization systems into three distinct components.

Therefore, in a localization system, we want to solve the following problem: Given a multihop network and a set of beacon nodes with their known positions, we want to find the position (e.g., latitude, longitude) of regular nodes based on available information. For more information about the localization systems in WSNs, the reader can refer to Chapter 11, in which the localization systems are divided into three distinct *Components* (Figure 18.1):

1. *Distance/Angle Estimation.* This component is responsible for estimating information regarding the distances and/or angles between two nodes. Known techniques used in this component include received signal strength indicator (RSSI), time [difference] of arrival (ToA/TDoA), number of hops, or the angle of arrival (AoA).
2. *Position Computation.* This component is responsible for computing a node's position based on available information about the distances/angles and positions of reference nodes. Some techniques used to compute a position include trilateration, multilateration, or triangulation.
3. *Localization Algorithm.* This is the main component of a localization system. It determines how the available information will be manipulated in order to allow most or all of the nodes of the WSN to estimate their positions. It is a distributed and usually multihop algorithm. Some known algorithms include the ad hoc positioning system (APS) [12], and the directed position estimation (DPE) [6].

To be deployed in hostile environments, WSNs need a *secure localization system*, in which we need to solve the localization problem but must also be aware that we are in the presence of *compromised nodes*—malicious nodes or network nodes that have been corrupted by a malicious code—and/or in the presence of a *compromised environment*, where hostiles can change the environment's characteristics and may also have physical access to nodes.

18.3 FRAGILITY OF CURRENT LOCALIZATION SCHEMAS

A number of attacks and hostile techniques can used to compromise a localization system. Also, when used in WSNs, these systems become more vulnerable due to a number of resources saving techniques. Thus, as a result, localization systems in WSNs can be attacked in a number of different ways. In the last section, the localization systems were divided into different components and we also showed how these components are strongly connected. Any small misbehavior in any of these components can greatly affect the whole functioning of a localization system. For instance, a malicious erroneous distance estimation can cause a position miscomputation, which will be propagated to the localization algorithm and will probably cause a major localization error for the sensor nodes. Due to this strong relationship between them, any of these components can be used to attack a localization system, making these systems very fragile and hard to secure. In the next sections, we discuss each of the localization systems' components, showing to which type of attack each is vulnerable.

18.3.1 Attacks on Distance/Angle Estimation

As mentioned before, distance estimations can be made based on signal strength, time of arrival, or hop count analysis. In the first case, an easy way to generate erroneous estimations is by making a compromised node send a packet with a *greater or reduced transmission power* in order to make neighboring nodes think it is nearer or farther away than it really is. In the second case, the *transmission time of a packet can be delayed*, causing problems to both ToA- and TDoA-based systems. Hop-count-based distance estimations can be confused by compromised nodes that *advertise wrongly computed hop counts*. In fact, since it is a multihop algorithm, hop count estimations can also be affected by attacks to the localization algorithm (Section 18.3.3).

When the nodes are secured but they are located in a compromised environment, both signal strength and time of arrival techniques can be targeted by *changing the physical medium*—for example, by introducing noise, obstacles, or even smoke. Also, angle of arrival-based systems could be compromised by deploying magnets in the sensor field.

18.3.2 Attacks on Position Computation

Positions computation are highly dependent on the received positions from the reference node as well as on the estimated distances. To compute a position, a node needs at least three known positions and three distance estimations. Any attack on distance estimations, as shown in the previous section, has the main goal of affecting the position computation. However, some attacks can affect the position computation directly by advertising wrong known positions. A wrong advertised position can lead to an erroneous position computation even when the distance is correctly estimated. In this case, a compromised node can not only send its own *packet with a wrong position* but also send additional packets as if it were different nodes in different locations. In a compromised environment, a GPS signal can be jammed, making it erroneous or even not possible for beacon nodes to estimate their positions using their GPS receivers.

18.3.3 Attacks on the Localization Algorithm

While attacks on distance estimation and position computation components are attacks specific only to localization systems, the third component of the localization systems, the localization algorithm, shares the same kind of vulnerabilities associated with other distributed systems. It happens because the localization algorithm is a distributed and usually multihop algorithm executed by all nodes of the WSN. Some of the attacks in which distributed systems are usually vulnerable include the *Sybil*, the *replay*, and the *wormhole* attack.

- *Sybil Attacks.* In this type of attack, a malicious node makes it appear that it is a set of different nodes and starts sending erroneous information. This erroneous information can be distance estimations, positions, number of hops, or nonexisting nodes or beacons. Figure 18.2a illustrates this attack when node 6 claims to be also nodes 12 to 15.
- *Replay Attacks.* In a replay attack, a compromised node stores a received packet (from a beacon node, for instance) and then resends the same packet later. Since it is a copy of the original packet, neighboring nodes wrongly deduce that the malicious node is the node that sent the original packet (Figure 18.2b). In this case, since the distance estimation will be done based on the compromised node while the position in the packet will be based on the original node, the position computation will be affected. Both signal strength and time-based distance estimations are affected, since the packet sent by the compromised node will have a different signal strength and different propagation time.
- *Wormhole Attacks.* In this case, the information received by one malicious node on one side of the network is sent and replicated by another malicious node on the other side of the network. The multihop path between these two attackers is a wormhole in the sense that packets arriving on one side are transported and received on the other side of the network, appearing as if it came from a

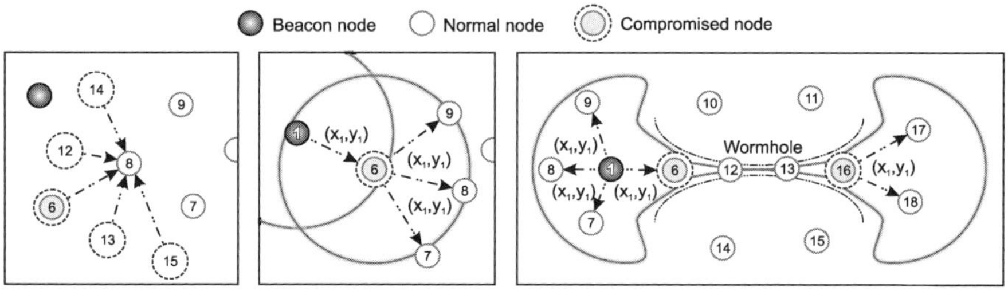

Figure 18.2. Attacks to the localization algorithms: (a) Sybil, (b) replay, and (c) wormhole.

neighboring node. This attack is illustrated in Figure 18.2c. This type of attack can greatly disrupt an insecure localization system by putting totally different and erroneous reference points in the position computations.

18.3.4 Examples in Current Insecure Solutions

In the following, we cite some known proposed localization systems for WSNs and we show how they can be vulnerable to a number of security attacks.

- *Ad Hoc Positioning System (APS) [12].* This is a distributed, multihop localization system that computes the average size of a hop and uses this information in the multilateration process. In this localization system, compromised beacon nodes can report wrong positions and a wrong average hop size, causing an erroneous position computation for the nodes. Compromised regular nodes can perform a Sybil attack to act like beacons, and they can also perform a replay attack or a wormhole attack. In these last two cases, the number of hops will be incorrect, which will cause a wrong estimation of the average size of a hop and result also in an erroneous position computation for the nodes.
- *Recursive Position Estimation (RPE) [8].* In RPE, estimated positions are broadcasted to help other nodes estimate their position as well. A similar algorithm is the Directed Position Estimation (DPE) [6], which uses a recursion with a direction. This kind of algorithm is more subject to security problems, because the miscomputation or the hostile advertisement of a single wrong position will be used in the computation of another position that will in turn be used to compute another position, until all nodes have an incorrect estimation. As an example, a hostile node can perform a replay attack by replicating a received position. Neighboring nodes will estimate their distance to the hostile node while using the position in the replicated packet, which will start an error propagation that will increase at each localization round.
- *Localization with a Mobile Beacon (MBL) [23].* in the MBL, once the nodes are deployed, the mobile beacon travels through the sensor field broadcasting messages that contain its current coordinates. When a regular node receives more than three messages from the mobile beacon, it computes its position, using a probabilistic approach, based on the received coordinates and on the RSSI distance estimations. A good aspect of this algorithm is that it is not vulnerable to some of the attacks performed in distributed algorithms such as the wormhole attack. It happens because network nodes do not need to exchange packets among them. They only need to receive packets from the mobile beacon. On the other hand, a compromised node can start sending packets as if it were the mobile beacon (Sybil or replay) and thereby mislead surrounding nodes. Also, a new hostile mobile beacon can be introduced in the network or the current mobile beacon can be compromised in order to attack the whole network.
- *Point-In-Triangulation Test.* The APIT algorithm [7] uses triangles formed by three beacon nodes, and a node decides if it is inside or outside these triangles

by comparing its signal strength measurements with the measurements of its neighbors. In the presence of a wormhole, a node can always think that it is outside the triangles, since it will receive packets from distant nodes that seem to be coming from neighbors.

18.4 TECHNIQUES FOR A SECURE LOCALIZATION

A number of security techniques have been proposed in recent years to enforce the localization systems and provide the secure positioning of nodes in hostile and military applications of WSNs. Most of these solutions achieve security by using cryptography, detecting and blocking compromised nodes or information, making statistical decisions, or filtering the positions used in computations.

18.4.1 Security through Cryptography

Cryptography is the first line of defense in most secure protocols. Since most security attacks are performed by a malicious node trying to pass as an entity it is not or by changing the values in data packets, these problems can be easily solved through cryptography by using *authentication* and/or *message integrity* checks. When using cryptography techniques, it is also possible to provide position *Confidentiality*, preventing malicious nodes and entities from gathering network information that could be used in other attacks or to search for uncovered areas in the sensor field (since the intruder would have access to the nodes' positions).

The good news is that cryptography can be used to protect against externally deployed hostile nodes that could execute any of the attacks cited. However, in the presence of compromised nodes in the local network, the attackers gain access to locally stored keys and passwords, which compromises the whole functioning of any cryptographic system. For this reason, most secure localization algorithms use noncryptographic security techniques, as will be shown in the next sections, and rely on cryptography as a second line of defense. This is the case in the HiRLoc [24], SeRLoc [25], and ROPE [26] localization algorithms, in which efficient cryptography is used to secure beacon transmissions. In SPINE [27], cryptography can be used to make an authenticated distance estimation, while it is used in reference 28 to assist the detection of malicious beacon nodes. In most cases, it is assumed that network nodes can establish pairwise secret keys.

Due to the limited availability of processor and memory resources in sensor nodes, cryptography is avoided in some works. However, it is a gray area, since, in most cases, if we need secure localization, we will also need secure media access, secure data routing and transfer, secure time synchronization, and a lot of other secure algorithms, all of which could take advantage of the same stored keys and implemented cryptographic algorithms. In these cases, cryptography can be a very plausible solution. Also, cryptography techniques can be used to provide a layer of security for all three localization system components (Distance/Angle Estimation, Position Computation, and Localization Algorithm), explained in Section 18.3, by providing authentication and integrity checks of the exchanged packets. However, it does not protect against compromised environment where it is possible to introduce obstacles, smoke, and so on, in the sensor field. In this last case, misbehavior detection, as shown in the next section, could be a possible solution.

18.4.2 Misbehavior Detection and Block

There are a number of cases where it is not possible to guarantee the security of the nodes and communications such as when the cryptography is compromised or when the environment is compromised. In these cases, the next layer of defense would be to make the nodes themselves detect and block the compromised factors. One way to make this happen and to defend against a possible attack is to observe the behavior of the nodes over time and decide whether to trust them. These techniques can be used mainly to protect the Position Computation component, since information gathered from untrusted nodes can simply be ignored when computing the position of the nodes.

For instance, in Liu et al. [28] it is proposed a set of techniques for detecting malicious beacon nodes. One technique detects malicious beacon nodes by comparing the distance estimated by using the location information provided by these beacon nodes and the distance estimated by means of the signal (e.g., RSSI, TDoA, AoA). Another technique evaluates the round-trip time (RTT) between two neighbors, based on the observation that the replay of a (malicious) beacon signal introduces extra delay. Then, the base station (or sink node) uses such information about malicious beacons to reason about the suspiciousness of each beacon node, and then it filters out malicious beacon nodes accordingly.

Srinivasan et al. [29] extends the techniques proposed by Liu et al. [28] by using a continuous scale and a reputation- and trust-based mechanism. The result is the Distributed Reputation-based Beacon Trust System (DRBTS), which is a distributed security protocol for excluding malicious beacon nodes. In DRBTS, each beacon node monitors its neighborhood for suspicious beacon nodes and provides information by maintaining and exchanging a Neighbor-Reputation Table, in such a way that other sensor nodes can choose trustworthy beacons based on a voting approach.

18.4.3 Robust Position Computation

Robust position computation can be used in order to increase the security of a localization system. This technique can be used when both cryptography and misbehavior detection is compromised. In this case, a way to deal with malicious nodes is to accept that they will be present in the network and propose robust position computations that can still work in the presence of bogus information. This is done mostly by using statistical and outlier filtering techniques. In these cases, it is assumed that the benign nodes outnumber the malicious ones. These techniques are used to protect against (or be robust to) attacks on the Position Computation and Distance/Angle Estimation components.

A technique proposed by Li et al. [30] uses the principle behind the least squares data fusion technique to propose an adaptive least squares and least median squares position estimator. The idea is to use the least squares in the absence of attacks and the least median squares in the presence of attacks, since the latter alternative tolerates up to 50% outliers and still provides correct estimates. The authors show that the use of traditional Euclidean distance is not robust to intentional attacks against base stations, and they introduce robustness to fingerprinting localization by means of a median-based distance metric.

Liu et al. [31] propose a method that uses the MMSE (Minimum Mean Squared Estimation), which is a data fusion technique for obtaining improved estimations, to identify and remove malicious location information. In this method, sensor locations are estimated using the MMSE-based method. Then, the method verifies whether the estimated location can be estimated from a set of consistent location references. If not, the most inconsistent reference is identified and removed, and the node location is estimated again. It repeats this process until all inconsistent references are discarded. The mean square error of the distance measurements is used as an inconsistency indicator.

A second method proposed by Liu et al. [31] is a voting-based location estimation technique. In this method, the sensor field is quantized into a grid of cells, and each reference node votes on the cells to which an unknown node may belong. Then, the method selects the most voted cell(s) and uses the "center" of these cell(s) as the estimated location. Voting results can be refined interactively to improve accuracy. This method requires few resources and is suitable for current resource constrained sensor nodes.

18.4.4 Location Verification

If all previous techniques fail to provide the required security, it is still possible to check the computed positions using redundant information available on the nodes and in the network. This is the core of some proposed works that focus on the reliability of the final position computations rather than on avoiding or detecting compromised nodes and attacks.

For instance, LAD [32] uses deployment knowledge, with a group-based deployment model, to check whether the computed positions of the nodes are consistent with the known model and observations.

In reference 33 an algorithm is proposed for in-region verification, in which a certain node can check whether another node really is inside the particular region that it claims to be. The proposed protocol, called Echo, uses known physical properties of both radio-frequency and ultrasound to compute distances and check whether a node really can be inside the claimed region. These techniques can be used to provide a layer of security for all three localization system components, since they only verify the result of the overall localization system.

18.4.5 Secure and Simple Algorithms

Localization systems are vulnerable mostly due to the number of components available to be attacked. Another way to secure a localization system is to use simple, less dependable localization algorithms, such as GPS-free, range-free, and/or one hop algorithms.

One example is SeRLoc [25], in which beacon nodes are equipped with a set of higher-power-directional antennas. These nodes send a packet using an asymmetric transmission that contains their position and the sector of the antenna in which the packets are sent. Because it is a range-free single-hop localization algorithm, it is protected against attacks aiming at altering range measurements and against regular compromised nodes. However, it does not protect against wormholes, which are avoided by checking network properties such as sector uniqueness and communication range.

A technique similar to SeRLoc is used in HiRLoc [24], which has a greater accuracy but an increased computational and communication complexity. Techniques like these can be used to protect the Localization Algorithm Component.

Another simple and secure algorithm is the already-mentioned Localization with a Mobile Beacon (MBL) [23]. Although MBL is not designed to be a secure algorithm and the authors do not mention anything about security in their work, the algorithm used in the MBL is quite simple, where a mobile beacon walks the network and broadcasts its position information to near nodes. In this case, regular nodes only need to listen to the beacon node and they never exchange messages among themselves. By being a simple single-hop algorithm, this localization system is secured against a number of distributed attacks such as the wormhole attack.

18.4.6 Comparison of Current Solutions

A widely known fact in network security is that there is no system that is totally safe and secure. There will always be weak points and the question is simply whether they are acceptable. In WSNs, this issue becomes a little more complicated due to resource limitations. In this case, we need to decide on the required level of security, which is application-dependent, and how many resources can be spent in providing these levels of security. Depending on this cost–benefit analysis, we can decide which solution or what security techniques will be used to secure the WSN. In Table 18.1, we compare each of the studied proposals, showing which type of security they use as well as some observations about them and their potential weaknesses. As we can see, most security proposals rely on some kind of lightweight cryptography as a second line of defense combined with other security techniques such as misbehavior detection, robust position computation, location verification, and simple algorithms combined with extra hardware.

18.5 CONCLUSIONS

In this chapter, localization systems were studied under the viewpoint of security. We showed how an insecure localization system can be attacked in a number of ways to compromise the whole functioning of a WSN and thus lead to incorrect military plans and decision-making. First, the localization systems were divided into three different

TABLE 18.1. Secure Localization Systems Comparison

Algorithm	Cryptography	Misbehavior Detection	Robust Pos. Computation	Location Verification	Simple Algorithms	Observation
HiRLoc [24]	Encryption and authentication of beacons' communication. Global preloaded keys.	—	—	—	Yes	Requires extra hardware (directional antennas) in beacon nodes.
SeRLoc [25]	Encryption and authentication of beacons' communication. Global preloaded keys.	—	—	—	Yes	Requires extra hardware (sectored antennas) in beacon nodes. Doesn't consider hostile beacons.
ROPE [26]	Encryption and authentication of beacons' communication. Global preloaded keys.	—	—	Beacons verify distances.	Yes	Requires extra hardware (directional antennas) in beacon nodes.
SPINE [27]	Symmetric or public-key cryptography for authenticated distance estimations.	—	Verifiable multi-lateration.	—	—	Nanosecond clocks. Uses ultrasound. High number of beacons.
DRBTS [29]	Encryption using a network-wide group key.	Reputation- and trust-based	—	—	—	Benign observations must be the majority. Dense network.
LAD [32]	—	—	—	Deployment knowledge	—	Requires deployment knowledge.
Echo [33]	—	—	—	Physical propagation of sound/RF.	—	Majority benign. No hostile beacons. Uses Ultrasound.
Li et al. [30]	—	—	Robust statistical methods.	—	—	Benign observations must be the majority.
Liu et al. [28]	Beacon packets are authenticated using shared pairwise keys.	Distances comp. and RTT	—	—	—	Requires redundant beacon nodes.
Liu et al. [31]	Authentication with pairwise key establishment.	—	Voting-based	—	—	Benign beacons must be the majority.

components: distance/angle estimation, position computation, and localization algorithm. We then went over each of these components, showing several techniques that could be used to compromise them and, consequently, the whole localization system when in the presence of compromised nodes and/or a compromised environment. After that, we showed some examples of current insecure localization systems and how they could be easily confused by some of the studied attacks. Finally, some of the techniques used by current secure localization systems were studied in order to show how to perform localization in the presence of hostile nodes and compromised environments.

18.6 EXERCISES

1. Cite the main differences in providing security in (a) wired networks, (b) wireless ad hoc networks, and (c) wireless sensor networks.
2. Why are localization systems for WSNs so hard to secure? Cite the main challenges in providing security in each of the three components of the localization systems: (a) distance estimation, (b) position computation, and (c) localization algorithm.
3. Cryptography is not so easy to be implemented in a WSN. Cite and explain the main challenges in providing a WSN with a cryptographic system.
4. Cite the main techniques to provide a security layer in a WSN. Show the pros and cons of each technique.
5. Choose a different nonsecure localization system (not studied in this chapter), explain how it works, show how it is possible to be attacked, and finally, show possible solutions to secure this chosen localization system.

BIBLIOGRAPHY

1. A. Boukerche. *Handbook of Algorithms for Wireless Networking and Mobile Computing (Chapman & Hall/Crc Computer & Information Science)*, Chapman & Hall/CRC, Boca Raton, FL, 2005.
2. I. F. Akyildiz, W. Su, Y. Sankarasubramaniam, and E. Cyirci. Wireless sensor networks: A survey. *Computer Networks*, **38**(4):393–422, 2002.
3. D. Estrin, L. Girod, G. Pottie, and M. Srivastava. Instrumenting the world with wireless sensor networks. In *International Conference on Acoustics, Speech, and Signal Processing (ICASSP 2001)*, Salt Lake City, UT, June 2001, pp. 2033–2036.
4. M. Ilyas and I. Mahgoub. *Handbook of Sensor Networks: Compact Wireless and Wired Sensing Systems*, CRC Press LLC, Boca Raton, FL, 2004, Chapter 20.
5. G. J. Pottie and W. J. Kaiser. Wireless integrated network sensors. *Communications of the ACM*, **43**(5):51–58, 2000.
6. H. A. B. F. Oliveira, E. F. Nakamura, A. A. F. Loureiro, and A. Boukerche. Directed position estimation: A recursive localization approach for wireless sensor networks. In *14th IEEE*

International Conference on Computer Communications and Networks, San Diego, October 2005, pp. 557–562.

7. T. He, C. Huang, B. M. Blum, J. A. Stankovic, and T. Abdelzaher. Range-free localization schemes for large scale sensor networks. In *MobiCom '03: Proceedings of the 9th Annual International Conference on Mobile Computing and Networking*, ACM Press, New York, 2003, pp. 81–95.
8. J. Albowicz, A. Chen, and L. Zhang. Recursive position estimation in sensor networks. In *The 9th International Conference on Network Protocols*, November 2001, pp. 35–41.
9. L. Doherty, K. S. J. Pister, and L. E. Ghaoui. Convex position estimation in wireless sensor networks. In *IEEE Conference on Computer Communications 2001*, Vol. 3, Anchorage, AK, April 2001, pp. 1655–1663.
10. A. Savvides, C.-C. Han, and M. B. Strivastava. Dynamic fine-grained localization in ad-hoc networks of sensors. In *7th ACM/IEEE International Conference on Mobile Computing and Networking*, Rome, Italy, 2001, pp. 166–179.
11. Y. Shang, W. Ruml, Y. Zhang, and M. P. J. Fromherz. Localization from mere connectivity. In *Proceedings of the 4th ACM International Symposium on Mobile Ad Hoc Networking and Computing*, Annapolis, MD, 2003, pp. 201–212.
12. D. Niculescu and B. Nath. Ad hoc positioning system (aps). In *IEEE Global Communications Conference (GlobeCom '01)*, San Antonio, TX, November 2001, pp. 2926–2931.
13. N. Bulusu, J. Heidemann, D. Estrin, and T. Tran. Self-configuring localization systems: Design and experimental evaluation. *Transactions on Embedded Computing Systems*, **3**(1):24–60, 2004.
14. H. A. B. F. Oliveira, E. F. Nakamura, A. A. F. Loureiro, and A. Boukerche. Error analysis of localization systems in sensor networks. In *13th ACM International Symposium on Geographic Information Systems*, Bremen, Germany, November 2005, pp. 71–78.
15. A. Boukerche, H. A. B. F. Oliveira, E. F. Nakamura, and A. A. F. Loureiro. A Voronoi approach for scalable and robust dv-hop localization system for sensor networks. In *Computer Communications and Networks, 2007. ICCCN 2007. Proceedings of 16th International Conference on*, 2007, pp. 497–502.
16. A. Boukerche, H. A. B. F. Oliveira, E. F. Nakamura, and A. A. F. Loureiro. Localization systems for wireless sensor networks. *IEEE Wireless Communications—Special Issue on Wireless Sensor Networks*, 2007, pp. 25–30.
17. H. A. B. F. Oliveira, E. F. Nakamura, A. A. F. Loureiro, and A. Boukerche. Localization in time and space for sensor networks. In *AINA'07: 21st IEEE International Conference on Advanced Information Networking and Applications*, Niagara Falls, Canada, May 2007, pp. 539–546.
18. A. Boukerche, H. A. B. F. Oliveira, E. F. Nakamura, and A. A. F. Loureiro. Vehicular ad-hoc networks—A new challenge for localization. *Elsevier Computer Communications*, 2008, to appear.
19. A. Boukerche, H. A. B. F. Oliveira, E. F. Nakamura, and A. A. F. Loureiro. A novel location-free greedy forward algorithm for wireless sensor networks. In *Proceedings of the 2008 IEEE International Conference on Communications (ICC 2008)*, Beijing, China, May 2008, IEEE CD-ROM.
20. A. Boukerche, H. A. B. F. Oliveira, E. F. Nakamura, and A. A. F. Loureiro. Towards an integrated solution for node localization and data routing in sensor networks. In *ISCC '07:*

12th IEEE Symposium on Computers and Communications, Aveiro, Portugal, July 2007, pp. 449–454.

21. B. Karp and H. T. Kung. Gpsr: Greedy perimeter stateless routing for wireless networks. In *Proceedings of the 6th Annual International Conference on Mobile Computing and Networking*, Boston, MA, 2000, pp. 243–254.
22. Y. Yu, R. Govindan, and D. Estrin. Geographical and energy aware routing: A recursive data dissemination protocol for wireless sensor networks. Technical Report CSD-TR-01-0023, Computer Science Department, UCLA, 2001.
23. M. L. Sichitiu and V. Ramadurai. Localization of wireless sensor networks with a mobile beacon. In *Proceedings of the 1st IEEE International Conference on Mobile Ad hoc and Sensor Systems (MASS 2004)*, Fort Lauderdale, FL, October 2004, pp. 174–183.
24. L. Lazos and R. Poovendran. Hirloc: High-resolution robust localization for wireless sensor networks. *IEEE Journal on Selected Areas in Communications*, **24**:233–246, 2006.
25. L. Lazos and R. Poovendran. Serloc: Secure range-independent localization for wireless sensor networks. In *Proceedings of WiSe'04*, ACM Press, New York, 2004, pp. 21–30.
26. L. Lazos, R. Poovendran, and S. Capkun. Rope: Robust position estimation in wireless sensor networks. In *Proceedings of IPSN*, April 2005, pp. 324–331.
27. S. Capkun and J.-P. Hubaux. Secure positioning of wireless devices with application to sensor networks. In *Infocom'05*, Miami, March 2005, pp. 1917–1928.
28. D. Liu, P. Ning, and W. Du. Detecting malicious beacon nodes for secure location discovery in wireless sensor networks. In *25th ICDCS*, Washington, DC, 2005, IEEE Computer Society, pp. 609–619.
29. A. Srinivasan, J. Teitelbaum, and J. Wu. Drbts: Distributed reputation-based beacon trust system. In *2nd IEEE DASC*, 2006, pp. 277–283.
30. Z. Li, W. Trappe, Y. Zhang, and B. Nath. Robust statistical methods for securing wireless localization in sensor networks. In *IPSN '05: Proceedings of the 4th International Symposium on Information Processing in Sensor Networks*, Los Angeles, CA, IEEE Press, New York, 2005, p. 12.
31. D. Liu, P. Ning, and W. Du. Attack-resistant location estimation in sensor networks. In *IPSN '05: Proceedings of the 4th International Symposium on Information Processing in Sensor Networks*, Los Angeles, CA, IEEE Press, New York, 2005, p. 13.
32. W. Du, L. Fang, and P. Ning. Lad: Localization anomaly detection forwireless sensor networks. In *19th IPDPS*, Washington, DC, 2005, IEEE Computer Society, p. 41.1.
33. N. Sastry, U. Shankar, and D. Wagner. Secure verification of location claims. In *WiSe'03*, New York, 2003, ACM Press, New York, pp. 1–10.

INDEX

Algorithms and Protocols for Wireless Sensor Networks, Edited by Azzedine Boukerche

WILEY SERIES ON PARALLEL AND DISTRIBUTED COMPUTING

Series Editor: Albert Y. Zomaya

Parallel and Distributed Simulation Systems / Richard Fujimoto

Mobile Processing in Distributed and Open Environments / Peter Sapaty

Introduction to Parallel Algorithms / C. Xavier and S. S. Iyengar

Solutions to Parallel and Distributed Computing Problems: Lessons from Biological Sciences / Albert Y. Zomaya, Fikret Ercal, and Stephan Olariu (*Editors*)

Parallel and Distributed Computing: A Survey of Models, Paradigms, and Approaches / Claudia Leopold

Fundamentals of Distributed Object Systems: A CORBA Perspective / Zahir Tari and Omran Bukhres

Pipelined Processor Farms: Structured Design for Embedded Parallel Systems / Martin Fleury and Andrew Downton

Handbook of Wireless Networks and Mobile Computing / Ivan Stojmenović (*Editor*)

Internet-Based Workflow Management: Toward a Semantic Web / Dan C. Marinescu

Parallel Computing on Heterogeneous Networks / Alexey L. Lastovetsky

Performance Evaluation and Characteization of Parallel and Distributed Computing Tools / Salim Hariri and Manish Parashar

Distributed Computing: Fundamentals, Simulations and Advanced Topics, *Second Edition* / Hagit Attiya and Jennifer Welch

Smart Environments: Technology, Protocols, and Applications / Diane Cook and Sajal Das

Fundamentals of Computer Organization and Architecture / Mostafa Abd-El-Barr and Hesham El-Rewini

Advanced Computer Architecture and Parallel Processing / Hesham El-Rewini and Mostafa Abd-El-Barr

UPC: Distributed Shared Memory Programming / Tarek El-Ghazawi, William Carlson, Thomas Sterling, and Katherine Yelick

Handbook of Sensor Networks: Algorithms and Architectures / Ivan Stojmenović (*Editor*)

Parallel Metaheuristics: A New Class of Algorithms / Enrique Alba (*Editor*)

Design and Analysis of Distributed Algorithms / Nicola Santoro

Task Scheduling for Parallel Systems / Oliver Sinnen

Computing for Numerical Methods Using Visual C++ / Shaharuddin Salleh, Albert Y. Zomaya, and Sakhinah A. Bakar

Architecture-Independent Programming for Wireless Sensor Networks / Amol B. Bakshi and Viktor K. Prasanna

High-Performance Parallel Database Processing and Grid Databases / David Taniar, Clement Leung, Wenny Rahayu, and Sushant Goel

Algorithms and Protocols for Wireless and Mobile Ad Hoc Networks / Azzedine Boukerche (*Editor*)

Algorithms and Protocols for Wireless Sensor Networks / Azzedine Boukerche (*Editor*)